The Chemistry of Mind-Altering Drugs

History, Pharmacology, and Cultural Context

The Chemistry of Mind-Altering Drugs

History, Pharmacology, and Cultural Context

Daniel M. Perrine

American Chemical Society
Washington, DC 1996

Library of Congress Cataloging-in-Publication Data

Perrine, Daniel M., 1943–
The chemistry of mind-altering drugs: history, pharmacology, and cultural context / by Daniel M. Perrine.
p. cm.

Includes bibliographical references and index.

ISBN 978-0-8412-3253-2
1. Psychotropic drugs. 2. Drugs of abuse. 3. Psychopharmacology. I. Title.
[DNLM: 1. Central Nervous System Agents—pharmacology. 2. Street Drugs—pharmacology. 3. Street Drugs—history. 4. Substance Abuse—ethnology. QV 76.5 P458c 1996]

RM315.P435 1996
615'.788—dc20
DNLM/DLC 96–28908
for Library of Congress CIP

Reproduced by ACS Books from camera-ready copy supplied by the author.

Printed in the United States of America
Printed on acid-free paper.

in memoriam

WALTER N. PAHNKE, M.D., Ph.D.

(1931-1971)

utriusque arcana eadem clavi

reserari posse

ausus est sperare

Contents

Preface

Of the more than 10 million chemical substances identified by *Chemical Abstracts*, this book is concerned with the tiny fraction, at most a few hundred, that have the astonishing capability of affecting human consciousness.

Because of this focus, this book of necessity explores much more than chemistry. In addressing so broadly human a topic as the mind itself and the drugs which affect it, it becomes impossible to understand the effects of such drugs outside the total context, and especially the cultural context, of their use. For instance, there is little meaning that can be assigned even to the term "drug" without specifying the cultural set and setting. An Afghan grandfather smoking a pipe of opium with the family after the evening meal, members of the Native American Church drinking peyote tea during their worship services, a once suicidal and bulimic young woman summoning the courage to live with the help of Prozac, Hasidic Jews reeling in drunkenness at Purim, or an urban American prostitute taking a "crack break"—all are clearly under the influence of a "drug," but just as clearly the drug effects take on profoundly different meanings in the different human settings. And so it is that the pages of this book will be found to contain much material which is not chemical or pharmacological but historical, anthropological, sociological—even literary, philosophical, theological, and religious. Chemists of an older generation may recognize that what has finally emerged is an expansive working in a genre once common but now not widely employed: the *descriptive chemistry* of some very interesting chemicals, in this case those that affect the human mind.

As such, it is hoped that the book will prove interesting and informative not only to chemists, but to the more general audience of science-conscious laypersons, students of the health sciences, educators, or science professionals in disciplines other than organic chemistry—all of whom as citizens may well be concerned with the social and ethical implications of the use of mind-altering drugs in our society, whether the drug in question be Prozac or PCP. To this end, as an aid to the many people for whom the arcane pentagons and hexagons of organic chemistry are either a hazy memory from their distant college days or a forbidding forest never entered, the book includes a series of four Appendices which comprise a simplified introduction to the language of organic chemical structures. For in the last analysis, it is in this language alone that there can be found the definition of what any drug entity is.

Those of us who teach organic chemistry may find a further aspect of this book to be of pedagogical value: the outline syntheses of representative drug types which can be found in each chapter. Otherwise dry discussions of the Friedel-Crafts or Mannich reactions can be made more interesting for the typical premedical student by using as examples of these reactions the syntheses of Valium and Prozac (or THC and psilocybin).

Comments and suggestions are welcome. Even the most carefully scrutinized reference works, I have found, contain their inevitable share of errors: in the course of writing this book, I have discovered several in the venerable pages of the *Merck Index*. Where Titans limp, mere mortals can but crawl. I would appreciate any errors being called to my attention by mail, fax, phone, or internet at the addresses given below.

Finally, I owe several expressions of heartfelt thanks: to Barbara Tansill, Cheryl Shanks, Barbara Pralle, and Randy Frey, editors at the American Chemical Society whose suggestions and encouragement have been so helpful over the several years of this book's gestation; to Loyola College, which generously provided a Faculty Development Grant for the summer of 1994 enabling me to travel to the Netherlands to study drug policy there; to Lee Tawney, assistant to Baltimore Mayor Kurt Schmoke, who provided me with an introduction to the health policy officials of Rotterdam and Amsterdam; to Dr. Hans van Mastrigt, officer for Addiction Policy for the Rotterdam Municipal Health Service, who extended himself far beyond the call of duty and well past the midnight hours to escort me into the demimonde of Rotterdam's drug users; and to Richard Yensen and Donna Dwyer of Baltimore's Orenda Institute, who spent many hours sharing with me their extensive knowledge and experience of the use of drugs in psychotherapy.

Dan Perrine
July 1996

Loyola College Chemistry Department
Knott Hall 468
4501 N. Charles St.
Baltimore, MD 21210-2699
Voice: 410-617-2717; Fax: 410-617-2803; E-mail: DMP@loyola.edu

1

Mind and Molecule: Neurotransmission in Context

> Some people have maintained, in my hearing, that they had been drunk upon green tea; and a medical student in London, for whose knowledge in his profession I have reason to feel great respect, assured me, the other day, that a patient, in recovering from an illness, had got drunk on a beef-steak.
>
> —De Quincey[1]

> *O the mind, mind has mountains; cliffs of fall*
> *Frightful, sheer, no-man-fathomed. Hold them cheap*
> *May he who ne'er hung there. . . .*
>
> —Hopkins[2]

Introduction

This chapter presents important background information essential to understanding the book as a whole. The focus is on the *psychoactive* (mind-altering) effects that the chemicals found in a few dozen plants, powders, and potions have on human behavior and perception—even to the point of changing or controlling people's lives. If it were not for this amazing power, the chemicals themselves would be no more interesting than the millions of other combinations of carbon, hydrogen, and oxygen atoms found in nature.

Throughout history, the power that many psychoactive drugs have exerted over the behavior of human beings has been variously ascribed to gods or demons, and such drugs have been believed to exert their influence over consciousness by acting on the heart, spleen, or liver. The idea that consciousness has its physiological basis in the functioning of the brain rather than some other organ is actually a relatively new conclusion of modern science. And so we look at the brain to understand the biological effects of these drugs, and this means understanding something about *neurotransmission*, the technical term for the process whereby the billions of nerves in the brain exchange signals with each other.

We can then understand something of how psychoactive drugs work: they exert their effects by mimicking, amplifying, blocking, or otherwise altering the effects of those messenger molecules called neurotransmitters (NTs), which the brain itself uses for signaling. There are only a few dozen of these chemicals, and we will see that a relatively small subset is invoked to explain (to the extent it can be explained) the first subject of this chapter, that of *drug dependency* or addiction.

Finally, the overall effect of any drug can be dramatically different depending on a number of pharmacological variables such as the mode of ingestion and the drug's concentration. This chapter will explain a number of these variables and some ways in which pharmacologists measure them.

The Human Context: Drug Use and Dependency

The Prevalence of Drug Use in American Society

The use of drugs that alter alertness, mood, perception, or behavior has been widespread in virtually every society, and often predates recorded history. In our own Western culture we are of course most familiar with caffeine, alcohol, and nicotine.[3] But opium and marijuana have been used in Eastern cultures for millennia, and a number of hallucinogenic botanicals can be shown from archeological records to have been in use for thousands of years by hundreds of native cultures in Africa and the Americas.

Here in the United States during the last decade of the 20th century, the most widely used psychoactive agents are caffeine, alcohol, and nicotine. About 80% of adult Americans consume beverages containing caffeine, but the number who experience discomforting symptoms and struggle to quit their habit is difficult to ascertain.[4] A close second to caffeine is alcohol; studies typically indicate that about 75% of adult Americans employ alcohol to some degree. There are regional and subpopulation differences that are probably less illuminating than curious. For instance, most urban Jews and Episcopalians drink, but few rural Baptists; most bartenders, writers, and surgeons drink, but few accountants, mail carriers, or carpenters. "Heavy" drinkers (more than two drinks a day) constitute about 13% of U.S. men and 3% of U.S. women. The third most widely consumed drug is likewise legal: about a third of adult Americans have used tobacco in recent years, and about a fourth are regular users.[5]

Most of the remaining psychoactive drugs are illegal. A recent national study of the use of these drugs (chiefly marijuana, cocaine, heroin, hallucinogens, and prescription sedatives and stimulants) from the National Comorbidity Survey showed that 51% of the population has used one or more of these drugs at some time in their lives; 15% used them in the past 12 months. The authors conclude that "drug use and dependence are highly prevalent in the general population."[6]

In economic terms, overall yearly expenditures by Americans on mind-altering drugs other than caffeine as of 1990 were approximately as follows: $50 billion on alcohol, $40–50 billion on cocaine, $35 billion on tobacco, $30 billion on prescription and over-the-counter (OTC) psychotropics, $25 billion on marijuana, and about $20 billion on all other illegal drugs combined. In total, this amounts to more than $200 billion. In comparison, during the same period Americans spent about $450 billion for medical care, $250 billion for nondrug recreation, and $100 billion for automobiles. In health terms, during the same period in the United States there were about 400,000 deaths from tobacco use, 100,000 from alcohol, 5,000 from overdoses of prescription drugs, and 3,500 from all illegal drugs combined (except marijuana, which is credibly claimed never to have caused a single death in all of history).[7] Using drugs is truly as American as apple pie.

Drug Use by the Individual

Drug use is one of the most human of behaviors, involving all the complex interactions of individuals with their history, culture, and society. In light of this, one should keep the following factors in mind when trying to understand anyone's interaction with a drug: (1) the *individual*, the particular human being, both as a unique biological organism with a possibly idiosyncratic response to a given chemical substance and as a unique personality and psychology; (2) the particular mental *set* of the person taking the drug, which often has a dramatic influence on its effects; (3) the *setting* in which a person takes a drug, which can range from a religious ceremony to a rock concert to an assisted suicide; and (4) the *pharmacology* of the drug itself.

None of these four elements can be completely isolated from the others—not even the last. The chemical structure can be considered by itself, and if the drugs we are talking about were diuretics to enhance kidney function, their pharmacology could be considered by itself. But of course they are not; they are psychoactive drugs, and the other three elements are always involved in trying to explain what that means.

Chemical/Substance/Drug Addiction/Dependence/Abuse

By combining each of the three adjectives with each of the three nouns in this heading, nine different phrases can be generated to describe a phenomenon difficult to delineate. A famous jurist said of pornography, "I cannot define it, but I know it when I see it";[8] Thomas à Kempis in the 15th century said of sorrow for his sins, "Opto magis sentire quam scire eius definitionem (I would rather experience it than know its definition)."[9] And so it may be of our subject. There are implicit value judgments associated with all the choices: "drug addiction" seems too harsh, "substance dependence" too euphemistic.

Everything but gods and ghosts consists of chemicals and "substances" (indeed, the latter term in the legal context of the Controlled Substances Act can include not only chemicals but toads and toadstools); and it is certainly as true of humans as of any other living beings that we are inextricably involved in an endless ecological web of mutual interdependence with the surrounding animate and inanimate worlds. From the abstract impersonal view of physical and biological science, there is nothing that distinguishes a chemical from a food or a drug—these distinctions are ultimately functional, defined by their use within the warp and woof of individual and social human life. The dependence we all have on such chemical substances as oxygen, water, vitamins, and food is much like the dependence of diabetics on insulin, the dependence of a patient undergoing surgery on anesthetics and morphine, or a street junkie's dependence on mainline heroin. What distinguishes these various dependencies in our minds is largely the evaluation we give them as good or bad. And the realm of such valuations is ultimately not that of the physical, biological, or even the health sciences, however great the part each of these sciences has to play in contributing to an informed judgment.

In the final analysis, it is impossible to draw the line between drug use and abuse without entering into the realm of ethics and value, without asking questions as timeless as those of the Greek philosophers: What is the good life? What is the truly human? To these questions the scientific method cannot offer the final answer, even assuming that there could be a final answer. Not only is it impossible to obtain universal agreement on such questions; it may even be one of the highest achievements of human society precisely to foster and protect a whole spectrum of different judgments about them. Should we pity, patronize, or penalize human beings who want nothing more each day for the rest of their lives than to doze in a drug-induced stupor? Should we execute drug runners and profiteers? Should we imprison a housewife who takes Valium or a successful lawyer who parties with cocaine on Saturday nights?

One standard definition of drug dependence is that of the American Psychiatric Association and the World Health Organization. They list nine criteria for dependence, which is further distinguished as *mild* when three of these criteria are met, *moderate* when five are met, and *strong* when seven or more are met. The criteria are:

- Taking the drug more often or in larger amounts than intended.
- Unsuccessful attempts to quit; persistent desire, craving.
- Excessive time spent in drug seeking.
- Feeling intoxicated at inappropriate times, or feeling withdrawal symptoms from a drug at such times.
- Giving up other things for the drug.
- Continued use, despite knowledge of harm to oneself and others.
- Marked tolerance in which the amount needed to satisfy increases at first before leveling off.
- Characteristic withdrawal symptoms for particular drugs.
- Taking the drug to relieve or avoid withdrawal.

The Merck Manual of Diagnosis and Therapy treats Drug Dependence (subtitled "Substance Use Disorders; Drug Addiction; Drug Abuse; Drug Habituation") in the section on Psychiatric Disorders, and has this to say:

> A single definition for drug dependence is neither desirable nor possible. The term **drug dependence of a specific type** emphasizes that different drugs have different effects, including the type and hazard of the dependence they produce. **Addiction** refers to a style of living that includes drug dependence, generally both physical and psychological, but mainly connotes continuing compulsive use and overwhelming involvement with a drug. Addiction additionally implies the risk of harm and the need to stop drug use, whether the addict understands and agrees or not.[10]

The *Merck Manual* goes on to say that drug abuse is only definable in terms of societal disapproval: usually, the term implies a use of drugs (1) for recreation, (2) as an escape from life's problems, or—in the long-term user—(3) to avoid the discomfort of withdrawal from the drug. The first two uses more or less establish *psychologic dependence*, and with some drugs this can be the only factor involved in establishing an intense craving and compulsive use. The last use implies *physical dependence*, which is an adaptation of the body to the nearly constant presence of the drug. The existence of physical dependence is defined by the correlative development of a *withdrawal syndrome*, an abnormal physiologic change that will occur upon abrupt discontinuation of the drug. There is no necessary correlation between the intensity of craving an individual may felt for a given substance and the presence or absence of physical dependence and its withdrawal syndrome: for instance, the *Merck Manual* joins the mainstream of traditional medical opinion in ascribing no physical dependence or tolerance to the use of cocaine, because only minor physiological symptoms attend its abrupt withdrawal.[11] However, numerous ex-users have testified to an overwhelmingly addictive craving for this drug, at least in its injected, snorted, or smoked forms. By contrast, phenobarbital, which is a slow-acting depressant drug rarely used for recreational purposes, can cause fatal epileptiform seizures if suddenly withdrawn from a longtime user.

The critical reader may object to these traditional distinctions. There is every reason to expect, and growing evidence to prove, that any intense craving has its origins in some neurotransmitter function in the brain; and the same can be said for such traditional "mental" phenomena as "habit" and "learning." But at this stage of knowledge of neurobiology, most of these higher functions are more or less black boxes.

Finally, either physical or psychologic dependence can be accompanied by *tolerance*, which is the need to progressively increase the dosage to obtain the same effect. Again, tolerance is quite different with different drugs. Amphetamine use in many people provokes a sort of vicious circle of ever-increasing tolerance, with ever-increasing drug intake, until the person enters a toxic psychosis or coma; for some, alcohol acts similarly. Nicotine, caffeine, and the opioids usually cause an initial tolerance, which then plateaus at a stable equilibrium where constant daily dosage elicits the desired effect. With all the drugs just mentioned, there is an increased craving that feeds the increasing tolerance. On the other hand, the classic psychedelics (LSD, mescaline, psilocybin) cause a very different sort of tolerance: a person who has taken one of these drugs will be essentially unresponsive to equivalent doses of any of them for the next few days, but after a few days of abstinence will once again be susceptible to their influence. However, there is almost never any desire to take these drugs on a daily basis, and so users rarely discover the sort of tolerance they cause.

Often in the case of cocaine, amphetamines, and alcohol, and almost always with nicotine, caffeine, and the opioids, the need to increase the dosage for the same effect is accompanied by the desire to attain the same effect, and it is these drugs which by and large are taken daily or not at all. Cocaine is subtly different from amphetamines. Large large doses of cocaine or the amphetamines can cause a toxic psychosis. However, the increased consumption shown by the many users who binge on cocaine until exhaustion sets in is not based on any increased tolerance but simply on an intense craving to repeat the experience, coupled perhaps to a dread of the letdown when the brief period of euphoria wears off.

What Is the Most Addictive Drug?

This is a difficult question to answer because there are significant predisposing conditions in the development of drug dependency other than the pharmacologic effects of the drug itself. These include varying cultural patterns, peer pressures, and individual psychological affects such as low self-esteem, social alienation, and feelings of impotence and helplessness. Opioid drugs like morphine are powerfully addictive to some individuals in some settings, yet hospitalized patients given days or weeks of full doses of these drugs rarely become addicted. Even patients with chronic pain problems requiring long-term administration usually are not addicts in the aforementioned *Merck Manual* definition (i.e. their life is not characterized as overwhelmingly involved with the use of the drug), although they may develop tolerance and physical dependence. A U.S. Surgeon General's report issued in February 1992 indicated that American medical practice, presumably out of excessive concern with the danger of inducing addiction, consistently underutilizes pain medication, causing needless suffering and even permanent injury. (One startling example is the practice of using no anesthetic or pain medication when major surgery is carried out on newborn infants.) Finally, drugs differ so much in their intrinsic pharmacology that any attempt to rank them by addictivity can be like comparing apples and oranges.

Nonetheless, we will consider two classifications of drugs according to their relative liability to induce dependence. The first is the U. S. Comprehensive Drug Abuse Prevention and Control Act of 1970, which established five "schedules" of drugs for law enforcement purposes:

- *Schedule I.* No prescriptions can be written for drugs in Schedule I, which are said to have a high potential for abuse, no accepted medical use, and a lack of accepted safety. Examples of drugs in this category are heroin (Chapter 2), marijuana (Chapter 7), and hallucinogens such as peyote, psilocybin, and LSD (Chapter 6).
- *Schedule II.* Drugs with an accepted medical use and a high potential for abuse. Prescriptions for these drugs must be handwritten (as opposed to telephoning), and refills cannot be renewed without a new prescription. Examples of drugs in this category are opioid narcotics like morphine, codeine, and Percodan (Chapter 2), and amphetamine stimulants like Ritalin, cocaine, and Benzedrine (Chapter 4).
- *Schedule III.* Drugs with an accepted medical use and moderate or low potential for physical dependence or high levels of psychological dependence. Prescriptions can be phoned in to a pharmacy by a doctor, and can be renewed every 6 months and refilled up to five times. Examples of drugs in this category are combination analgesic-narcotic preparations like Tylenol with codeine, and paregoric (Chapter 2).
- *Schedule IV.* These drugs are said to have low potential for physical dependence and psychological dependence. Prescription restrictions are as with the Schedule III drugs. Drugs in this category include the benzodiazepine hypnotics and anxiolytics (Chapter 3).
- *Schedule V.* These drugs have a low potential for abuse; often a prescription is not needed. Included are cough syrups with codeine (Chapter 2).

Although the U.S. Scheduled Drug Classification sounds quite authoritative, it originated in an effort to assign criminal penalties to the use, distribution, and sale of illegal drugs. There are some glaring inconsistencies; for instance, the crude plant material from *Cannabis sativa* (marijuana) is a Schedule I drug and should therefore have no legitimate medical use. However, about 13 patients in the United States were provided with marijuana cigarettes from government research plots for treatment either of glaucoma or of nausea associated with chemotherapy from 1976 to 1992 (the program was closed by the Bush administration); furthermore, the most active ingredient in the plant, the chemical responsible for its euphoriant effects, tetrahydrocannabinol, is a Schedule II drug marketed as Marinol for use as an antinauseant.

Additionally, there are several notable omissions from the list of scheduled drugs: alcohol, nicotine, and caffeine are all addictive drugs with little significant medical use (except that caffeine is used in some nonprescription analgesics and in some prescription drugs for migraine, the moderate social use of alcohol is arguably as effective an anxiolytic as the benzodiazepines, and nicotine patches have been shown to improve the symptoms of colitis). Alcohol, nicotine, and caffeine have as great a potential for addiction and are as capable of creating as great a physical dependence with a correlative withdrawal syndrome as many or most of the scheduled drugs. The difference in their treatment is probably chiefly cultural: these three drugs have been in extensive use in Northern Europe and North America for several centuries—in the case of alcohol, for several millennia.

A better evaluation of the relative addictive potential of a drug has been obtained from a survey of health professionals involved with the treatment of addicts. Responders were asked to prescind from social and economic pressures such as drug availability or acceptability, and to evaluate only the inherent addictive potential of each drug. Table 1.1 shows the average composite ranking of each drug on a 100-point scale from least (lowest score) to most (highest score) addictive-liable.

Table 1.1 Relative Addictivity of Common Drugs[12]

Nicotine	99	Alcohol	81	Marijuana	22
Smoked methamphetamine	98	Heroin	80	Ecstasy	20
Crack	97	Nasal amphetamine	80	Psilocybin (mushrooms)	19
Injected methamphetamine	92	Nasal cocaine	71	LSD	16
Valium	83	Caffeine	70	Mescaline (peyote)	15
Seconal	82	PCP	57		

The results are somewhat surprising. According to these respondents: (1) the most addictive drug known is not only not scheduled, it can be purchased without a prescription by anyone over the age of 18; (2) cocaine is about as addictive as coffee or tea; and (3) alcohol is about as addictive as heroin. Of course addictive here means only how difficult it is to shake the habit, not how much damage it may do to a person's life.

More recently, rating scales have been developed that are more nuanced, recognizing the several factors that contribute to the overall likelihood of a drug to become a "problem drug." One of these ranks the six most commonly used drugs—caffeine, alcohol, nicotine, marijuana, cocaine, and heroin—according to five categories: *withdrawal* (the extent and severity of physical withdrawal symptoms, if any); *reinforcement* (a measure of the substance's potential in both human and animal tests for inducing repeated self-administration); *tolerance* (to what extent increasing amounts of the substance must be taken to maintain satisfaction, and at what level of consumption equilibrium is finally reached); *dependence* (a measure of how difficult it is for a person habituated to its use to discontinue, and the degree to which persons experience craving for the substance even when they know it is doing them harm); and *intoxication* (the extent to which those who use the substance are mentally or physically incapacitated).

Two experts in the field of addiction, Dr. Jack E. Henningfield of the National Institute on Drug Abuse (NIDA) and Dr. Neal L. Benowitz of the University of California at San Francisco, were asked to rank the six drugs within each of these five categories, using 1 to indicate the most serious degree and 6 the least (a value in italics indicates a tie rating).[13] Tables 1.2 and 1.3 summarize their respective positions.

Table 1.2 Ranking of Drug Addictivity (Henningfield)

SUBSTANCE	WITHDRAWAL	REINFORCEMENT	TOLERANCE	DEPENDENCE	INTOXICATION	Σ
NICOTINE	3	4	2	1	5	15
HEROIN	2	2	1	2	2	9
COCAINE	4	1	4	3	3	15
ALCOHOL	1	3	3	4	1	12
CAFFEINE	5	6	5	5	6	27
MARIJUANA	6	5	6	6	4	27

Table 1.3 Ranking of Drug Addictivity (Benowitz)

SUBSTANCE	WITHDRAWAL	REINFORCEMENT	TOLERANCE	DEPENDENCE	INTOXICATION	Σ
NICOTINE	*3*	4	*4*	1	6	18
HEROIN	2	2	2	2	2	10
COCAINE	*3*	1	1	3	3	11
ALCOHOL	1	3	*4*	4	1	13
CAFFEINE	4	5	3	5	5	22
MARIJUANA	5	6	5	6	4	26

Notable is the two experts' exact agreement as to the rankings of the drugs with regard to liability to induce dependence (col. 5), with nicotine first and marijuana last. Both experts likewise agree that the withdrawal from alcohol is most problematic (indeed, unlike any of the other drugs considered, it can be fatal); and that alcohol, not heroin, is the drug causing the most intoxication—that is, clouding of reason, memory, and judgment. With both authorities, cocaine is the most reinforcing of all the drugs.

In both tables, a final column sums the rankings for each drug so as to give a rough measure of the relative overall likelihood of each drug to create problems in the life of the user. It can be seen that both authorities rank heroin as the most problematic and marijuana as the least. Indeed, Henningfield's rankings tie marijuana with caffeine, and Benowitz considers marijuana less problematic than caffeine.

Is Food an Addictive Substance?

Paraphrasing Abraham Lincoln, there are certainly some substances (cocaine, alcohol, heroin, tobacco) that will addict some of the people all of the time—but there are probably no substances that will addict all of the people all of the time. There are many reports of individuals who have tried Schedule I substances once and never used them again. Of every 10 people who light up a first cigarette, 9 will have trouble quitting, compared with 2 of 10 first-time cocaine users. In the final analysis, the definition of addiction must lie not in the physical dependence a substance may cause, nor in the intensity of any withdrawal syndrome, but in an overall (value) judgment of the extent to which the interaction of the individual and the substance disproportionately and inappropriately dominates his or her life. Food provides an excellent example of a "substance" that to the great majority of "users" is not at all addictive in the sense just defined. Yet we all quite obviously are physically dependent on food, and the "withdrawal syndrome," hunger, is something we encounter daily without panic, even with pleasure. Most of us are "social eaters," and derive enhanced enjoyment from food when we eat together. Nonetheless, to a significant minority of the population food has all the attributes of a drug.[14] Some suffer from the problems of bulimia and anorexia nervosa; there are others, the morbidly obese, who simply spend their entire day eating, to the point that they lose their employment, their family, and their health. Is food an addictive substance? Ordinarily not; but it can be. Sometimes one particular food can be addictive. The *New England Journal of Medicine* has reported a case of a man addicted to eggs:

> An 88-year-old man who lived in a retirement community was in generally excellent health with no significant atherosclerosis. He had never smoked, and he never drank excessively. His consumption of 20–30 eggs/day was verified; his physician and friends testified he had eaten this many for well over 15 years. He always soft boiled the eggs and ate them throughout the day, keeping a careful record, egg by egg, of each egg eaten. The nurse at the retirement home confirmed that about 2 doz eggs were delivered to him each day. A psychiatrist and a psychologist described this habit as compulsive behavior; efforts to modify it had all failed. The patient said, "Eating these eggs ruins my life, but I can't help it."[15]

Is Exercise Addictive?

Jogging can, indeed, be addictive—even fatally so. Two Finnish runners are described in the *Journal of the American Medical Association*. Both had aortic stenosis and were told that jogging was threatening their lives. Both refused to quit. One died suddenly at age 22, while continuing his 120–210 km/week regimen in preparation for his second marathon. The other, 46, continued to run frequent marathons despite fainting and suffering bone fractures and mild paralysis from head injuries. But he was still alive in 1992. The Finnish doctors reporting the cases conclude that "the mechanism of this addiction to jogging, which is dangerous for those with unhealthy hearts, is unknown, but endurance exercises increase serum β-endorphin."[16]

Are Shopping, Gambling, and Russian Roulette Addictive?

Moreover, there are some behaviors that do not involve interaction with any substance external to the body, and yet have all the other characteristics of a drug addiction. Compulsive shopping is one, compulsive gambling another. Several antidepressant drugs have proved helpful in treating these disorders, which seems to substantiate the likelihood that they involve addictions to behaviors that release endogenous NTs. Fluvoxamine has been shown to be quite effective in helping compulsive shoppers,[17] while clomipramine and carbamazepine seem to reduce the craving in the brain for gambling.[18] Another sort of "gambling"—with one's life rather than with one's property—is Russian roulette. The late author Graham Greene relates in the first volume of his autobiography, *A Sort of Life*, that for many years in his late teens and early twenties he compulsively played this "game." These were not suicide gestures, according to Greene, but attempts to relieve his overwhelming boredom. (Boredom is one of the reasons drug users frequently adduce for beginning their habit; escape from the meaninglessness of life is another—significantly, the second volume of Greene's autobiography is titled *Ways of Escape*.) Greene's description of his toying with suicide makes frequent reference to drugs, and he remarks that the game lost its interest when the drug parallel deteriorated from adrenaline to aspirin:

> Boredom seemed to swell like a balloon inside the head; it became a pressure inside the skull: sometimes I feared the balloon would burst and I would lose my reason. . . . I can remember very clearly the afternoon I found the revolver in the brown deal corner cupboard in a bedroom which I shared with my elder brother. . . .
>
> I had been reading a book (I think Ossendowski was the author) which described how the White Russian officers, condemned to inaction in southern Russia at the tail end of the counterrevolutionary war, used to invent hazards with which to escape boredom. One man would slip a charge into a revolver and turn the chambers at random, and his companion would put the revolver to his head and pull the trigger. The chance, of course, was five to one in favor of life....
>
> There was no hesitation at all: I slipped the revolver into my pocket, and the next I can remember is crossing Berkhamsted Common toward the Ashridge beeches. Perhaps before I had opened the corner cupboard, boredom had reached an intolerable depth. The boredom was as deep as the love and more enduring—indeed it descends on me too often today. For years, after my analysis, I could take no aesthetic interest in any visual thing: staring at a sight that others assured me was beautiful I felt nothing. I was fixed, like a negative in a chemical bath. Rilke wrote, "Psychoanalysis is too fundamental a help for me, it helps you once and for

> all, it clears you up, and to find myself finally cleared up one day might be even more helpless than this chaos."
>
> Now with the revolver in my pocket I thought I had stumbled on the perfect cure. I was going to escape in one way or another. . . . This was not suicide, whatever a coroner's jury might have said: it was a gamble with five chances to one against an inquest. The discovery that it was possible to enjoy again the visible world by risking its total loss was one I was bound to make sooner or later. I put the muzzle of the revolver into my right ear and pulled the trigger. There was a minute click, and looking down at the chamber I could see that the charge had moved into the firing position. I was out by one. I remember an extraordinary sense of jubilation, as if carnival lights had been switched on in a dark drab street. My heart knocked in its cage, and life contained an infinite number of possibilities. It was like a young man's first successful experience of sex—as if among the Ashridge beeches I had passed the test of manhood. . . .
>
> This experience I repeated a number of times. At fairly long intervals I found myself craving for the adrenaline drug, and I took the revolver with me when I returned to Oxford. . . . The revolver would be whipped behind my back, the chamber twisted, the muzzle quickly and surreptitiously inserted in my ear beneath the black winter trees, the trigger pulled. Slowly the effect of the drug wore off—I lost the sense of jubilation, I began to receive from the experience only the crude kick of excitement. It was the difference between love and lust. And as the quality of the experience deteriorated, so my sense of responsibility grew and worried me. . . . It was back in Berkhamsted during the Christmas of 1923 that I paid a permanent farewell to the drug. As I inserted my fifth dose, which corresponded in my mind to the odds against death, it occurred to me that I wasn't even excited: I was beginning to pull the trigger as casually as I might take an aspirin tablet. I decided to give the revolver—since it was a six-chambered—a sixth and last chance. I twirled the chambers round and put the muzzle to my ear for a second time, then heard the familiar empty click as the chambers shifted. I was through with the drug.[19]

Later, Greene describes using alcohol, and still later a dentist's anesthetic as alternative escapes from boredom and from disappointment in love. Oddly enough, it is boredom with the "drug experience" itself which finally leads him to quit—and a similar story could be told by many adolescents who have experimented with marijuana, alcohol, or cocaine only to find them finally a more boring alternative than life itself, even with all its pains.

Are God, Sex, and Work Addictive?

One can be addicted to traditionally "virtuous" behaviors as well. Religious professionals are well acquainted with people who abuse religious ritual and ideation in what can only be described as an addictive manner—indeed, an Anglican priest has written a book entitled *When God Becomes a Drug*,[20] and a group of psychologists have published *Toxic Faith: Understanding and Overcoming Religious Addiction*.[21]

When some experts were asked to apply the standard psychiatric definition of addiction (a positive response to three of nine test questions such as: Do you curtail important social, occupational, or recreation activities because of X? Do you use or do X despite persistent problems caused thereby? Do you suffer cravings, anxiety, depression, jitters when unable to obtain X?) to such substance usage or behaviors as *eating chocolate, jogging, shopping, sex, work, watching television,* or *mountain climbing*, many of these proved to be just as addictive as marijuana and much more so than LSD.[22]

Polydrug Users

Intravenous drug users in the United States and in the Netherlands are very likely to be users of many drugs sequentially or simultaneously: alcohol, tobacco, marijuana, cocaine, heroin. . . . Nor is this an entirely new phenomenon. At the close of the 19th century, the writer Aleister Crowley and his friend Allan Bennett were members of the Order of the Golden Dawn, a group of theosophists interested in mysticism and magic whose members included William Butler Yeats.[23] Crowley wrote of Bennett that "his cycle of life was to take opium for about a month, when the effect wore off, so that he had to inject morphine. After a month of this he switched to cocaine, which he took until he began to 'see things,' and was then reduced to chloroform."[24] Bennett, who had been trained as an analytical chemist, eventually became a Buddhist monk in Sri Lanka.

In the drug culture of the 1960s, celebrities were notorious for the simultaneous use of a cornucopia of drugs. A frequently cited antihero of this sort of behavior is Lenny Bruce, whose "usual number" was to smoke a few joints as a warm-up to an IV injection of 12 Dilaudid pills (an opioid) mixed with an ampule of Methedrine (an amphetamine) followed by a "fistful" of mescaline tablets (a psychedelic) washed down with a "chocolate Yoo-Hoo"—and possibly a few double scotches.[25] Somewhat miraculously, Bruce lived to his 40th birthday.

The characters in the stories of the current novelist Thom Jones are perhaps typical in their almost lovesick enthusiasm for drugs—any drugs:

> The Dramamine had hit her fast behind all that wine. She picked up four more Stelazine tablets and swallowed those with more wine. . . . She looked at her pills: her cache of Librium, glossy black-and-green capsules —500 or more; Valiums in blue; Xanax all pearly white; red-and-gray Darvons; Ludiomils in good-morning-sunshine orange; tricolored Tuinal in red, redder and baby blue . . . multicolored Dexedrine spansules, passionate purple Parnate; there were her Nembutals, and the sea-green, let's-do-the-job-up-right Placidyl gel caps (Baby Dills)—the pills and capsules suddenly became an object of immense beauty, a treasure. It was better than unearthing a pirate captain's sea chest . . . better because pills did things. Drugs altered sensations. They could alter the worst sensation—permanently.[26]

Some Misconceptions

How is it that people start using illegal drugs? Part of the prevailing mythology is that there are drug "pushers" out there, usually in school playgrounds seductively proffering their wares as bait to the unwise. This may occasionally happen, but it is actually prima facie quite unlikely: most children will not take candy from strangers, let alone drugs. Furthermore, the overwhelming majority of drug users trace their first use to peer settings, where they became interested because they had trusted friends who already used drugs. Many people try drugs in settings like these and find the results—even with pharmacologically powerful drugs like heroin and cocaine—unpleasant or, even if pleasurable, incompatible or undesirable in terms of their overall lifestyle and life choices. Only a small percentage of those who experiment with drugs, usually in their adolescence or early adulthood, actually go on to become habitual users. And of these, some use only occasionally and some after a few years discontinue any use. But some become compulsive users. Why? Certainly the objective pharmacology of the substance itself, isolated from social and psychological set and setting, plays a role. This is supported by animal tests in which rats and monkeys will continue to self-inject cocaine, heroin, and alcohol even to the point of convulsions, exhaustion, and death.[27] But it is certainly not an all-determinative role. There is also the factor of nature versus nurture.

Nature versus Nurture

At one time, it was thought that there was such a thing as an addictive personality type. However, careful studies of the psychological profiles of alcoholics and addicts have shown that they do not differ significantly from the ordinary population. Studies of a presumed genetic predisposition to alcoholism (the addiction most extensively studied) are conflicting, but there is a residue of evidence (e.g., identical twins raised in separate families) that seems unassailable. Thus, if there is no addictive personality, there is probably an addictive physiology: in this view, the addict is probably self-medicating some sort of neurochemical deficit. Surely the case of Dr. William Halsted, the first head of surgery at Johns Hopkins Hospital, fits this pattern: despite having an outstandingly successful career and a happy marriage, and despite his acute awareness of the risk cocaine and morphine use posed to both of these, and finally despite extraordinary efforts to quit and extraordinary help from such loyal friends as Dr. William Osler, he was never able to eliminate the use of intravenous morphine—although he was able to quit injecting cocaine and to reduce his morphine habit to a minimum daily dose that escaped easy detection and allowed him to function without any obvious incapacity. (More detail about Halsted can be found in Chapters 2 and 4.) What is stunning about his case is why he found it impossible to quit a habit that many others with far less external support and structure have been able to drop with no great difficulty. The most likely explanation points to something in his physiological makeup.

Although appeals to genetics and idiopathic neurochemistry seem to refer the question to a black box, drug use can be an understandable response to a sufficiently harsh or deprived environment. Thousands of U.S. troops in the Vietnam war became heavy users of smoked and intravenous heroin, which was cheap and readily available. But this seems to have been mostly a response to the acute stress of the situation; the overwhelming majority of them had no trouble quitting the habit after their return home.

A recent study of 95 male adoptees separated at birth from their biological parents[28] confirmed several previous studies by isolating two independent genetic factors involved as predictors of drug/alcohol abuse (a biological parent who was either alcoholic or exhibited antisocial personality disorder) and one independent environmental factor (a disturbed adoptive parent, e.g., one who was an alcohol/drug abuser or had other psychiatric disorders or marital/legal problems).

Does Long-Term Use of a Drug Cause Permanent Need for the Drug?

That is: once an addict, always an addict? Perhaps so, at least with some people and some drugs. Some people experience craving for nicotine years after giving it up, some people after years of high levels of coffee consumption claim it is absolutely impossible for them to experience normal mental functioning without caffeine, and there are many authorities who consider long-term opioid addicts unable to function normally without a minimal daily maintenance dosage. Neurobiological data support this impression: in detoxified addicts, unlike controls, the plasma levels of ACTH, an important hormone of the adrenal cortex, do not increase in response to naloxone, an opioid antagonist.[29] Also, long-term addicts may relapse without experiencing any conscious craving for the drug; the drug-seeking and drug-administering behavior has become an automatized learning that corresponds to a craving which can "exist at a physiologic level in the absence of conscious awareness."[30] Correlative to this is the phenomenon demonstrated in recent studies of the physiology of pain: after months of experiencing pain in a certain part of the body, a kind of "kindling" (like that long known to occur from epileptic seizures) takes place whereby the person will continue to experience the pain long after its physiological cause is gone.

Great strides have been made in prying open the black box of genetics and neurochemistry alluded to as "nature" in this section. Much is now known about the normal functioning of the mass of intertwined nerves we call the brain, and much of this information illumines how drugs affect the mind. The next section explores the *molecular biology* of brain functioning—that is, how the neurons of the brain communicate with each other at the molecular level to provide the substratum for what we experience as consciousness: thought, memory, and feelings. By entering into this substratum, psychoactive molecules can affect our feelings: from pain to pleasure, from desire to despair.

The Biological Context: Neurotransmission

Neurons and Neurotransmission

Nothing is so obvious or immediate in our experience as the phenomenon we call consciousness. Nor is anything so mysterious or obscure when we try to understand it. It is the starting point for profound systems of religious belief, of theology, and of philosophy. But whatever else it may be, consciousness is also dependent on the biological functioning of the brain, which is a vast and intricate network of more than 10 billion nerve cells (neurons) with several thousand times as many interconnections among them. This network is what physiologists call the central nervous system (CNS); the network's continuation outside the brain and spinal cord is referred to as the peripheral nervous system, or simply the periphery. Although this total system is obviously of enormous complexity both in numerical magnitude and in its final effects, there is little doubt where and how psychoactive drugs exert their effects on it. This is because all the neurons of the CNS interact with each other in only one way: by releasing signaling chemicals called neurotransmitters (NTs) at their points of interconnection, called synapses. All the psychoactive drugs we will consider in this book act by mimicking or modifying the release of NTs.

To understand this, we must consider the structure and functioning of the neuron. Neurons are cells specialized for receiving, conducting, and transmitting signals. All neurons are divided into three parts. The central feature of the neuron is the cell body, which contains the nucleus of the cell, and outside of which certain more complicated processes such as protein synthesis (including the synthesis of peptide NTs) cannot take place. Extending from the cell body are two types of filamental processes specialized in receiving and transmitting signals from the cell to other cells that, in the case of the CNS, are other neurons. One of these constitutes the second part of the neuron: a single long process called the *axon*, which can traverse several centimeters of the brain (or even a meter in the periphery) and which extends from each neuron and conducts signals from the cell to more distant neurons. At the end of the axon, multiple branchings like the roots of a plant occur so that the axon potentially communicates with many thousands of other neurons. In most cases, as we shall see, this transmission occurs via the release of specific chemical substances, NTs, from the termini of the axon. (Peptide NTs must be transported down the length of the axon from their locus of synthesis in the cell body, but the smaller nonpeptide NTs can be synthesized from common amino acids by means of enzymatic reactions throughout the axon). Dendrites constitute the third part of the neuron. These are multiple smaller outbranchings (*dendron* is Greek for tree) from the cell body whose chief function is to receive the signals from the axons of other neurons. (However, dendrite-to-dendrite communication is also possible.) There can be as many as 100,000 dendrites on a single neuron. The locus of interaction between two neurons is, as we have said, the synapse.

Signaling between neurons at the synapse is almost always chemically transmitted, but the signal within the neuron itself consists of an electrochemical *action potential* radiating down the axon (and continually amplified along the way) at a speed of about 100 m/sec. While speedy enough, this is obviously not the speed of light or even the speed of electric transmission in a copper wire. The neuron with its extended axon does bear some resemblance to an electric wire, including a myelin sheath that functions as an insulator. But a wire transmits any electric potential, however small, whereas neurons only transmit an action potential when they experience an overall stimulus (most often by NTs, occasionally by direct neuron-to-neuron electrical synapses) that exceeds a certain threshold strength. The action potential then discharges in an all-or-nothing fashion down the axon.

How is this action potential generated and propagated? For details, the interested reader will need more specialized works, but roughly speaking, the process is as follows.[31] By means of a sodium/potassium ion pump that drives sodium out of the neuron and potassium into the neuron, combined with pores that leak potassium slowly out of the cell along its concentration gradient, the interior of the neuron ordinarily maintains, relative to the aqueous saline osmotic surroundings of the bloodstream, (1) an overall excess of anions over cations and hence a negative electrical potential; (2) an excess of potassium ions; and (3) a deficit of sodium and chloride ions. (The first two of these may seem somewhat contradictory; one must remember the large number of phosphate and small organic acid anions.) Although the difference in potential inside and outside the neuron is only about 50-100 mV, the charges are held apart by a very thin plasma membrane of about 5 nm, thus producing a significant voltage gradient of about 10^5 V/cm. The neuron is thus a sort of biological capacitor, and the action potential consists of a momentary pulse of depolarization. This depolarization is initiated via stimulation by an adjacent neuron interacting at the synapse (see next section), and it is propagated by the opening of *voltage-gated Na^+ channels*. These channels selectively allow an inrush of sodium ions (about 8,000 ions/msec) into the neuron when they are triggered to open upon detecting a depolarization in an adjacent part of the neuron. The action potential is thus propagated in one direction down the axon. At the end of the axon, the depolarization pulse triggers the release of an NT at the synapse.

Neurotransmission at the Synapse

While it is possible for neurons to communicate depolarization directly via *gap junctions* (small channels connecting cytoplasmic fluid) at *electrical* synapses in specialized organs, all the synapses connecting neurons in the CNS communicate by *chemical* synapses. Such synapses involve no direct electrical connection between the neurons. When the action potential reaches the terminus of the axon, the presynaptic neuron releases an NT from *synaptic vesicles* on the membrane surface into the small space between the neurons, the *synaptic cleft*. (This happens by *exocytosis*, a cunning process whereby the vesicle merges transiently with the presynaptic cell membrane; the integrity of the cell membrane is never lost.) Upon release from the presynapse, the NT rapidly diffuses across the very narrow synaptic cleft and binds to *receptors* on the surface of the postsynaptic cell that are very selectively designed to accommodate the specific NT's exact molecular shape and structure. (The word *bind*, in the context of all that follows concerning receptor interaction, means such intermolecular interactions as hydrogen bonding and van der Waals forces whereby a large protein structure wraps around a small NT and in so doing undergoes a change in its own conformation; there are generally no covalent bonds broken or formed.)

Most NTs interact primarily with the postsynaptic receptor (R-1 in Figure 1.1); however, *autoreceptors* on the presynaptic neuron can, by a feedback mechanism, either increase (R-2, +) or decrease (R-2, –) the release of NTs. (An initial sharp increase can be followed by an abrupt decrease, thus producing a step-function response in the postsynapse.) Other presynaptic receptors (R-3) respond to NTs from adjacent neurons to either increase or decrease release of NTs. Finally, there are receptors (R-4) on the presynapse for other neuromodulators such as steroids and prostaglandins; binding at these receptors can also modulate release of NTs.

Binding of an NT to the postsynaptic receptor can induce a response by one of two major mechanisms. The first provides the most rapid response; it consists of a receptor that controls the opening of a transmitter- or ligand-gated ion channel. There are two major types of transmitter-gated ion channels corresponding to the two major types of neurotransmission. *Excitatory* neurotransmission is mediated by cation ion channels: upon binding of NT to receptor, the channel opens to allow a selective rush of cations (most commonly Na^+, but also K^+ and Ca^{++} ion channels) into the second neuron and cause a

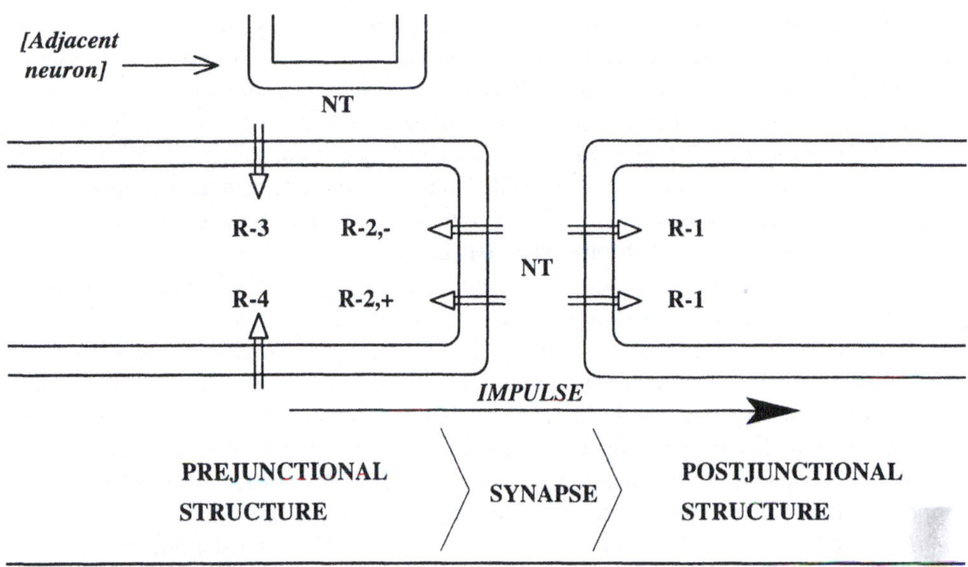

Figure 1.1 Receptor sites at the synapse.

localized depolarization. If sufficient channels open in a short enough time, an action potential will be initiated in the second neuron. Acetylcholine, glutamate, and serotonin are three such excitatory NTs. On the other hand, neurotransmission can be *inhibitory*. Glycine and γ-aminobutyric acid (GABA) are two such inhibitory NTs. In this case, the NT binds to a receptor located on a transmitter- or ligand-gated Cl^- channel. These channels are selective for the transit of chloride ions that, like sodium ions, are more concentrated in the extracellular medium. The inrush of chloride anions into the postsynaptic neuron causes hyperpolarization of the neuron, and thus inhibits the likelihood of its firing.

The second type of response induced by NT binding at the postsynaptic receptor is less direct, but it can be much more nuanced in its final effect. In this very common system, the receptor is said to be *G-protein-coupled*.[32] The receptor in this case is not part of an ion channel but is a complex structure that extends through the cell membrane into the cytoplasm, where it is able to be bound to a G-protein (so called because the protein is itself bound to a molecule of guanosine triphosphate, GTP). When the NT or other (e.g., drug) ligand links to the extracellular receptor site on the receptor protein in the synapse, its extension into the cytoplasm alters, exposing a binding site within the cell for this G-protein. Soon a stray G-protein, which ordinarily carries a molecule of guanosine diphosphate (GDP), comes along and hops on board this site. By taking a seat in the receptor protein, it loosens its grip on the GDP molecule, which wanders off; but the vacant site is now ready to take on a molecule of GTP. When a GTP hops on, the G-protein leaves its seat on the receptor complex. G-protein with a GTP molecule, as opposed to one with GDP, is able to bind to a molecule of adenylate cyclase in such a way as to activate this enzyme. Adenylate cyclase begins churning out numerous molecules of cyclic adenosine monophosphate, cAMP, from adenosine triphosphate (ATP).

This is a powerful second messenger, a major intracellular signaling molecule in the postsynaptic cell that, depending on the cell, can initiate a variety of processes within the cell. This whole process allows for a diverse response from different target neurons to the same external NT at the synapse; it also involves an automatic amplifying process, since one extracellular NT signal has produced multiple intracellular cAMP messenger molecules. Still further diversity is possible because G-proteins can also couple to other intracellular messenger systems, chiefly calcium (released from intracellular storage sites into the cytosol) and inositol triphosphate (IP_3).

Still another feature of G-protein-coupled receptors is that the binding site for the NT, unlike the relatively superficial site for ligand-gated ion channel receptors, is buried much more deeply into a pocket within the transmembrane domain, where there are possibilities for multiple interactions between NT and receptor. Consequently, the binding affinity of ligand for receptor in G-protein-coupled receptors is typically several orders of magnitude higher than for ligand-gated ion channel receptors. And the same NT can couple to the same receptor site (cloned) in two different systems and yet produce exactly opposite effects in the target neuron. This is because there are G-proteins capable of *inhibiting* cAMP as well as G-proteins that activate cAMP (or calcium, or potassium channels, or sodium channels, etc.)

Agonists, Antagonists, Partial Agonists, Inverse Agonists

Most of the drugs to be discussed in this book are known to act by binding like the endogenous NTs to one or more of the types of receptor described previously. If the drug binds like the NT and evokes the same response, it is said to be an *agonist* at the receptor (the NT itself is, of course an agonist ex officio). But some drugs, called *antagonists* or *blockers*, are known to bind to the receptor site even better than the native NT, and yet elicit no response. How this can happen in view of the mechanisms discussed previously is not always known. Presumably, the drug can fit well enough to occupy the receptor site but not well enough to alter the conformation of the receptor protein in a way similar enough to the natural NT to cause an ion channel to open, engage a G-protein, and so forth.

If an agonist and antagonist simply compete with one another so that the measured response is a function only of their relative concentration (dose ratio), they are said to exhibit *competitive antagonism*. With other drugs the agonist concentration can be increased indefinitely and yet the effect of the antagonist drug cannot be overcome. Such a drug is said to be a *noncompetitive antagonist*. This can happen because the site occupied by the antagonist is not really the same as that occupied by the agonist; no amount of agonist will succeed in displacing the antagonist because they are going to different places. An instance of this is the drug *tubocurarine*, which is derived from a South Amerindian arrow poison and which causes paralysis by blocking the acetylcholine receptor (AChR) at the receptor site between a motor neuron and a muscle. Even in the presence of increasing amounts of nicotine, a powerful agonist at the same AChR, the paralysis cannot be entirely abolished. It turns out that the tubocurarine has two effects: one is a competitive antagonism of the AChR site, but the other is a noncompetitive blockage of the ion channel itself that is gated by the AChR.[33] One further possibility is that of a chemical substance that functions as a *partial agonist*. A drug that has partial agonist activity can act on the same receptors at the same concentration as a full agonist and yet only evoke a partial response. Finally, there can be *inverse agonists* at a receptor. The best known instance of this is a subtype (below) of the GABA receptor, the $GABA_A$ receptor, where at least three sites other than the site for GABA itself are known to surround the chloride ion channel that is gated endogenously by GABA. One of these sites is the *benzodiazepine* receptor site. This is where such familiar drugs as Valium and Halcion (see Chapter 3) bind to cause their typical effects: sedation and alleviation of anxiety. Occupancy of the benzodiazepine site by a typical benzodiazepine agonist such as Valium modulates the GABA ion channel so that it more efficiently allows chloride ion to enter the neuron, thus hyperpolarizing the neuron and generally exercising an inhibitory, sedative effect on the CNS. But a class of drugs exemplified by DMCM[34] has an opposite effect, causing anxiety and epileptic seizures in experimental animals.

Receptor Subtypes

Receptors for the same NT are not necessarily identical, but can be differentiated into subclasses. For instance, it has long been known that cholinergic receptors can be divided into the N_1 and N_2 receptors, which are responsive to nicotine (the alkaloid from tobacco) and the M_1 and M_2 receptors, which are responsive to muscarine (a toxic alkaloid from the mushroom *Amanita muscaria*). Similarly, there are four subclasses of adrenergic receptors: α_1, α_2, β_1, and β_2. Successful drugs have been developed by exploiting these differences: *propranolol*, an antihypertensive and anti-angina medicine, is a β-adrenergic blocker; anti-ulcer drugs like *cimetidine* block the H_2 histamine receptors of the stomach, while ordinary OTC antihistamines found in allergy and cold medications block the H_1 receptors of the upper respiratory tract. There is constant research to develop agents that selectively agonize or antagonize further subclasses of these and other receptors, particularly the dopaminergic and serotonin receptors.

Receptors are often characterized for which there are no known endogenous agonists, only to have the endogenous substances be discovered at a later time: this has been the case with the opioid receptors most strikingly, and later in the case of the benzodiazepine and the cannabinoid receptors. The benzodiazepine receptor is responsible for the action of such common drugs as Valium and Halcion (see Chapter 3) and with the barbiturates and alcohol is associated with the GABA neurons. Although an endogenous peptide NT has been isolated that binds to the benzodiazepine receptors, displacing benzodiazepines, it has been found to have the properties of an inverse agonist. That is, in contrast to the actions of the benzodiazepines, it induces (instead of suppressing) anxiety, conflict, and seizures.[35] The cannabinoid receptor is agonized by the active ingredient in marijuana, Δ^9-tetrahydrocannabinol. Only recently, an endogenous agonist of the cannabinoid receptors has been discovered, *anandamide* (see Chapter 7).

Finally, receptors are protein complexes, and as such they are continuously synthesized and degraded by the cell, with half-lives of from days to weeks. If continually stimulated they become hyposensitive (*downregulated*); if not stimulated or blocked by antagonists they become hypersensitive (*upregulated*).

Still More Complexities. As one of the pharmacologists who developed Prozac remarked to Dr. Peter Kramer, "if the human brain were simple enough for us to understand, we would be too simple to understand it."[36] In any case, let us consider two of the many facts that mar the pristine landscape just portrayed: (1) *Dale's law* (1935) postulated that one and only one NT is released from a given neuron's axonal termini. This seems like a tidy and reasonable hypothesis. However, it now seems that, while Dale's law is probably true for the monoamine NTs (no more than one per neuron), a neuron may release one monoamine NT from its termini that is accompanied by a second neuropeptide or a prostaglandin NT or hormone.[37] What the effect of this might be in terms of psychopharmacology or behavior is not yet known; (2) although the termini of axons are the usual places where NTs are released, they can also be released from dendrites. For example, dendrites in the cells of the substantia nigra release dopamine, which then diffuses over the general region to act on axons and dendrites of GABAergic and dopaminergic neurons in other regions of the basal ganglia. The dendrites from one cell can also communicate directly with the dendrites of an adjacent cell.

Drugs at the Synapse

A given drug can affect the NT signal at the synapse via at least one or more of the following mechanisms: (1) it can increase or decrease NT *synthesis*; (2) it can inhibit or enhance NT *transport* to the presynaptic storage vesicles; (3) it can modify the *storage* of the NT in the vesicles; (4) it can modify the *release* of NT from the vesicles; (5) it can accelerate or retard the *degradation* of NT within the synapse; (6) it can block *reuptake* of NT from the synapse; (7) it can directly mimic a NT and act as an *agonist* at the receptor; (8) it can *block* the receptor; and (9) it can affect the *up-* or *downregulation* of the receptors themselves.

The action of a given drug on a receptor has been likened to that of a key in a lock: a specific key (the agonist drug) can fit in a lock (the receptor) and turn the lock (induce a physiological response); continuing the analogy, an antagonist drug fits in the lock but cannot turn and thus effects no physiological response while also, by blocking the keyhole (the receptor site), preventing an agonist drug from having any effect.

As knowledge of the complexity of the drug-receptor interaction has grown, this model has appeared increasingly inadequate. In particular, the receptor is far from being a rigid socket into which the agonist drug must exactly fit; rather, both agonist and receptor can assume a new conformation in their interaction with each other. And just because a given drug interacts with the same receptor protein as a known NT does not mean it is interacting at exactly the same site in the protein as does the endogenous NT; there are often several different adjacent sites for several different classes of drug interaction.

Perhaps a more adequate analogy from everyday life would be the interaction of the human hand (receptor) and a door handle (agonist) in the process of opening a door. Think for a moment of the enormous variety of door, cabinet, and drawer knobs and handles. The human hand has to assume a variety of quite different conformations in order to use all these devices, yet we do so with little or no inconvenience. Analogously, the quite varied molecular structures with opioid or anxiolytic properties can still be interacting at the same receptor. If you are right-handed, try opening and closing doors for a day using your left hand. With many doors you will find there is no great problem, while with others, you will find it awkward; analogously, some agonist-receptor interactions are achiral, others chiral.

Imagine an anthropologist from another galaxy whose only knowledge of the human hand was an extensive collection of doorknobs, steering wheels, keyboards, computer mice, handles, and cranks. How well could he deduce the shape and structure of the human hand? Molecular pharmacologists are in much the same position when they attempt to deduce the shape of a receptor from the structures of the various drugs that interact with it. But the task grows easier from year to year: molecular biologists are now often able to isolate the genetic code for a given receptor, clone the receptor, and use computer models of the receptor and drug to predict the most "comfortable" (lowest energy) fit of the drug within the receptor.

Let us now consider the exact structure of the dozen or so small NT molecules that are the natural (endogenous) signaling messengers in the brain.

The Chemical Context: Neurotransmitters

A Prefatory Caution

With all that is understood about receptors and neurotransmission, one might expect that much light could be shed on the human phenomenon of drug use and abuse. But the darkness is still quite extensive. It is rightly considered to be a significant breakthrough when the effects of a psychoactive drug on human thought and behavior can be traced to the drug's interaction with a class or subclass of NT receptor types in the brain. However, a little reflection, a few steps back from the tree to view the forest, and a prudent skepticism sets in. As the authors of a highly regarded text on neuropharmacology caution:

> What is enormously difficult to comprehend is the contrast between the action of a drug on a simple neuron, which causes it either to fire or not to fire, and the wide diversity of central nervous system effects, including subtle changes in mood and behavior which that same drug will induce. . . At the molecular level, an explanation of the action of a drug is often possible; at the cellular level, an explanation is sometimes possible; but at a behavioral level, our ignorance is abysmal. There is no reason to assume, for example, that a drug that inhibits the firing of a particular neuron will therefore produce a depressive state in an animal: there may be dozens of unknown intermediary reactions involving transmitters and modulators between the demonstration of the action of a drug on a neuronal system and the ultimate effect on behavior.[38]

Nonetheless, very much has been discovered about the interaction of many psychoactive drugs with the functioning of the NT systems. A brief summary of the most important of these systems and a description of the chemical structures involved will be helpful in understanding much of the material about specific drugs in the chapters that follow.

The Major Neurotransmitters

***Acetylcholine (ACh,* 1-1,** *Figure 1.2).* Acetylcholine is the major NT of the neurons controlling muscular activity in the periphery; it is also the NT of many neurons in the CNS (particularly the basal ganglia and motor cortex). Neurons using ACh as their NT are often collectively called *cholinergic*. The stimulation of receptors by ACh is terminated by the destruction (hydrolysis) of ACh by *acetylcholinesterase* (*ACE*), which is present on neurons and neuromuscular junctions. "Nerve gases" inactivate ACE so that muscles are uncontrollably stimulated by ACh, with resultant respiratory failure. The opposite effect is caused by curare (tubocurarine), which is a competitive neuromuscular blocking agent: it binds to ACh receptor sites at the postjunctional membrane, competitively blocks the transmitter action of ACh, and thereby causes paralysis.

Alzheimer's and Acetylcholine. Autopsies of Alzheimer victims show depleted stores of ACh and increased accumulation of a natural brain protein, β-amyloid. Research at the National Institutes of Health seems to show that these phenomena are related: the protein causes leakage of choline through cell membranes.[39] The single drug presently given FDA approval for treatment of Alzheimer's is a cholinesterase inhibitor, *tacrine* (tetrahydroaminoacridine), marketed as Cognex. A mixture of hydrogenated ergot alkaloids, *ergoloid mesylates*, has also shown some benefit: it probably acts by increasing activity at multiple NT systems including the AChR.

***Dopamine (DA,* 1-7,** *Figure 1.3).* DA, epinephrine, and norepinephrine are collectively referred to as *catecholamines* because they are amino derivatives of catechol (1,2-dihydroxybenzene). DA is the NT of some peripheral nerve fibers and of many CNS neurons (particularly in the substantia nigra, midbrain, hypothalamus). Its formation begins with the amino acid tyrosine (**1-5**, itself formed from phenylalanine, **1-4**), which is taken up by dopaminergic neurons and converted by tyrosine hydroxylase to 3,4-dihydroxyphenylalanine (*DOPA*, 1-6); DOPA is decarboxylated (by aromatic L-amino acid decarboxylase) to DA, which is stored in vesicles. After release, DA interacts with dopaminergic receptors and is then pumped back by active reuptake into the presynaptic neurons.

The Mesotelencephalic Dopamine System. Evidence has accumulated (particularly from experiments with laboratory animals trained to self-administer abusable drugs) that the major mediator of reward-reinforcing mechanisms in the mammalian brain is a system of dopaminergic neurons in the ventral tegmental area.[40] It is likely that all habit-

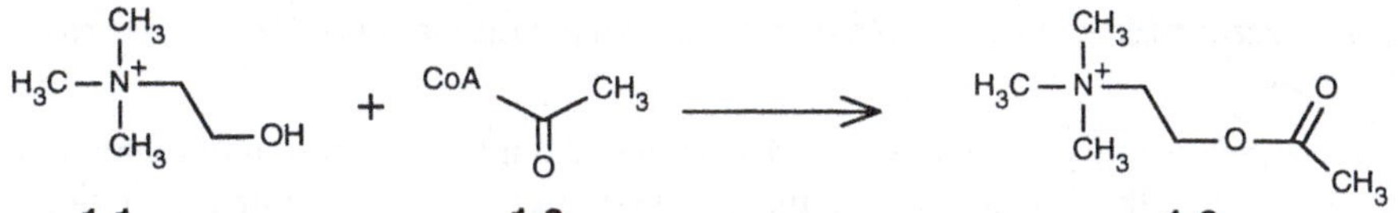

CH_3, H_3C-N^+, CH_3, OH + CoA, CH_3, O ⟶ CH_3, H_3C-N^+, CH_3, O, O, CH_3

1-1 **1-2** **1-3**

Figure 1.2 Biosynthesis of acetylcholine.

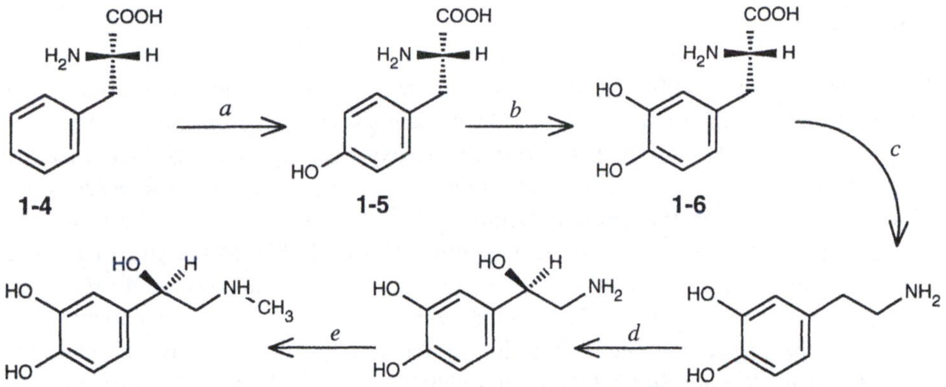

a: phenylalanine hydroxylase; b: tyrosine hydroxylase; c: dopa decarboxylase; d: dopamine β-hydroxylase; e: phenylethanolamine-N-methyltransferase

Figure 1.3 Biosynthesis of the catecholamines.

forming drugs either directly affect this system as DA agonists, or by enhancing DA release, or by acting on other neurons responding to different NTs—which, however, synapse upon this DA system to activate it. It is known that cocaine blocks reuptake of DA from the synapse, and that amphetamines trigger its release into the synapse; and these two drugs are among the most reinforcing known. In the case of depressant drugs such as the barbiturates, benzodiazepines, and opiates, the same system is indirectly affected by lowering its threshold of activation.[41] Even the reinforcing effects of nicotine, alcohol, caffeine, PCP, and the cannabinoids are believed to be finally traceable to indirect activation of this system. The hypothesis is probably the best explanation to date of hitherto puzzling results such as how opioid antagonists like naltrexone, or DA partial agonists like bromocriptine[42] can suppress craving for alcohol, which acts directly at the GABA neurons rather than at the opioid or dopamine systems.[43]

Dopamine, Schizophrenia, and Parkinson's. Antipsychotic drugs such as chlorpromazine (Thorazine) and haloperidol (Haldol) are antagonists at dopamine receptors, particularly the D_2 subset in the nucleus accumbens and frontal cortex; it is believed that their efficacy is owed to this property, and this is to some degree substantiated by the phenomenon of an acute psychotic state that can result from very high doses of cocaine or amphetamines and that is often indistinguishable from natural schizophrenia. However, schizophrenia is a very poorly understood disease, and the efficacy of all drugs used to treat it at this time leaves much to be desired (see Chapter 5).

If schizophrenia can be thought of as due to an excess of dopaminergic stimulation, Parkinson's disease is known to be caused by inadequate dopaminergic transmission in the part of the brain known as the substantia nigra; the symptoms of the disease (e.g., frozen facial muscles and a shuffling gait) can be partially offset by administering DOPA along with another drug, *carbidopa.* The DOPA enters the CNS and is transformed there to DA while the carbidopa, which cannot penetrate the CNS, blocks transformation of

DOPA to DA in the periphery, thus inhibiting the unwanted effects of excess dopamine outside the brain.

***Norepinephrine (NE,* 1-8)**. NE is the NT of most sympathetic nerves and many CNS neurons in parts of the brain such as the locus ceruleus and hypothalamus. NE is made from tyrosine, like DA, but the DA is further hydroxylated (by dopamine-β-hydroxylase) to NE, which is stored in vesicles. After release from the vesicles and interaction with the adrenergic receptor, the action is terminated by reuptake of NE into the presynaptic neuron. NE is also metabolized via monoamine oxidase (MAO) to inactive metabolites. The amphetamines (see Chapter 4) act indirectly at DA, NE, and serotonin presynapses to trigger the release of these NTs into the synapse. NE has also been known as noradrenaline; the prefix *nor* implies that the nitrogen is missing a methyl or other alkyl group (German, *N ohne Radikal*, "nitrogen without the radical," is one possible etymology). If a methyl group is attached to the nitrogen of NE, *epinephrine* (**1-9**), also known as *adrenaline*, is formed. This is the hormone released from the adrenal glands under conditions of physical or psychological stress (e.g. the "flight or fight" reaction).

***γ-Aminobutyric acid (GABA,* 1-10**, Figure 1.4). GABA is a major inhibitory NT in the CNS (basal ganglia, cerebellum). It is derived from glutamic acid, **1-12**, via glutamic acid decarboxylase. After interaction, it is restored to the presynaptic neurons by active uptake. Drugs acting principally at the GABA synapses are the barbiturates, the benzodiazepines (e.g., Valium), and alcohol (see Chapter 3).

***Glycine* (1-11), *Glutamic Acid* (1-12), *Aspartic Acid* (1-13)**. These are the more important of several amino-acid NTs used by CNS neurons in the cortex, cerebellum, and spinal cord. One subset of the glutamic acid receptor class is known as the *N-methyl-D-aspartate (NMDA*, **1-14**) receptors because they selectively respond to this nonendogenous ligand. The relationship of the NMDA receptor subtype to two other receptor types, the *PCP* and *sigma receptors*, has been quite confusing. It seems that the PCP receptor site is allosterically coupled to the NMDA site, and is probably located within the ion channel controlled by the NMDA receptors.[44] PCP and several related drugs bind to this site and block the increase in nerve depolarization that is induced by NMDA. This is potentially of clinical usefulness, because much of the damage from stroke or other ischemic events in the brain (and quite likely from amyotrophic lateral sclerosis, "Lou Gehrig's disease") is caused by flooding of adjacent cells by glutamate from dying neurons; NMDA neurons may also be involved in epileptic seizures. It was once believed

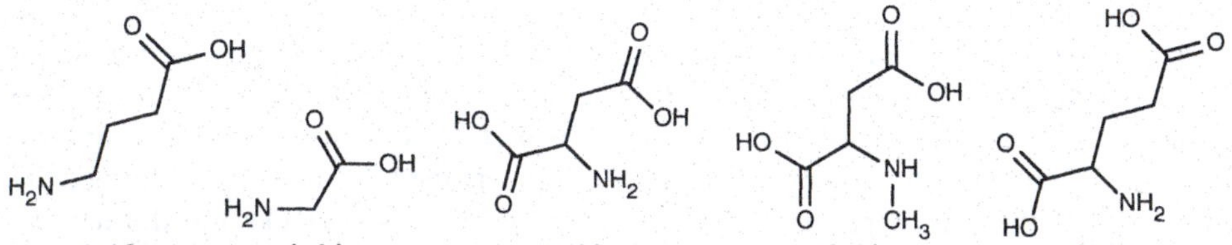

1-10 **1-11** **1-12** **1-13** **1-14**

Figure 1.4 Some amino-acid NTs and the receptor agonist NMDA.

that the PCP receptors and the sigma opioid receptors (see Chapter 2) were the same. Indeed, many compounds that bind to the sigma sites induce behavioral effects quite similar to those caused by PCP and often bind to PCP sites as well as sigma sites. However, it is now known that the anatomical distribution of these sites in rodent brain at least is quite different, and drugs have been created that selectively bind at the sigma opioid sites; *DTG (1,3-di-ortho-tolylguanidine)* is one such selective sigma receptor ligand.

***Serotonin (5-HT,* 1-17**, Figure 1.5). Serotonin is also commonly referred to as *5-HT*, an abbreviation of its chemical synonym, *5-hydroxytryptamine*. It is the NT of many CNS neurons in a rather restricted area, mostly in the raphe nucleus. It is produced in the body from tryptophan (**1-15**) via hydroxylation (tryptophan hydroxylase), and decarboxylation (aromatic L-amino acid decarboxylase). It is metabolized via MAO to 5-hydroxyindole-acetic acid (**1-19**). In the pineal gland, serotonin is transformed into *melatonin* (**1-18**), a hormone involved in skin pigmentation and the circadian light-dark (sleep/wake) cycle.

The number of receptor subtypes for 5-HT seems to grow yearly. Drugs that act principally at the serotonin synapses are the newer class of antidepressants, the selective serotonin reuptake inhibitors (SSRIs) such as *fluoxetine* (Prozac, Chapter 5); a drug used to suppress appetite, *fenfluramine* (Chapter 4), which although structurally very similar to amphetamine, probably acts in part by releasing serotonin into the synapse;[45] the antimigraine drug *sumatriptan*, which is a serotonin agonist; and psychedelics such as *LSD* (Chapter 6).

Endorphins, Enkephalins. *β-endorphin* is a polypeptide containing 31 amino acids. It is the NT of many CNS neurons (hypothalamus, amygdala, periventricular thalamus, locus ceruleus). β-endorphin and a number of other peptides including the *enkephalins* interact with the peptidergic (opioid) receptors, the same receptors that are stimulated by opium, heroin, and so forth (Chapter 2). They have received no little notoriety because of frequent claims in the popular press that joggers and other exercise enthusiasts are

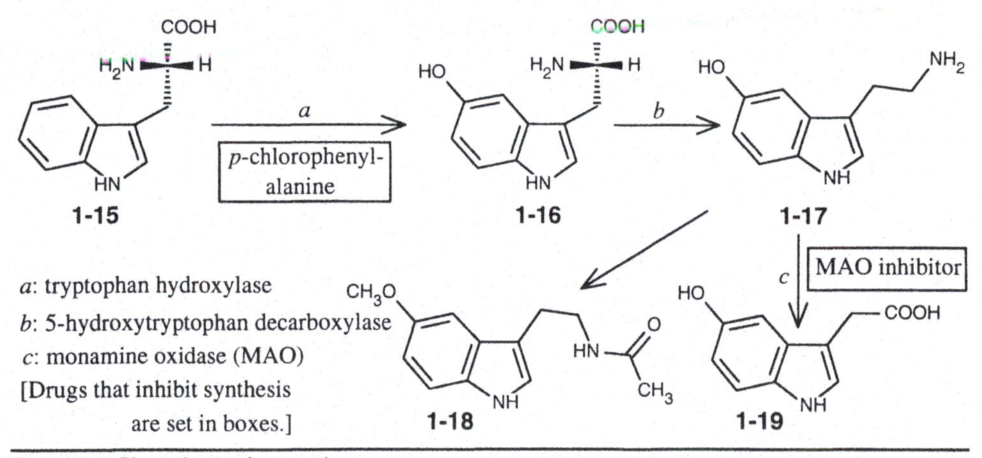

Figure 1.5 Biosynthesis of serotonin.

addicted to these endogenous opioids. As we shall see in the next chapter, there is very likely a role played—at least in some individuals—by the endorphins in behaviors like compulsive exercise, anorexia nervosa, and bulimia.

Histamine (**1-20**, Figure 1.6). Evidence that this amine has an effect on mental functioning can be provided by anyone who has ever taken an antihistamine allergy, cold or sleep medication (Benadryl, Contac, Sominex). Nonetheless, progress in understanding the function of histamine in the CNS has been slow. It is known that there are three types of histamine receptors in the CNS with a high density in the hypothalamus. They may be involved in thermoregulation and in food and water intake. There are H_1 receptors in the reticular formation, and these may be responsible for the sedating effect of common antihistamine drugs. In the periphery, the antiulcer medications *Tagamet* and *Zantac*, which have recently been made available OTC, are blockers of a second subtype of histamine receptor, H_2, found in the stomach lining; the function of these receptors is to stimulate the release of hydrochloric acid when food is eaten.

Adenosine (**1-21**). Adenosine is a "mysterious modulator . . . found at virtually every synapse that has been examined. . . exhibiting a variety of behavioral effects ranging from evoking premature arousal in hibernating ground squirrels to anticonvulsant activity."[46] It has been described as a neuromodulator[47] rather than a NT, but well-defined receptors exist and have been cloned.[48] *Caffeine* and *theophylline* (see Chapter 4) act as antagonists at the adenosine A_1 receptors.

Brain Gases. Some of the most unusual NTs are not known at this point to be extensively involved with addictive behavior. But as the role of memory in learned behavior is better understood at the molecular level, they may prove to be critical. What makes them so unusual is that these NTs are very small molecules indeed: *nitric oxide (NO)* and *carbon monoxide (CO),* both of which are ordinarily encountered in the environment as poisonous gases contributing to air pollution. NO is generated from arginine, an amino acid, by the recently cloned enzyme nitric oxide synthase[49] and is responsible for vasodilation as well as being antitumorigenic and antimicrobial.[50] CO is perhaps the crucial NT enabling the possibility of memory. Storing information in the brain seemingly must involve the permanent alteration of some neurons (like the off/on bits of computer memory storage). Memory takes place largely in the hippocampus of the brain, and it seems to involve the phenomenon of *long-term potentiation*: when a pre-

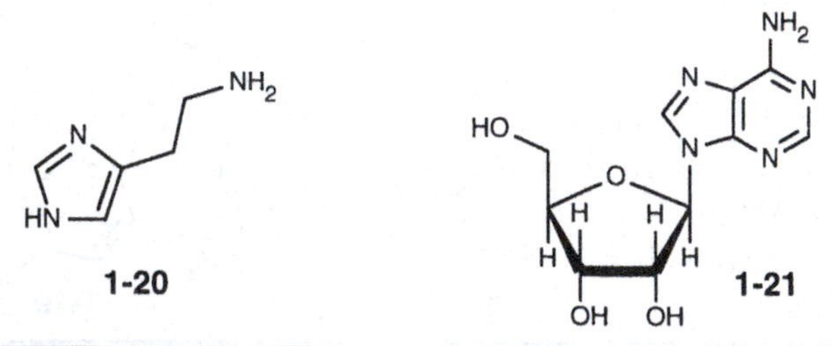

Figure 1.6 Histamine and adenosine.

synaptic NMDA neuron fires the same way several times in succession so that it is firing when the postsynaptic membrane is already depolarized, the postsynaptic membrane sends a messenger NT in a retrograde manner back to the presynapse, which then becomes "potentiated" more or less permanently (i.e., it will "remember" by responding very strongly to similar future stimuli). The retrograde NT is believed to be CO or possibly NO.[51] Naturally, as gases, these NTs cannot be stored in membrane vesicles; rather, each is synthesized as needed.

We have considered the human phenomenon of addiction, and gone to the new science of molecular biology to understand how the chemicals we call drugs interact within the brain's own mechanisms to contribute to this phenomenon. Next we briefly look to a much older discipline, pharmacology, for some simplified notions (with apologies to any pharmacologist readers) of how a drug's potency and effectiveness are measured, and for an understanding of how the varied methods of formulating and ingesting a drug can significantly alter its final effect on the mind.

The Pharmacological Context: Measuring Drug Activity

How Drugs Are Developed

To provide a context for becoming acquainted with some useful concepts used in pharmacology, we will follow the development of an imaginary drug from its first discovery through its commercial marketing as a prescription pharmaceutical to its use "on the street" as an abused drug. Suppose we were scientists working for the large multinational pharmaceutical company Dandy Drugs, and we had just succeeded in isolating a new chemical from a species (also imaginary) of African marigold known botanically as *Pseudopelargonium africanum* L., *Geraniaceae*. We believe that the chemical is an analgesic (pain reliever) because several native African tribal folk healers have traditionally employed the leaves and flowers of the plant for this purpose. After it is purified and its structure is determined, we first check *Chemical Abstracts* to be sure that it has not been reported before. At this point, supposing its structure is simple enough, we might try to synthesize it to provide a cheaper source or to be able to vary its structure and improve its properties. Academic chemists would then publish a report of its isolation and structural identification, but we chemists at pharmaceutical companies are interested in a viable commercial product, so we would hold back on publishing this information until we are sure of its medical use and have patented it (or until we are sure it is commercially useless).

Toxicity

As a first step, we want to have some idea of the overall toxicity of the drug. Of course, it would be nice if the substance had no toxicity whatsoever, but this is impossible. Even water is toxic if you drink enough of it.[52] Indeed, as Paracelsus observed long ago: "All substances are poisons; there is none that is not a poison. The right dose differentiates a poison and a remedy."[53] So it helps to find the upper dose limit above which our geranium chemical begins to kill. The simplest (if somewhat crude) estimate of toxicity is the LD_{50} (lethal dose for 50% of a population); this is the dose that, when administered to, say, 100 rats will result in 50 of the rats dying. It is usually measured in units of milligrams of drug per kilogram of animal (mg/kg). The assumption is that the effects of the drug are linearly proportional to the animal's body weight, so that a dose of 10 mg given to a 0.25-kg rat would have the same effect as a 20-mg dose given to a 0.50-kg rat. More to our purposes, the expectation is that one can extrapolate from rats to humans: if a drug has an LD_{50} equal to 30 mg/kg of rat, it should have the same LD_{50} per kg of human. The average (male) human weighs 70 kg, so a dose of this drug that would be expected from the results of the rat studies to kill 50% of a population of 70-kg human males would be $30 \times 70 = 2100$ mg = 2.1 kg.

Some standard assumptions are made in carrying out this sort of test: usually a consistent breed (Sprague-Dawley, Wistar) and sex (male) of rats is used; and the rats are all of the same age, frisky young adults. The latter is important because rats are like humans; as they grow older, everything slows down, including their rate of metabolizing drugs. With some drugs, this can be quite significant. For instance, the LD_{50} for the oral ingestion of alcohol in *young* rats is 10.6 g/kg; in *old* rats it is only 7.06 g/kg.[54] Extrapolating to our species, this means that if you are a 70-kg youngish human, you have roughly even odds of surviving a binge drinking episode in which you toss down about 742 g of alcohol (the equivalent of 1.8 L of 80 proof scotch); if you have avoided such youthful excesses and have survived to your 50s or 60s, your risk increases and you have an equal chance of killing yourself with only 1.2 L.

As can be seen from the example just given, the extrapolation to humans seems roughly credible; 1.8 L certainly sounds like a near-lethal amount of scotch. But of course, there are many problems with this measure of toxicity, not the least of which is the number of animals that must be killed to determine it. With many drugs, there can also be very significant variations in toxicity between animal species; and of course in particular between animals and humans. (A well-known example: morphine controls severe visceral pain in rats, cats, horses, and humans. In rats and humans, higher doses result in sedation and somnolence; but in cats and horses, higher doses cause restlessness and excitement.[55]) Furthermore, the test only measures acute, short-term toxicity: after receiving certain levels of drugs, the test animals will look fine for 48 hours, but they will nonetheless all die within 2 weeks from delayed toxic effects. Still more often, a dose that does no harm administered once on an acute basis can have a very high chronic toxicity that only becomes apparent when it is given regularly for months or years. Depending on the purpose of the drug this may or may not be relevant: antibiotics are given for a few weeks at most, anorexiants (appetite suppressants) may be taken for a lifetime.

For all these reasons, the requirement that a pharmaceutical company report the LD_{50} of new drug entities has been dropped by most countries' regulatory agencies, to be replaced by some sort of "limit" test on animals of the largest dose (proportionately, mg/kg) likely to be ingested by a human being. After a drug is marketed for human use, and is more widely used, accidental or deliberate overdoses inevitably if lamentably occur, some of them fatal, and this provides direct data on human toxicity. For instance, the antipsychotic chlorpromazine (see Chapter 5) has been in use for almost half a century, and there is at least one report of 26 g proving fatal in an adult.[56] This can be compared to the oral LD_{50} for chlorpromazine in rats, which is 225 mg/kg. If the adult weighed a standard 70 kg, extrapolating from the rat toxicity gives a 50% likelihood that a (70 × 0.225 g) 15.75-g dose would prove fatal to a human. Inasmuch as the anonymous person in the report took nearly twice that dose, it is not surprising that the result was fatal, and the rat data seem to be quite reliable as a ballpark figure for human toxicity.

For all its limitations, the LD_{50} is still of use in this book because it is often the only estimate of toxicity available for "underground" drugs, and because the acute ingestion of an extremely large dose is a phenomenon that often happens with such drugs.

The Name Game

But back to our story. Suppose our new chemical entity seemed to have a reasonably low level of toxicity and in preliminary tests shows some analgesic activity. It is time to give it a name. Of course the chemical structure provides a name, but these names are inordinately lengthy and essentially useful only for systemic structural identification. At this point in the development of the drug, the chemical will have been given an in-house code name such as *DDC-111* (for Dandy Drug Company's 111). Dandy Drugs will then negotiate with the United States Adopted Names (USAN) Council to provide a nonproprietary (i.e., not patented) name everyone can use for the drug. The name usually reflects bits and pieces of the nomenclature for the chemical structure: in the case of DDC-111, USAN gives us the name *chlorthipromine*. In the United States, only the Secretary of Health and Human Services has the legal authority to designate an official nonproprietary or *generic* name that must thereafter appear on all manufacturer's labels, but the Secretary nearly always follows the usage of the USAN. (The USAN in turn coordinates its naming with the World Health Organization's International Nonproprietary Names Committee so as to ensure there are not two different drugs being marketed with the same name.) Now that Dandy Drugs has a generic name, they can give chlorthipromine their own exclusive proprietary name, preferably one that will have a catchy ring to it for Dandy Drugs' chief customers, the prescribing physicians. After due consultation with its marketing and advertising experts, Dandy selects *Gerain*®, reflecting the drug's origins in African geraniums. They may also be thinking ahead to when they might want to advertise: "Need a fix to stop your pain? Ask your doctor for Gerain!" Or: "Pain? One drug's germane—Gerain!" Note that the proprietary name is capitalized and the generic name is not. (A real-world example of this multiple layer of names is Eli Lilly's revolutionary discovery of the serotonin reuptake inhibitor *LY-110140*, with structural name N-methyl-3-(4-trifluoromethylphenoxy)-3-phenylpropylamine, generic name *fluoxetine*, and proprietary names *Prozac, Adofen, Fluctin, Fluoxeren, Fontex, Foxetin,* and *Reneuron.*)[57]

We see that fluoxetine, a relatively new drug, already has at least 7 proprietary names. This is why there has been a movement in recent years among physicians and health workers to try to use the generic (i.e., official nonproprietary) name as much as possible. The problem is that for reasons such as euphony in different languages and price differentials in different markets, international pharmaceutical companies market the same drug in different countries with different proprietary names. And after the 20-year period during which the originating company has exclusive rights to market the product, other companies produce generic versions of the drug using *their* proprietary names. Older drugs that have been used for many years in many countries have by now acquired a plethora of proprietary names. Chlorpromazine, for instance, has at least 35: *Hebanil, Hibanil, Hibernal, Klorpromex, Largactil, Largaktyl, Megaphen, Promacid, Chloractil, Chlorazin, Sonazine, Marazine, Propaphenin, Taroctyl, Thorazine, Torazina, Clordelazin, Chlorderazin, Chlorpromados, Contomin, Esmind, Fenactil, Novomazina, Promactil, Prozil, Plegomazin, Sanopron, Aminazine, Ampliactil, Amplictil, Promazil, Chlor-Promanyl, Proma, Elmarin,* and *Wintermin.*[58]

Measuring Drug Efficacy

Dandy Drugs has tested the toxicity of its new drug Gerain in mice, and has found the LD_{50} is about 700 mg/kg. Now they want to test how well Gerain suppresses pain. It is compared to morphine in a test where mice treated with morphine versus Gerain versus control (saline) are placed on a hot plate and the temperature increased at a steady rate. The temperature at which the animal first licks its paw is noted. If Gerain is like morphine, this temperature will be significantly higher than is the case with the control population. Gerain proves to be less effective than morphine at the same dose range, but the LD_{50} of Gerain is almost twice that of morphine, and so a larger dose can safely be used. It turns out that when Gerain is administered at 0.60 mg/kg it is as effective as morphine at 0.20 mg/kg.

Of course, this test cannot be done on just one mouse; the results above are those done on a population of mice and the median effective dose (ED_{50}), is calculated analogously to the LD_{50}. Gerain is not as potent as morphine, since it takes about three times as much to have the same effect. But its toxicity is also lower than morphine's (the LD_{50} for morphine is mice is about 225 mg/kg), so it is relatively safer to use. This ratio of risk to benefit is called the *therapeutic index* (T.I.), and one way of quantifying the TI is to define it as the ratio of the LD_{50} to the ED_{50}. The greater the T.I., the safer the drug. In the case of Dandy Drug's Gerain, the T.I. is $700/0.60 \cong 1200$. The T.I. for morphine is $225/0.20 \cong 1125$, so the larger dose of Gerain is likely to be safer. If all else were equal (of course it never is), Gerain could well replace morphine for controlling pain, and be a very valuable drug!

Some Statistical Assumptions

Obviously, toxicity or efficacy could not be reliably measured if a drug were tried on only one mouse. Mice are like people, and have considerable variation. How many trials are needed to be reasonably certain that the results are reliable and not just a chance occurrence? There are two types of errors that can conceivably occur, called type

I and type II errors. A type I error, or false positive, occurs when a difference in efficacy between drug A and drug B is found, but in reality there is no difference. A type II error, or false negative, is the opposite: no difference is observed, but in reality there is one. There are statistical tables that summarize a great deal of mathematical theory in this area, and using these tables and various algorithms, the number of trials needed for a desired degree of certainty with regard to each type of error can be calculated. The *significance* of a test is the probability that a type I error has occurred. If drug A and drug B have been shown to be different at the 0.05 level of significance, this means that the likelihood that a false positive error has occurred is 5 in 100, or 1 in 20. For drug trials in medical studies, 0.05 is usually considered an adequate level of significance.

The second kind of error, a false negative, is not usually as strictly accounted for. The *power* of a test is the probability that a type II error has *not* occurred (note that power is defined negatively). Usually trials with a power of 0.80–0.90 (i.e., an 80–90% likelihood that such an error has not occurred, or a 10–20% likelihood that it has) are considered acceptable.

Human Trials

Dandy Drugs is pretty happy with the results so far, and wants to see how their new drug works in human beings. In the United States, it must now file an Investigational New Drug application with the FDA, presenting much more detailed studies on the safety of the drug in animals than we have sketched here. If the FDA grants the application, Dandy Drugs can now test the drug on humans. The first experiments are called Phase I studies. In these, Gerain will be tested in normal volunteers not experiencing the problem or illness (in this case, pain) that Gerain is intended to treat. The tests are primarily to see how safe and how well tolerated the new drug is, and what dosage is likely to be appropriate in humans. If all goes well, Gerain moves to Phase II trials, where it is used on patients (who have given their informed consent) suffering from pain. At this point it is important that we consider whether the trial is *open, placebo-controlled, single-blinded,* or *double-blinded.*

Particularly in the case of psychoactive drugs in human beings, whether these are antidepressants, analgesics, or hallucinogens, the knowledge possessed by either the patient or the person administering the drug can have an extraordinarily dramatic effect on the outcome. A person who has been suffering from chronic pain for years, pain that is intractable to ordinary analgesics, and who is told that he or she is about to be given a new and much more effective drug but is actually given an inert substance like sugar (a placebo), is likely to experience a significant reduction in pain. This is particularly the case when the administrator believes that the new drug is going to work. When this factor is not taken into account, but when both patient and administrator are aware they are getting a new drug, the experiment is called an open trial. When the drug is randomly administrated with a look-alike placebo, it is said to be placebo-controlled but single-blinded; if neither the administrator nor the patient knows which drug sample is the placebo and which is the drug, it is double-blinded. Sometimes the side effects of the drug are noticeable (flushing, sweating, and so forth.), which may give a clue to patient or administrator as to when a placebo is being used; in this case it is sometimes possible to use an *active* placebo—a substance that is inert as far as the effect being tested goes but has a harmless side effect like the real drug. (For instance, the vitamin nicotinic acid causes flushing but otherwise is inert.) Finally, to control for idiosyncratic reactions to a drug, clinicians can use a *crossover* regimen, where each patient is randomly selected to receive the drug for a given period, then the placebo, or vice versa.

If the Phase II trials show that the drug is efficacious, safe, and well-tolerated, larger Phase III trials are conducted. Assuming that Gerain works well, Dandy Drugs will apply to the FDA for a New Drug Application. If it is granted on the basis of the Phase III trials, the drug can be marketed commercially. Many hitherto unnoticed problems can arise, of course, when a new drug is actually used in this much larger population, and physicians are asked to fill out Adverse Reaction Reports on any that arise. On the basis of these so-called Phase IV results, the drug may be withdrawn or the dosage altered.

Much to the dismay of Dandy Drugs, it turns out that significant numbers of patients like the effects of the Gerain very much and begin to take more than they are prescribed, apply to more than one physician for additional prescriptions, and so on. The drug is given a Schedule II rating because of its abuse potential. Word gets out on the street about Gerain's properties, and soon a thriving black market exists in stolen or "diverted" capsules. Even more interesting, specimens and seeds of the African geranium are smuggled into the United States, and are soon being grown under hydroponic conditions in New York City lofts. The leaves are rolled and smoked in water pipes; this process is called "G-bonging." Or the red, orange, and purple flowers are dried and eaten; these are known as "G-rainbows." Users of G-rainbows say the experience is quiet, profound, and mellow; but G-bongers say they get a more exciting flash or rush, and scorn G-rainbowers as "wussies."

A few pharmacologic facts can explain why the G-bongers may indeed be getting a more intense rush. The mental effect of any psychoactive drug depends not on how much drug is in the stomach or intestines, but on its concentration in the brain. And the brain concentration depends in part on how the drug is taken—whether smoked, eaten, injected, and so forth.

Drug Absorption

How does a drug get into the brain? How fast does it get there? How long does it stay? How does it get out of the brain and the body? These are questions from *pharmacokinetics*, and we can only touch on a few generalizations here. Drug absorption is most precisely defined as the process whereby a drug moves from the site of administration into the plasma, the largely aqueous fluid that constitutes the blood along with the red and white blood cells suspended within it. The most important variable in how a drug reaches the plasma is its route of administration, and the main possibilities are the following:

- *Injection*, using a hypodermic syringe (needle). The injection can be intravenous (IV), into a vein; intramuscular (IM), into a large muscle such as the thigh or shoulder; or subcutaneous (SC), just under the skin ("skin-popping" in the jargon of the street user). All these modes of injection are collectively referred to collectively as parenteral (which literally means by any route other than oral). IV injection and inhalation (below), introduce a drug very rapidly into the bloodstream and provide the quickest drug effect possible; IM injection, insufflation (below), and SC routes are intermediate; the oral route is the slowest of all.

- *Inhalation*, either as a gas or an aerosol of small droplets or particles. Because freshly oxygenated blood from the lungs is pumped directly to the brain, without transiting the liver, this can provide a quicker rush than even that from IV injection. With drugs like nicotine and cocaine, which are rapidly metabolized, this can become a preferred route of ingestion.
- *Insufflation*. Drugs can be "snorted" or, more technically speaking, taken by nasal insufflation. There is a fairly extensive area of mucous membrane in the upper respiratory tract through which many drugs can be easily absorbed.
- *Oral*. Ingestion of a drug by mouth is much slower because the drug must first pass through the stomach into the small intestine. It can also be a less effective means of drug intake (whether the drug is a pharmaceutical or illicit) because many drugs are metabolized or decomposed by stomach enzymes or by the stomach's very acid environment. (For instance, the first penicillin, penicillin G, had to be administered IM because it was unstable and decomposed in stomach acid. Later modifications like ampicillin were designed to overcome this problem and could be taken orally.) Still another problem with oral ingestion of some drugs is what is called *first-pass metabolism*: drugs or nutrients absorbed by the gastrointestinal tract first enter the portal circulation, a network of blood vessels that carries them directly to the liver. In the liver are numerous enzymes capable of chemically modifying many foreign substances, including many drugs; the products of these transformations are usually inert materials sent on to the kidneys for elimination. But often they are toxic or have an altered biological activity. In the case of some psychoactive drugs, first-pass metabolism can reduce the amount of drug that reaches the brain. As an example, as much as 70–80% of orally ingested cocaine is metabolized by the process of first-pass hepatic (liver) biotransformation.[59]
- Other routes less commonly employed are *sublingual* (chewing tobacco), *transdermal* (nicotine, scopolamine, and fentanyl patches); and *rectal* (some children's aspirins).

In our imaginary scenario, the G-bongers get a quicker and more intense high from their drug because the smoke they inhale is absorbed directly into the bloodstream from the lungs, which have a very large surface area. G-bongers probably feel the first effects within a few minutes, and the effects will reach a peak rapidly. The G-rainbowers, on the other hand, will feel no effects for perhaps 45 minutes, and the drug concentration in the brain will gradually rise to a plateau and then as gradually wear off.

As we make our way through the diverse drug types discussed in this book, we will encounter drugs like heroin that are almost exclusively taken IV or by smoking because the rush is prized by users. Other drugs, notably cocaine, are commonly taken by IV injection, smoking, or insufflation. All of these drugs could be taken effectively by mouth, but this is rarely done. On the other hand, psychedelic substances like LSD, and of course alcohol and caffeine, are almost always taken orally (though the active agent in all of these could be injected), and this is probably because the desired effect is a relatively lengthy experience of several hour's duration rather than an intense high.

Drug Distribution: The Blood–Brain Barrier

Once a drug has gotten into the plasma and is circulating in the blood, it can be distributed to most parts of the body. For instance, a bacterial infection in the foot can be treated with an oral antibiotic such as ampicillin; but severe bacterial infections of the brain may have to be treated by injecting the antibiotic *intrathecally*, that is, directly into the CNS. Many drugs are distributed into the brain from the bloodstream either poorly or not at all. Why is this? We have passed over a critical issue in our discussion of drug absorption thus far. The sort of drugs we are concerned with, psychoactive drugs, will only have an effect if they can get from the plasma into the brain; but to do this they must pass through a special partition in the body called the blood-brain barrier (BBB). This is a layer of cells with unusually tight junctions between them such that, generally speaking, only very small molecules such as water, oxygen, and carbon dioxide can get through. It is also permeable to most nonionic or nonpolar molecules, those that can generally be described as *lipophilic* ("fat loving"). Lipophilic substances dissolve readily in such organic solvents as ether, chloroform, or benzene, while substances composed of ionic molecules (bearing a positive or negative electric charge) are insoluble in these solvents and dissolve in water (they are *hydrophilic*, "water loving").

Obviously, the drugs described in this book cross the BBB or they would not have any effect on the mind. Indeed, the overwhelming preponderance of psychoactive drugs are amines. Amines be can be either lipophilic or hydrophilic depending on their environment: In an acidic environment, the amine nitrogen atom binds to a positively charged hydrogen ion and they become hydrophilic; in a basic environment they lose this charge and become lipophilic. Most of the major classes of psychoactive drugs (the opioids, cocaine, nicotine, the psychedelics), as well as many of the pharmaceuticals used in treating disease, were first discovered as amines naturally occurring in plant materials, called *alkaloids*. And one of the reasons these were the first drugs discovered is that it was relatively easy to isolate them from plants known from traditional use to have medicinal activity: all that often had to be done was to percolate the plant material in strong acid, make the filtered solution basic, extract the solution with ether and evaporate the ether. The illicit refining of opium and coca is carried out today by fairly unskilled workers in much the same way. In the buffered environment of plasma (pH = 7.4), amines exist as an equilibrium mixture of their hydrophobic and hydrophilic forms; on reaching the BBB, the hydrophobic form rapidly crosses the barrier and enters the brain. A few of the drugs in this book are not amine bases (alcohol; THC, the active principle in marijuana; the barbiturates; some of the benzodiazepines); all of these are, however, quite lipophilic and easily cross the BBB.

Drug Redistribution or Storage: Body Fat

Some unusual effects can occur in the relatively rare cases when a drug is very highly lipophilic, particularly when the drug is taken by IV or inhalation. In obese persons, fat may constitute as much as 50% of body weight; and even in persons near starvation it is almost 10%. This can be used to advantage in some cases. The barbiturate thiopental ("truth serum") is very lipid-soluble. When it is administered IV for anesthesia, it crosses the BBB very rapidly to cause loss of consciousness within 30–60 seconds. Consciousness returns almost as rapidly when injection of the drug into the vein is stopped, increasing the margin of safety in using the drug. One might assume that the thiopental is removed from the brain by rapid metabolism, but this is not actually the case. Rather, the drug first enters the brain because it is richly perfused by the bloodstream, but once the drug ceases to be injected, equilibrium favors its being quickly sopped up by (relatively poorly perfused) body fat, in which it is even more soluble than it is in the brain.

A similar process is at work with marijuana smokers; the active ingredient THC (see Chapter 7) is very lipophilic and quickly enters the brain. But experienced users find they can titrate the drug effects quite reliably by adjusting their rate of puffing; as with thiopental, the THC exits the brain fairly rapidly and enters the body fat. This is why several studies have shown that heavy users of marijuana can have detectable traces of THC metabolites in their urine even several weeks after their last use of the drug.

Drug Metabolism and Excretion

Most drugs are eventually removed from the body in two steps, called Phase I and Phase II. Phase I biotransformations occur mainly in the liver and serve to oxidize, reduce, or hydrolyze the drug molecule so that it has a convenient "handle"—such as a hydroxy group, OH, or an H_2N—for the the Phase II step. In the Phase II step, also called the *conjugation*, this function is condensed with one of several strongly hydrophilic conjugates, most commonly β-glucuronic acid.[60] Glucuronic acid is a polyhydroxy compound similar to glucose in structure, and the resulting conjugate is water-soluble, does not cross the BBB, is usually pharmacologically inert, and is rapidly eliminated by the kidneys. Drug molecules that already have these convenient OH or NH handles may undergo conjugation directly. An example is *acetaminophen (Tylenol*, **1-23**, Figure 1.7), which is conjugated directly to *uridine diphosphoglucuronic acid* (UDP), **1-22**, by means of its phenolic oxygen to form the elimination product **1-24**. (Chemists will note that UDP, an excellent leaving group, is positioned in the axial α-orientation so as to facilitate S_N2 attack forming the more stable equatorial β-glucuronide.)

Often more than one path of Phase I transformation is followed, and not infrequently one or more intervening step in a sequential oxidation or reduction leads to a structure that is itself pharmacologically active; these are called *active metabolites*, and in some cases, notably the benzodiazepines, can lead to extending the effects of a drug long after the parent substance has disappeared. In still other cases, the drug is itself relatively inert, and only on biotransformation is an active structure formed. Aspirin, for example, is this sort of *prodrug*, most of its activity coming from its metabolite salicy-

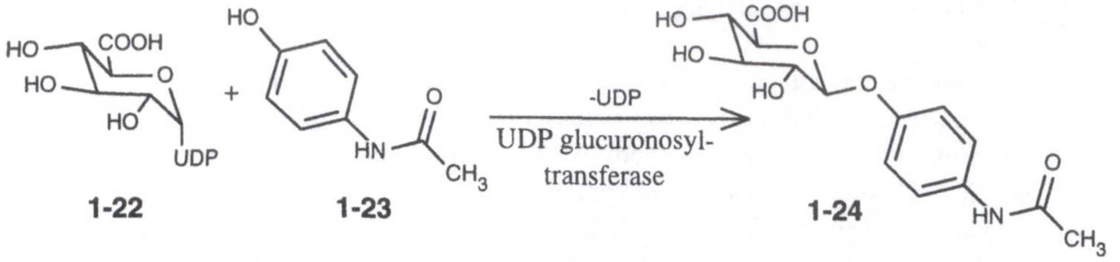

Figure 1.7 Metabolic transformation of acetaminophen to form the β-*O*-glucuronide.

late; and so is another 19th-century product of the Bayer company, heroin, which crosses the BBB better than its more hydrophilic parent molecule, morphine, but which is changed in the brain back into the active parent drug.

One common measure of the duration of action of a given drug is the *half-life* or *half-time*, which is the time required for the concentration of a drug in the plasma to drop to one-half its initial value (chemists will recognize this as an instance of first-order kinetics). If half of whatever amount of the drug that is present is eliminated every half-life, then nearly 94% of the drug is cleared from the system after 4 half-times, and the usual assumption is that the drug is totally excreted sometime between 4 and 5 half-times. As an example, a careful study of cocaine metabolism in volunteers using radioactive cocaine showed the half-time for IV use to be about 80 minutes. A rough calculation gives $80 \times 5 = 400$ min = 6–7 hours, and we would expect that plasma levels would be virtually zero at this time. Indeed, the plasma concentration at 6 and 8 hours had dropped from an initial 180 ng/mL to 4 and 2 ng/mL or 2 and 1%, respectively. In this study, cocaine was no longer able to be detected in the plasma after 11 hours.[61] The picture is often more complicated, of course. The assumption made in using the half-time measure is that increasing the dosage will not increase the rate at which the drug is metabolized (although, of course, it increases the steady-state concentration); this is not always the case. Furthermore, there are often long-lived metabolites of the original drug (in the study of cocaine cited previously, for example, the primary metabolite of cocaine in humans, *benzoylecgonine*, could still be detected at 30 hours). And some very lipophilic drugs, as we saw above, can be sequestered into body fat and be slowly leached from there over an extended period. A dramatic example is the ultrashort-acting anesthetic thiopental: in average individuals the half-time for this drug is about 11.5 hours; in the morbidly obese, it is increased to nearly 28 hours.[62]

But now it is time to leave the imaginary problems of the Dandy Drug Company and the simpler world of generalities to encounter the specific drugs that are the subject of this book. We will start in the next chapter with the opioid drugs (morphine, heroin, and so forth.) because they are among the most ancient and most powerful of mind-altering drugs.

References and Notes

The following abbreviations are used for frequent sources in these References/Notes:

BBN: The Biochemical Basis of Neuropharmacology, 6th ed.; Cooper, J. R.; Bloom, F. E.; Roth, R. H.; Oxford: New York, 1991.

DE95: *Drug Evaluations: Annual 1995*; American Medical Association Division of Drugs and Toxicology, Department of Drugs; Bennett, D. R., Ed.; AMA: Chicago, 1995.

G&G8: *Goodman and Gilman's The Pharmacological Basis of Therapeutics*, 8th ed., Gilman, A. G.; Rall, T. W.; Nies, A. S.; Taylor, P., Eds.; Pergamon: New York, 1990.

G&G9: *Goodman and Gilman's The Pharmacological Basis of Therapeutics*, 9th ed., Hardman, J. G.; Limbird, L. E., et al., Eds.; McGraw-Hill: New York, 1996.

JAMA: *The Journal of the American Medical Association.*

MBC: *Molecular Biology of the Cell*; Alberts, B.; Bray, D.; Lewis, J.; Raff, M.; Roberts, K.; Watson, J. D.; Garland: New York, 1994.

MCGR: *Medicinal Chemistry: The Role of Organic Chemistry in Drug Research*, 2nd ed.; Ganellin, C. R.; Roberts, S. M., Eds.; Academic Press: San Diego, CA, 1993.

MI11: *The Merck Index*, 11th ed.; Budavari, S.; O'Neil, M. J.; et al., Eds.; Merck: Rahway, NJ, 1989. References are given by monograph number.

MI12: *The Merck Index*, 12th ed.; Budavari, S.; O'Neil, M. J.; et al., Eds.; Merck: Rahway, NJ, 1996. References are given by monograph number.

MM16: *The Merck Manual of Diagnosis and Therapy*, 16th ed.; Berkow, R.; Fletcher, A. J., Eds.; Merck: Rahway, NJ, 1992.

NEJM: *The New England Journal of Medicine.*

OP: *Opium and the People: Opiate Use in Nineteenth-Century England*, Berridge, V.; Edwards, G.; Yale University Press: New Haven, CT, 1987.

PHP: *Pocket Handbook of Psychiatric Drug Therapy*; Kaplan, H. I.; Sadock, B. J.; Williams & Wilkins: Baltimore, 1993.

SA: *Substance Abuse: A Comprehensive Textbook,* 2nd ed.; Lowinson, J. H.; Ruiz, P.; Millman, R. B.; Langrod, J. G., Eds.; Williams & Wilkins: Baltimore, 1992.

1. De Quincey, T., *Confessions of an English Opium-Eater*, Ticknor, Reed, & Fields: Boston, 1853 [reprint of the 1822 edition], p. 73.
2. Hopkins, G. M., "No worst, there is none," *The Poems of Gerard Manley Hopkins*, 4th ed., Gardner, W. H.; MacKenzie, N. H., Eds., Oxford: New York, 1970, p. 100.
3. For a brief social history of drug use in the United States, see Buchanan, D. R., "A social history of American drug use," *Journal of Drug Issues*, **1992**, *22*, 31-52.
4. Greden, J. F.; Walters, A., "Caffeine," *SA*, pp. 357-370.
5. Goodwin, D. W., "Alcohol: clinical aspects," *SA*, pp 144-151. Winick, C., "Epidemiology of alcohol and drug abuse," *SA*, pp. 15-29.
6. Warner, L. A.; Kessler, R. C., et al., "Prevalence and correlates of drug use and dependence in the United States," *Arch. Gen. Psychiatry*, **1995**, *52*, 219-229. Cited material p. 219.
7. Statistics are taken from Scriven, P., *The Medicine Society*, Michigan State University Press: East Lansing, 1992, pp. 1-2.
8. Former Supreme Court Justice Potter Stewart in the 1964 case Jacobellis vs. Ohio, 197.
9. À Kempis, T., *Imitatio Christi*, I, iii.
10. *MM16*, p. 1549.
11. Indeed, before the crack and cocaine crisis of the 1980s, the definition of an *addictive* drug was one that caused a withdrawal syndrome and therefore physical dependency. Cocaine dependence and even compulsion was well known; it was described as *cocainism* in the older literature. Cocaine was feared as an enslaving drug, whether injected or taken orally—this is evident from the literature surrounding such phenomena as Vin Mariani, Coca-Cola, and the use of cocaine by Dr. Halsted of Johns Hopkins (see Chapter 4). It was denounced as "habit forming" and "demoralizing." But technically speaking, it was not considered addictive as was heroin or the other opioids, or the barbiturates. The much broader use of the term addiction as it is now employed has resulted in some lack of clarity, and some have (vainly) called for reverting to the older usage: Akers, R. L., "Addiction: the troublesome concept," *Journal of Drug Issues*, **1991**, *21*, 777-793.
12. Franklin, D., "Hooked: Not everyone becomes addicted. How come?" *Health*, **1990**, *4*, 38.
13. Hilts, P. J., *The New York Times*, 2 August 1994, C3.
14. It is noteworthy that *SA* includes a chapter on eating disorders: Krueger, D. W., "Eating disorders," *SA*, 371-379.
15. *NEJM*, **1991**, *324*, 896-899.
16. Partanen, J.; Nieminen, M. S., "Noncritical aortic stenosis in two men unable to quit running marathons—well, one quit," *JAMA*, **1992**, *267*, 511.
17. McElroy, S. L.; Keck, P. E.; et al., "Kleptomania, compulsive buying, and binge-eating disorder," *Journal of Clinical Psychiatry*, **1995**, *56* (Suppl. 4), 14-27.

18. Clomipramine: Hollander, E.; Frenkel, M.; et al., "Treatment of pathological gambling with clomipramine," *Am. J. Psychiatry*, **1992**, *149*, 710-711. Carbamazepine: Holloway, M., "Trends in pharmacology: Rx for addiction," *Scientific American*, 1991, *264*, 95.

19. Greene, G., *A Sort of Life*, Simon and Shuster: New York, 1971, pp. 120, 128-132, 157-158.

20. Booth, L., *When God Becomes a Drug*, Jeremy P. Tarcher: Los Angeles, 1991.

21. Arterburn, S.; Felton, J., *Toxic Faith: Understanding and Overcoming Religious Addiction*, Oliver Nelson: Nashville, 1991.

22. Franklin, D.; op. cit., ("Hooked . . .), 38.

23. *The Oxford Companion to English Literature*, 5th ed., Oxford Press: Oxford, M. Drabble, Ed., p. 243.

24. Crowley, A., *The Confessions of Aleister Crowley*, Symonds, J.; Grant, F., Eds., Jonathan Cape: London, 1969, p. 180. Cited in *OP*, p. 224.

25. Cited in Maisto, S. A.; Galizio, M.; Connors, G. J., *Drug Use and Misuse*, Holt, Rinehart, & Winston: Fort Worth, TX 1991, p. 17.

26. Jones, T., "I need a man to love me," *Cold Snap: Stories*, Little, Brown: Boston, 1995, pp 180. Jones seems to have some acquaintance with drugs himself, since he credits in his acknowledgments, "Wyeth/Ayerst Laboratories and Stuart Pharmaceuticals for further expanding that narrow channel of joy by manufacturing Effexor and Elavil; drugs so good they feel illegal." *Ibid.*, p. 228.

27. But nothing is ever as simple as it seems. Some have argued that the unnaturally stressed situation of these test animals—often immobilized in a cage with a permanent intravenous catheter, something analogous in human terms to solitary confinement on Death Row—may play a significant part in motivating them to take the drug. Even M. S. Gold, who strongly stresses the brute pharmacology of these drugs as inducing their compulsive use, admits that "conditions of housing and stress can influence how rapidly animals learn to self-administer these drugs and whether they will learn to self-administer marginally effective doses." Gold, M. S.; Miller, N. S.; et al., "Cocaine (and crack): neurobiology," *SA*, 222-235; quoted material p. 232.

28. Cadoret, R. J.; Yates, W. R.; et al., "Adoption study demonstrating two genetic pathways to drug abuse," *Arch. Gen. Psychiatry*, **1995**, *52*, 42-52.

29. Gold, M. S.; Pottash, A. L., et al., "Evidence for an endorphin dysfunction in methadone addicts: lack of ACTH response to naloxone," *Drug Alcohol Dep.*, **1981**, *8*, 257-262.

30. Bauer, L. O., "Psychobiology of craving," *SA*, pp 51-55. Cited material p. 54. Bauer cites two studies that showed, one for alcohol and one for tobacco, that addicted subjects most likely to relapse were those who, when exposed to drug cues, exhibited such responses as pupillary dilation or slowed heart rate, despite any conscious craving for the drug.

31. The relevant chapters of the following texts are very helpful: (a) *MBC*. (b) Darnell, J.; Lodish, H.; Baltimore, D., *Molecular Cell Biology*, Scientific American (Freeman): New York, 1990. (c) De Robertis, E. D. P.; De Robertis, Jr., E. M. F., *Cell and Molecular Biology*, Lea & Febiger: Philadelphia, 1987.

32. A review coauthored by one of those (AGG) later to receive the Nobel Prize for his work on G proteins: Linder, M. E.; Gilman, A. G., "G Proteins," *Scientific American*, **1992**, *July*, 56-65.

33. Gibb, A. J., "Receptor pharmacology," *MCGR*, pp. 52-53.

34. (a) Nogrady, T., *Medicinal Chemistry: A Biochemical Approach*, 2nd ed., Oxford: New York, 1988, p. 233. DMCM is methyl 4-ethyl-6,7-dimethoxy-β-carboline-3-carboxylate and one of several β-carbolines which act as inverse agonists: (b) Triggle, D. J.; Langs, D. A., "Ligand-gated and voltage-gated ion channels," *ARMC* **1990**, *25*, 225-234. (c) Browne, L. J.; Shaw, K. J., "New anxiolytics," *ARMC*, **1991**, *26*, 1-10.

35. Cooper, J. R.; Bloom, F. E.; Roth, R. H., *The Biochemical Basis of Neuropharmacology*, 6th ed., Oxford: New York, 1991, pp. 419-420.

36. Quoted in Kramer, P. D., *Listening to Prozac*, Viking: New York, 1993, p. 134. A Gödelian (or Orwellian?) paradox is involved here: on observing a Black Hole, it is perhaps best to reverse direction and pursue another perspective.

37. Leonard, B. E., *Fundamentals of Psychopharmacology*, Wiley: New York, 1992, p. 16-17.

38. Cooper, J. R.; Bloom, F. E.; Roth, R. H., *The Biochemical Basis of Neuropharmacology*, 6th ed., Oxford University Press: New York, 1991, p. 4.

39. Galdzicki, Z., et al., *Brain Res.*, **646**, 345 (**1994**).

40. (a) Wise, R. A., Rompre, P-P., "Brain dopamine and reward," *Annu. Rev. Psychol.*, **1989**, *40*, 191. (b) Wise, R. A., "Action of drugs of abuse on brain reward systems," *Pharmacol. Biochem. Behav.*, **1980**, *13* (Suppl. 1), 213.

41. Gardner, E. L., "Cannabinoid interaction with brain reward systems—the neurobiological basis of cannabinoid abuse." In: *Marijuana/Cannabinoids: Neurobiology and Neurophysiology*, Murphy, L.; Bartke, A., Eds., CRC Press: Boca Raton, FL, 1992, pp. 275-336.

42. Lawford, B. R., et al., "Bromocriptine in the treatment of alcoholics with the D_2 dopamine receptor A1 allele," *Nature Medicine*, **1995**, 337-341.

43. An excellent summary of this hypothesis by one of its principal contributors (including nearly 500 references) can be found in Gardner, E. L., "Brain reward mechanisms," *SA*, 70-99. For other hypotheses, see: Koob, G. F., "Drugs of abuse: anatomy, pharmacology, and function of reward pathways," *Trends in Pharmacological Sciences*, **1992**, *13*, 177-184.

44. Johnson, G.; Bigge, C. F., "Recent advances in excitatory amino acid research," *Annual Reports in Medicinal Chemistry*, **1991**, *26*, 11-22. Newman, A. H., "Irreversible ligands for drug receptor characterization," *Annual Reports in Medicinal Chemistry*, **1990**, *25*, 271-280.

45. Nogrady, T., *Medicinal Chemistry: A Biochemical Approach*, 2nd ed., Oxford: New York, 1988, pp. 203-204.

46. *BBN*, pp. 129-30.

47. Nogrady, T., *Medicinal Chemistry: A Biochemical Approach*, 2nd ed., Oxford: New York, 1988, p. 223.

48. Gibb, A. J., "Receptor pharmacology," *MCGS*, p. 45.

49. Geller, D. A., et al., "Molecular cloning and expression of inducible nitric oxide synthase from human hepatocytes," *Proc. Nat. Acad. Sci.* **1993**, *90*, 3491-3495.

50. Snyder, S. H.; Bredt, D. S., "Biological roles of nitric oxide," *Scientific American*, **1992**, *(May)*, 68-77.

51. *MBC*, pp. 544-547.

52. Especially for infants. The FDA is recommending new labeling that will warn parents of the risk for water intoxication if bottled water is used to replace infant formula. The problem is lack of sodium, or hyponatremia. Between 1984 and 1992, one Wisconsin hospital had 27 cases of infants experiencing seizures because their parents fed them bottled

water: "Hyponatremic seizures among infants fed with commercial bottled drinking water—Wisconsin, 1993," *Morbidity and Mortality Weekly Report*, **1994**, *Sept. 9; 43*, 641-643.

53. Cited (in English) by Klaassen, C. D., "Principles of toxicology," *G&G8*, pp. 49-61. This dictum of Paracelsus (a.k.a. Theophrastus Bombastus von Hohenheim, 1493-1541) is frequently quoted as part of the oral tradition in pharmacology and toxicology, but the source is never given. I seem to recall seeing the phrase in Latin, but I could not track it down. In any case, it was part of Paracelsus' rebellion against the traditional dogmatic medicine of Galen to write almost exclusively in German (although the works frequently had Latin titles). His "device" or motto (which *is* in Latin) expresses this very well: *Alterius non sit, qui suus esse potest* (Let he who can be his own man follow no other). At any rate, I was able to find two sentences expressing his famous toxicological principle from his *Defensiones*. The first is quite close to the English given by Klaassen: *Alle ding sind gift und nichts ohn gift; alein die dosis macht das ein ding kein gift ist.* (All things are poison and there is nothing without a poison[ous characteristic—side effect?]; only the dose determines whether something is not a poison.) A second variant is more aphoristic and includes a rhyming couplet: *Es ist nicht zu vil noch zu wenig, der das mittel trift der entpficht kein gift.* (Neither too much nor too little; who hits the mean has taken nothing ill.) References to the original sources in Paracelsus's works can be found in Pagel, W., *Paracelsus: An Introduction to Philosophical Medicine in the Era of the Renaissance*, 2nd ed., Karger: Basel, 1982, p. 363.

54. *MI11*, 3716 (monograph number), p. 594.

55. The *Merck Index* says that morphine is "contraindicated in cats; unreliable in horses," *MI11*, 6186 (monograph number), p. 988. But according to the *Merck Veterinary Manual, 7th ed.*, Fraser, C. M., Bergeron, J. A., et al., Eds., Merck: Rahway, NJ, 1991, p. 1387, the drug should be used in horses and cats to control deep visceral pain, with the precaution of not using higher than the recommended doses.

56. *PHP*, 245.

57. *MI12*, 4222, p. 709.

58. *MI11*, 2186, p. 338. Some of these names are for the free base, some for the hydrochloride, one for the maleate salt. It is interesting to speculate on the etymology of some of these names: the original observations of chlorpromazine's effects in animals and humans suggested a "hibernation" (psychotics were "chilling out" in a more current usage), and this is probably why there are names like Wintermin, Hibernal, and so forth.

59. (a) Gold, M. S.; Miller, N. S.; et al., "Cocaine (and crack): neurobiology," *SA*, pp. 222-235. (b) Verebey, K.; Gold, M. S., "From coca leaves to crack: the effects of dose and routes of administration in abuse liability," *Psychiatr. Ann.*, **1988**, *18*, 513-520.

60. For an extensive treatment of drug deactivation and elimination with numerous examples, see Silverman, R. B., *The Organic Chemistry of Drug Design and Drug Action*, Academic Press: San Diego, CA, 1992, pp. 287-351.

61. Jeffcoat, A. R.; Perez-Reyes, M.; et al., "Cocaine disposition in humans after intravenous injection, nasal insufflation (snorting), or smoking," *Drug Metab. Disp.* **1989**, *17*, 153-159.

62. Jung, D., "Thiopental disposition in lean and obese patients undergoing surgery," *Anesthesiology*, **1982**, *56*, 269-274. Cited in *DE93*, 177.

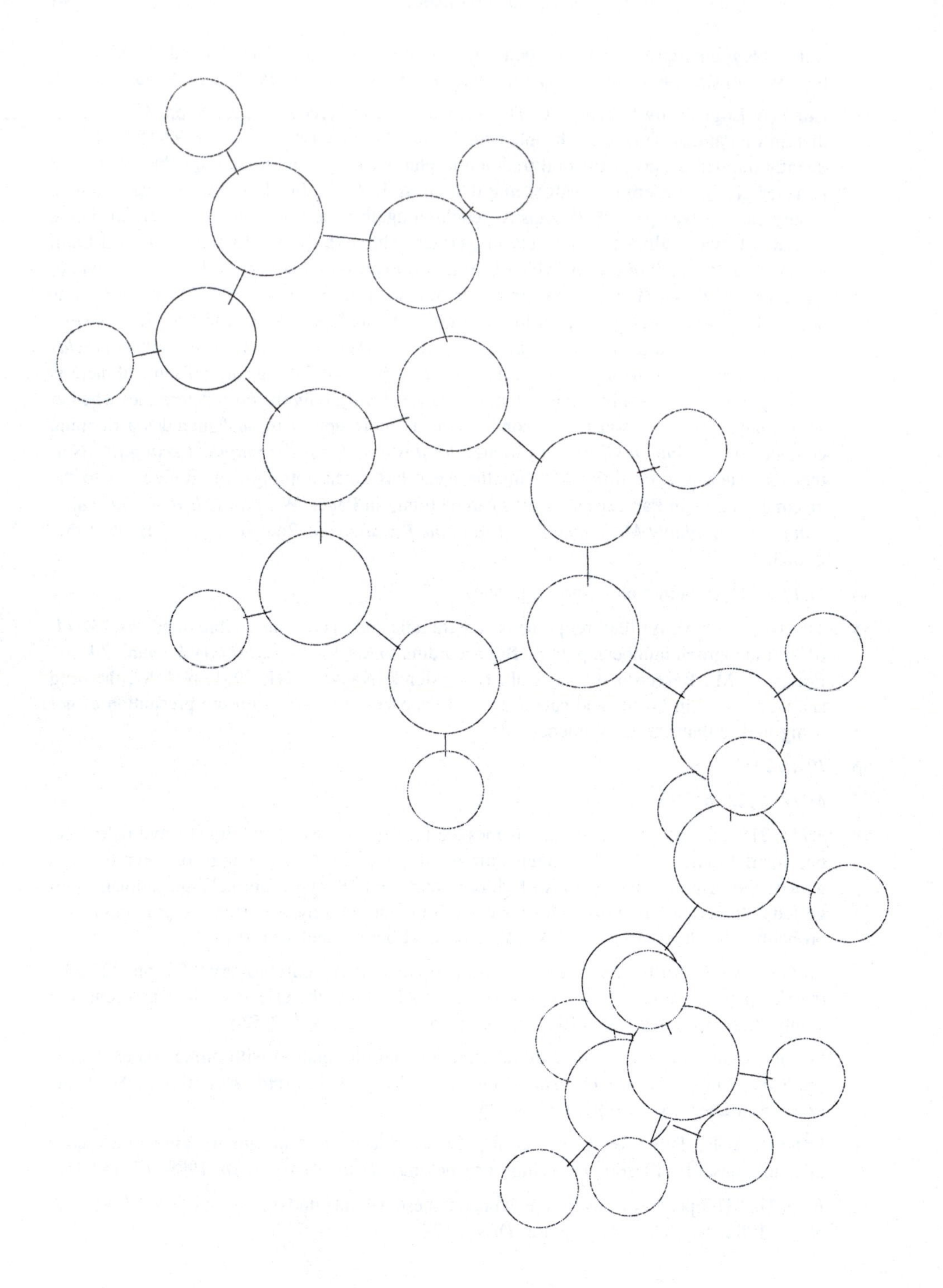

2

Opium and the Opioids

Thou'rt slave to Fate, Chance, kings and desperate men,
And dost with poison, war, and sickness dwell;
And poppy, or charms can make us sleep as well,
And better than thy stroke. Why swell'st thou then?
—John Donne[1]

Time is no more, I pause and yet I flee.
A million ages wrap me round with night.
I drain a million ages of delight.
I hold the future in my memory.
—Arthur Symons[2]

Introduction

Opium and alcohol are two of the oldest drugs known to mankind. Almost every culture has used them extensively, and almost every culture has outlawed one or the other at some time in its history because of the problems accompanying their use. In this chapter we explore the use of opioids beginning with the use of poppy resin in early medicine and the extraction of morphine from opium in the 19th century. Since then, an enormous number of varied chemical structures have been designed that show morphinelike activity of one sort or another. For convenience—and because this pattern largely reflects chemical structure and historical development as well—we will first consider the opioids found in nature (chiefly in the poppy plant, but also in the neural systems of mammals); then those that have been obtained semisynthetically by modifying structures derived from nature (chiefly morphine and etorphine); and then the many simpler structures that are completely synthetic. Following this, we look at a smaller class of drugs that act by blocking the opioid receptors and that have been finding an increasing range of interesting applications including the treatment of alcohol and drug addiction. This will lead us finally to describe some of the programs that have been developed over the years to help opioid drug users live with or control their addiction.

A comment needs to be made regarding the terms "opioid" and "opiate." In the past, some authorities used the first term exclusively to describe morphine, codeine, and any drugs derived semisynthetically from these or other opium alkaloids, while reserving the second term for synthetic drugs and the endorphin NTs found in mammalian brain. This eventually proved to be quite arbitrary, and we will follow more recent parlance that favors using "opioid" inclusively to mean any substance interacting with the brain's opioid receptors.[3]

Natural Opioids

The Opium Poppy

The opium poppy, *Papaver somniferum*, is an Old World plant native to Asia Minor, but it has been successfully cultivated on every continent, including Europe and the United States. Even in Peru's inaccessible highlands, former coca growers have successfully turned to growing these beautiful red, yellow, and white flowers; the heroin obtained from processing these plants has proved to be much more profitable than cocaine. The plant produces an abundance of small black seeds that are used as a major ingredient of most birdseed, as well as a common garnish on rolls. Poppyseeds can also be ground into flour, and the oil obtained by expressing them can be used for cooking and for the manufacture of paints, varnishes, and soaps. The seeds have only a very small amount of morphine in them, but someone eating a Polish *galotchki* or Bohemian *buchtva* (pastries stuffed with a black poppyseed jam) can test positive in a gas chromatography-mass spectroscopy (GC-MS) test for morphine, the final metabolite of heroin. Recently, a 26-year-old woman at the University of Michigan was astonished when a GC-MS showed traces of morphine and codeine in her urine; she denied any use of opiates. The problem turned out to be a lemon poppyseed muffin the woman had eaten 5 hours before the urine test. Six days later she tested negative for opiates, ate a second muffin, and in 5 hours was again positive. The Michigan scientists concluded that "positive results on such tests must be handled with caution, and their use for screening purposes must be seriously questioned."[4]

Opium poppies were widely grown in the United States during the first few centuries European settlement, but their cultivation was criminalized in this century. The analgesic effects were valued in colonial times, when settlers would dissolve the resin in whiskey to provide relief from coughing or pain. Thomas Jefferson grew opium poppies at Monticello: his historic garden was maintained, and seeds from its plants, including the poppies, were sold at the gift shop at the Thomas Jefferson Center for Historic Plants—until a drug bust (which did not involve poppy seeds) at the nearby University of Virginia in 1991. Realizing that if Thomas Jefferson were around to practice his now criminal gardening habits, he would be a felon liable to mandatory imprisonment in a federal jail, the board of directors ripped up the plants and burned the seeds. But, whether in ignorance or disregard of the law, some amateur gardeners continue to grow the opium poppy today, prizing its spectacular blooms.[5]

Because of a famous poem by John McCrae, the poppy became the symbol of that 11th-hour, 11th-day, 11th-month truce ending the slaughter of World War I:

In Flanders fields the poppies blow
Between the crosses, row on row....
We are the Dead. Short days ago
We lived, felt dawn, saw sunset glow,
Loved and were loved, and now we lie
In Flanders fields....
To you from failing hands we throw
The torch; be yours to hold it high.
If ye break faith with us who die
We shall not sleep, though poppies grow
In Flanders fields.[6]

Part of the poignancy of the poem derives from the association of the poppy flower with sleep: it has been the symbol of the dead and sleep since antiquity, and the Latin *somniferum* of its botanical name means "bearer of sleep."[7] Notwithstanding Dorothy's experience in *The Wizard of Oz*, when she falls asleep in a field of poppies, the flowers are no more narcotic than the seeds. But if, after the poppy has flowered, several small incisions are made in the unripe seed capsule, a milky latex flows out that can be collected and dried: the resulting yellow- to brown-colored paste is raw opium.

Opium

The power of opium to relieve pain and induce sleep was probably known by the Sumerians (ca. 4000 B.C.); it was accurately described by Theophrastus, the "father of botany" (ca. 300 B.C.); and in the first century A.D. Dioscorides wrote a description of the process of incising poppy capsules. Opium was probably introduced into India and China by Moslem traders sometime in the 14th century. The two Opium Wars of the mid-19th century were an effort by the British Empire to preserve and expand the market in China for British Indian opium against the Chinese government's desire to curtail domestic addiction. An ironic compounding factor in the conflict was the burgeoning market in England for Chinese tea, which had become popular as a more healthful alternative to traditional beer, ale, and spirits.

The active ingredients in opium, being nitrogenous alkaloids occurring in their freebase form, are significantly less soluble in water than in alcohol; an advance in opium processing occurred in the 1500s, when Paracelsus first concocted *laudanum*[8] by extracting opium into brandy, a process that in effect produced a more concentrated alcohol solution (tincture) of morphine. By the 19th century, laudanum and opium itself were freely available in England at any chemist's (drugstore) or grocery store. Use was extraordinarily common among the rural peasantry in the Fens area centered around the Isle of Ely, probably in part because of the frequency of malarial "ague" in the as yet incompletely drained marshes, but perhaps also because, in the words of the writer Thomas Hood (1799–1845), "the Fen people in the dreary, foggy, cloggy, boggy wastes of Cambridge and Lincolnshire had flown to the drug for the sake of the magnificent *scenery*."[9] A physician in the area likewise hypothesized that "their colorless lives are temporarily brightened by the passing dreamland vision afforded them by the baneful poppy."[10] The British Medical Association estimated that half the British imports of opium, about 30,000 lb per year, were consumed in the sparsely populated Fens area—and this use was supplemented by generous use of poppy tea brewed from homegrown poppies weaker in alkaloid content than the Indian product.[11]

We are more familiar with two famous writers of the period, Samuel Taylor Coleridge and Thomas De Quincey, who acquired a lifelong addiction to laudanum, which they first used because of severe pain.[12] Coleridge began taking laudanum in about 1796 to relieve the pain of rheumatism in his knees. By 1802 he was heavily addicted, and by 1808 his drug use had led to complete estrangement from his family. He became nonetheless one of the most influential figures in the English romantic movement, and what may be his best-known work, "Kubla Khan," was written in 1797 in a dreamlike trance under the influence of opium:

> *In Xanadu did Kubla Khan*
> *A stately pleasure-dome decree:*
> *Where Alph, the sacred river, ran*
> *Through caverns measureless to man*
> *Down to a sunless sea....*[13]

In one of the most well-known literary portrayals of addiction, De Quincey's *Confessions of an English Opium-Eater,* the author asserts that "it was not for the purpose of creating pleasure, but of mitigating pain in the severest degree"[14] that he began to use laudanum. But it seems to be the pleasures, admittedly quite intellectual, which kept him using it, until with growing tolerance he finally reached an enormous habit of 8,000 drops (about 20.8 g or perhaps 2–3 g morphine)[15] of laudanum a day. He attributes a heightening of his mental powers to its use, which he felt was much to be recommended over that of alcohol:

> Whereas wine disorders the mental faculties, opium ... introduces amongst them the most exquisite order, legislation, and harmony. Wine robs a man of his self-possession; opium greatly invigorates it.... Wine constantly leads a man to the brink of absurdity and extravagance; and, beyond a certain point, it is sure to volatilize and to disperse the intellectual energies; whereas opium always seems to compose what had been agitated, and to concentrate what had been distracted.... A man who is inebriated ... is often ... brutal; but the opium-eater ... feels that the diviner part of his nature is paramount; that is, the moral affections are in a state of cloudless serenity; and over all is the great light of the majestic intellect.[16]

De Quincey protests that, far from nodding off in a stupor, as was then and is now the stereotypical image of an opioid user, he was quite alert and of heightened sensibility. He would often attend the opera under the influence of opium, and found his experience of the music peculiarly enhanced:

> Now opium, by greatly increasing the activity of the mind, generally increases, of necessity, that particular mode of its activity by which we are able to construct out of the raw material of organic sound an elaborate intellectual pleasure. . . . It is sufficient to say, that a chorus, etc., of elaborate harmony, displayed before me, as in a piece of arras-work, the whole of my past life—not as if recalled by an act of memory, but as if present and incarnated in the music; no longer painful to dwell upon, but the detail of its incidents removed . . . and its passions exalted, spiritualized, and sublimed.[17]

(It is curious how similar this description of enhanced appreciation of music and the significance of music is to what others have said of their experiences using marijuana or psychedelics.) But De Quincey found there was a price for all this; eventually he became so tormented by wild, fantastic dreams that sleep was nearly impossible:

> Whatsoever things capable of being visually represented I did but think of in the darkness . . . they were drawn out, by the fierce chemistry of my dreams, into insufferable splendor that fretted my heart. For this, and all other changes in my dreams, were accompanied by deep-seated anxiety and gloomy melancholy, such as are wholly incommunicable by words. I seemed every night to descend—not metaphorically, but literally to descend—into chasms and sunless abysses, depths below depths, from which it seemed hopeless that I could ever reascend. . . . Space swelled, and was amplified to an extent of unutterable infinity. . . . I sometimes seemed to have lived for seventy or one hundred years in one night. . . . The minutest incidents of childhood, or forgotten scenes of later years, were often revived.[18]

And there were nightmares worthy of Stephen King on acid:

> I was stared at, hooted at, grinned at, chattered at, by monkeys, by parakeets, by cockatoos. I ran into pagodas, and was fixed, for centuries, at the summit, or in secret rooms: I was the idol; I was the priest; I was worshipped; I was sacrificed. I fled from the wrath of Brahma through all the forests of Asia: Vishnu hated me; Shiva laid wait for me. I came suddenly upon Isis and Osiris: I had done a deed, they said, which the ibis and crocodile trembled at. I was buried, for a thousand years, in stone coffins, with mummies and sphinxes, in narrow chambers at the heart of eternal pyramids. I was kissed, with cancerous kisses, by crocodiles; and laid, confounded with all unutterable slimy things, amongst reeds and Nilotic mud.[19]

Life became less than fun, and De Quincey was driven to reduce his habit gradually to about 250 drops of laudanum a day. He found he was never able to get below this dosage without suffering enormously painful withdrawal symptoms that seemed never to go away. At this maintenance dosage he was able to write prolifically and publishably—but the dreams never quite stopped.

Laudanum could prove worse than addictive: in 1862, the poet and painter Dante Gabriel Rossetti returned from a night on the town with his fellow poet Swinburne to find his invalid wife dead from a probably deliberate overdose. In a passion of grief and remorse, he buried with his wife's body the only copy of his most recent poems. (However, six years and a new love later, he had the poems exhumed and published.)

In the United States, according to David Courtwright, opium addiction increased steadily throughout the 19th century due to two influences. The first was opium smoking among Chinese immigrants and a small number of Anglo Americans who learned the habit from their associations with the Chinese underworld. The second much larger group acquired the dependence iatrogenically (from being given the drug by a physician), and most of these people were upper-class women like Mrs. Henry Lafayette Dubose in Harper Lee's *To Kill a Mockingbird*, or Mary Tyrone in Eugene O'Neill's *Long Day's Journey into Night*. The white male addicts were mostly the significant proportion of the physician population that managed to habituate itself.[20] For a variety of reasons, this population declined and the character of the typical opioid addict changed until by World War II it was identified in the public mind with "lower-class urban males, down-and-outs like Frankie Machine, the hustling, poker-dealing junkie of Nelson Algren's *Man With the Golden*

Arm."[21] By then the form of the drug was no longer opium or morphine but, increasingly, heroin.

In present-day medicine, tincture of opium, essentially the same as laudanum, is still occasionally prescribed as a treatment for diarrhea, and a purified mixture of opium alkaloids (below) is also available for intramuscular or subcutaneous injection—for there are yet those who believe that the "natural" mixture, just as it comes from the hands of God, Mother Nature, and Brother Papaver, is superior in stilling pain with fewer side effects than any product of human artifice. But this has never been subjected to any probative test.[22] Another form in which opium is still employed is in *paregoric,* a more dilute solution of powdered opium, anise oil, benzoic acid, camphor, and glycerin in alcohol, which is sometimes prescribed to treat diarrhea. The extraneous ingredients have been carefully selected for their horrendous taste, and this appears to succeed in discouraging all but the most dedicated abusers, since the mixture is able to be sold as a Schedule III drug.

The Opium Alkaloids

About 25% of the mass of crude opium consists of alkaloids (Figure 2.1). Since these structures have a basic amine function, they can be extracted into a solution of aqueous hydrochloric or sulfuric acid; the solution can then be filtered to remove plant fibers and other insoluble material. When the filtrate is made basic, the alkaloids come out of solution and can be extracted into an organic solvent immiscible with water, such as ether, chloroform, or even gasoline. Upon evaporation of the organic solvent, the so-called "free base" is obtained. This process was discovered in the early 19th century, and it is simple enough to be carried out today by unskilled workers in clandestine processing laboratories.

There are over 20 alkaloids in opium, the five most abundant being *morphine*, 2-1

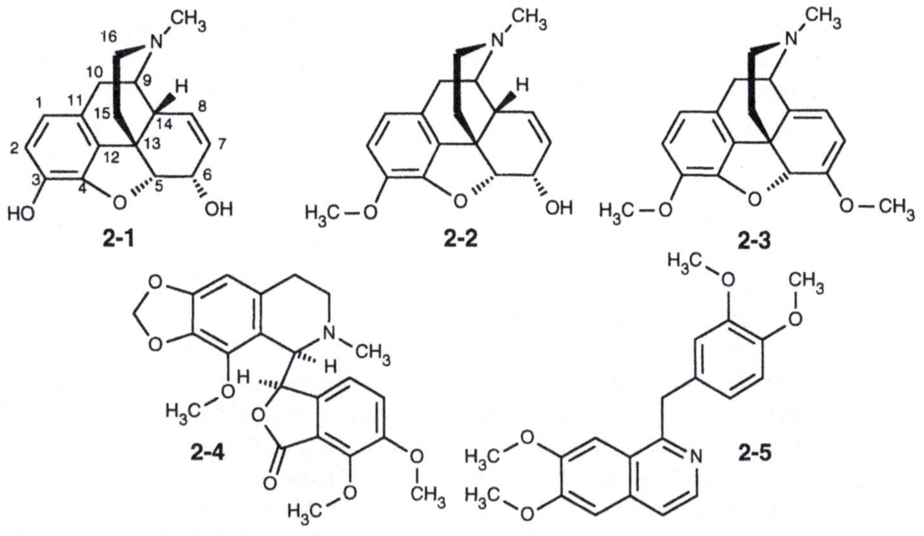

Figure 2.1 The opium alkaloids.

(10-15%); *codeine*, 2-2 (1-3%); *thebaine*, 2-3 (1-2%); *noscapine*, 2-4 (4-8%); and *papaverine*, 2-5 (1-3%). Of these, only morphine and codeine are opioids and have the property of dulling pain (more technical terms for substances having this activity include *analgesic, analgetic,* and *antinociceptive*). While thebaine has a structure quite similar to morphine (all five of these structures are biosynthetically related, as we will see below), it is not an analgesic; rather, it causes strychninelike convulsions. Nonetheless, chemical modifications of the thebaine molecule produce some of the most useful opioid drugs (etorphine, buprenorphine). Noscapine also lacks narcotic activity; although it found some use in the past as an antitussive (cough suppressant), there are far more effective drugs (such as dextromethorphan, used in OTC medications like Robitussin DM) now available for this use. Papaverine has no narcotic action, but it is a smooth muscle relaxant, and it has recently been used either alone[23] or in conjunction with phentolamine[24] (an alpha-adrenergic blocking agent) to treat male impotence. The drugs are injected with a fine gauge needle into the base of the penis. In one double-blind crossover study, more than 80% of the patients had erections with papaverine.[25]

Morphine

Morphine, **2-1**, was first isolated in pure form from opium latex in the first decade of the 19th century. The discovery was made virtually simultaneously by Jean-Francois Derosne, Armand Séguin, and Friedrich Wilhelm Sertürner; however, recognition of the importance of their achievement did not come for more than 10 years, when Gay-Lussac published a translation of Sertürner's work in the prestigious *Annales de Chimie*, accompanying it with an editorial predicting that morphine was only the first of a new class of alkaline plant derivatives containing nitrogen, for which Gay-Lussac proposed the name *alkaloid*.[26] It was Sertürner who named this first alkaloid after Morpheus, the Greek god of dreams.[27]

Morphine is still widely used in modern medical practice (and is, therefore, classified as a Schedule II drug under the Controlled Substance Act). However, it must usually be given parenterally (intravenously or intramuscularly); it is much less potent orally because of rapid metabolism in the intestinal wall and liver. It is still the preferred drug for the treatment of severe pain accompanying myocardial infarction (heart attack). No small part of the efficacy of morphine lies in its ability to reduce psychological anxiety as well as physical pain. When given intravenously, the maximum analgesic effect occurs in about an hour, and continues for 3-4 hours. Morphine can also be prescribed in the form of tablets for oral use; in this form the drug can be used in a home environment for control of the pain of terminal cancer. But dosages have variable effects when taken orally, and must be tailored to the individual's metabolism. The total effect of the drug is diminished by about one-sixth, but proportionately larger doses provide as good relief from pain as with IV administration, and the analgesic effect is more prolonged.[28]

The Mystery of Pain. Pain, like love, is a many-splendored thing, and almost as poorly understood. Pain is even like the weather: everyone talks about it, but nobody does anything about it. Ronald Melzack, research director of the Pain Clinic at the Montreal General Hospital, has been doing something about it. He proposed in 1981 that there are two kinds of pain—treated by morphine, the prolonged, *tonic* kind that persists long after the injury, and the acute, sharp *phasic* pain like that of a freshly cut finger, which is briefly intense and then falls away to be replaced by the tonic variety. Morphine and the opioids are usually used for tonic pain, but it turns out that tolerance does not develop to morphine's effects on this sort of pain. Traditionally, morphine and other drugs have been tested for their capacity to induce tolerance (and, as the usual assumption goes, addiction) by observing their effects on phasic pain in animals; and tolerance does develop to morphine's effects on phasic pain. This seems to explain in part why patients in hospital settings given injections of narcotics for months simply do not become opioid addicts, and why many cancer patients do not develop tolerance to the opioids given for their pain. Melzack has also found that there are, at least in rats, genetic strains in which naloxone, an opioid antagonist, actually enhances morphine analgesia, and speculates that susceptibility to addiction may have a genetic component in some people.

One corollary of all this may be that people with chronic pain are being deprived of the drugs that could help them because of fear that they will become addicted to ever-increasing levels of opioids, when it may be that pain sufferers (and perhaps some heroin addicts as well) could lead fairly normal lives if only they were provided with a constant maintenance level of morphine.[29] It has long been argued that chronic pain is inadequately treated in American medical practice due to a prevailing cultural bias that one should grit one's teeth, stiffen one's upper lip, and bear with the pain of life without recourse to the "crutch" of drugs,[30] an attitude that has been labeled "pharmacological Calvinism."[31] In 1994, a national pain survey sponsored by the American Medical Association confirmed this impression: nearly one in five adults suffers from pain lasting 6 months or more, and 44% of this group said that the medication given by their doctors did not help.[32] A report of the same year on the management of pain in patients with cancer of the Agency for Health Care Policy and Research, a branch of the Department of Health and Human Services, concurs: "Undertreatment of cancer pain is common because of clinicians' inadequate knowledge of effective assessment and management practices, negative attitudes of patients and clinicians toward the use of drugs for the relief of pain, and a variety of problems related to reimbursement for effective pain management."[33]

According to this report, the sequence of pharmacological agents to be used for cancer pain is first acetaminophen or nonsteroidal antiinflammatory agents like aspirin, ibuprofen, or naproxen, then mixtures of these drugs with codeine or hydrocodone. When these no longer control pain, recourse should be had to the "third step of the analgesic ladder. At this stage drugs such as codeine or hydrocodone are replaced by more potent opioids, usually morphine, hydromorphone, methadone, fentanyl, or levorphanol," with morphine "the most commonly used" because of its availability, well-known profile of side effects, and relatively low cost.[34] They reassure the practicing physician that "tolerance and physical dependence are common and predictable consequences of long-term administration of opioid agents," but that this should not be "confused with psychological dependence ('addiction') manifested as drug abuse. . . . There should be no ceiling or maximal recommended dose for full opioid agonists: very large doses of morphine (e.g., several hundred milligrams every 4 hours) may be needed for some patients with severe pain."[35] Patients with stable disease usually require only moderate increases in medication. A suddenly increasing need for pain medication, they suggest, is often a sign that the cancer has begun to progress rapidly, or perhaps that the end is near.

Death by Morphine Drip. As morphine has been used for millennia to take away the pain of disease, so it has been and continues to be used to take away the "disease" of life. Not only those who have abandoned hope, but some of the greatest philosophers, poets, and religious leaders, men like Siddhartha Gautama, Socrates, and Jesus, have come to view death as an eventually desirable part of life:

Come lovely and soothing death,
Undulate round the world, serenely arriving, arriving,
In the day, in the night, to all, to each,
Sooner or later delicate death.

Prais'd be the fathomless universe,
For life and joy, and for objects and knowledge curious,
And for love, sweet love—but praise! praise! praise!
For the sure-enwinding arms of cool-enfolding death.

Dark mother always gliding near with soft feet,
Have none chanted for thee a chant of fullest welcome?
Then I chant it for thee, I glorify thee above all,
I bring thee a song that when thou must indeed come, come unfalteringly.[36]

But except for some rare, wise, and deliberately iconoclastic individuals like Walt Whitman, whose words we read above, the approach of our dark mother, death, is generally not spoken of in American culture.[37] What is a physician to do when a patient needs more and more morphine, even when a dose level is finally reached at which death will be caused through respiratory failure? Traditional ethical theory has held that, by the "principle of double effect," if the primary intent is to relieve pain, even though the secondary, anticipated, but "unintended" consequence is death—then it is legitimate to administer a lethal dose of morphine (assuming, of course, that patients or their proxies have given their consent to this procedure). The morphine is not given as a single large dose, but in a slow intravenous infusion called a "morphine drip" that, over the course of 6–12 hours, results in the unfaltering visitation, via respiratory failure, of Whitman's "lovely and soothing death."[38]

Drugs Complementing Morphine's Effects. If only morphine caused less euphoria and less respiratory depression—and relieved more pain—then it would be an ideal drug, many a physician must have mused. Surprisingly, this seems to be just what happens when morphine is administered along with *verapamil*, a coronary vasodilator and calcium channel blocker used to treat hypertension. This could be considered a phenomenon begging for discovery, since both drugs are given to treat myocardial infarction. But the anecdotal evidence has been nailed down: first in rodents (where it was shown that calcium channel antagonists enhance opiate-induced analgesia and antagonize respiratory depression produced by morphine) and now in humans. Morphine-verapamil interactions were studied in 12 experienced polydrug users with histories of heroin abuse by using a double-blind, crossover study design. Verapamil significantly potentiated the ability of morphine to elevate the pain threshold while it significantly reduced morphine-elevated scores of euphoria as measured on the Morphine-Benzedrine Group subscale of the Addiction Research Center Inventory.[39]

It has long been known that the toxicity of morphine and heroin is lowered by simultaneous administration of small amounts of amphetamines, while the analgesic and anxiolytic effects are augmented; several antidepressant drugs, and some of the antihistamines have similar synergetic effects. In the case of the terminally ill particularly, these effects can be used to advantage to lower the dose of morphine needed to control pain.[40]

Morphine Analgesia Outside the CNS. Thus far we have discussed opioid analgesia from the traditional perspective, which assumes that it takes place exclusively within the CNS. But there is now evidence that, at least in some types of trauma, there are opioid receptors located on sensory nerves outside the CNS at the site of inflammation that respond to endogenous β-endorphin and met-enkephalin produced by immune cells in response to the trauma. They also respond to morphine. Stein at Johns Hopkins University injected morphine into the joint during arthroscopic knee surgery and found that the drug diminished postoperative pain.[41] The morphine was as effective at the site as a local anesthetic, was reversed by naloxone, an opioid antagonist, and lasted up to 48 hours after injection.[42]

The Relative Analgesic Efficacy of Opioid Drugs. If for no other reason than that it is the oldest known opioid drug, morphine is still felt to be the paradigm against which all other such drugs are measured. In the following sections, we will be looking at a score or so of other opioids, some of which have specialized uses (e.g., in veterinary medicine or general anesthesia). In these cases, there is often only a rough estimate of analgesic efficacy relative to morphine, one that has usually been obtained from animal studies. In order to compare the relative effectiveness and potency of the handful of opioid drugs most often used in human medicine, Table 2.1 lists the approximate equianalgesic doses of these drugs, which are used to relieve moderate-to-severe pain in adults and children weighing 50 kg or more,[43] in descending order of potency (with allowance made for methadone and levorphanol, which have longer half-lives). The most common ways of administering analgesic opioids are the oral and parenteral routes (IV, IM, or SC), and these are also compared in Table 2.1.

Codeine

The name of this alkaloid derives the Greek word κωδεια, for the head of the poppy. Codeine is found in small amounts with morphine in opium, and in the past this was the principal source of the drug. But nowadays, useful amounts of codeine are ordinarily obtained semisynthetically from morphine by methylation of the phenolic hydroxide. It is used to relieve mild to moderate pain; while less effective than morphine as an analgesic, it is also thought to be less liable to establish addiction. Unlike morphine, codeine is rarely used intravenously because of its greater liability to trigger the release of histamine; on the other hand, when taken orally, it is not as rapidly inactivated by first-pass hepatic metabolism as morphine. About 90% of ingested codeine is metabolized by conjugation with glucuronic acid (Chapter 1); the remaining 10% is *O*-demethylated to form morphine, and it is likely that codeine's analgesic efficacy (but probably not its antitussive effect) actually comes from the morphine thus produced (i.e., codeine is really a prodrug). The enzyme that effects this biotransformation is called *sparteine oxygenase*, and about 7%

of the U.S. population displays a genetic deficiency in this enzyme. Similarly, *hydrocodone* (**2-34**) probably owes its efficacy to its biotransformation to *hydromorphone* (**2-33**). "Thus, for 7% of the people in the United States," cautions Marcus Reidenberg of Cornell Medical College, "codeine, and probably hydrocodone, acts as a placebo and has little or no analgesic activity. We should think of this when patients in pain say they are not getting relief from either of these two drugs."[44]

Table 2-1 Relative Potency of Opioid Drugs

Opioid	Oral	Parenteral
Meperidine	300 mg every 2-3 hours	100 mg every 3 hours
Morphine	30 mg every 3-4 hours	10 mg every 3-4 hours
Methadone	20 mg every 6-8 hours	10 mg every 6-8 hours
Hydromorphone	7.5 mg every 3-4 hours	1.5 mg every 3-4 hours
Levorphanol	4 mg every 6-8 hours	2 mg every 6-8 hours
Oxymorphone	—	1 mg every 3-4 hours

But they may still get alleviation of cough. Codeine is more effective than morphine as an antitussive (a relatively small oral dose of codeine, 10 or 20 mg, although insufficient to suppress pain, is nonetheless effective in suppressing cough), and this is probably due to the direct interaction of unmetabolized codeine at a distinct set of receptors. Supporting evidence for this is the powerful antitussive action of a drug used in Europe, though not in the U.S.—*pholcodine*. The structure of pholcodine is that of codeine but with the 3-methyl group replaced by a much longer radical, the morpholinoethyl group, which renders it resistant to sparteine oxygenase. Because it is not biotransformed to morphine, it has a long half-life and no opioid-like action.[45]

Codeine controls pain better, often having a synergetic (greater than merely additive) effect, when it is used in combination with nonopioid analgesics such as acetaminophen or aspirin. A Schedule II drug when used alone, in mixtures like these codeine is Schedule III.

Endorphins

In the early 1970s, several groups (Solomon Snyder and Candace Pert at Johns Hopkins University, Lars Terenius in Uppsala, and Eric Simon at New York University) simultaneously reported evidence for stereospecific binding sites for opiate drugs in mammalian brain.[46] Such sites were soon found to be widespread in all vertebrates, and this led to the search for endogenous ligands—that is, chemical agents naturally occurring in the brain that bind to these sites. Soon Hughes and Kosterlitz in Aberdeen, Scotland (soon followed by Snyder's group in Baltimore) had isolated two such endogenous ligands for these receptors: they were small peptides that the Scottish group called *enkephalins*.[47] Within 10 years a number of other such peptides had been discovered, and since then at least a dozen have been found. These are now collectively designated as *endorphins* (endogenous morphine) and include the original two, *leu-enkephalin* and *met-enkephalin,* each of which contains 5 amino acid residues; *dynorphin,* with 17 residues; and *β-endorphin,* with 31. Various fragments of dynorphin and β-endorphin are also active and are given different Greek letters to distinguish them.

The opioid peptides are formed by cleavage from three much larger proteins with much larger names: preproopiomelanocortin (POMC, also called adrenocorticotropic hormone-β-lipotropin precursor), preproenkephalin A, and preproenkephalin B (also called preprodynorphin). These precursor proteins each have about 260 amino acid residues. POMC is formed in the pituitary and is cleaved not only to form β-endorphin but several other seemingly unrelated biologically active peptides: adrenocorticotropic hormone (ACTH) and the melanocyte-stimulating hormones. Proenkephalin originates in the adrenal cortex and contains one copy of leu-enkephalin and several of met-enkephalin; prodynorphin has been isolated from the brain, the spinal cord, pituitary, adrenal, and reproductive organs and contains one copy of dynorphin. There is considerable redundancy and interrelatedness between the opioid peptides. For instance, the first four amino acids (tyrosine-glycine-glycine-phenylalanine) of the two smallest, met-enkephalin (**2-6,** Figure 2.2) and leu-enkephalin, **2-7**, are identical. Moreover, the first five residues of dynorphin are those of leu-enkephalin, and the first five residues of β-endorphin are those of met-enkephalin.

When injected directly into the brain (intraventricularly), the pharmacological activities of the endogenous opioids have proved to be surprisingly similar to those of morphine and the synthetic opioids, despite the vast difference in chemical structure. Like morphine, the effects of the endogenous opioids are reversed by opiate antagonists such as naloxone (and probably by an endogenous peptide antagonist, *cholecystokinin*[48]). Thus far, the hope of finding an ideal, nonaddicting analgesic among these peptides has not been fulfilled; it was shown very early that tolerance and physical addiction develop to pharmacologically effective doses of the opioid peptides just as with morphine and the synthetic opiates; furthermore, cross-tolerance develops between these classes of opioids.[49]

How these very different structural classes—the rigid, pentacyclic morphine alkaloids on the one hand and the linear, flexible peptide chains on the other—both act with almost indistinguishable effect at the same specific and selective set of receptors, remains something of a mystery. Many proposals have been made, but there is little agreement except concerning the tyramine residue in the enkephalins, which corresponds to the *p.*-hydroxyphenylethylamine fragment in morphine. Most hypotheses, such as that of the "β-bend" revolve around conformations induced in the peptide chain by intramolecular hydrogen bonding.[50]

***Hunger's Heroin High*.** According to one theory, patients with anorexia nervosa or bulimia are actually addicted to the release of their own endorphins.[51] Put at its baldest, this amounts to the claim that anorexia is

> perpetuated by an addiction to starvation .... Anorexics are compulsive, often centering their lives on fasting and other anorexic behaviors just as drug addicts center their lives on getting their next 'fix.' They typically deny they have a problem, or ... report feeling driven by an almost demonic force beyond their control.[52]

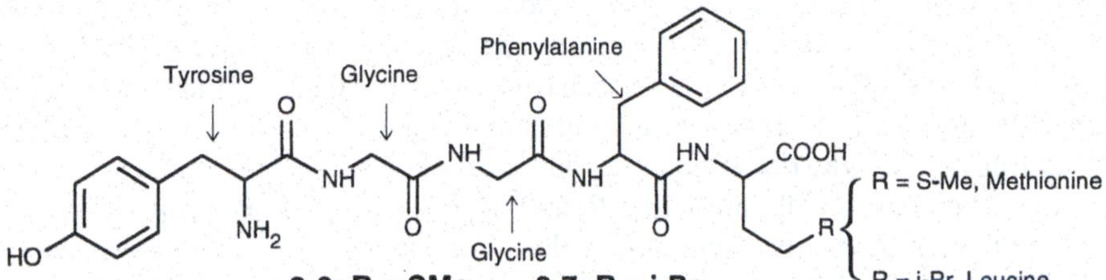

Tyrosine Glycine Phenylalanine Glycine

2-6: R = SMe **2-7: R = i-Pr**

R = S-Me, Methionine

R = i-Pr, Leucine

Figure 2.2 The enkephalins.

What motivates this hypothesis is the well-established fact that fasting induces the release of endogenous opioids. In one animal experiment substantiating this, 24-hour deprivation from food caused analgesia in rats equivalent to morphine in a standard pain test (tail-flick), and the analgesia was blocked by naloxone. Furthermore, when food was alternately provided freely for 24 hours, then withdrawn for 24 hours over a course of 10 days, the animals in effect developed tolerance, with the analgesia greatly reduced, and they simultaneously developed cross-tolerance to morphine.[53] The pattern is suggestively like the feast-or-famine, binge-or-bust characteristics of many people with eating disorders or addictions. In fact, in a human study comparing binge eaters with normal controls, the opioid antagonist naloxone significantly reduced total caloric intake from snacks on the part of binge eaters, but not in controls, and the reduction was particularly pronounced for sweet high-fat foods like cookies or chocolates.[54] Moreover, from one-quarter to one-half of those who have anorexia nervosa or bulimia also have a history of substance abuse.[55]

The relationship is supported by success in using opioid antagonists to treat eating disorders. When naltrexone was given to 10 bulimics who had not improved with antidepressant therapy, 7 of the 10 had an at least 75% improvement in their symptoms and maintained this improvement for several months.[56] In a second study, it was found that the effect was dose-related, with higher doses of naltrexone needed than are needed to block exogenous opioids like heroin.[57]

An elaboration of this theory comes from the observation that mice respond to morphine by hyperactivity and reduced feeding, whereas most species, including rats and normal humans, respond in an exactly opposite manner, with slowed activity and increased intake of food. Could it be that people with anorexia nervosa have an atypical opioid system like that found in mice?[58] Or the problem could be that bulimics do not produce adequate amounts of endorphins and are fasting in order to trigger their release: two studies showed that bulimics with normal weight had significantly lower levels of β-endorphin than nonbulimics of normal weight.[59]

In what may be a related phenomenon, several studies have shown that the terminally ill are much more comfortable and die in greater peace if they are not given any more food or water than they ask for, which is often little or nothing. "Starving," the researchers report, "seems to ease the death of such patients because dehydration lessens consciousness, promotes sleepiness and diminishes pain"[60]—all typical effects of the endogenous opioids. "This is, of course, how people have always died, before there were hospitals with IVs and feeding tubes," said Dr. Groth-Juncker, one of the researchers.[61]

Jogging Junkies—That "Natural" High. None of what was said above with regard to anorexia or bulimia necessarily contradicts the importance of social and cultural influences on eating disorders as on all "addictive" behavior. Why are there so many more anorectic women than men? Most likely, because our culture itself is so addicted to a paradigm of the beautiful woman who doesn't have an ounce of fat (about as reasonable as not having an ounce of bone). We all know from Renaissance art that the beautiful women of the 17th century were pleasantly chubby—only the lower classes were so poor and overworked that they couldn't get fat.[62] But in our culture it is almost a rite of passage for a young woman to start dieting as soon as she reaches puberty. The males in our culture would probably get just as hooked on their endorphins if they had any motivation for a really determined diet, but they are all too busy taking steroids and pumping iron— or jogging. There have been conflicting results from studies attempting to show that endorphins play some role in the dedication of the jogging, iron-pumping, or otherwise aerobic types. Endorphins almost certainly are involved; the problem has been to detect them, since they are produced in small quantities and rapidly metabolized. The best evidence for the origins of the sense of well-being gained by "sweataholics" after they do their thing was a double-blind, placebo-randomized crossover study in which participants in an aerobics class were given either a placebo or the opioid blocker naltrexone before exercising, then evaluated for mood afterwards by several standardized questionnaires. There was indeed a significant tendency for the nonblocked exercisers—those who had taken placebo—to experience a calm, pleasant, relaxed mood while the naloxone-treated subjects were still relatively fretful.[63] But, as with the eating disorders, there is probably more involved, and more NTs involved, than just the endorphin system.

The situation is also fuzzy with regard to acupuncture, electrical nerve stimulation, and hypnotism, all of which can produce genuine analgesia, but by mechanisms that are still little understood. There are conflicting results in studies assessing whether naloxone can block these analgesic effects, and the most likely conclusion is that opioids are partly involved.[64] Two recent examples of research in this area are (1) in China, a group of 32 patients with muscle spasticity from spinal injury experienced amelioration of their spasticity when treated with high-frequency electrical stimulation, an effect that was partly reversed by a high dose of naloxone;[65] and (2) at UCLA, a double-blind test of dental pain showed that electrical stimulation raised the pain threshold to a small but significant degree and that the effect was partially but not completely blocked by naloxone.[66]

Opioid Receptor Subtypes

About the same time that the endorphins were being discovered, other workers were noticing some important differences in receptor response between morphine and some other opioids. It was shown that three distinctly different behavioral syndromes could be obtained from administration of morphine, *ketazocine*, (**2-8**, Figure 2.3), or N-*allylnormetazocine*, **2-9**, to experimental animals; that none of these three drugs was able to substitute for either of the others in suppressing withdrawal symptoms in dogs who had been treated chronically with any of them; and finally, somewhat later, that the vas deferens of the mouse was much more responsive to enkephalins than it was to morphine. This seemed to show that there were at least four subtypes of opioid receptor, and they were designated μ for *m*orphine,

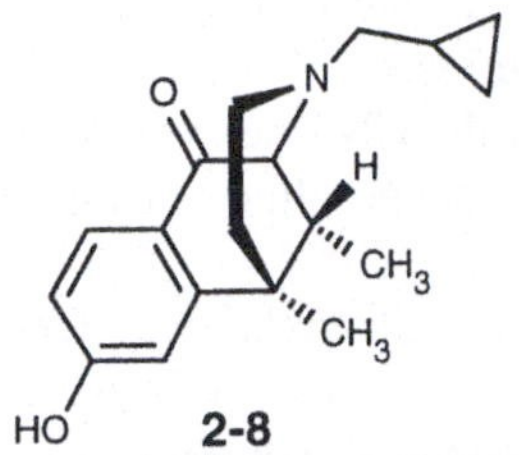

HO 2-8 HO 2-9

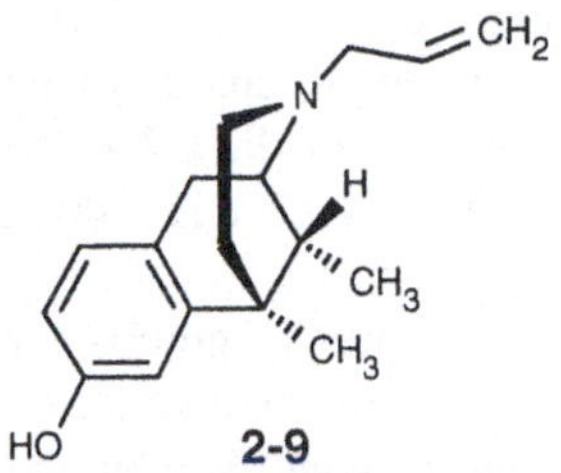

Figure 2.3 Ketazocine and *N*-allylnormetazocine.

κ for *k*etazocine, σ for *S*KF 10,047 (which was the Smith, Kline, and French code for *N*-allylnormetazocine), and δ for *d*eferens. However, the σ receptor is no longer considered to be a true opioid receptor, and its function remains unclear[67]; many drugs like SKF 10,047, while resembling morphine in low doses, in larger doses produce dysphoric, anesthetic, and psychotomimetic effects. They are now thought to interact at both σ receptors and another nonopioid class, the PCP receptors. (PCP is phencyclidine, or "angel dust," discussed in Chapter 7.) Many of the effects ascribed to the σ receptors by the original workers were probably caused by action of SKF 10047 at PCP receptors.[68]

Many claim that opiates that do not act at the μ receptor, but exclusively at the δ or κ receptor, can produce analgesia without the undesirable side effects of euphoria, respiratory depression, constipation, tolerance, or dependence. Agonists highly selective for the κ receptors now available, and it appears that in contrast to the μ- and δ-selective agonists, these induce aversive reactions and are *not* self-administered by experimental animals;[69] additionally, some appear to be neuroprotective.[70] Other such agonists tested in humans, while producing analgesia, caused dysphoria rather than euphoria.[71] Although there seems to be promise in pursuing these leads, at this writing none of these agents has final approval in any country for human therapeutic use. Thus far, the idea of an analgetic that does not produce respiratory depression, etc., has been most closely approximated not by drugs that act exclusively at one of these receptor subtypes but by those, like buprenorphine, that have both agonist and antagonist action.[72]]

Structure-Activity Relationships for the Opioids

One of the most useful methods of studying drug effects is to chemically modify an active structure, producing a new substance, and then observe what differences there are in pharmacological activity between the old and new substance; these are called structure-activity relationships (SAR) for the drug. Morphine, as the prototype analgesic and the longest-known drug of this class, has had many such alterations made to its structure. Some of the more dramatic effects are the following:[73]

- Blocking the phenolic 3-OH of morphine (see **2-1**, Figure 2.1) by esterification (as in heroin) or etherification (as in codeine) decreases CNS depressant effects and increases antitussive activity.

- Blocking the alcoholic 6-OH, or oxidizing it to the ketone, as in oxycodone, increases the euphoriant and CNS depressant effects.
- Hydrogenation of the C7–C8 bond enhances the CNS depressant effect.
- Breaking the C4–C5 ether linkage or opening the piperidine ring markedly decreases activity.
- Replacing the methyl group on the nitrogen atom with allyl ($CH_2{=}CHCH_2$—), cyclopropylmethyl, cyclobutylmethyl or similar groups transforms an agonist into a mixed agonist–antagonist, as in buprenorphine, butorphanol, nalbuphine, and pentazocine, or into a pure antagonist, as in naloxone and naltrexone.
- The five chiral centers of morphine (carbons 5, 6, 9, 13, and 14) are critical to its activity. The mirror image of morphine has been prepared from the alkaloid sinomenine, and it has no analgesic or addictive properties.[74]

In the early part of this century, consideration of which fragments of the morphine structure were essential for activity led to the formulation of the "morphine rule," which states that "any structure that has a phenyl group attached to a quaternary carbon that in turn is two atoms distant from an amine nitrogen" will have morphinoid activity.[75] Most of the drugs considered in this chapter will roughly follow this rule, although those in the fentanyl class are a prominent exception. More recently, efforts to describe the typography of the opioid receptor have led to the postulate that the *N*-methyl-3-phenylpiperidine moiety of the morphine molecule is essential for its activity, and that the opioid receptor must be characterized by the following three features:[76]

- A flat surface allowing van der Waals binding to the hydroxyphenyl ring (this corresponds to the hydroxyphenyl moiety of the tyrosine residue in either met- or leu-enkephalin);
- A negatively charged site able to bind the positively charged nitrogen; and
- A cavity that can accommodate the ethylene bridge (C15–C16 in morphine) that projects from the plane containing the phenyl ring and the nitrogen atom.
- From SAR of the tremendously enhanced activity of such powerful agonists as *etorphine* (below), it can be concluded that there is an additional lipophilic pocket, not accessed by morphine, into which the —C(OH)MePr chain extending from carbon 7 fits, and which corresponds to the phenylalanine ring found in either met- or leu-enkephalin. (When the —C(OH)MePr chain of etorphine is replaced with —C(OH)MeBz, receptor binding is still further enhanced.)[77]

However, the fact that there are still other substances that show powerful morphinelike activity at the opioid receptors and have structures seemingly quite different from any suggested by the descriptions above "suggests that our knowledge of the opiate-receptor topography is still rudimentary."[78] Two examples shown in Figure 2.4 are (+)-*S-tifluadom*,[79] **2-10**, which is an effective opioid analgesic acting primarily at the κ receptor, and *etonitazene*, **2-11**, which acts at the μ receptor with an analgesic activity approximately 1,000 times that of morphine. It can be seen that tifluadom looks like a benzodiazepine—a class of compounds that normally shows not analgetic but anxiolytic and hypnotic effects; while etonitazene, a benzimidazole, has no strong resemblance to any existing class of drugs. (Nonetheless, appearances notwithstanding, it is quite possible to rationalize the fit of these two structures to the receptor site previously.)

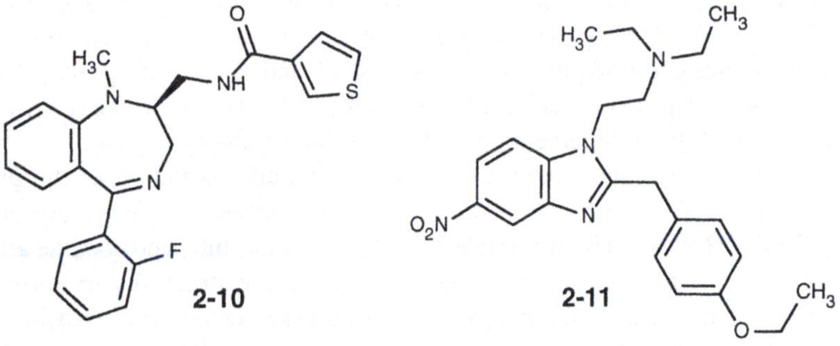

Figure 2.4 Two anomalous opioids.

Endogenous Synthesis of Morphine in Living Organisms

How is morphine made by the poppy plant? Research using radioactively tagged amino acids shows that the opioids are ultimately synthesized from two molecules of *tyrosine*. One molecule of tyrosine is first transformed by oxidation and decarboxylation into *dopamine* (see Chapter One, Figure 1.3), and the dopamine is then combined with a second unit of tyrosine to form *tetrahydropapaveroline* (**2-12**, Figure 2.5). This is transformed through several steps to *reticuline*, **2-13**.[80] Here there is a branching, and some of the reticuline goes on via *O*-methylation and oxidation to become papaverine (**2-5**, Figure 2.1), while the rest undergoes what chemists will recognize as an oxidative phenolic coupling (responsible for such diverse phenomena as the browning of toast and sliced apples) after an unexpected gymnastic contortion that would seem plausible only to an enzyme. (Follow the arrows of **2-13** in a counterclockwise twirl to produce the equivalent structure, **2-14**, with its arm twisted—which then couples intramolecularly to form *salutaridine*, **2-15**, whose new chiral carbon is the magic quaternary carbon of the morphine rule.) After this, things

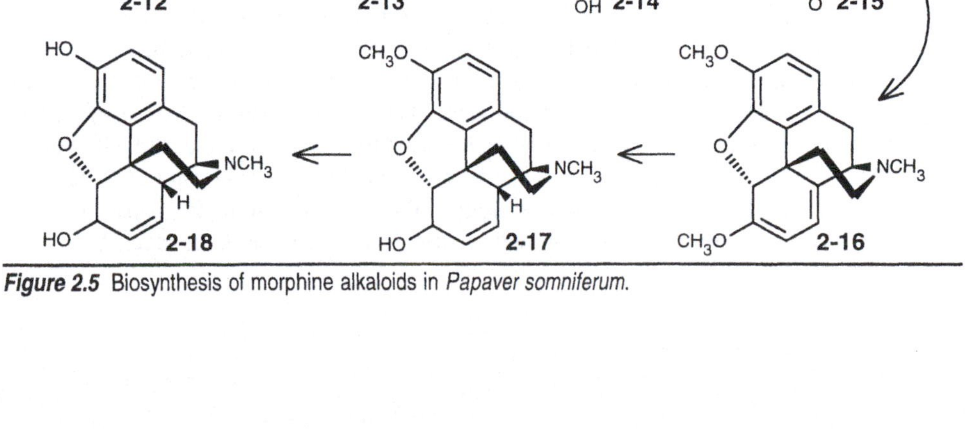

Figure 2.5 Biosynthesis of morphine alkaloids in *Papaver somniferum*.

are relatively straightforward: the lone hydroxy group of salutaridine adds to the enone to make the furan bridge and we have (give or take a redox hydrogen) *thebaine*, **2-16**, which is sequentially transformed to *codeine*, **2-17** and finally *morphine*, **2-18**. (This sequence may seem a little unexpected, but it is well established: radioactive tyrosine shows up first in thebaine, then codeine, and only last in morphine.)[81]

Phylogenetically, one would not expect so unique a structure as morphine to be formed outside the *Papaveraceae*—indeed, it is only formed in a few representatives of the poppy family. But there is now ample evidence that morphine and codeine are naturally produced by mammals[82] (and cockroaches[83]) and that these structures are formed—albeit in far smaller quantity than in the poppy—by the same sequence, from tyrosine via reticuline![84] In 1986, Spector's group showed salutaridine and thebaine were converted to codeine and morphine in rat brain, intestine, liver, kidney, and blood.[85] In 1987, Goldstein (whose earlier work inspired the successful search for the endorphins) showed that the key biosynthetic step generating salutaridine from reticuline took place in vitro and in vivo in rat liver,[86] and this has recently been shown to take place by means of a very regio- and stereoselective mammalian (cytochrome P-450) enzyme.[87]

What exactly this means functionally in mammalian physiology is not obvious, in that the quantity of these opioids produced seems small. But it may play a role in the long noted, if anecdotal, greater sensitivity of some people to pain. "I am a lone lorn creetur, . . . and I feel it more than other people," was the constant refrain of Mrs. Gummidge in *David Copperfield*.[88] Perhaps she did. Perhaps she was in part such a "lone lorn creeter" because she did not express the *CYP2D6* gene and hence was a poor metabolizer of sparteine/debrisoquine oxygenase. Her endogenous codeine stayed lone and forlorn in her brain, unable to undergo *O*-demethylation to morphine, unable to provide her with comfort in her distress. Whatever the case with Mrs. Gummidge, a recent study from Denmark showed that when a group of 94 extensive metabolizers was compared with 82 poor metabolizers in three tests of pain, there was a significantly greater peak pain rating for the poor metabolizers when exposed to an "ice-water cold pressor" test than for the extensive metabolizers. The authors conclude that greater sensitivity to some types of pain "may be related to an inherited defect in endogenous synthesis of morphine via *CYP2D6* in the brain."[89]

Exogenous Synthesis of Morphine by Human Organisms

We move now from the mindless synthesis of morphine in the brain to its more thoughtful synthesis in the laboratory. As we have seen, the structure of morphine is quite intricate, and it was not unravelled correctly until 1923, more than a century after its isolation, by Robert Robinson and John Gulland.[90] Another 30 years was to elapse before the first successful synthesis was achieved, and it represented the work of more than a decade (1945–1956) by Gates, Doering, and Woodward. The synthesis (Figure 2.6), started with a substituted anthracene, **2-19**, which, after elaboration to **2-20**, was condensed with butadiene, **2-21**, in a classic Diels-Alder reaction to form the phenanthrene ring of **2-22**. The final yield of morphine, **2-23**, was on the order of 0.002%, after some 25 steps, including the use of a relay derived from opium and resolution of the racemic product.[91]

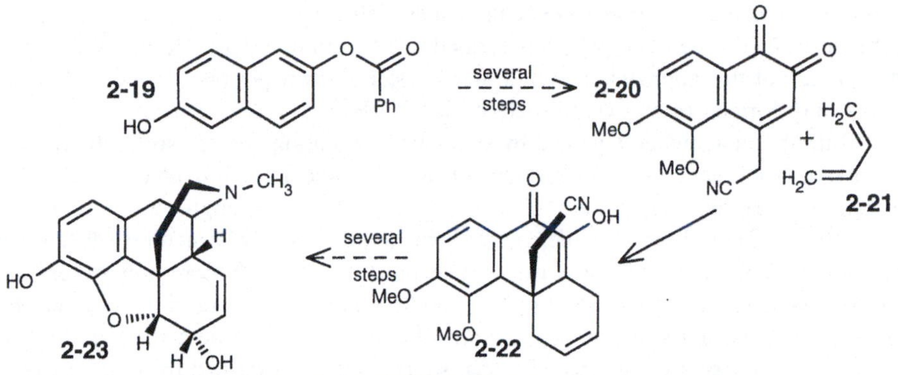

Figure 2.6 Synthesis of morphine by Gates and Woodward.

There are a total of five rings in the morphine system, and this first synthesis rather naturally built them up one at a time. A more recent synthesis of Fuchs and Toth[92] (Figure 2.7) uses a convergent approach, first building up two halves of the molecule, and two of the ring systems, from isovanillin, **2-24**, and 2-allylcyclohexane-1,3-dione, **2-26**. These are then strung together by an ether linkage that will eventually become the oxygen of the tetrahydrofuran ring. The key step in the synthesis is the tandem intramolecular alkylation process in which an excess of *n*-BuLi converts **2-25** to **2-28**. The cleverness of this step, which is actually a three-reaction cascade, rivals nature's arm-twisting of reticuline. The process is initiated by the greater stability of aryl lithiums over alkyl lithiums like *n*-BuLi, which leads to an immediate transformation of the phenyl bromide moiety to phenyl lithium. Now this anion is poised invitingly over a highly polarized electron-deficient alkene, to which it promptly bonds, forming the critical quaternary carbon center and displacing the anion to the sulfoxide-substituted (and stabilized) terminus, where it in turn is perfectly aligned for an S_N2 displacement of the second bromine. In one smooth process,

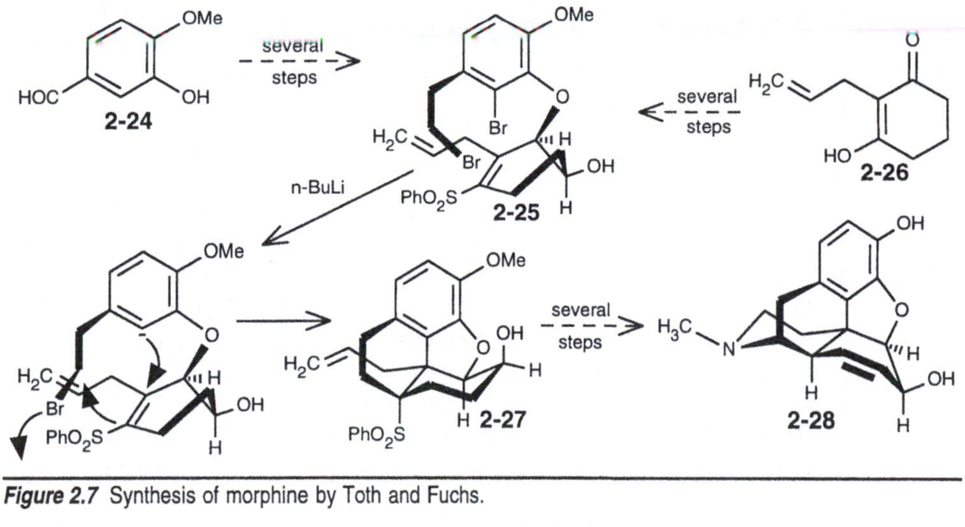

Figure 2.7 Synthesis of morphine by Toth and Fuchs.

two more of morphine's rings have been created; left for last is the least encumbered of all, the unstrained piperidine, which is formed in a few more steps. Despite the elegance and economy of this approach, it is still much easier to grow poppies; racemic morphine, **2-28**, is only formed in an overall yield of about 1%.[93]

But, of course, what is gained by synthetic explorations of this sort, which include the work of Nobel laureates like Woodward, is a rich understanding of the idiosyncratic chemistry of the opioid ring system. This makes it much easier to engineer relatively minor modifications of morphine, codeine, and thebaine to produce new structures (the so-called semisynthetic opioids discussed in the next section) that may be more potent, less toxic, or both. And often along the way to a total synthesis of a natural product like morphine, a much simpler partial structure is encountered that turns out to be a successful drug in its own right. These are the structures that should not be disdained for being synthetic. After all, there is nothing so synthetic as nature herself, and our best contrivances have come only from imitating that painted lady who seems to spend most of her time imitating us.

Semisynthetic Opioids

Modifications of Morphine

Heroin. To Heinrich Dreser and Felix Hoffman of the Friedrich Bayer Laboratories can be attributed the ambivalent credit for the introduction and enthusiastic promotion, in the late 1890s, of two of the all-time best-selling drugs of the 20th century: aspirin and heroin.[94] Both were promoted in the belief that, by forming the acetyl ester of a substance found in nature, its pharmacological profile would be modified for the better. Salicylic acid could be obtained from the bark of willow trees (*salices* in Latin, whence the name of the acid), long known for its ability to lower fever, or from *Spirea ulmaria* (whence a*spirin*). Aspirin was acetylsalicylic acid and, unlike its parent acid, it did not cause nausea or vomiting. It was indeed a great improvement over nature. Dreser believed that the same process of acetylation carried out (this time twice) on morphine to produce diacetylmorphine had also produced an improved, even a "heroic" drug, having in particular none of the addiction liability of its parent morphine: hence the Bayer trade name of heroin (diamorphine, **2-29**, Figure 2.8). Heroin can be made easily by reacting morphine with any number of acetylating agents, the simplest and cheapest probably being acetic anhydride, (**2-30→2-29**); and this seems to be how it is usually made nowadays in illicit drug laboratories.[95] As it turns out, Dreser's masking of both hydroxy groups in heroin presumably allows it to cross the blood-brain barrier (BBB) more rapidly as the freebase; on the other hand, the hydrochloride salt is much more soluble in water than morphine hydrochloride, and this allows a more concentrated solution for injection. For whatever reason, it is a fact that heroin produces relief from pain more promptly than morphine. Heroin itself, like codeine, is a prodrug and inactive as such. But in the brain, the phenolic 3-acetyl ester is hydrolyzed, producing 6-monoacetylmorphine, which actively binds to the opioid receptors; the 6-monoacetylmorphine is itself rapidly hydrolyzed to morphine, which is of course an active opioid (and the end metabolite excreted in the urine).[96] That the "rush"

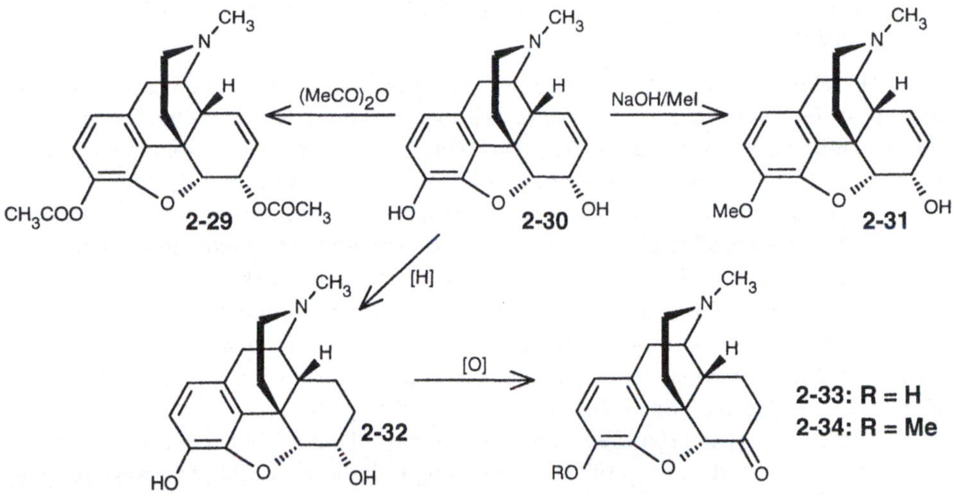

Figure 2.8 Synthesis of heroin, codeine, hydromorphone, and hydrocodone.

from injected heroin is perceived as greater than that from an equivalent amount of morphine may be as much a factor of its much greater solubility as of its more rapid crossing of the BBB.

Nonmedical Use of Heroin. The stereotype of a jobless, impoverished, inner-city user is a very small part of the story. If only jobless and impoverished people bought and used heroin and cocaine, it would not be a multibillion-dollar industry, and the drug lords would be as poor as country churchmice. "There are enormous numbers of people in all walks of life who have integrated heroin use with their lives," says Dr. Robert B. Millman, director of drug abuse programs at New York Hospital-Payne Whitney Psychiatric Clinic. He says that many people find they can experiment with it or use it on weekends without developing a compulsion to take it more frequently or in greater amounts.[97] Here is the abbreviated story of one divided soul anonymously interviewed by *The New York Times.* He is a 44-year old New York executive with a salary of about $200,000, who lives in a condo on the Upper East Side, plays tennis in the Hamptons, and vacations with his wife in Europe, but unknown to his professional colleagues has been using heroin for 20 years. He is trying to quit now, and taking naltrexone to reinforce his motivation, but only because of his wife's insistence:

> "The drug is an enhancement to my life," he said. "I see it as similar to a guy coming home and having a drink of alcohol. Only alcohol has never done it for me. . . . In my heart, I really don't feel there's anything wrong with using heroin. But there doesn't seem to be any way in the world I can persuade my wife to grant me this space in our relationship. I don't want to lose her, so I'm making this effort. I wish I believed in the effort more."
>
> The businessman developed [an] on-again, off-again cycle [of heroin use]. "It's like hide-and-seek," he said of his system. . . . "Once you're clean and straight, you're in control again. Once you're in control it's OK to indulge again. For the first two weeks after I've stopped, it does not occupy my thoughts in an overwhelming sense. But by the third week it is creeping in there

> and it just gets to the point where I want it. I say: 'It's time. I've been good enough. I want my reward.'"[98]

He takes the drug by scooping it up to his nose and inhaling; he used to inject the drug but suffered an overdose some 10 years previous to this interview. One further feature of this gentleman's story (that the *New York Times* reporter says is corroborated from the director of a Cornell University drug treatment program the man attended) that may or may not bear on his ability to "control" his habit is that his withdrawal symptoms are relatively mild—a few sore muscles, chills, some insomnia. Nor did his first experience with heroin sweep him off his feet; instead, he says he had to "actually work at liking it."[99]

Businessman X probably represents a far higher proportion of heroin and cocaine users than most citizens imagine. He gets his heroin through a middle-class source who has supplied him for years. His "ultimate jeopardy" is discovery, and he says he could never risk going to the "South Bronx or wherever" to get drugs.

Others are not so privileged. The Johns Hopkins University School of Hygiene and Public Health has studied the profile of those using Baltimore's needle-exchange program, which has outlets in areas comparable to New York City's South Bronx. Clients average 39 years of age and more than half are high-school dropouts: 93% are unemployed, 91% inject drugs at least daily, and 90% use both heroin and cocaine.[100] Some typical stories:

> "All day, you're in and out of stores, stealing, then running back and forth to the drug man," said Pam Day, who is fighting her heroin and cocaine addictions. "At the end of the day, your arm is sore, your vein is almost collapsed. You're run down, beat down. You've barely eaten anything."
>
> Joan White, 50, developed a sophisticated shoplifting network to pay for her $40- to $80-a-day habit. The West Baltimore woman took special orders from customers who wanted her to steal a certain baby outfit or dress from Hecht's [a local department-store chain] and then deliver the merchandise to them. Ms. White said she tried to get help for her 30-year cocaine and heroin addiction but failed because of the shortage of treatment slots, even when it was part of her probation for stealing. "The judge is threatening you to get help, and I'm still out there," said Ms. White.
>
> She looks back fondly on her four years in prison. She didn't wake up sick and without money. She could function. She could eat. "I had really gotten to like myself," Ms. White said, smiling as she remembered life without drugs.[101]

China White. The street term "China white" originally referred to a high-grade, smokeable or snortable heroin that could be 40–90% pure, unlike the brown, highly diluted heroin of the '60s and '70s, which was often almost 97% inert sugar (or worse). China white originally came from opium poppies grown in the Golden Triangle, where Myanmar (formerly Burma), Thailand, and Laos intersect, and was marketed exclusively through ethnic Chinese clans in Hong Kong (the 14K), Taiwan (the United Bamboo), and California (the Wah Ching).[102]

But the term China white has come to be used to describe any high-purity heroin, whether from the Golden Triangle, the Golden Crescent (Afghanistan, Pakistan, and Iran), Mexico, or–most recently–Peru and Columbia. Additionally, several synthetic opioid drugs like *fentanyl* (below), which is much more potent than heroin and has either been diverted from hospitals or manufactured in clandestine labs, and some designer drugs like 3-methyl fentanyl, have been sold on the street as China white.[103] In any case, the high purity of the drug allows it to be smoked or snorted.

What is common to all of these products is the users and the usage: increasingly middle-class nouveau addicts, among them many women, who will not inject drugs because of their quite reasonable fear of overdose or infection, but will snort or smoke China white. And there are many crack users who turn to China white to mellow off the sharp edges of a cocaine high. There are two common methods of smoking freebased heroin: the usual one is to use a pipe, often equipped with a small wire screen and useful for either heroin or crack. A more exotic Oriental custom is "chasing the dragon," a process in which the heroin is placed on a piece of aluminum foil and heated from below with a cigarette lighter. As the vaporized drug drifts into the air, it is sucked up with a straw inserted in the user's nose.[104] As a matter of fact, there is a higher safety margin with either of these methods: persons smoking or snorting heroin are more likely to nod off into a stupor before reaching their fatal limit, whereas an injected dose of the same drug could easily be lethal. And, of course, there is no risk of AIDS.

Medical Use of Heroin. Although the import or manufacture of heroin is illegal in the United States, where it is a Schedule I drug, it is prescribed in the United Kingdom, Belgium, Canada, Iceland, Malta, the Netherlands, and Switzerland for very severe pain (usually of terminal illness). Used intravenously, heroin produces peak analgesia a few minutes earlier than does morphine, presumably because of its more efficient transition of the BBB. However, the advantages are probably overall quite marginal.[105] In Britain, the Royal Brompton Hospital and St. Christopher's Hospice have long used an empirically developed mixture of alcohol, heroin, chloroform, and cocaine, commonly called "Brompton's Cocktail," for easing the pain of the terminally ill. While at first blush this sort of concoction may seem frighteningly toxic, it appears to help those in their terminal agony, who have often become highly tolerant to most opioid drugs, and the mixture has a basis in pharmacological fact: we have seen that the amphetamines have a synergetic effect on opioid-induced euphoria and analgesia, and this is probably true of cocaine as well: in mice, μ-agonist opioids reduce the toxicity of cocaine, and this effect is blocked by naloxone.[106]

In Canada, a bitter controversy has waxed and waned since 1979, when Dr. Ken Walker, writing a health column "The Doctor Game" under the pseudonym of "W. Gifford-Jones, M.D.," advocated that Canada allow the medical use of heroin (as England has for nearly a century) because it is a superior painkiller, as well as being more soluble and hence better for injections.[107] Eventually, there was such a public outcry at the seeming indifference of the Canadian bureaucrats that the law was changed and since 1984, heroin can legally be prescribed in Canada. But as recently as 1991, the strictures on its use were so severe (it could be obtained only from Toronto and only if accompanied by a police escort) that it was receiving very little use.

As for the objective scientific benefits of heroin over morphine (or other opioids)—well, like so much of the debate about addictive drugs, this is less a debate about evidence than it is about sociocultural values, and real or (probably mostly) imagined fears. There is no question that heroin (hydrochloride) is about twice as potent as morphine on a weight/weight basis, and it is many times more soluble in water than morphine hydrochloride, which allows a much smaller volume of IM injectate to be required for a terminal cancer patient with very little muscle mass left.[108] But similar advantages could be cited for hydromorphone (hydrochloride), which is only about a third less soluble than heroin and about eight times as potent as morphine. Furthermore, heroin is a prodrug; its effects, psychological and antinociceptive, are almost entirely due to morphine, its active metabolite.[109]

Dr. Walker's anger at the medical "old boy network," as he calls it, is understandable, since the opposition to using heroin is almost entirely sociomoralistic. "What's in a name?" asks Juliet. In the case of heroin, it is everything: doctors are appalled at the thought of employing a substance about which there is so lush a mythology of superstitious dread and cultural revulsion. As the present drug of choice of the outcast street addicts, heroin is in our culture what the smoked opium of the dread Chinese dens was at the turn of the century. If it were only called something else (Gerain for Pain!) and were said to be as "new and·improved" as the latest detergent, there would doubtless be no problem with its taking a modest place in the ranks of the other opioid drugs in the physician's armamentarium. It is neither as uniquely dreadful as its enemies fear, nor as uniquely useful as its friends wish, nor as uniquely desirable as its addicted communicants believe. The real problem of cancer patients suffering unnecessary pain comes from the broader fear that physicians and the culture at large have in employing any "narcotic" drug whatsoever—and, when they finally do use narcotics, their unwillingness to use them in adequate amounts.

Hydromorphone (Dilaudid) and Hydrocodone. Hydromorphone, **2-33**, is used for the same indications as morphine, but it is both more potent (by a factor of about 7, compared mg to mg; Table 2.1) and more soluble in water (as the hydrochloride salt). Consequently, a much smaller volume needs to be injected, and this can be less painful for the patient—particularly in the case of persons suffering from terminal cancer, when a great deal of muscle mass has often been lost. It can also be taken orally. The duration of its action is a little shorter than with morphine. Hydrocodone, **2-34**, is used only in combination, either with acetaminophen (*Lortab, Vicodin*) in analgesic preparations, or in antitussive mixtures with homatropine (*Hycodan*), or with pseudoephedrine (***Tussend***). Hydrocodone is about three times as potent as codeine in suppressing cough. Both of these compounds are made from morphine in a few short steps: morphine is first hydrogenated with one of the usual catalysts (platinum or palladium metal), selectively reducing the isolated double bond to form **2-32**, *dihydromorphine (Paramorphan)*, which has had some use as an analgesic in the past. Oxidation of the cyclohexyl alcohol gives hydromorphone, **2-33**, and the phenolic hydroxide can be easily methylated by the same procedure used to synthesize codeine from morphine, giving hydrocodone, **2-34**.

Modifications of Thebaine by 1,4-Addition

Oxycodone (Roxicodone) and Oxymorphone (Numorphan). Some of the most powerful opioid drugs have been developed by various additions to the 1,3-cyclohexadiene system of thebaine, which, as was mentioned earlier, is itself a convulsant of no pharmacological value occurring only in small amounts in the usual opium poppy, *Papaver somniferum*. However, the dried latex from a related poppy, *P. bracteatum*, can contain as much as 26% thebaine,[110] and so it is readily available for chemical modification. Thus, when thebaine is treated with peroxides (hydrogen peroxide[111] or, in an improved synthesis,[112] *m*-chloroperbenzoic acid), the elements of hydrogen peroxide add in a 1,4 manner to the cyclohexadiene ring; on hydrogenation oxycodone, **2-36** (Figure 2.9), is formed. Oxycodone is similar to morphine, both in its effectiveness in controlling pain and in the duration of its effect. However, unlike morphine it is a more reliable oral drug; about 50% of its parenteral effect is retained when it is given orally.

Oxymorphone, **2-37**, can be synthesized by hydrolysis of the methyl ether of oxycodone. As an analgesic, it is about 10 times as potent as morphine on a mg/mg basis, but unlike morphine, it has little or no cough-suppressant activity.

Nalbuphine (Nubain). The *N*-methyl group of morphine or oxymorphone can be removed easily (using, e.g., the classical van Braun reaction with cyanogen bromide) and replaced with another alkyl group. As previously discussed, a well-known feature of the SAR for morphine is that if the replacement group is allyl, cyclopropylmethyl, or cyclobutylmethyl, the new structure usually acquires some antagonistic properties. In the case of nalbuphine, **2-38**, substitution of the *N*-cyclobutylmethyl group for the methyl group of oxymorphone (and reduction of the ketone group to an alcohol) results in a drug that is a mixed agonist-antagonist. It seems to be effective in controlling moderate to severe pain (although in one study with rhesus monkeys, there was a startlingly large variation in individual response),[113] and exhibits a plateau effect in causing respiratory depression, which is not further increased with dosages larger than 30 mg. One hoped-for goal in manufacturing mixed agonist-antagonist drugs like nalbuphine is to curtail diversion into the street market of illicit users. But it is notoriously difficult to predict the populace's taste in these matters. Perhaps nalbuphine owes its good fortune to an approximately 10-fold greater antagonist personality than pentazocine; on the other hand, compared to pentazocine, it has a much reduced likelihood to produce psychotomimetic effects.[114] (Whether the latter is viewed as an asset or liability on the street is hard to say,[115] but thus far nalbuphine remains an unscheduled drug under the Controlled Substances Act.)

Modifications of Thebaine by the Diels-Alder Reaction. The Oripavines

From 1960-1980, a group at Reckitt and Colman led by Kenneth W. Bentley[116] developed a series of very useful drugs by reacting the cyclohexadiene ring of thebaine with a

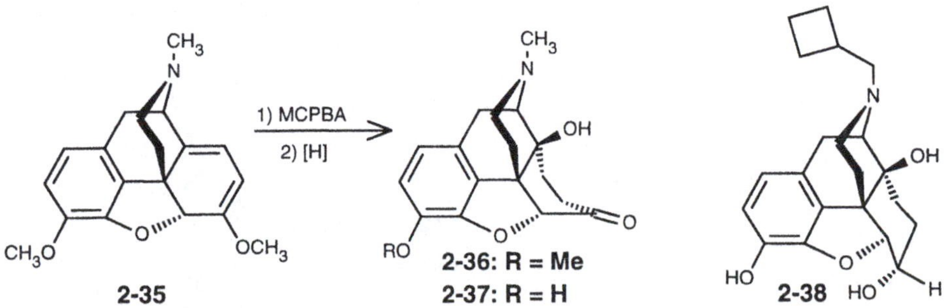

CH_3 N CH_3O O OCH₃ **2-35** 1) MCPBA 2) [H] CH_3 N OH O RO O **2-36: R = Me** **2-37: R = H** N OH HO O HO H **2-38**

Figure 2.9 Oxycodone, oxymorphone, and nalbuphine.

dienophile in a classical Diels-Alder reaction. Later in the synthesis, the phenolic methyl ether is cleaved (a process that produces a much more dramatic increase in potency than in the case of the analogous codeine–morphine transformation), and these compounds have therefore come to be called *oripavines* after the alkaloid, O^3-demethylthebaine (i.e., thebaine minus the methyl group on the 3-oxygen), that occurs in *P. orientale* and *P. bracteatum*. The name is not very descriptive, in that the operative structural feature is not the O^3-methyl ether but rather the transformation of the cyclohexadiene system of thebaine and oripavine into a bicyclo[2.2.2]octane ring with an extended alkanol substituent. The discovery of the oripavine series led to a further elaboration of the characteristics of the opioid receptor.

Etorphine. Etorphine (**2-41**, Figure 2.10) is a semisynthetic opioid derived from thebaine, an alkaloid found with morphine in opium. Its extreme potency (in some animal tests 10,000 times that of morphine, but see below for human trials) is used to advantage in the immobilization of large wild animals: enough can be dissolved in about 1 mL to stupefy an elephant, and this volume of solution can easily be put into a hypodermic dart small enough not to damage the animal. In human trials with nondependent subjects, etorphine was effective in doses as low as 25 μg (0.025 mg). Doses of 50 and 100 μg produced pupillary constriction and morphinelike euphoria; etorphine was estimated to be overall about 500 times as potent as morphine, but with a very rapid onset and short duration of action. In morphine-dependent subjects, etorphine suppressed abstinence but for a shorter period than morphine.[117] This makes it and LSD far and away the most potent drugs to be encountered in this book; as we shall see in Chapter 6, LSD requires about 10 μg to produce mild euphoria and loosening of inhibitions and about 50 micrograms for its typical psychedelic effects.[118] The Federal Controlled Substance Act requires that etorphine, carfentanil and diprenorphine be stored in a safe or steel cabinet "equivalent to a U.S. Government Classification V security container." They can be dispensed only in "reason-

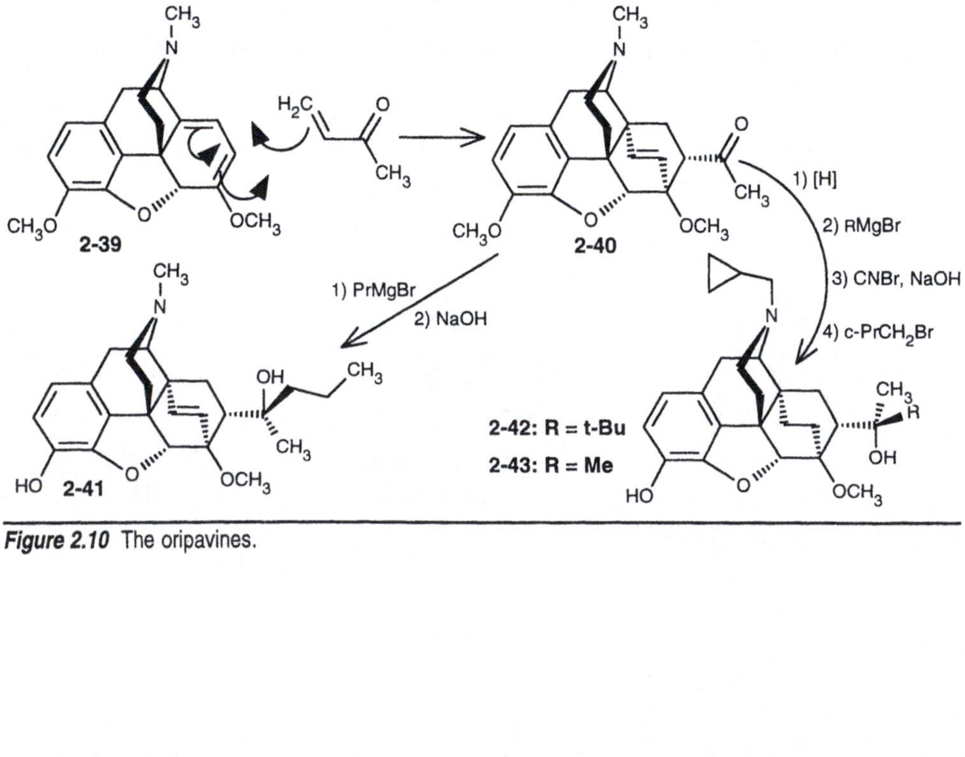

Figure 2.10 The oripavines.

able amounts" to veterinarians engaged in zoo and exotic animal practice, wildlife management programs, or research.[119] A mixed preparation of etorphine and *acepromazine* (*Vetranquil*, a phenothiazine tranquilizer whose structure is that of chlorpromazine with a acetyl group replacing the chlorine) is marketed as *Immobilon* for use in large animal management.

It was originally hoped that the extreme potency of etorphine would extend only to its analgesic effects and that it could be used in quite small doses to suppress pain without concomitant suppression of respiration. Unfortunately, this proved not to be so. In a double-blind comparison with morphine for analgesia in the treatment of postoperative abdominal pain, the dosage of etorphine that adequately controlled pain, although 25 times smaller (0.56 mg/70 kg vs. 10.5 mg/70 kg), caused comparable respiratory depression.[120] Indeed, very tiny amounts can cause death from respiratory paralysis in humans.

Buprenorphine. Buprenorphine (Buprenex, **2-42)** is claimed to represent the most successful effort to date toward the achievement of that perhaps chimeric goal: a drug that controls severe pain but has no tendency to induce addiction and is safe even in overdose. Buprenorphine acts at the μ, κ, and δ receptors with a different proportion of partial agonist activity at each. Buprenorphine binds to the μ receptor, much more strongly (i.e., with much higher affinity) than does morphine, but it induces a maximum level of agonist response that is well below that of morphine (i.e., it has lower intrinsic activity at this receptor). This means that a much lower dose of buprenorphine is needed to provide a level of analgesia equal to that of morphine for moderate to severe pain, but there are very intense levels of pain that only morphine can suppress. At the κ and δ receptors, buprenorphine also has an extremely high affinity but an even lower proportion of intrinsic activity: at this receptor subtypes, it exerts nearly pure antagonist activity. As a consequence, it has an overall bell-shaped dose-response curve: at very high dose levels, undesirable opioid effects such as respiratory depression and inhibition of gastrointestinal motility (constipation) do not become more pronounced, but actually begin to recede. This makes it a much safer and more tolerable drug in overdose than morphine.

Another advantageous feature of buprenorphine is that, while it is very lipophilic and crosses the BBB rapidly, its onset is nonetheless not rapid, but gradual. This is because the kinetics of its interaction with the receptor is relatively slow—probably because its rigid and convoluted shape requires considerable adjustment of the receptor for binding to take place. But once bound it has a long duration of action, and the reverse process of dissociation from the receptors is of course equally slow. The effect of this in the human recipient is that there is little euphoric "rush" on administration of the drug, and any withdrawal symptoms are gradual.[121]

One of the drawbacks of buprenorphine is that 95% of an oral dose is metabolized in the intestines or on first pass through the liver. But it can be sufficiently well absorbed (>50%) and conveniently administered as a sublingual lozenge.[122]

Buprenorphine in Addiction Therapy. Buprenorphine seems to be as effective as methadone in either suppressing the symptoms of withdrawal from opioid addiction (usually heroin) or in providing a maintenance substitute for more toxic opioids. Several studies have found it superior to clonidine[123] and about equal to methadone[124] in opioid withdrawal or maintenance programs. The much wider margin of safety for buprenorphine in overdose compared to heroin—especially street heroin, and the use of a sublingual candylike lozenge rather than a potentially infecting hypodermic needle are two significant improvements in public health from use of maintenance buprenorphine. And its slow onset generally results in an alert, responsive user who can lead a normal, productive life. An advantage over methadone is that buprenorphine need only be given every other day, allowing patients to visit the clinic less frequently without the risk of diversion associated with take-home doses.[125]

Studies using buprenorphine to help control the craving users experience withdrawing from cocaine or from combined cocaine and heroin addiction are puzzling. At least in low maintenance dosage, buprenorphine has not helped withdrawal from cocaine; and in the case of combined addiction, while withdrawal from heroin was helped, withdrawal from cocaine was not.[126] Cocaine itself seems to reduce opioid withdrawal symptoms when buprenorphine or methadone is given in low doses (hence, paradoxically, use of low-dose methadone may exacerbate cocaine use in addicts withdrawing from both drugs). But with higher doses of buprenorphine or methadone, cocaine use seems to trigger increased opiate withdrawal symptoms resulting in an aversive reaction when cocaine abuse is attempted.[127] Thus, the higher regimen may represent one approach towards medicating cocaine addiction. (It is intriguing to speculate that Dr. William Stewart Halsted discovered this phenomenon in his own regard while trying to break his cocaine addiction; see Chapter 4.)

One might think that there would be a strong motive for pharmaceutical companies to invest in research to identify drugs like buprenorphine that would be helpful in overcoming the widespread problem of drug dependency, but the opposite is the case. The history of methadone shows why. As we have seen, methadone is one of the most useful drugs available for treating the pain of cancer; it was developed just for this purpose as a long-lasting analgesic by Eli Lilly. But once it became associated with heroin addiction, many doctors refused to prescribe it, and the FDA gave it a higher scheduling when addicts diverted it for street sale. Finally, the benevolent intervention of the government provided funding to dispense methadone to treatment programs free of charge (with the requirement that the manufacturer sell it at very low profit),[128] so that "Eli Lilly has not made a penny off methadone," says Mary Jeanne Kreek, a drug-abuse researcher at Rockefeller University.[129]

Buprenorphine in Depression. One of the most interesting uses of buprenorphine is in treating the depression of those who have not been helped by standard antidepressant medication (see Chapter 5). In the 19th century and until the 1950s, when the first antidepressant drugs were discovered, patients with suicidal depression were often treated with morphine simply because it was one of the very few psychotropic drugs available at the time: any resulting addiction was considered, quite understandably, as a better alternative than a successful suicide.[130] (As we shall see, methadone has likewise been shown to benefit refractory schizophrenics.) Now there is the possibility of treating depression with newer opioids like buprenorphine, with its partial agonist activity, safer profile, and much reduced abuse liability. This was the reasoning that led a group at Harvard to study the use of buprenorphine to treat depressed patients who were either unresponsive to or intolerant of conventional antidepressant agents. In an open trial, 10 such subjects with unipolar, nonpsychotic, major depression were treated with buprenorphine. While 3 were unable to tolerate the drug because of its side effects, the remaining 7 completed 4 to 6 weeks of treatment, with 4 achieving complete remission of symptoms, and two showing improvement.[131]

Diprenorphine. Diprenorphine (Revivon, **2-43**) is a pure antagonist, and is used to reverse the powerful effects of etorphine in cases of accidental overdose. Although it could hardly be considered a likely drug of abuse (it would cause immediate and massive withdrawal symptoms if taken by an addict), its storage is regulated by the federal government as strictly as etorphine and carfentanil.

Synthesis of etorphine, buprenorphine, and diprenorphine. The synthesis of these oripavine structures constitutes one of the most creative applications of the Diels-Alder reaction. Diels-Alder adducts are known for their complex piling up of intertwined ring systems in a manner that seems to fly in the face of entropy, and the addition of a sixth ring to the five of thebaine certainly follows this pattern. In etorphine and buprenorphine, 6 rings have been constructed with only 20 atoms—a Medusa-like tangling (0.30 rings/atom) exceeding even that of strychnine, which boasts 7 constructed with 24 (0.29 rings/atom). There are some further marvels about these syntheses: not only is the crucial Diels-Alder reaction of thebaine, **2-39**, with methyl vinyl ketone stereo- and regiospecific, bolting three new chiral centers onto thebaine's scaffold of three, but the subsequent addition of a Grignard reagent to the ketone adduct **2-40** to create yet another chiral center is also stereospecific, being controlled by an intramolecular complex between the organometallic reagent and the oxygen atom of the C-6 methoxy group.[132] This meant that the highly chiral opioid receptor could be catered to almost at will, and the group at Reckitt and Colman exploited this to the full, etorphine being one of the most potent of several dozen alternatives tested. Nonetheless, the path from adduct **2-40** to etorphine, **2-41**, is achieved in two simple steps: addition of propyl MgBr and alkaline hydrolysis of the phenolic C-3 ether function. For the synthesis of buprenorphine, **2-42**, and diprenorphine, **2-43**, the isolated double bond is first hydrogenated, and the product then treated with either the *tert*-butyl or methyl magnesium bromide. The *N*-methyl group is removed by treatment with cyanogen bromide followed by basic hydrolysis (which simultaneously cleaves the phenolic ether), and its replacement by a cyclopropylmethyl group results in the partial or complete antagonist activity of buprenorphine or diprenorphine, respectively, depending on the nature of the R group at the quaternary carbinol.

Synthetic Opioids

Morphinans

Logically enough, the first attempts to improve or modify the pharmacological properties of the opioids consisted in some relatively simple modifications of the structure of morphine and codeine, the two most useful natural opioids, followed more recently by the complex elaboration of the structure of thebaine, which has led to etorphine and buprenorphine. The other possibility was to start from scratch and attempt the synthesis of a simplified, stripped-down version of morphine that, it could be hoped, would nonetheless possess some of its desirable properties. Many successful ventures were made in this direction, and some of the molecular structures produced bear only a fleeting resemblance to their original inspiration. Our brief review of this extensive field starts with the morphinans, which is the class of structures bearing the closest resemblance to the natural opioids they emulate.

As can be seen from the structure of *levorphanol* (**2-48**, Figure 2.11), the ring system is almost identical to that of morphine shorn of the oxygen bridge that forms its furan ring.

The morphinan system is synthesized in a few simple steps from easily available, and easily modified, starting materials. The methoxybenzyl Grignard **2-44** adds to the pyridinium quaternary salt, **2-45**, to form **2-46**. Catalytic hydrogenation can be nuanced so that the double bond adjacent to the nitrogen is selectively reduced to give the cyclohexene amine **2-47**. This is nicely constructed to invite an intramolecular Friedel-Crafts cyclization that, after cleaving the methoxy ether, gives equimolar amounts of the mirror-image pair **2-48** and **2-49**. This racemic mixture is called *racemorphan* (**2-51**). When the optical isomers are separated, the enantiomer that rotates light in a counterclockwise sense is called *levorphanol (Levo-Dromoran,* **2-48**), while its mirror-image, **2-49**, is known as *dextrorphan.*[133] Levorphanol, whose structure corresponds to the chirality of natural morphine, codeine, and thebaine, was found to be an opioid analgesic with about four to eight times the potency of morphine after IM injection. Like morphine, it has the usual complement of opioid side effects and abuse liability (it is a Schedule II drug), but its mirror-image partner, dextrorphan—because its right-handed chirality makes it a total misfit for the exclusively left-handed gloves of the endogenous receptors—is totally inert in any test of analgesia, euphoria, or abuse liability. But the opioid receptors affecting the cough reflex are apparently quite different in structure and not so demanding as to their chirality: dextrorphan still showed considerable activity in quieting coughs, and when the methyl ether was restored to dextrorphan to make it resemble codeine, **2-51** turned out to be an even more potent antitussive drug.

Dextromethorphan. The synthesis of dextromethorphan (DM, **2-51**), a widely used and effective cough suppressant, shows how useful the synthetic tinkerings of medicinal chemists can be. Precisely because most of the common synthetic procedures in organic chemical laboratories, unlike the enzymatic processes of plants and animals, do *not* result

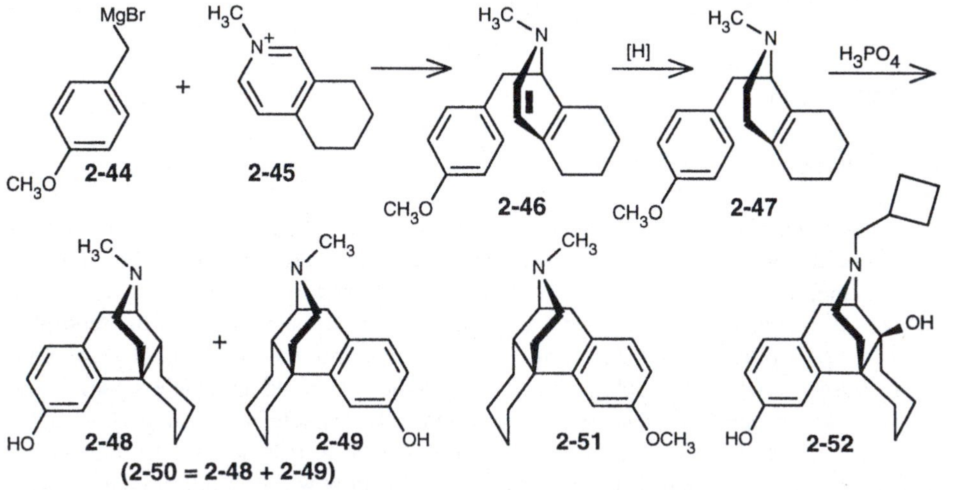

Figure 2.11 The morphinans.

in exclusively the right- or left-handed configurations of chiral molecules but a 50-50 mixture of both, a complete separation of the addictive and antitussive properties of the opioids was discovered. Dextromethorphan has no addictive, analgesic, or sedative actions and causes no respiratory depression (at least with the ordinary doses needed to suppress coughs). There are a few mild side effects, affecting only a small percentage of those using it: some dizziness, nausea, or drowsiness. Except for cases of very severe coughing, it seems to be as effective as codeine. After many uneventful years as a prescription drug, the FDA allowed it to be sold OTC, and it is now widely available in a multitude of generic formulations: syrups, lozenges, capsules; unmixed, with an expectorant (usually *guaifenesin*), or in "multisymptom" cocktail mixtures with antihistamines, decongestants, and alcohol. But following more than a decade of unproblematic OTC use, some problems have emerged from people who deliberately take dextromethorphan in megadoses for the psychic effect it seems to give them.

"Robo Weekends." Among U.S. high school and college student populations, dextromethorphan products are often ingested in quantities many times the recommended dose—e.g., the entire contents of an 8-oz. bottle of Robitussin Maximum Strength—to induce a "Robo high." There are relatively few references to this phenomenon in the literature. In one case, a 22-year-old male U.S. army private claimed to experience an intense high by drinking a 4-oz bottle of Robitussin DM every 4 hours throughout the day.[134] From Germany, there is a report of three persons taking as much as 1,500 mg DM daily to achieve enhanced visual and auditory perception.[135] Two doctors reported in 1993 from Charleston, SC, that they had treated two teenage males, a 16-year-old and a 14-year-old, both from middle-class backgrounds, who were brought to hospital emergency rooms in a psychotic state after taking large quantities of Robitussin DM. Both were admitted users of alcohol, both tested positive for marijuana, and one had a history of seizure disorder treated by carbamazepine. One of them said that many of his friends use Robitussin DM to "get high," and that he regularly drank about 24 oz of the syrup a week.[136] An "epidemic" of adolescent and teenage abuse of DM in Utah in the 1980s led to a voluntary policy of stocking these medications behind the pharmacy counter.[137] There were two reports from Sweden of death from DM overdose, one probably deliberate, the other apparently accidental; in 1986, Sweden removed DM products from the OTC category and reinstated them in the prescription drug category.[138]

Most recently, in 1995, two physicians in the Emergency Medical Division at the University of Utah reported the case of a gentleman with the questionable distinction of consuming "the highest daily dose [of dextromethorphan] for the longest duration yet reported in the world's English-language literature."[139] He claimed to have consumed three or four 12-oz bottles of Robitussin DM several times a week for the first 5 years of his habit and then daily for the most recent 2 years; often, if not always, he followed these DM binges with a six-pack beer chaser. The Utah doctors report the following psychological symptoms of DM intoxication: euphoria; increased perceptual awareness; altered time perception; feelings of floating; tactile, visual, and auditory hallucinations; paranoia; and disorientation. Similar psychic effects were reported on the part of another DM user in Watertown, NY (however, the patient was also diagnosed with chronic paranoid schizophrenia).[140]

It is difficult to evaluate these cases, because they are so small in number and so varied in circumstance. There are conflicting reports as to whether DM overdose is or is not reversed by naloxone.[141] There are claims that DM is helpful in alleviating the symptoms of opioid withdrawal.[142] Anecdotally, the response to these large doses of DM seems to be idiosyncratic, with a few claiming a euphoria they compare to that from psychedelics or marijhuana, while most others have no response at all or are nauseated.[143]

It is well known that dextromethorphan undergoes *O*-demethylation to form *dextrorphan*, **2-49**, via the same *CYP2D6* enzyme system that converts codeine to morphine.[144] And it has been shown in rat brains that dextrorphan binds neither to the same sites as dextromethorphan nor to the sigma opioid site, but predominantly to the same site on the *N*-methyl-D-aspartate (NMDA) receptor as TCP (an analogue of PCP or angel dust; see Chapter 7).[145] This may account for the fact that only some of those who experiment with "robo weekends" derive any satisfaction other than nausea and cough relief—the psychotropic effects may derive from dextrorphan, and only extensive metabolizers would be expected to produce this metabolite.

Butorphanol. The cyclopropylmethyl group on the nitrogen atom of butorphanol (Stadol, **2-52**) lends the structure mixed agonist-antagonist properties. While it can control severe pain with a potency three to six times greater than morphine, there is a ceiling to this beneficial effect, and very intense pain requires a pure agonist like morphine. On the other hand, when former opioid addicts were given butorphanol, they identified it as a barbiturate and expressed indifference or even dislike; thus far, it has not been classified under the Controlled Substances Act.

A further interesting feature of butorphanol is its potential to be used as an intranasal spray. Many illnesses make oral pain medication impractical or impossible, and an intranasal analgesic of the potency of butorphanol can bypass the risks of attempting hypodermic use in a home setting. When taken orally instead of as an IM injection, the onset of maximum pain relief is delayed from about 30 minutes to an hour or more, but the duration of its effect is also proportionately prolonged, and pain relief can last as long as 6 hours instead of 4.

Methanobenzazocines, Methanobenzazonines, Methanobenzocyclodecenes

Even the most avid nomenclature enthusiast might be dismayed at these jaw-breaking terms. A picture is worth a thousand polysyllabic words, and the reader is directed first to Figures 2.11 and 2.12. The structure of *pentazocine* is **2-58**, and comparison of its structure and synthesis with the structure and synthesis of levorphanol, **2-48**, shows that the structures are very similar, except that pentazocine has been further simplified relative to morphine by the elimination of one more ring, the pendant cyclohexane ring of levorphanol. Only two methane stumps of this ring remain. The synthetic pathway to pentazocine is likewise essentially the same as that leading to levorphanol, except for a final tagging of an isopentenyl group on the nitrogen.[146] What complexity has been lost in the realm of structure, however, seems to have been recovered in the ethereal world of nomenclature, for pentazocine has a very solemn name: (2α,6α,11*R**)-1,2,3,4,5,6-hexahydro-6,11-dimethyl-3-(3-methyl-2-butenyl)-2,6-methano-3-benzazocin-8-ol. It is the underlying ring system of a benzene ring fused to an octagon of atoms, one of them nitrogen, as shown in the second, equivalent, structure of **2-58**, which is called a "benzazocin," the one-carbon

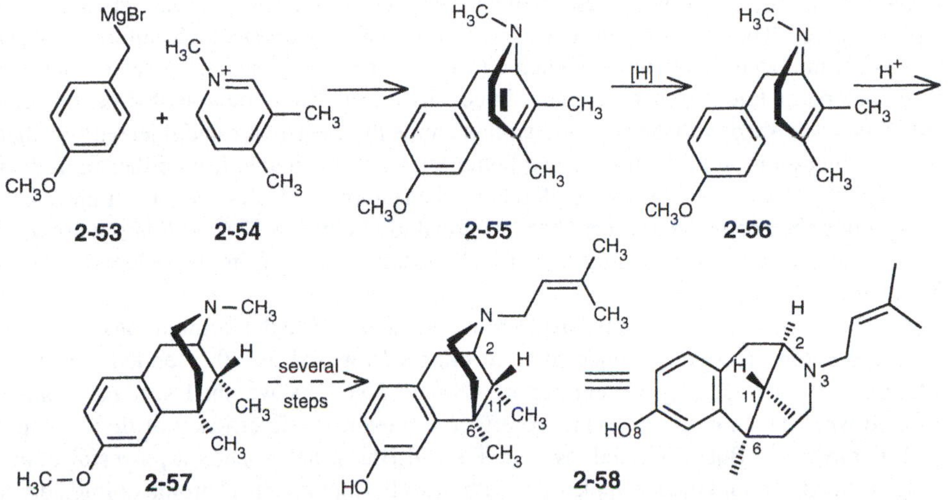

Figure 2.12 Synthesis of pentazocine.

bridge between atoms 2 and 6 being the "methano." Pentazocine and *cyclazocine* (**2-59**, Figure 2.13) are both methanobenzazocines; later variations on these structures broadened the eight-atom ring to a nine-atom ring in *eptazocine*, **2-60** (which is a *1,6-methano-4-benzazonin*), and to a 10-atom ring, with the nitrogen moved to a position on the methano bridge, in *dezocine*, **2-61** (which is a *5,11-methanobenzocyclodecene*).[147]

All of these drugs act pharmacologically as mixed agonist–antagonists. In the case of the two benzazocines, pentazocine and cyclazocine, this is clearly because the methyl group on the nitrogen of morphine has been replaced by either a (substituted) allyl or cyclopropyl methyl group, both of which have long been known to confer antagonistic properties. The SAR for eptazocine and dezocine is less obvious, but in both cases the critical nitrogen atom has been subtly jogged from its position relative to the rest of the molecule; for instance, it is exactly seven atoms from the phenolic hydroxide by any of three pathways in morphine and the benzazocines, whereas it is six to eight atoms away in eptazocine and dezocine. There are other subtle chiral and steric alterations as well.

Pentazocine. Pentazocine (Talwin, **2-58**, Figure 2.12) was was the first mixed agonist-antagonist to be marketed. It appears to be an agonist at the κ and σ receptors and a weak antagonist at the μ. It is about 0.2 times as potent as morphine on a weight/weight basis, and about one-third as active orally as it is parenterally. While it relieves moderate pain, it is less effective than morphine in severe pain. For instance, morphine is still preferred over pentazocine to relieve the extreme pain of myocardial infarction (heart attack). The adverse effects of pentazocine are the same as those of morphine, but it causes less severe respiratory depression.

The dependence liability for pentazocine was at first thought to be essentially eliminated because of its mixed agonist-antagonist character, and it is certainly less attractive to the typical addict than pure agonists like morphine or heroin. Indeed, a common side effect of the drug is dysphoric hallucinatory episodes, which most people find unpleasant or alarming. But in these matters the taste of street addicts has proved to be unpredictable. They used it, and managed to discover a pharmacologic interaction that had gone unnoticed by mainline medicine (a result of the empirical drug-salad approach of many street addicts: mix everything, inject anything). They found that if pentazocine tablets were crushed with the common antihistamine *pyribenzamine* and injected together, the combination produced a unique and better buzz not obtainable from either ingredient by itself.[148] Because of the color of the pyribenzamine capsules and the trade name, Talwin, of pentazocine, this process became popularly known as *T's and Blues*, a nuisance that led some states to raise pentazocine to a Schedule II rating from the federal Schedule IV category.

The manufacturer attempted a different approach. Since the formulations of Talwin that were being diverted and injected were intended for oral use, they added 0.5 mg of naloxone to every 50 mg of pentazocine (*Talwin Nx*). As we shall see, naloxone is inactive when taken orally, so it has no effect on the analgesic efficacy of the Talwin if used as directed. But if injected, as with T's and Blues, naloxone is a powerful opioid antagonist, and should induce unpleasant withdrawal symptoms in a chronic opioid addict, or at least block any opioid effects from the pentazocine. Perhaps the manufacturer did not add enough naloxone, because it continued to be diverted and abused as before; the frequency has declined, however, due to more abundant street supplies of higher quality and cheaper heroin.[149] Another phenomenon of more recent development is injecting a mix of powdered Talwin and Ritalin (methylphenidate, a drug used for attention deficit disorder discussed in Chapter 4).[150]

Cyclazocine. Cyclazocine (**2-59**, Figure 2.13) was initially used as a long-lasting analgesic but eventually was abandoned for this use because of the frequent occurrence of hallucinations, paresthesias (a sort of sensory hallucination involving tingling, burning, and itching of the skin), and other dysphoric phenomenon. While there were several apparently successful trials of cyclazocine as an aid to recovering narcotics abusers, it has lost ground to newer drugs like buprenorphine that have fewer unpleasant side effects.[151]

Eptazocine and Dezocine. Eptazocine (Sedapain, **2-60**) is licensed in Japan for the

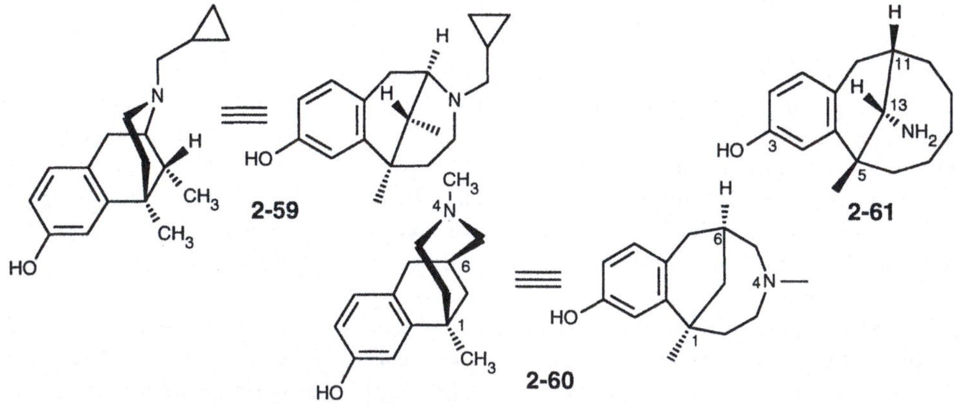

Figure 2.13 Cyclazocine, eptazocine, and dezocine.

treatment of postoperative pain; it is thought to have little or no abuse potential because after chronic administration to test animals there was no cross-tolerance with either pentazocine or morphine.[152] However, as with all such promising beginnings, time will tell.

Dezocine (Dalgan, **2-61**) is an injectable mixed agonist–antagonist analgesic that seems to be as effective as morphine in postoperative and cancer pain. Like eptazocine, dezocine also is believed by its manufacturers to have a very low abuse potential; the FDA and DEA seem to agree, since it has not been scheduled under the Controlled Substances Act.[153] The drug is thought to owe this wholesome balance of properties to a combination of a high affinity for μ and δ receptors and a low affinity for κ receptors.[154] However, its unscheduled career may be short-lived: in a test on nondependent drug abusers, it exhibited a morphine-like subjective response.[155]

Phenylpiperidines, Fentanyls, Phenylpropylamines

If the cyclohexane ring of pentazocine is discarded, there are only two rings left from morphine's original five: the phenyl ring and the six-atom ring containing the alkaloid nitrogen, the piperidine ring. Several successful analgesic drugs, most notably *meperidine*, have been developed that retain only this residual phenylpiperidine resemblance to morphine; additionally, there are some very effective antidiarrhetics based on this structure, such as *loperamide*, which are all the more useful for lacking any euphorigenic or dependence liability. By a still further modification of the phenylpiperidine structure, introducing an amide nitrogen between the phenyl ring and the piperidine ring, still another class of opioids many times more powerful than morphine has been synthesized, the paradigm of which is *fentanyl*. We will look at the individual properties of each of these drugs in the following section.

Meperidine. Meperidine (Pethidine, Demerol, **2-65**, Figure 2.14), like its close cousin methadone, was discovered in Germany in 1937 at laboratories of the I. G. Farbenindustrie (a conglomerate formed in the 1920s by the merger of a number of German chemical corporations including Hoechst and Bayer). Looking for atropine-like compounds that would act as antispasmodics to treat bladder and intestinal spasms, a chemist named Schaumann studied a series of compounds synthesized seven years earlier by Eisleb. When he administered the compound to a cat, he saw to his surprise that the cat's tail flipped backwards into a rigid S shape—the Straub reaction, which is a typical response of this species to opioids.[156] On further testing, meperidine was shown to have analgesic properties, being about one-eighth as potent as morphine. Because its structure was so different from that of morphine, it was, of course, at first believed that it would not induce tolerance or addiction. By now the reader is familiar with this fragile hope eternally springing in the breast of every medicinal chemist; as usual, it was to be dashed on the rocks of grim experience: meperidine proved not significantly different from morphine in this regard. Unlike morphine, however, meperidine has no effect on the cough reflex.

Meperidine (known in Britain as pethidine, and generally better known by its trade name Demerol) is still widely used today as an obstetric and anesthetic premedication; it is considered preferable to morphine because it is effective in a shorter time and is metabolized more quickly, which allows for greater flexibility of administration. When used

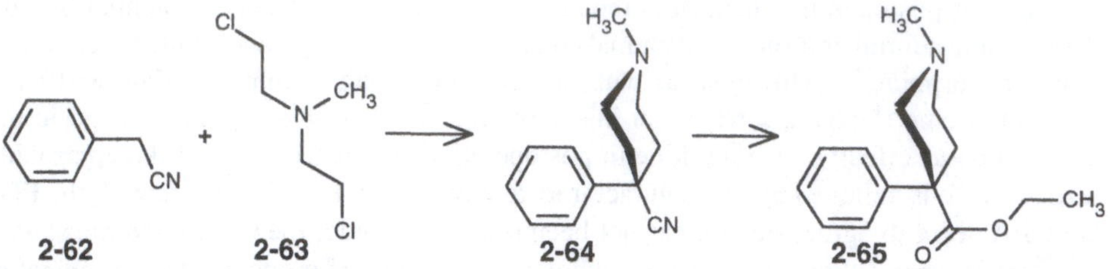

Figure 2.14 Synthesis of meperidine.

as an obstetric analgesic, it is seems to depress the respiration of the fetus less than does morphine. And—no small benefit for the comfort of an immobilized patient—it is much less constipating than morphine.

Meperidine is metabolized in the body to normeperidine—that is, meperidine without the *N*-methyl group. This metabolite has no analgesic activity; rather, it induces seizures. In the ordinary uses of meperidine, this problem does not occur, since normeperidine is excreted very rapidly by the kidneys. However, meperidine is so widely available in hospitals that it is frequently diverted from its intended users by opioid-dependent doctors and nurses to supply their own habits. Such is the seduction of opportunity that many of these hospital personnel become daily users, and can even consume so much that the accumulated normeperidine causes them to suddenly fall into an epileptiform seizure, thereby betraying their habit to their colleagues.[157]

Synthesis of Meperidine and the Prodines. Meperidine is easily synthesized by forming the anion at the α position of phenylacetonitrile, **2-62**, through addition of two moles of base: the anion displaces the twin chlorine substituents of **2-63**, and the resulting phenylcyanopiperidine **2-64** is then hydrolyzed to the acid and esterified with ethanol to form meperidine, **2-65**.

Meperidine represents one of the simplest structures in existence to exhibit opioid activity, and it was only natural that thoughtful chemists would attempt to make variations on this structure. One hypothetical variant of meperidine, the same structure but with the ester function simply reversed, is just as easy to make, perhaps even easier. Reacting 1-methyl-4-piperidine, **2-66** (Figure 2.15), with one of the cheapest and most easily formed of Grignard reagents, phenyl magnesium bromide, forms alcohol **2-67**. A simple esterification with propionic anhydride produces the reversed ester of meperidine, **2-68**. This is just what a group at Hoffman-La Roche did in 1947 (following in part the work of a Danish group a few years before) and found that **2-68** was 25 times more potent than meperidine and three times more potent than morphine.[158] They made several dozen other variations on this theme, finally settling in 1950 on *alphaprodine (Nisentil)* for patenting and marketing. Alphaprodine is **2-68** with an additional methyl group at the indicated 3-position [159] One name for the simpler lead structure, **2-68**, is *desmethylprodine* (proline minus the methyl), but it has since become much more famous than its parent, and is now commonly referred to by the acronym *MPPP*, which stands for 1-methyl-4-phenyl-4-propionoxypiperidine. MPPP has earned its notoriety because in the 1980s it underwent an unwanted resurrection from dusty mummification in the chemical journals to become the first and thus far, if indirectly, the most lethal example of a "designer drug."

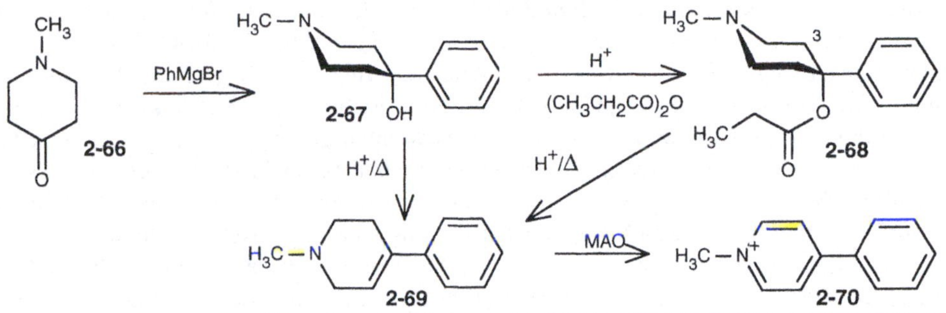

Figure 2.15 Synthesis of "reversed meperidine."

Designer drugs harken back to a less inquisitorial and at the same time more legalistic interpretation of the Controlled Substances Act. It is the "me too" phenomenon of the underground pharmaceutical industry. Just as with many legitimate drugs for arthritis, headache, or hemorrhoids, if a given chemical structure proves to be a runaway best-seller, competing pharmaceutical companies are understandably tempted to use the successful drug as a lead compound. In many cases, it is possible that tagging a single methyl group onto a rather humdrum stretch of alkyl chain in a given structure will affect its pharmacological properties very little one way or another (of course, in a critical area of the molecule the properties may be dramatically changed or even reversed by a single methyl group, and it is not always easy to predict with any confidence where the humdrum landscape of a molecule segues into the critical). Hence, as an example, if diacetylmorphine (heroin) is a Schedule I drug, then a daring entrepreneur, flitting on the fringes of a vast and profitable market in the illegal drug, could simply whip up a batch of di*propionyl*morphine, and market it as *Heroin II* or *Heroin-Pro*. It would probably not be very different—it might even be much better, since it could well cross the BBB all the better. In any case, since this new structure is not listed in the Controlled Substances Act, its manufacture and sale would be questionably ethical but perfectly legal; the anonymous producer could make a tidy profit and then, when the government finally got around to making the latest Heroin-Pro drug "scourge" illegal, slip off to Switzerland to enjoy a comfortable retirement. There are almost certainly a few people out there whose biographies roughly match this scenario. Of course, the regular marketers of ordinary heroin have almost as much firepower and money as the agents of the federal government, and they have as much or more cause to resent the intrusive competition. So there are probably as many ambitious entrepreneurs spending their retirement years in cement shoes just off the California coast. In the famous case of MPPP, however, the entrepreneurial chemist has probably passed to a better world after a short and unusually rapid demise from Parkinson's disease. And there, endowed with a wisdom no endorphins of this world can provide, he or she may be doing penance by listening over and over to a gentle admonition from the ghosts of Professors Schotten und Baumann: "Immer mit Kali, mein(e) Liebe(r)! Immer mit Kali!" Ah, yes. As far back as the 19th century, chemists found that you should always add alkali when carrying out an alkoxy-de-halogenation reaction, as we would call it today. Otherwise. . . . Well, otherwise, instead of MPPP, you might get something you don't want, like *MPTP*.[160]

Prodrug and Neurotoxin: MPTP and MPP$^+$. In 1982, four drug users in northern California, one female and three males, whose ages ranged from 26 to 42 years, tried out something being sold as "new heroin." After injecting 5 to 20 g of the drug over the course of a week or so, they suddenly developed symptoms of severe Parkinson's disease, including near total immobility, inability to speak intelligibly, a fixed stare, and constant drooling.[161] One of the four became so immobile that he could move only his eyes and required IV hydration because he was unable to swallow.[162] Neurologists at Stanford found that the "new heroin" was not heroin, but a crude mixture of about 96% inert material (this is typical of street drugs, which have been cut time and time again), a 0.3% trace of MPPP, and about 3.2% of what proved to be a byproduct of careless or clumsy MPPP production, *1-methyl-4-phenyl-1,2,5,6-tetrahydropyridine,* or *MPTP*, **2-69**.[163]

Something had gone wrong in our clandestine chemist's lab: either the wrong esterification process had been used, or there was excessive sulfuric acid used and the product was overheated.[164] In any case, the four drug users had made an unwilling and unwitting contribution to the understanding of Parkinson's disease; on investigation, it was discovered that a similar case had occurred several years previously when a chemistry graduate student who was used to making his own supply of MPPP took a few unspecified "shortcuts" in the lab and developed Parkinson's disease. At that time, MPTP was likewise suspected as the toxic agent, but when rats were tested with the substance, they did not develop any Parkinsonian symptoms, and so the cause of the student's disorder remained a mystery. When the graduate student died a few years later of a drug overdose, an autopsy performed on his brain showed extensive nerve cell destruction in the substantia nigra, a phenomenon also observed on autopsy of patients with idiopathic (i.e., ordinary) Parkinson's.[165]

Why had the rats not developed Parkinson-like symptoms when given MPTP? Within a year (and three victims later, one a nondrug-abusing pharmaceutical chemist whose job required him to produce batch upon batch of MPTP),[166] the Stanford group was able to show that rats are resistant to damage from this drug, but primates like squirrel monkeys develop Parkinsonism with destruction of neurons in the substantia nigra when given MPTP. Additionally, the damage can be prevented if they are first given a drug, *pargyline*, which is a monoamine oxidase (MAO) inhibitor (a class of drugs used for its antidepressant effect; see Chapter 5). MAO is widely distributed in the mammalian brain and will induce the in vitro conversion of MPTP to *MPP$^+$* (2-70), a process that can also be blocked by pargyline.[167] Presumably, MPTP is a sort of prodrug (a proneurotoxin in this case) that is converted to a species-specific neurotoxin, MPP$^+$.

Still later work published in 1985 (by which time two more industrial chemists who had recrystallized kilogram batches of MPTP were among the victims) showed that the variety of Parkinson's produced by MPTP can be distinguished from the ordinary, or idiopathic variety. MPTP-induced Parkinsonism destroys only the dopamine neurons in the substantia nigra, whereas the idiopathic variety destroys norepinephrine neurons as well.[168] Nonetheless, there is continued speculation as to whether environmental toxins of human or natural origin perhaps contribute to the development of idiopathic Parkinson's disease.[169] There is probably a genetic predisposing factor coupled to environmental insult: as we will see in Chapter 5, many antipsychotic drugs—among them *haloperidol* (**2-73**, Figure 2.16, a structure very similar to **2-67**—cause Parkinsonian symptoms in some, but not all, patients who take them.

Indeed, it is no coincidence that the drugs in Figure 2.16, *loperamide* (**2-71**), *diphenoxylate* (**2-72**), and *haloperidol* (**2-73**) appear to be deliberate elaborations either of the structure of desmethylprodine (loperamide and diphenoxylate) or of meperidine (haloperidol). They were all synthesized by the extremely productive Belgian chemist Paul Janssen in an effort to develop better analgesics based on the phenylpiperidine model. As it turns out, none of these drugs has analgesic properties, but the first two retain the opioid feature of slowing intestinal motility and are used as antidiarrhetics; the third unexpectedly turned out to be one of the most potent antipsychotics known.

Loperamide. Loperamide (Imodium, **2-71**) is a drug for one of those problems that are perhaps amusing when afflicting others but very much less than amusing when experienced oneself. Unlike Thomas à Kempis's compunction, diarrhea is something we would all rather be able to define than to experience. In double-blind tests, loperamide has been shown to be as effective as other opioids, such as codeine or paregoric, in controlling diarrhea; although codeine or paregoric are less expensive, they are also quite addictive. Like diphenoxylate, it was developed by Janssen as an analgesic, but it proved so poorly soluble that it was unsuitable for parenteral injection. Taken orally, 85% seems to remain in the intestinal tract; even that which enters systemic circulation seems not to cross the BBB well, and about 40% exits the body in the feces unchanged. For these reasons, it is considered fairly abuse-resistant, and recently it has been given OTC status in the United States by the FDA, much to the solace of those numerous anguished travelers we see so often on TV attempting to endure the third world terror of a nonrestroom-equipped bus. On a more serious level, loperamide can be lifesaving in infants with severe diarrhea and in patients with irritable bowel syndrome, colostomy, or Crohn's disease.

Diphenoxylate. Like loperamide, diphenoxylate (Lomotil, **2-73**) was developed by Paul Janssen as a potential analgesic. But it was found that the drug, taken orally, was subject to such rapid first-pass metabolism in the liver that it never reached the CNS. Furthermore, even the salts of this amine are so poorly soluble that it is impractical to attempt its parenteral use.[170] It is marketed as a mixture containing 2.5 mg diphenoxylate hydrochloride for every 0.025 mg atropine sulfate. Atropine is one of the active alkaloids in *Atropa belladonna* L. (deadly nightshade) and *Datura stramonium* L. It is a powerful anticholinergic and reduces mucus secretions, thereby providing symptomatic relief from

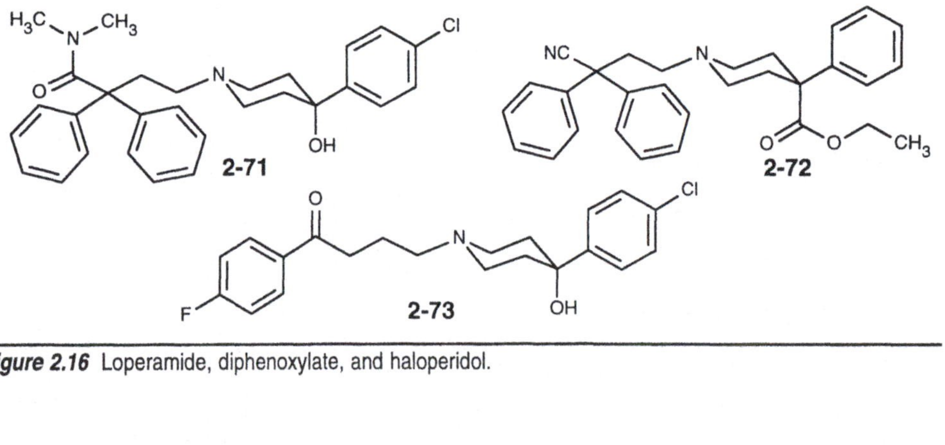

H_3C N CH_3 O N Cl OH **2-71**

NC N O O CH_3 **2-72**

O N Cl F OH **2-73**

Figure 2.16 Loperamide, diphenoxylate, and haloperidol.

diarrhea. It also makes injection or overdose of the mixture unlikely, allowing it to be marketed as a Schedule V drug under the Controlled Substances Act. If taken in very large doses, some of the substance will get past the liver and enter the CNS (additionally, one of the chief metabolites, difenoxin (the acid resulting from hydrolysis of the ethyl ester), is itself an active opioid. Doses as large as 40–60 mg have produced a morphine-like euphoria; and it is possible to take so much that there is respiratory depression and coma, reversible by naloxone.[171]

Fentanyl. Perhaps in an effort to overcome the poor solubility that made loperamide and diphenoxylate ineffective as analgesics, Janssen introduced a second nitrogen into the molecule between the phenyl ring and the piperidine ring. This can be done in a short, straightforward synthesis: phenethylpiperidone, **2-74**, forms a Schiff base with aniline that can be reduced to the diamine **2-75**. This is selectively acylated at the secondary amine with propionyl chloride to give fentanyl (Sublimaze, **2-76**, Figure 2.17).

Because of the introduction of the additional nitrogen atom, fentanyl no longer follows the morphine rule (as do meperidine, methadone, and propoxyphene) and were its historical development as an analogue of meperidine not known, one might be unlikely to guess from its structure that it was an opioid. But it is, and one of the most potent ever developed: on a weight/weight basis, fentanyl is 50-100 times as potent as morphine. Fentanyl takes effect much more rapidly than morphine and likewise its duration of action is only half or a third as long (provided that multiple doses are not given, in which case it can accumulate in lipid and muscle tissues and then only slowly or unexpectedly redistribute into the CNS). But this is used to advantage in one of the most common applications of fentanyl, general anesthesia. Fentanyl is powerful enough to induce anesthesia by itself, and should respiratory problems develop, its effects are quickly abolished by naloxone. It is also used in combination with neuroleptics (very powerful sedatives most of which also have antipsychotic properties) like haloperidol or, most commonly, *droperidol* (**2-82**, Figure 2.18), in what is called *neuroleptanalgesia.*

By now this has become a mantra, but here we go again: the high from this drug is so short that it was at first thought not to be addiction-liable. (Why do pharmaceutical companies think people would find a short high, as opposed to no high at all, so objectionable?) Of course, this has proved to be a very mistaken notion; there were soon instances of its diversion from hospitals for recreational use. And the synthesis of fentanyl is simple

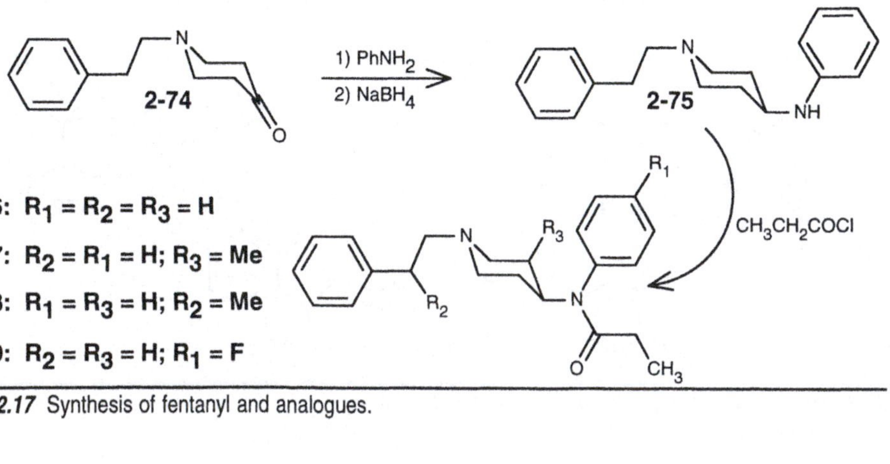

2-74 → 1) $PhNH_2$ 2) $NaBH_4$ → **2-75** → CH_3CH_2COCl

2-76: $R_1 = R_2 = R_3 = H$

2-77: $R_2 = R_1 = H$; $R_3 = Me$

2-78: $R_1 = R_3 = H$; $R_2 = Me$

2-79: $R_2 = R_3 = H$; $R_1 = F$

Figure 2.17 Synthesis of fentanyl and analogues.

enough that it has invited clandestine laboratory production, so that it and several of its designer-drug analogues have been sold on the street as China white (although, as previously mentioned, this term properly refers to highly purified heroin from the Golden Triangle).[172] The problem with fentanyl and its analogues is that they are too good: it is hard to dilute them sufficiently, and street users accustomed to low-potency heroin unwittingly overdose on them. In 1992, about 30 deaths due to accidental overdose from fentanyl occurred in the Baltimore region alone.[173]

Careful studies of fentanyl substantiate that it produces an opioidlike high, although there seems to be a wide variation in response, with most subjects liking the initial effects of a moderately high dose but with some finding the overall experience after 3 hours disagreeable.[174] Fentanyl is available as a transdermal patch for pain control, and in 1994 there was a report of a 36-year-old male who heated and inhaled the contents of one such patch, immediately collapsing with a near-vanishing respiratory rate and blood pressure. Fortunately, he was in a hospital setting where naloxone could be promptly administered.[175] Indeed, fentanyl is becoming the medical professional addict's opioid of choice; not only is it readily available to anesthesiologists and other physicians, but it provides a quick, efficient high and is extremely difficult to detect in blood or urine. But the margin of safety is narrow, even in the hands of trained professionals: it is estimated that there have been more than 20 physician deaths nationwide from fentanyl overdose.[176]

Designer Drug Analogues of Fentanyl. Three designer-drug analogues of fentanyl appeared on the streets between 1981 and 1984: 3-methylfentanyl, **2-77**, alpha-methylfentanyl, **2-78**, and *p*-fluorofentanyl, **2-79**. With regard to potency, they are all significant achievements of the clandestine laboratory technician's art: 3-methylfentanyl (perhaps emulating the methyl group added to the piperidine ring in alphaprodine?) is about 3,000–6,000 times as potent as morphine and has a duration of action like that of heroin (thereby eliminating the problem of too short a high).[177]

All of these substances have more or less promptly been given a place of honor in the Schedule I category of the Controlled Substances Act; and they seem to have disappeared from the scene since 1984. Perhaps Chemist X repented after so many overdoses (or retired to a chateau on Lake Geneva). Meanwhile, Paul Janssen had been designing his own variations on fentanyl.

Sufentanil (Sufenta, **2-80**), ***Alfentanil*** (Alfenta, **2-81**), ***Droperidol*** (Inapsine, **2-82**). All of these closely related structures in Figure 2.18 were developed by Janssen. Sufentanil and alfentanil differ from fentanyl in that one of the aromatic rings has been varied and an additional methoxymethyl group has been tagged on. They are even more potent than fentanyl and used exclusively for induction of general anesthesia. Sufentanil is about 10 times as potent as fentanyl and about 700 times as potent as morphine; alfentanil is still more potent and even more short-acting, so that its effects dissipate only about 20 minutes after administration. This makes it quite useful as an anesthetic for short procedures.

Finally, there is droperidol, a Janssen creation whose structure is a hybrid of haloperidol, **2-73**, and the fentanyl class. It is an extremely powerful neuroleptic, producing an "altered state of awareness"[178] and sedation that many find unpleasant. It also exercises a potent antiemetic effect, and it is consequently used with fentanyl for surgery in which subsequent nausea would be particularly dangerous.

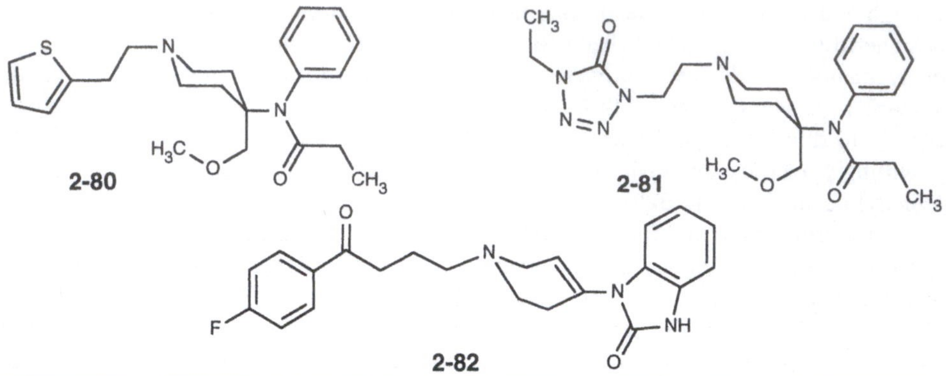

Figure 2.18 Opioids used for general anesthesia.

Phenylpropylamines

We are finally coming to the end of our stroll through the forest of opioid drugs. The piperidine ring is gone now, though not its all-essential amine nitrogen, and there is almost nothing left of the morphine structure but the phenyl ring and its three-carbon attachment to the nitrogen—phenylpropylamine. And yet, with naught but these few bones of the morphinoid skeleton to build on, two very widely used analgesics have nonetheless fleshed out a life of considerable zest and interest.

Methadone. Methadone (Dolophine, **2-83**, Figure 2.19) was originally prepared at I. G. Farbenindustrie as a potential antispasmodic in the same batch of chemicals as meperidine. The structure of methadone has been drawn to emphasize what similarity to morphine is left. As can be seen, while it lacks a piperidine ring, it still follows the morphine rule: a phenyl ring (two for good measure) attached to a quaternary carbon and, two carbons away, the amine nitrogen. (One can speculate that perhaps the lack of a piperidine ring is to some extent supplied by an enol form of the ketone, stabilized by resonance with the two phenyl groups, which could form a hydrogen bond with the amine nitrogen and produce a conformation of the nitrogen similar to the piperidine ring in morphine.) Relative to its distant ancestor morphine, methadone is not only cyclically but chirally depauperate. The number of chiral carbons in methadone has been reduced from morphine's five to a single chiral center—and not, as one might have supposed, the quaternary carbon, because that carbon bears two identical phenyl substituents. Rather, it is the methyl-substituted carbon immediately adjacent to the nitrogen that makes methadone a chiral molecule. Methadone is marketed as the inexpensive *d,l* racemic mixture, but it has been long known that the *l* form is responsible for almost all the analgesic activity (as well as the less desirable features of causing respiratory depression and tolerance/addiction). On the other hand, as with so many opioid pairs (levorphanol and dextrorphan, above; and the isomers of propoxyphene, below), the *d* form is still an effective antitussive, emphasizing the apparent chiral neutrality of the opioid cough receptor.[179]

As we saw in Table 2.1, methadone is about 1.5 times as potent as morphine, and its duration of analgesia is prolonged by about the same amount. Orally, it is about half

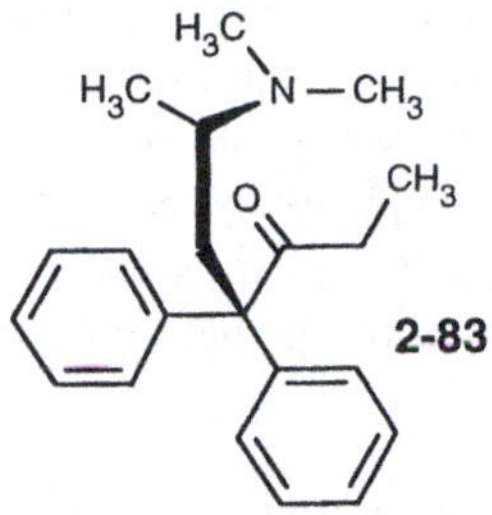

Figure 2.19 Methadone and propoxyphene.

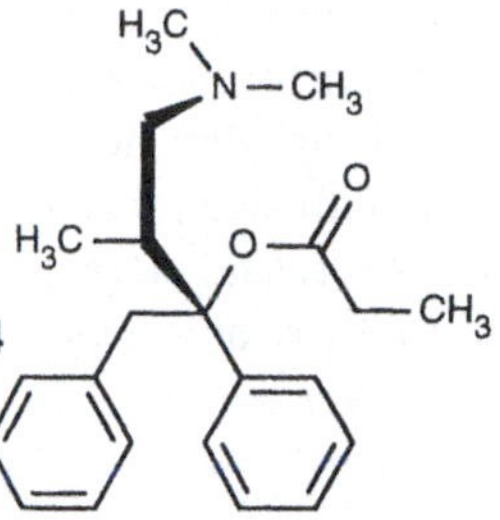

as potent as it is parenterally, but it is far more reliably absorbed when taken orally than is morphine. Tolerance and physical dependence develop with methadone as with morphine, and like morphine it is a Schedule II drug. But withdrawal from methadone is less intense and more gradual than withdrawal from morphine or heroin. Furthermore, the cross-tolerance between methadone and other μ-agonists is such that patients maintained on doses of 60 to 100 mg of methadone orally report little or no euphoria from a dose of heroin up to 25 mg. It is features like these that have led to its use as an oral medication in opioid maintenance and detoxification programs.

Methadone for Schizophrenia. Health-care personnel who staff methadone-dispensing stations get to know their clients well after many visits and are often struck by the seemingly positive results methadone has on the so-called ambulatory schizophrenic segment of the street population. It has been speculated that a good number of these people take up heroin use as an effort to self-medicate their illness—and perhaps find the more stabilizing long-term effects of methadone more effective. In any case, there has been at least one controlled trial of the use of methadone on refractory schizophrenics (those not responding to traditional antipsychotic drugs): there was an initial lessening of psychotic symptoms, but there seemed eventually to be a rebound due perhaps to some form of tolerance.[180] The opposite side of the coin is that withdrawal from opioids may precipitate a psychosis that might otherwise have lain dormant or been suppressed by the drug.[181]

Propoxyphene. Propoxyphene (Darvon, **2-84**) is even more distant in structure from morphine. It is a variation on the methadone structure, with one of the phenyl groups replaced by a benzyl, the ketone turned into a propionate ester, and the methyl group shifted one atom down the chain away from the nitrogen. Propoxyphene has two chiral centers and exists as a pair of diastereomers, which are referred to as the α and β forms. The usual form of the marketed drug, Darvon, is actually the dextrorotatory isomer of α-propoxyphene; most of the activity of propoxyphene as an analgesic is owed to this chiral form.[182] The mirror image of Darvon, the levorotatory isomer of the α form, is called *levopropoxyphene (Novrad)*, and as might be expected from the case of levorphanol and methadone, has no opioid analgesic properties but has been used as an antitussive. (Novrad, if you haven't noticed this, is Darvon spelled backwards.)

Although Darvon is an opioid, its ability to control pain is not really something to write home to Mother Morphia about: it is estimated that 65 mg of Darvon is no more effective than 650 mg of aspirin or acetaminophen. Unlike these latter drugs, propoxyphene has no antiinflammatory or antipyretic action, and unlike codeine (and despite the activity of its mirror image) has little or no antitussive activity. It is probably slightly less liable to "induce" dependency than codeine, but abuse certainly occurs. Alone or in combination with alcohol and other CNS depressants, propoxyphene has caused numerous deaths from overdose, often indeliberately. Many health experts consider it to be overprescribed, since there are several other drugs or combinations of drugs that will control pain better with lower overall risk.

Opioid Antagonists

As we mentioned in the previous discussion of morphine SAR, replacement of the *N*-methyl group of morphine with an allyl or cyclopropylmethyl group endows a structure with antagonistic properties. We now consider a set of drugs that have found wide use for a variety of sometimes quite unexpected uses. All are powerful opioid antagonists, able to reverse the effects of even the most powerful opioid agonists if given in sufficient dosage. The most commonly used, *naloxone,* **2-86** (Figure 2.20), and *naltrexone*, **2-87**, are synthesized from oxymorphone in a few short steps that replace the methyl group with allyl and cyclopropylmethyl, respectively. The third opioid antagonist we will consider is a more recent development, *nalmefene*, **2-88**, and is synthesized from naltrexone via the Wittig reaction.

Naloxone. Naloxone (Narcan, **2-86**) acts very rapidly to reverse the effects of opioid agonists, and for this reason it is the drug of choice to treat respiratory depression caused by opioid overdose, as might occur when using fentanyl or similar agents in surgery for anesthesia or when patients are brought into a hospital emergency room comatose from suspected heroin overdose. It will terminate coma caused by pure agonists like heroin, or mixed agonists–antagonists like pentazocine, and will arrest the convulsions caused by chronic use of meperidine. But it has no effect whatsoever on respiratory failure induced by barbiturates (see Chapter 3). Naloxone itself has no opioid effects and is not a scheduled drug. Naloxone acts very quickly; usually an effect is seen within 2 minutes. It is also rapidly eliminated, and 60% of the amount of naloxone needed to initially reverse opioid overdose symptoms must be continuously readministered hourly until the agonist opioid causing the problem has cleared the body. And the administration must be parenteral: IV, IM, subcutaneous or, if possible, sublingual. The drug is inert when taken

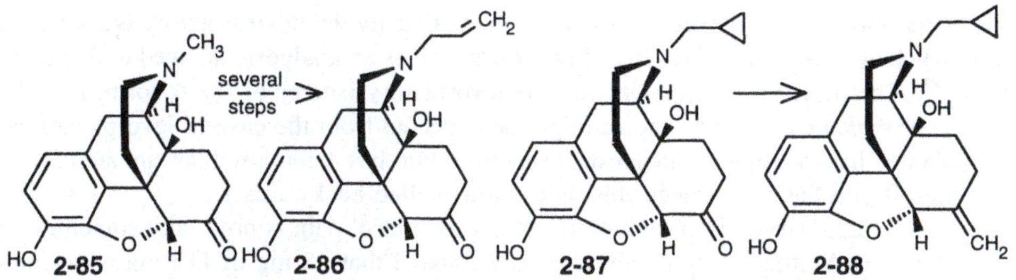

2-85 —several steps→ 2-86 2-87 → 2-88

Figure 2.20 Synthesis of naloxone and nalmefene.

though the oral-intestinal route because it is so extensively degraded by first-pass metabolism.

Some intriguing uses have been made of this limitation. We have seen that oral pentazocine has been formulated mixed with naloxone to discourage its diversion for use as an IV recreational drug. A similar technique has been applied successfully in maintenance clinics. In these settings, it is usually customary for patient addicts to take their daily oral methadone Monday through Friday in a solution of orange juice under the direct observation of a clinic staff person. However, the patients are given two day's supply to take home for Saturdays and Sundays when the clinic is closed. A temptation emerges in that less reform-minded or recidivist fellow addicts will offer the recovering addict money or other drugs in exchange for the weekend dose. The active drug is then extracted and injected for a better high.

To reduce the incidence of this problem, doctors at the University of Vienna added naloxone to the methadone mixture on the weekend. If the ingredients in this mixture are extracted and injected, a severe withdrawal syndrome lasting 15 to 30 minutes ensues because of the antagonist effects of the naloxone. However, if the mixture is taken as it should be, orally, the naloxone is ineffective and the addicts get their desired (moderate) high.[183]

Naltrexone. Naltrexone (Trexan, **2-87**) is quite similar to naloxone in its ability to competitively antagonize opioid drugs, but it is surprisingly different in its pharmacokinetics. Naltrexone is very effective orally and it has a very much longer duration of action. An oral dose suppresses the effects of opioids for 48 to 72 hours, and "insulates" or "immunizes" the recipient to any injected opioid agonist like heroin. This can be helpful to the strongly motivated patient, but, obviously, the temptation is simply not to take the drug. There is an biodegradable bead form of the drug under study that would be implanted subcutaneously and last for a full month; this would allow recovering addicts to summon their supererogatory resolve only with the frequency of a new moon.[184] No one would dispute that methadone maintenance would be easier, if less noble. The motivation to discontinue drug use and to "immunize" oneself from the effects of opioids by taking naltrexone is also, naturally enough, conditioned by what alternatives to drug use life has to offer. In treatment programs with large numbers of street addicts, only 10% expressed an interest in using naltrexone, and only a small number of this group stayed in treatment more than 4 weeks.[185] On the other hand, the best results with naltrexone were from groups of physicians attempting to overcome a drug habit, when over 74% completed at least 6 months of treatment and were opiate-free and practicing medicine at a 1-year follow-up.[186] The powerful carrot-stick appeal of loss of licensing on the one hand with the social acceptance and monetary rewards of practicing medicine on the other; and the contrast of this incentive system to that of the street addict—sober and homeless versus pain-free and homeless—will probably be evident even to those among us who are not rocket scientists.

Most studies indicate that when naltrexone is taken by individuals with no history of opiate use there is usually no effect whatsoever—which seems surprising, since naltrexone definitely blocks endogenous peptide opioid functioning as well as any exogenous opiate drugs (of course OTC antidiarrheal medications like loperamide and cough-suppressing drugs like dextromethorphan are blocked and ineffective in persons taking naltrexone). One of the few effects on some (male) individuals were incidents of inappropriate penile erections.[187]

This effect, while somewhat indelicate, was not entirely unexpected. As long ago as 1563, Garcia d'Orta wrote that, contrary to popular belief, opium did not improve sexual function but often led to impotence. William S. Burroughs, gives concordant testimony in *Junkie*, written under the penname of William Lee in 1953:

> Junk [heroin] short-circuits sex. The drive to non-sexual sociability comes from the same place sex comes from, so when I have an H[eroin] or M[orphine] shooting habit I am non-sociable. If someone wants to talk, O.K. But there is no drive to get acquainted. When I come off the junk, I often run through a period of uncontrolled sociability and talk to anyone who will listen.[188]

It is now well known that chronic injection of opiate drugs in men causes loss of libido and sexual potency,[189] and that use of opiates by women can depress libido and cause amenorrhea and infertility. Indeed, on the hypothesis that some instances of male impotence were due to an excess of endogenous endorphins, a group of Italian scientists administered naltrexone to a random group of impotent patients. The naltrexone patients showed significant improvement over placebo, and the authors hypothesized that "an alteration in central opioid tone is present in idiopathic impotence and is involved in the impairment of sexual behavior."[190]

***Nalmefene* (2-87).** Nalmefene is still in the experimental stages of development. It combines many of the advantages of naloxone and naltrexone, since it can be taken both orally and parenterally, and it seems to be more potent with a greater duration of efficacy than either oral naltrexone or parenteral naloxone. In a pilot study, it was compared with naloxone for emergency treatment of opioid overdose, and was found to be just as effective and potentially safer than naloxone because of its longer (4- to 8-hour) half-life.[191] It has shown promise in a variety of uses: the treatment of interstitial cystitis;[192] the amelioration of the symptoms of posttraumatic stress disorder (on the assumption that the emotional numbing characteristic of this disorder is due to endogenous opioid release);[193] the further improvement in psychological status of schizophrenic patients already stabilized by antipsychotic drugs;[194] and in reducing the extent of neuron death in ischemia.

Both naltrexone and nalmefene have been shown to significantly reduce craving for alcohol and recidivism among recovering alcoholics attempting to maintain sobriety (see Chapter 3).

Treating Opioid Addiction

Opioid Addiction

Why do they do it? Over the last 150 years, many theories have been proposed as to why people become opioid addicted; roughly, these fall into three categories: (1) they start because of pain (disease, war wounds), but continue to avoid the agony of withdrawal; (2) they keep using them because the "high" (euphoria) feels good; and (3) addicts are attempting to self-medicate an endogenous psychological or physical deficiency. All of these factors may play a part in a given individual. Much current work focuses on how learned emotions and environmental cues re-evoke (1) the distress of withdrawal, (2) the memory of opioid euphoria, or (3) the relief given by opioids to endogenous dysphoria.

As we saw in Chapter 1, it is thought that this learning is a reinforced behavior mediated by dopaminergic neurons, probably in the nucleus accumbens of the brain. Interestingly, while μ-agonists increase activity in this system, κ-agonists have no effect or decrease dopamine release (yet the goal of a "nonaddictive" selective κ-analgesic has yet to be achieved) and amphetamines and cocaine, by releasing/inhibiting reuptake of dopamine, also cause dopamine release at the nucleus accumbens. Addicts frequently inject cocaine and heroin simultaneously ("speedballing") because it intensifies the euphoria; opioid analgesia, as we have seen, is also synergized by cocaine (as in the Brompton cocktail, used in the United Kingdom for terminal cancer pain).

Opioid Tolerance

Physicians prescribing opioids to patients with terminal cancer have observed that tolerance to these drugs varies considerably from person to person: some persons never need more than the minimum initial dose, while others continually require more and more.[195] Some long-term abusers of opioids, amphetamines, and barbiturates can come to tolerate astonishingly large drug doses that would be fatal to a neophyte and still function normally in their usual activities. But the tolerance is often uneven over the entire spectrum of pharmacological effects: some heroin users become tolerant to the euphoric and respiratory depressant effects and yet have constricted pupils and severe constipation.[196] On the other hand, people given even large amounts in a normal hospital setting for acute pain usually show no tendency to develop dependence. We saw some evidence in Chapter 1 that there was a genetic component to the phenomenon of addiction, and some such trait may underlie the considerable idiosyncratic variation from individual to individual in the tendency of opioids to establish tolerance. Eric Simon, professor of psychiatry and pharmacology at NYU Medical Center, writes:

> No disease involving genetic alterations of the opioid peptides or their receptors has yet been discovered. However, the author is willing to predict that such diseases will eventually be found. Hereditary insensitivity to pain is a candidate for such a genetic defect, though the evidence is still sparse. Addictive disorders could yet prove to involve genetic or environmentally induced changes in the endogenous opioid machinery.[197]

The notion of treating the problems of opioid addiction encompasses many related problems such as overdose, withdrawal, and opioid maintenance or abstinence. We have discussed elements of all of these in passing; here we will try to outline more completely some approaches to these issues,[198] moving from the easier to the more difficult.

Treatment of Opioid Overdose

Overdose, which can easily occur because the concentration of uncontrolled black-market product is completely unknown, leads to depression of the respiratory centers and consciousness: breathing becomes extremely shallow, blood pressure drops to the nearly undetectable, and the person is completely unresponsive to any stimuli. (These are more or less the characteristics that are deliberately evoked in a surgical setting when opioids like alfentanil are used for general anesthesia.) If a person experiencing overdose symptoms is given intravenous naloxone or nalmefene, a dramatic reversal of these symptoms will occur almost immediately.

Treatment of Opioid Withdrawal (Detoxification)

The chief withdrawal symptoms that come from abruptly discontinuing the long-term high-dosage use of opioids like heroin or morphine are diarrhea, intestinal cramps, nausea, vomiting, and a profuse cold sweating with piloerection (hair standing on end).[199] It is the conjuncture of these last two that inspired some poetic wag long ago to give the whole phenomenon the appellation "going cold turkey"—since the person undergoing opioid withdrawal feels to the touch just like a well-refrigerated, uncooked Thanksgiving turkey.[200]

But while withdrawing from opioids is obviously no fun, it is considerably less hazardous than being the Thanksgiving turkey, for—unlike the sudden untapered withdrawal from barbiturates, which is often fatal; or from alcohol, which is occasionally fatal—opioid withdrawal is never life-threatening. And the intensity of the opioid withdrawal symptoms can be moderated considerably by use of any of a number of pharmacological regimens:

Methadone. The most common way of easing the agony is to replace a short-acting IV drug like heroin with a slow, orally active drug like methadone, and once it has been substituted, to slowly taper the dosage of methadone until the patient can comfortably dispense with it. Buprenorphine, as previously discussed, is used in an analogous manner.

Clonidine (Catapres). Clonidine is an α_2-adrenergic receptor agonist usually prescribed (either orally or as a transdermal patch) to lower blood pressure, but which has been found helpful in several studies in reducing the physical symptoms of opioid withdrawal (hypertension, tachycardia, sweating, runny nose).[201] One of the side effects of clonidine is pronounced sleepiness and sedation (less frequently there are reports of hallucinations), and considerable tolerance develops to its effects: the rebound hypertension from too abrupt withdrawal from the drug can be quite dangerous. Even tapered withdrawal occasionally activates dormant psychosis or mania; and there is at least one instance of a death attributed to a street user's being abruptly cut off from the drug when undergoing detox from his opioid habit.

Oddly enough, clonidine has emerged in some areas as a street drug in its own right. Street users say, "It's a downer. It cools you out. Like a muscle relaxant." It may be that heroin users so often suffer from involuntary withdrawal symptoms (street heroin is often excessively dilute or unavailable) that they find clonidine stabilizing. Or it may simply be a case of monkey see, monkey do. A Baltimore drug counselor says "as typical with addicts, they put two and two together and get five. They see it being used in detox. They either use it themselves or see others nod out on it. So they think people on clonidine are getting high. Then it perpetuates into being a street drug."[202]

While clonidine relieves many of the physical symptoms of opioid withdrawal, it does not appear in double-blind tests to be as effective in suppressing illicit opioid use as methadone or buprenorphine. It has been used with naltrexone to help patients quickly quit methadone use;[203] the problem may be that they often change their minds. For those with sufficient motivation, the use of naloxone, with its long duration of action and oral dosage form, can provide the helpful reinforcement of "perceived unavailability—the cognitive understanding of the patient that any opioid administered will be ineffective and that any money spent on them will have been wasted."[204]

Is Opioid Addiction Curable?

This is and probably always will be a disputed issue. In the (perhaps) better-understood area of alcohol addiction, groups like Alcoholics Anonymous (AA) hold the conviction, central to their treatment philosophy, that alcoholism is incurable but abstinence is possible—hence the paradoxical use of the phrase "recovering alcoholic." One physiological interpretation of this condition would be that there has been some permanent alteration in brain chemistry whereby recovering alcoholics can no longer use alcohol moderately, although they can hope to be abstinent with all the helps provided by AA. In the case of opioid addiction, there is a longstanding opinion that holds that it also is incurable—but in the more radical sense that it is impossible for the addict to be abstinent. As the editors of *Consumer Reports* conclude from their exhaustive study:

> No effective cure for heroin addiction has been found—neither rapid withdrawal nor gradual withdrawal, neither the drug sanitariums of the 1900s, nor long terms of imprisonment since 1914, nor Lexington since 1935, nor the California program since 1962, nor the New York State program launched in 1966, nor the National Addiction Rehabilitation Administration program, nor Synanon since 1958, nor the other therapeutic communities. Nor should this uninterrupted series of failures surprise us. *For heroin really is an addicting drug.*[205]

And first-hand experience seems to concord: "Once a junky, always a junky," is the testimony of William S. Burroughs.[206] Fortunately, for many people, maintenance levels of opioid drugs do not deprive them of the ability to lead a normal or even a very successful life, as witness the careers of Thomas De Quincey and Dr. William Stewart Halsted. This is because (unlike the pattern with many alcoholics but like the pattern with a few cigarette smokers) many opioid addicts who are able to obtain an adequate maintenance level of the drug may be able to develop tolerance to its soporific effects while experiencing little or no interference with their intellect or judgment and no need to increase their daily dosage. As the ludicrously solemn primness of Sir Clifford Allbutt's 1909 *System of Medicine* puts it:

> Opium is used, rightly or wrongly, in many oriental countries, not as an idle or vicious indulgence, but as a reasonable aid in the work of life. A patient of one of us took a grain [65 mg] of opium in a pill every morning and every evening for the last fifteen years of a long, laborious, and distinguished career. A man of great force of character, concerned in affairs of weight and of national importance, and of stainless character, he persisted in this habit, as being one which

> gave him no conscious gratification or diversion, but which toned and strengthened him for his deliberations and engagements.[207]

Treatment: Methadone Maintenance

Methadone maintenance programs are controversial. (Indeed, there is almost as much passion evoked by the word *methadone* as there is by *heroin*. Sometimes the passion becomes paranoia: it has been claimed that methadone is called *Dolophine* out of homage to Adolf Hitler.)[208] There seem to be two irreconcilable views of the matter. On the one hand, there are those who do not see any fundamental difference between prescribing methadone for an opioid-dependent person and prescribing insulin for a diabetic. On the other hand, there are those who feel that replacing an intensely addictive narcotic with a mildly addictive one is an unconscionable compromise of the principle that people should not tolerate themselves or others being enslaved by narcotics. In the words of the author of a distinguished text in medicinal chemistry, the programs are nothing more than "the ethically reprehensible substitution of heroin addiction with methadone addiction."[209]

Methadone maintenance programs were developed in 1963 by Drs. Vincent Dole and Marie Nyswander, working for the Rockefeller Institute in New York City.[210] The program aimed not at total abstinence but at providing a minimum daily "non-euphoric" dose of the methadone that could block the high from any additional injected heroin while affording addicts a stable physiological plateau from which they could, it was hoped, lead reasonably functional and normal lives. The overall goal was "to reduce illicit drug consumption and other criminal behavior and secondarily to improve productive social behavior and psychological well-being."[211]

Although many states have established methadone maintenance programs, they are by no means universally available. And, over time, due in no small part to the controversial nature of the programs, layers of federal and state legislative micromanagement have proliferated that probably do not enhance the effectiveness of the programs:

> In many states, the original Dole-Nyswander protocol has been altered to make abstinence the priority.... the Drug Enforcement Administration (DEA) oversees the security and dispensing of the medication. These minimum standards regulate admissions, staffing patterns, record keeping, treatment planning, service provision, storage, facility standards, frequency of visits and of urine testing, and dose limitations. Some states interdict altogether the practice of methadone maintenance. Some place a ceiling on the maximum dose, making it impossible to produce a narcotic blockade or to remove narcotic craving. Some prohibit take-home medication, and others place a limit on time in treatment before a patient must be withdrawn from methadone. Such restrictions present physicians with serious dilemmas. Instead of being able to rely on their professional judgment and clinical experience, they are often forced to make medical decisions that are independent of the needs of the patients.[212]

For admission to methadone treatment, federal standards mandate a minimum of 1 year of addiction and require current evidence of addiction. A physician may legally prescribe an opioid drug to treat an addict only if the existence of a physiologic dependence can be proved.[213] This can be difficult because of individual variations in tolerance, and because—due to the frequent availability on the street of only low-grade heroin—most addicts have not been able to maintain a high enough intake of heroin to establish anything but psychological dependence.

In an editorial for the *Journal of the American Medical Association*, Dr. James Cooper of the Medical Affairs Branch of the Drug Abuse and Mental Health Administration concludes that all these restrictions taken together amount to the imposition of "ineffective methadone doses and the premature termination of methadone maintenance treatment."[214] A national study from the Institute for Social Research at the University of Michigan substantiates this judgment and concludes that such policies are "counterproductive both for reducing illicit drug use and for preventing HIV infection among needle users."[215] Cooper in effect sees this underutilization of methadone as one instance of the general cultural unwillingness to utilize opioid drugs.[216] But this will probably have little weight with those who are convinced that it represents no great moral improvement to substitute a legal use of narcotic drugs for an illegal one.

Treatment to Support Abstinence

When heroin addicts arrive in a hospital emergency room suffering from an overdose or some complication of their habit, the short-term goal of detoxifying them without excessive withdrawal symptoms is accomplished relatively easily. As we have seen, buprenorphine, methadone, or clonidine can suppress or ameliorate the withdrawal syndrome. Withdrawal is not the difficult problem; abstinence is. The rate of recidivism from enforced withdrawal in hospital settings is extremely high. Oddly enough, what would seem to be the enforced withdrawal of a prison setting still rarely works, because it has been shown time and time again that it is as easy or easier to obtain drugs within most prison systems as it is outside; withdrawal is not a problem there because it never takes place.[217]

Therapeutic Communities. Probably the most successful programs for those relatively rare individuals who are highly motivated to become abstinent are the therapeutic communities such as Synanon, Daytop Village, and Phoenix House. But these are long-term (about 1-1.5 year) residence centers; they are consequently quite expensive; and the selectivity they engage in whereby only the highly motivated participate leads one to suspect that they are saving the saved. (Of course the opposite criticism is leveled against methadone maintenance programs: that they condone the corrigible.)

Chemotherapy for Drug Addiction? Ibogaine? 12-Hydroxyibogamine?

Ibogaine is an alkaloid with a complex indole-isoquinuclidine structure (see Chapter 6) isolated from the plant material of an African shrub, *Tabernanthe iboga* Baill., which is used by tribes in Central Africa in pubertal initiation rites and during hunting expeditions to combat sleep and hunger.[218] Despite its reputation as a hallucinogen (in initiation rites it produces an altered state lasting 48 hours in which participants encounter a universal ancestor named Bwiti),[219] it displays only weak affinity at 5-HT (serotonin) receptors, which are the usual locus of hallucinogenic effect. Nor does it seem to be perceived as LSD in animals trained to recognize that classic psychedelic,[220] but it does exhibit significant activity at the κ-opioid receptors.[221] And Dr. Deborah Mash of the Department of Neurology at the University of Miami School of Medicine, who has permission for the only FDA-sanctioned human experiments, has found that ibogaine and its primary metabolite, 12-hydroxyibogamine, act at both NMDA and serotonin receptor sites.[222] Ibogaine therapy may be something quite new, or it may represent a psychedelic catalyzed process of psychological insight; we will see (Chapter 6) that LSD was used by several psychotherapy groups in the 1960s for treatment of alcohol dependency.

Of course, in the United States ibogaine is a Schedule I drug. But human ibogaine testing has been going in the Netherlands, Israel, and other countries, and many recipients of the drug who were heroin and cocaine addicts claim it has dissolved their craving for drugs and allowed them to be abstinent. There were numerous animal trials in the last decade in which ibogaine was indisputably shown to significantly slow learned responses to opioid and cocaine self-administration.[223] In enormous doses, however, it can cause neuronal degeneration,[224] and hence caution is called for.

Opium is one of mankind's most ancient companions, and the use of opioid drugs in and out of medicine is still extensive. We now turn to another of mankind's longtime companions, alcohol. Alcohol is somewhat like the opioids in that it is a depressant drug, causing sleep or even coma in sufficiently large doses. But it is, of course, in many ways quite different from the opioids. We will consider with alcohol a group of drugs (e.g., benzodiazepines, barbiturates, kava) that are believed to act primarily, as does alcohol, at a common site in the CNS, the GABA receptors.

References and Notes

The following abbreviations are used for frequent sources in these References/Notes:

ARMC: *Annual Reports in Medicinal Chemistry*; Academic Press: New York.

CMC: *Comprehensive Medicinal Chemistry: The Rational Design, Mechanistic Study, and Therapeutic Application of Chemical Compounds*, Hansch, C.; Sammes, P. G.; Taylor, J. B., Eds.; Pergamon: Oxford, 1990. The **bold** number following *CMC* indicates the volume number (1-6); this is followed by the page number.

DE95: *Drug Evaluations: Annual 1995*; American Medical Association Division of Drugs and Toxicology, Department of Drugs; Bennett, D. R., Ed.; AMA: Chicago, 1995.

G&G8: *Goodman and Gilman's The Pharmacological Basis of Therapeutics*, 8th ed., Gilman, A. G.; Rall, T. W.; Nies, A. S.; Taylor, P., Eds; Pergamon: New York, 1990.

G&G9: *Goodman and Gilman's The Pharmacological Basis of Therapeutics*, 9th ed., Hardman, J. G.; Limbird, L. E., et al., Eds; McGraw-Hill: New York, 1996.

L&ID: *Licit and Illicit Drugs: The Consumers Union Report on Narcotics, Stimulants, Depressants, Inhalants, Hallucinogens, and Marijuana—Including Caffeine, Nicotine, and Alcohol*, Brecher, E. W., and the Editors of Consumer Reports, Eds, Little, Brown: Boston, 1972.

MCGR: *Medicinal Chemistry: The Role of Organic Chemistry in Drug Research*, 2nd ed.; Ganellin, C. R.; Roberts, S. M., Eds.; Academic Press: San Diego, CA, 1993.

MI11: *The Merck Index*, 11th ed.; Budavari, S.; O'Neil, M. J., et al., Eds.; Merck: Rahway, NJ, 1989. References are given by monograph number.

MI12: *The Merck Index*, 12th ed.; Budavari, S.; O'Neil, M. J., et al., Eds.; Merck: Rahway, NJ, 1996. References are given by monograph number.

MM16: *The Merck Manual of Diagnosis and Therapy*, 16th Ed.; Berkow, R.; Fletcher, A. J., Eds.; Merck: Rahway, NJ, 1992.

NEJM: *The New England Journal of Medicine.*

OCDS: *Organic Chemistry of Drug Synthesis,* Lednicer, D.; Mitscher, L. A.; Georg, G. I.; Wiley: New York, 1977, 1980, 1984, 1990, 1995. The **bold** number following OCDS indicates the volume number (1-5); this is followed by the page number.

OP: *Opium and the People: Opiate Use in Nineteenth-Century England,* Berridge, V.; Edwards, G.; Yale University Press: New Haven, CT, 1987.

SA: *Substance Abuse: A Comprehensive Textbook,* 2nd ed.; Lowinson, J. H.; Ruiz, P.; Millman, R. B.; Langrod, J. G., Eds.; Williams & Wilkins: Baltimore, 1992.

1. Donne, J., "Death be not proud," *Holy Sonnets*, iii. *The New Oxford Book of English Verse: 1250-1950,* Gardner, H., Ed.; Oxford: New York, 1972, p. 197.
2. Symons, A., "The opium smoker," *Poems*, Heinemann: London, 1902, Vol. I, p. 3.
3. Thus the *Merck Index* and the *American Heritage Dictionary*, 3rd ed.
4. *JAMA*, **1991**, *266*, 3130.
5. Anne Raver of *The New York Times* describes how a lot of "old-fashioned gardeners" continue to be supplied with seeds for *Papaver somniferum* though private exchange or from "subversive seed catalogues" selling them under such code descriptions as "lettuce-leaf," "bread-seed" or "hens and chicks" poppy. "Sure, it could be a dangerous plant," says John Fitzpatrick, former director of the Jefferson garden, "but so is every fourth plant in the garden." The law, like 55-mph speed limits on interstate highways, is largely unenforceable. As Fitzpatrick says, "Opium poppy seeds last a long time. They could come in on the wind. And you would be a felon."
6. McCrae, J., *In Flanders Fields and Other Poems*, G. P. Putnam's Sons: New York, 1919.
7. Flanders Fields in Belgium, part of a campaign in which more than 300,000 soldiers, including more than 100,000 allies, lost their lives. The poppy in Flanders fields was probably not *P. somniferum*, despite Canadian physician McCrae's allusion, but *P. rhoeas* (also known as the Shirley poppy). According to Mary Gold in *The Baltimore Sun*, May 19, 1991, p. 18, "the picture of white crosses amid a sea of red poppies is not a myth. Many World War I burial fields of France and Belgium had lain barren for years, trampled and packed, riddled with trenches and bomb holes. Botanists were amazed when the new cemeteries burst forth with vibrant flowers that no one had planted—so amazed that they flocked to the sites to experiment and test. Not only did they find that poppyseeds had great longevity, but there were more than 2,500 of them per square foot of soil."
8. The rather odd word "laudanum" is probably an alteration of the Medieval Latin "ladanum," which was used to describe the resin from another plant, the rockrose, which the Greeks called ληδον.
9. T. Hood, *Reminiscences*, quoted in H. A. Page, *Thomas De Quincey: His Life and Writings,* John Hogg: London, 1877, pp. 233-234. Cited in *OP*, p. 45.
10. Vicars, G. R., "Laudanum drinking in Lincolnshire," *St. George's Hospital Gazette*, **1893**, *1*, pp. 24-26. Cited in *OP*, p. 45.

11. *OP*, p. 45.

12. William Ober, M.D., argues that John Keats, Francis Thompson, and George Crabbe were also addicted to laudanum: Ober, W. B., "Drowsed with the fume of poppies: opium and John Keats," Chapter 4 in *Boswell's Clap and Other Essays: Medical Analyses of Literary Men's Afflictions*, Harper Perennial: New York, pp. 119-136.

13. According to *The Oxford Companion to English Literature*, 5th ed., Oxford Press: Oxford, Drabble, M. Ed., p. 541, the imagery was inspired by the book he was reading as he fell asleep, *Purchas his Pilgrimage*, which had to do with the Khan Kubla and a palace he commanded to be built. As De Quincey noted, "If a man whose talk is of oxen should become an opium eater, the probability is that ... he will dream about oxen." (Cited by Hayter, A., *Opium and the Romantic Imagination*, Faber & Faber: London, 1968, p. 107; cited in *OP*, p. 51.)

14. De Quincey, T., *Confessions of an English Opium-Eater*, Ticknor, Reed, & Fields: Boston, 1853 [reprint of the 1822 edition], p. 17.

15. I use De Quincey's calculation that 8,000 drops contained about 320 grains (or 20.8 g) of opium. But, as he says, the crude opium varied "much in strength and the tincture still more." De Quincey, T., *Confessions . . .* , p. 90-91.

16. *Ibid.*, p. 69-71.

17. *Ibid.*, p. 76-77.

18. *Ibid.*, p. 110-111.

19. *Ibid.*, p. 118-119.

20. Should these be called cases of *ipso*iatrogenic addiction? Why most of the 19th century addicts were women is explored by Courtwright, and attributed largely to the fact that women were more likely to be treated by the medical profession because of the exigencies of childbirth. (It is a more complicated system, the female: note there is no gender equality to the medical specialties of gynecology and obstetrics.) But it is interesting to consider that in our late 20th-century culture, anorexia and bulimia are largely diseases found among women, and also, as will be mentioned later in this chapter, associated with some (perhaps sex-determined genetic) alteration in endogenous opioid function. In any case, extensive documentation and fascinating case studies can be found in Courtwright, D. T., *Dark Paradise: Opiate Addiction in America before 1940*, Harvard: Cambridge, 1982. It is Courtwright who draws the comparison between the typical addict and the characters in O'Neill's play (p. 1) and Harper Lee's novel (p. 42).

21. Courtwright, *op. cit.*, p. 1.

22. *DE95*, pp. 104-105.

23. Brindley, G. S. *Br. J. Psychiatry* **1983**, *143*, 332-337.

24. Lue, T. F.; Tanagho, E. A. *J. Urol.* **1987**, *173*, 829-836.

25. Gasser, T. C., et al., *J. Urol.* **1987**, *137*, 678-680.

26. Sneader, W., *Drug Discovery: The Evolution of Modern Medicines*, Wiley: New York, 1985, pp. 6-9.

27. It is usually said that Morpheus was the Greek god of dreams. But he had a limited to nonexistent press in older Greek literature, and iconography (Pausanias, *Description of Greece*, 2:20,3); indeed, he was largely the invention of the Latin poet Ovid (43 B.C.–17 A.D.). Ovid tells us that Morpheus is the son of Sleep (*Somnus*), and the nephew of Death (*Thanatos*)—the older generation is a pretty quiet lot. Morpheus gets his name because of

the power of dreams to give *form* (Gk μορφη) to "airy nothings" (*Brewer's Dictionary of Phrase and Fable*, revised by Ivor H. Evans, Harper: New York, 1970, p. 730). But it turns out that Morpheus is just one of several gods of dreams, each with a fairly restrictive job description. He can imitate only human forms that flit through mortals' dreams; there is a second god known as Icelos, or Phobetor, who is charged with the role of beasts, birds, and snakes. And a third, named Phantasos, probably has the most challenging shtick of all: rocks, streams, and trees (Ovid, *Metamorphoses*, xi, 592-650). If we look up this passage we find, sure enough, that the poppies are represented, too. As Ovid's hearers approach the cave where sluggish Somnus snoozes life away with brother Thanatos, we are told that *ante fores antri fecunda papavera florent* ("before the cavern's entrance abundant poppies bloom").

In any case, this shows that Sertürner probably had more than one possibility for christening morphine. The fact that he chose not to name it after the god of sleep, e.g. "hypnine, somnine," may indicate that to Sertürner, as to De Quincey, the most prominent effect of opium was not so much the sleep as the vivid dreams it induced.

28. *DE92*, 98.

29. Melzack, R., "The tragedy of needless pain," *Scientific American*, **1990**, *262 (2)*, February, 27-33.

30. One is irresistibly drawn to continuing this analogy, precisely because of how boldly it limps: perhaps Nature provides some with endogenous crutches, while others are obliged to inject them.

31. Klerman, G., "Psychotropic Hedonism vs. Pharmacological Calvinism," *Hastings Center Report*, **1972**, *2*, 1-3.

32. Kolata, G., "Study says 1 in 5 Americans suffers from chronic pain," *The New York Times*, 21 October 1994.

33. Jacox, A.; Carr, D. B.; Payne, R., "Special report: new clinical-practice guidelines for the management of pain in patients with cancer," *NEJM*, **1994**, *330*, 651-655. Cited material on p. 651. Full text of the guidelines is available from: Jacox, A.; Carr, D. B.; Payne, R., "Management of cancer pain. Clinical practice guideline No. 9, " Agency for Health Care Policy and Research: Rockville, MD, 1994 (AHCPR publication # 94-0592).

34. Jacox, A.; et al., *op. cit.*, p. 653.

35. Jacox, A.; et al., *op. cit.*, p. 653-654. It is curious that this rather antiquated distinction between "psychological" and "physical" dependence is employed, but the authors probably have in mind an audience that is relatively unsophisticated in matters of drug dependency. In the ordinary acceptance of the words, people suffering from the pain and terror of terminal cancer would seem to have every right in the world to be physically and psychologically dependent on opioid drugs. Again, the issue being mincingly circumambulated is ultimately a moral judgment, not a pharmacological one.

36. Whitman, W., "Memories of President Lincoln: When lilacs last in the dooryard bloomed" (#14). In: *Leaves of Grass*, 1891-92 ed. *Complete Poetry and Collected Prose*, Library of America: New York, pp. 464-465.

37. An interviewer opined to the Dalai Lama that the Buddhist culture of Tibet appeared much less fearful of death than Westerners. He replied that it seemed to him just the opposite: recently, he pointed out, there was a murder in Tibet and everyone was aghast; but thousands of murders occur yearly in the United States and no one seemed surprised. Not only that, but Americans appeared to have an ongoing loveaffair with violence, guns, and the instruments of war.

38. Whether this should be done when patients have *not* requested greater medication for pain, or expressed their informed consent that their life be ended, is another matter. The issue is discussed by Dr. Thomas A. Preston, a cardiologist and professor of medicine at the University of WashingtonPreston, T. A., "Killing pain, ending life," *The New York Times*, OP-ED Tuesday, 1 November 1994, A27.

39. Vaupel, D. B; Lange, W. R; London, E. D, "Effects of verapamil on morphine-induced euphoria, analgesia and respiratory depression in humans," *J. Pharmacol. Exp. Ther.*, **1993**, *267*, 1386-1394.

40. Jaffe, J. H.; Martin, W. R., "Opioid analgesics and antagonists," *G&G8*, p. 499.

41. (a) Stein, C., "The control of pain in peripheral tissue by opioids," *NEJM*, **1995**, *332*, 1685-1690. (b) Stein, C., "Peripheral mechanisms of opioid analgesia," *Anesth. Analg.*, **1993**, *76*, 82-91. (c) Stein, C.; Comisel, K.; et al., "Analgesic effect of intraarticular morphine after arthroscopic knee surgery," *NEJM*, **1991**, *325*, 1123-1126.

42. Khoury, G. F.; Chen, A. C. N.; et al., "Intraarticular morphine, bupivacaine and morphine/bupivacaine for pain control after knee videoarthroscopy," *Anesthesiology*, **1992**, *77*, 263-266.

43. Based on the Agency for Health Care Policy and Research report for the management of cancer pain: Jacox, A.; et al., *op. cit.,* p. 652.

44. Reidenberg, M. M., "Clinical pharmacology," (*Contempo* review) *JAMA*, **1995**, *273*, 1664-1665.

45. Findlay, J. W., "Pholcodine," *J. Clin. Pharmacol. Ther.*, **1988**, *13*, 5-17.

46. Some intra- and interteam controversy, in particular concerning the 1978 Laskar award, has somewhat sullied the pristine purity of this scientific milestone. For an excellent account, see Levinthal, C. F., "Days of discovery: 1971-1973," Chapter 4 of *Messengers of Paradise: Opiates and the Brain: The Struggle over Pain, Rage, Uncertainty, and Addiction*, Anchor Doubleday: New York, 1988.

47. A firsthand and nontechnical account: Snyder, S. H., *Drugs and the Brain*, Scientific American Library: New York, 1986.

48. Stanfa, L; Dickenson, A.; et al., "Cholecystokinin and morphine analgesia: variations on a theme," *TIPS*, **1994**, *15*, 65-66.

49. Wei, E.; Loh, H. H. "Physical dependence on opiate-like peptides," *Science*, **1976**, *193*, 1262-1263.

50. *CMC* **3**, 815-816 and references cited therein.

51. There is good evidence too, from the efficacy of antidepressant drugs in treating some anorectics, that mood disorders and/or the serotonin system may likewise be involved—see Chapter 5—but these are not necessarily mutually exclusive hypotheses; it is also quite likely that there is more than one etiology for this behaviorally defined syndrome.

52. Rustina, R., "Starvaholics? Anorexics may be addicted to a starvation 'high,'" *Scientific American*, November 1988, p. 36.

53. Davidson, T. L.; McKenzie, B. R.; et al., "Development of tolerance to endogenous opiates activated by 24-hour food deprivation," *Appetite*, **1992**, *19*, 1-13.

54. Drewnowski, A.; Krahn, D. D.; et al., "Taste responses and preferences for sweet high-fat foods: evidence for opioid involvement," *Physiol. Behav.*, **1992**, *51*, 371-9. Since these foods are easily as dangerous to the public health as many Schedule I drugs, the FDA should perhaps consider requiring naloxone to be added to all high-fat ice-cream and chocolate

candy.

55. Jonas, J. M.; Gold, M. S., "Naltrexone treatment of bulimia: clinical and theoretical findings linking eating disorders and substance abuse," *Adv. Alcohol Subst. Abuse*, **1987**, *7*, 29-37.

56. Jonas, J. M.; Gold, M. S., "Treatment of antidepressant-resistant bulimia with naltrexone," *Int. J. Psychiatry Med.*, **1986-7**, *16*, 305-309.

57. Jonas, J. M.; Gold, M. S., "The use of opiate antagonists in treating bulimia: a study of low-dose versus high-dose naltrexone," *Psychiatry Res.*, **1988**, *24*, 195-199.

58. (a) Buck, M.; Marrazzi, M. A., "Atypical responses to morphine in mice: a possible relationship to anorexia nervosa?" *Life Sci.*, **1987**, *41*, 765-73. A later study specified the behavior in four distinct mouse strains and found differential response to morphine vs. a selective κ-opioid agonist: (b) Marrazzi, M. A.; Lawhorn, J.; et al., "Effects of U50,488, a selective kappa agonist, on atypical mouse opiate systems," *Brain Res. Bull.* **1990**, *25*, 199-201.

59. (a) Waller, D. A.; Kiser, R. S.; et al., "Eating behavior and plasma β-endorphin in bulimia," *Am. J. Clin. Nutr.*, **1986**, *44*, 20-23. (b) Brewerton, D. D.; Lydiard, R. B.; et al., "CSF β-endorphin and dynorphin in bulimia nervosa," *Am. J. Psychiatry*, **1992**, *149*, 1086-1090. The second study measured levels of both β-endorphin, and dynorphin and found only the β-endorphin levels were abnormal.

60. McCann, R. M.; Hall, W. J.; Groth-Juncker, A., "Comfort care for terminally ill patients: the appropriate use of nutrition and hydration," *JAMA*, **1994**, *272*, 1263-1266.

61. "Study backs terminally ill refusing food, "*The New York Times*, 26 October 1994, A20. The authors of the study surmise that these effects come from decreased carbohydrate consumption and ketosis; the patients were given exogenous opioid medication *prn*, and serum endorphin levels were not monitored.

62. Anorexia nervosa seems like such a modern disease. Where were the anorectics before the 20th century? Maybe some have churches named after them: *St. Theresa, St. Catherine, St. Margaret Mary*. There are many saints in the old Roman *Martyrology* who were celebrated for eating nothing for months at a time but the wafer at Holy Communion. Without being reductionistic, there can be an exaltation associated with traditional religious fasting which may owe some of its character to the endorphins: "Meanwhile, his disciples urged him, 'Rabbi, eat something.' But he said to them, 'I have food to eat that you know nothing about.'" (John, 3:31-32)

63. Daniel, M.; Martin, A. D.; Carter, J., "Opiate receptor blockade by naltrexone and mood state after acute physical activity," *British J. Sports Medicine*, **1992**, *26*, 111-115.

64. Naloxone does seem to block the effects of *electro*acupuncture: (a) Wang, M.; Xu, W., et al., "Mu opiate receptor antagonist blocks electroacupuncture inhibition on noxious blood pressure response in rabbits," *Acupunc. Electrother. Res.*, **1994**, *19*, 3-9. But with hypnosis and ordinary acupuncture the results are ambiguous: (b) Moret, V.; Forster, A.; et al., "Mechanism of analgesia induced by hypnosis and acupuncture: Is there a difference?" *Pain*, **1991**, *45*, 135-140.

65. Han, J. S.; Chen, X. H.; et al., "Transcutaneous electrical nerve stimulation for treatment of spinal spasticity," *Chin. Med. J. (Engl.)*, **1994**, *107*, 6-11.

66. Simmons, M.S.; Oleson, T. D.,"Auricular electrical stimulation and dental pain threshold," *Anesth. Prog.*, **1993**, *40*, 14-19.

67. (a) Pechnick, R. N., "Effects of opioids on the hypothalamo-pituitary-adrenal axis." In: *Annu. Rev. Pharmacol. Toxicol.*, **1993**, *32*, 353-382 (pp 362f). (b) Quirion, R.; Chicheportiche, R.; et al., "Classification and nomenclature of phencyclidine and sigma receptor sites,"

Trends Neurosci., **1988**, *10*, 444-446.

68. *SA* 198, 293; *ARMC 27 (1992)* 53.

69. Woods, J. H.; Winger, G. "Behavioral characterization of opioid mixed agonist-antagonists," *Drug Alcohol Depend.* **1987**, *20*, 303-315.

70. (a) Cheng, C. Y.; Hsin, L. W; Schmidt, W. K; et al., *J. Med. Chem.*, **1994**, *16*, 3121-3127. (b) Baskin, D. S.; Widmayer, M. A; Schmidt, W. K., et al., "Evaluation of delayed treatment of focal cerebral ischemia with three selective kappa-opioid agonists in cats," *Stroke,* **1994**, *25*, 2047-2053.

71. Musacchio, J. M. "The psychotomimetic effects of opiates and the sigma receptor," *Neuropsychopharmacology*, **1990**, *3*, 191-200.

72. For a detailed summary of inter- and intraspecies variations in the behavioral effects of subreceptor specific agonists, see *CMC*, **3**, 814-815.

73. Korolkovas, A., *Essentials of Medicinal Chemistry*, 2nd ed., Wiley: New York, 1988, p. 240.

74. Goto, K.; Yamamoto, I. *Proc. Jpn. Acad.*, **1954**, 769.

75. *OCDS*, **1**, 78-79.

76. Korolkovas, *op. cit.*, p. 249-251.

77. Lewis, J. W., "Discovery of buprenorphine, a potent antagonist analgesic." In: *Medicinal Chemistry: The Role of Organic Chemistry in Drug Research,* Roberts, S. M.; Price, B. J., Eds., Academic Press: London, 1985, p. 131.

78. Tollenaere, J. P., "Structure-Activity Relationships of Morphinomimetics," in *Clandestinely Produced Drugs, Analogues and Precursors: Problems and Solutions*, Klein, M., Sapienza, F., McClain, Jr., H., Eds.; U.S. Dept. of Justice Drug Enforcement Administration: Washington, 1989, p.. 131.

79. Petrillo, P.; Amato, M.; Tavani, A. *Neuropeptides (Edinburgh)*, **1985**, *5*, 403.

80. There are some less-than-obvious twists and turns. For references to the original research, see Torssell, K. B. G., *Natural Product Chemistry: A Mechanistic and Biosynthetic Approach to Secondary Metabolism*, Wiley: New York, 1983, pp. 274-281.

81. Nakanishi, K. *Rec. Adv. Phytochem.* **1975**, *9*, 283.

82. Largely from the work of Spector's group over more than a decade; representative papers are: (a) Gintzler, A. R.; Levy, A.; Spector, S., *Proc. Natl. Acad. Sci. U.S.A.*, **1976**, *73*, 2132. (b) Kodaira, H.; Lisek, C. A., et al., *Proc. Natl. Acad. Sci. U.S.A.*, **1989**, *86*, 716.

83. (a) Martin, D., "Roach queen retires," *The New York Times*, 9 February 1995, a biographical sketch of Dr. Berta Scharrer, pioneer in the study of the cockroach brain, wherein opioid receptors and even endogenous morphine are to be found: (b) Stefano, G.; Scharrer, B., *Neuropept. Immunoregul.* **1994**, 1-13. (c) Sonetti, D.; Ottaviani, E.; et al., "Microglia in invertebrate ganglia," *Proc. Natl. Acad. Sci. U. S. A.*, **1994**, *91*, 9180-9184.

84. Spector's group has also found morphine naturally occurring in toad skin (Oka, K.; Kantrowitz, J. D.; Spector, S., *Proc. Natl. Acad. Sci. U.S.A.*, **1985**, *82*, 1852-1854).

85. Donnerer, J.; Oka, K., et al., "Presence and formation of codeine and morphine in the rat," *Proc. Natl. Acad. Sci. U.S.A.,* **1986**, *83*, 4566-4567.

86. Weitz, C. J.; Faull, K. F.; Goldstein, A., "Synthesis of the skeleton of the morphine molecule by mammalian liver," *Nature*, **1987**, *330*, 674-677.

87. Amann, T.; Zenk, M. H., "Formation of the morphine precursor salutaridine is catalyzed by a cytochrome P-450 enzyme in mammalian liver," *Tetrahedron Letters*, **1991**, *30*, 3675-3678.

88. "I'd better go into the house, and die" (Dickens, C., *David Copperfield*, Chapter 3).

89. There was no difference in sensitivity to the other two types of pain. Sindrup, S. H.; Poussen, L., et al., "Are poor metabolizers of sparteine/debrisoquine less pain tolerant than extensive metabolizers?" *Pain*, **1993**, *53*, 335-9. With reference to Mrs Grummidge, the "ice water cold pressor test" sounds very much like the climate of London during much of the year.

90. Sneader, *Drug Discovery . . . op. cit.*, p. 74-75.

91. Gates, M.; Tschudi, G., *J. Am. Chem. Soc.*, **1956**, *78*, 1380. For a discussion of this synthesis see Fleming, I., *Selected Organic Syntheses: A Guidebook for Organic Chemists*, London: Wiley, **1973**, pp. 50-56.

92. Toth, J. E.; Fuchs, P. L. *J. Org. Chem.* **1987**, *52*, 473-475.

93. Tomas Hudlicky of Virginia Polytechnic Institute concludes that "an arbitrary limit of about ten operations can be placed on any total synthesis in order for it to be practical." If each step gives 75% product, the overall yield for 10 steps is 5.6%; if each of 10 steps gives a 90% yield, the overall yield is 34%. He also points out that the cost of morphine available from natural opium is about $250/kg, and concludes that "it is difficult to envision a chemical synthesis that could compete with this cost." Hudlicky, T.; Boros, C. H.; Boros, E. E., "A model study directed towards a practical enantioselective total synthesis of (-) morphine," *Synthesis*, **1992**, 174-178.

94. Heroin—like aspirin—had actually been synthesized some years before it was marketed by the Bayer company. It had been synthesized in 1874 by Wright and tested for pharmacological effects the same year by Pierce and in 1890 by Dott and Stockman. References to these publications as well as the details of Hoffmann and Dreser's firm belief in the power of acetylation as a "refining" and "detoxification" process can be found in: de Ridder, M., "Heroin: new facts about an old myth," *J. Psychoactive Drugs*, **1994**, *26*, 65-68. One of the many mysteries that the present writer has decided to leave for the reader to unravel is whether de Ridder (who should know, being privy to the hitherto closed archives of the Bayer Co.) or Walter Sneader, or perhaps neither, has the correct spelling of Felix Hoffman(n)—too often confused in most chemists' minds with a more noted chemist, August Wilhelm von Hofmann, who had nothing to do with aspirin or heroin. (We leave Roald Hoffmann and his orbital-symmetries, August Heinrich Hoffmann of *Deutschland über alles*, and E. T. A. Hoffmann of Offenbach's *Les contes d'Hoffmann* out of consideration altogether.)

95. Cooper, D. A., "Clandestine Production Processes for Cocaine and Heroin," in *Clandestinely Produced Drugs . . . op. cit.*, pp. 95-116.

96. Jaffe, J. H., "Opiates: clinical aspects," *SA*, 186-194.

97. Treaster, J. B., "Executive's secret struggle with heroin's powerful grip," *The New York Times*, 22 July 1992, A1, B4.

98. Treaster, J. B., "Executive's . . . " *op. cit.*

99. Treaster, J. B., "Executive's . . . " *op. cit.*

100. Cited in Sugg, D. K., "Syringe trade fights AIDS," *The Baltimore Sun*, 18 May 1995, 1A, 18A.

101. Sugg, D. K., "Addicts describe their agonizing, violent journey toward death on Baltimore streets," *The Baltimore Sun*, 18 May 1995, 18A.

102. *Newsweek*, 14 Oct **1991**.

103. Kram, T. C.; Cooper, D. A.; Allen, A. C. *Anal. Chem.*, **1981**, *53*, 1379A-1386A.

104. The author was privileged to observe this graceful procedure, which creates its own dance-like rhythm, at the *Perron Nul* ("Platform Zero"), an "open scene" where drug taking is tolerated under the careful observation of the Rotterdam police. Like many of the practices of street drug users, it probably has an experimental validation: there is almost certainly less destruction of freebased heroin or cocaine if it is ingested this way than if it is smoked in a pipe, with or without tobacco, where the much higher temperatures of the direct flame cause extensive pyrolysis. For a more extensive portrayal of this open scene see: Perrine, D. M., "The view from Platform Zero," *America*, 15 October 1994, *171 (11)*, 9-12.

105. Sawynok, J., "The therapeutic use of heroin: a review of the pharmacological literature," *Can. J. Physiol. Pharmacol.*, **1986**, *64*, 1-6.

106. *SA*, 189.

107. Walker, Ken, "How a medical journalist helped to legalize heroin in Canada," *J. Drug Issues*, **1991**, *21*, 141-146.

108. Kaiko, R. F.; Wallenstein, S. L.; et al., "Analgesic and mood effects of heroin and morphine in cancer patients with postoperative pain," *NEJM*, **1981**, *304*, 1501-1505. A crossover study on 166 cancer patients. The analgesic peak for heroin was about 20 min, and the peak mood improvement about 40 minutes earlier than for morphine; for doses of equal analgesic potency, there were "comparable improvements in various elements of mood, particularly feelings of peacefulness." The authors conclude that heroin has "no apparent unique advantages or disadvantages for the relief of pain in patients with cancer." Dr. Walker disputes this conclusion to some degree, and the interpretation given the results of this study by an editorial accompanying it: Lasagna, L, "Heroin: a medical me too," *NEJM*, **1981**, *304*, 1539-1540.

109. Inturrisi, C. E.; Max, M. B.; et al., "The pharmacokinetics of heroin in patients with chronic pain," *NEJM*, **1984**, *310*, 1213-1217.

110. *OCDS*, **2**, pp. 318-319; Sharghi, N.; Lalezari, L. *Nature*, **1967**, *213*, 1244.

111. Freund, M.; Speyer, E. *J. Prakt. Chem.* **1916**, *94*, 135.

112. Hauser, F. M.; Chen, T-K; Carroll, F. I. *J. Med. Chem.* **1974**, *17*, 1117.

113. Gerak, L. R.; Butelman, E. R., et al., "Antinociceptive and respiratory effects of nalbuphine in rhesus monkeys," *J. Pharmacol. Exp. Ther.*, **1994**, *271*, 993-999.

114. *DE95*, 114.

115. It is often quite difficult to know what is meant in the medical literature when drugs are said to have side effects described as "psychotomimetic," or "dysphoric," or even "euphoric." Probably this is unavoidable: as the use of drugs for "recreation" amply shows, not everyone likes the same cup of tea. Many people find the "euphoria" caused by the barbiturates or even alcohol to be extremely unpleasant and repellant; that is, they find it "dysphoric." "Psychotomimetic" can mean, from the view of the external observer, that the person is "acting crazy." Those of us who have lived on college campuses know that many in this age group enjoy acting "crazy," even though most of them eventually outgrow their relish for this amusement.

116. An interesting description of this research by one of the participants can be found in: Lewis, J. W. "Discovery of Buprenorphine, a Potent Antagonist Analgesic." In: Roberts, S. M., Price, B. J., Eds. *Medicinal Chemistry: The Role of Organic Chemistry in Drug Research*, Orlando, FL: Academic Press, **1985**, 119-142.

117. Jasinski, D. R.; Griffith, J. D.; Carr, C. B., "Etorphine in man, I. Subjective effects and suppression of morphine abstinence," *Clin. Pharmacol. Ther.*, **1975**, *17*, 267-272.

118. Klee, G. D.; Bertino, J., et al., "The influence of varying dosage on the effects of lysergic acid diethylamide (LSD-25) in humans," *Journal of Nervous and Mental Disease*, **1961**, *132*, 404-409. When Hofmann, the (accidental) discoverer of LSD, attempted to assay its properties by taking the lowest dosage he could conceive of as being active, 250 micrograms, he was actually employing a fairly walloping dose.

119. 21 C. F. R. 1305.16; 1301.75(d). See Nielsen, J. R., *Handbook of Federal Drug Law*, Lea & Febiger: Philadelphia, 1992, pp. 83, 104.

120. Macbeath, J. T.; Moore, B. A., "A double-blind clinical trial of the analgesic effects of R & S 218-M, a new potent analgesic for the relief or pain following abdominal surgery: Comparison with morphine sulphate," *Br. J. Anaesth.*, **1976**, *48*, 97-104.

121. Thus at any rate was the hope of those who developed the drug: Lewis, J. W., "Discovery of buprenorphine, a potent antagonist analgesic." In: *Medicinal Chemistry: The Role of Organic Chemistry in Drug Research*, Roberts, S. M.; Price, B. J., eds, Academic Press: London, 1985, pp. 119-142. Abrupt discontinuation of buprenorphine can elicit a characteristic opioid withdrawal syndrome, albeit of slow onset and only peaking after 5 days: San L.; Cami J., et al., "Assessment and management of opioid withdrawal symptoms in buprenorphine-dependent subjects," *Br. J. Addict.*, **1992**, *87*, 55-62.

122. Lewis, J. W., "Discovery of . . . " *op. cit.*, p. 140.

123. (a) Janiri, L.; Mannelli, P., et al., "Opiate detoxification of methadone maintenance patients using lefetamine, clonidine and buprenorphine," *Drug Alcohol Depend*, **1994**, *36*, 139-145. (b) Nigam, A. K.; Ray, R., et al., "Buprenorphine in opiate withdrawal: a comparison with clonidine," *J. Subst. Abuse Treat.*, **1993**, *10*, 391-4. However, another study found no difference in effectiveness between the drugs: (c) Cheskin, L. J.; Fudala, P. J., et al., "A controlled comparison of buprenorphine and clonidine for acute detoxification from opioids," *Drug Alcohol Depend.*, **1994**, *36*, 115-121.

124. At least when subjects were given higher doses of buprenorphine in response to cocaine or heroin compliance failures up to a maximum of 16 mg buprenorphine or 90 mg methadone. (50% of the subjects reached this maximum level.) Those patients who remained in treatment likewise showed a reduced incidence of cocaine-positive urines. (a) Strain, E. C.; Stitzer, M. L., et al., "Comparison of buprenorphine and methadone in the treatment of opioid dependence," *Am. J. Psychiatry*, **1994**, *151*, 1025-1030. An earlier study using a maximum of 6 mg buprenorphine compared with 65 mg methadone found methadone significantly superior to buprenorphine: (b) Kosten, T. R; Schottenfeld, R., et al., "Buprenorphine versus methadone maintenance for opioid dependence," *J. Nerv. Ment. Dis.* **1993**, *181*, 358-364. A recent study comparing 80 mg methadone with 8 mg buprenorphine also found the buprenorphine regimen "less than optimally efficacious": (c) Ling, W.; Wesson, D. R.; et al., "A controlled trial comparing buprenorphine and methadone maintenance in opioid dependence," *Arch. Gen. Psychiatry*, **1996**, *53*, 401-407.

125. (a) Amass, L.; Bickel, W. K., et al., "Alternate-day dosing during buprenorphine treatment of opioid dependence," *Life Sci*, **1994** *54*, 1215-1228. (b) Rosen, M. I.; Wallace, E. A., et al., "Buprenorphine: duration of blockade of effects of intramuscular hydromorphone," *Drug*

Alcohol Depend., **1994**, *35*, 141-149.

126. In a clinical study, buprenorphine suppressed the ability of 69% of experienced drug users to distinguish IV morphine from saline; but none had any difficulty distinguishing IV cocaine from saline: Teoh, S. K.; Mello, N. K., et al., "Buprenorphine effects on morphine- and cocaine-induced subjective responses by drug-dependent men," *J. Clin Psychopharmacol*, **1994**, *14*, 15-27.

127. Stine, S. M.; Kosten, T. R., "Reduction of opiate withdrawal-like symptoms by cocaine abuse during methadone and buprenorphine maintenance," *Am. J. Drug Alcohol Abuse*, **1994**, *20*, 445-458. This may explain the odd result of a double-blind study on a small number of detoxified inpatients dependent on both cocaine and heroin, which found that buprenorphine "significantly enhanced patients' ratings of cocaine-induced pleasurable effects" but this effect *lessened* from day 3 to day 5 of buprenorphine treatment: Rosen, M. I; Pearsall, H. R., "Effects of acute buprenorphine on responses to intranasal cocaine: a pilot study," *Am J. Drug Alcohol Abuse*, **1993**, *19*, 451-464. It also may explain the lessened cocaine usage among the population given high dose buprenorphine in the study cited above (Strain, E. C.; Stitzer, M. L., et al., "Comparison of buprenorphine and methadone in the treatment of opioid dependence," *Am. J. Psychiatry*, **1994**, *151*, 1025-1030).

128. An example of what Nobel laureate economist Milton Friedman somewhere calls the "invisible foot" of government.

129. Cited in "Lukewarm turkey: drug firms balk at pursuing a heroin-addiction treatment," *Scientific American*, March 1989, 32.

130. The German psychiatrist Emil Kraepelin suggested using laudanum to treat melancholia (the "opium cure"); he prescribed gradually increasing and then decreasing the drug so as to prevent addiction: Levinthal, C. F., *Messengers of Paradise: Opiates and the Brain: The Struggle over Pain, Rage, Uncertainty, and Addiction*, Anchor Doubleday: New York, 1988, pp. 148-149.

131. Bodkin J. A.; Zornberg G. L., et al., "Buprenorphine treatment of refractory depression," *J. Clin. Psychopharmacol.* **1995**, *15*, 49-57.

132. Lewis, J. W., "Discovery of buprenorphine, a potent antagonist analgesic." In: *Medicinal Chemistry: The Role of Organic Chemistry in Drug Research*, Roberts, S. M.; Price, B. J., Eds, Academic Press: Orlando, FL, 1985, pp. 119-142.

133. Leading references to the synthetic method given here can be found in: Schneider, O.; Grüssner, A., *Helv. Chim. Acta*, **1951**, *34*, 2211. See *OCDS*, **1**, 293.

134. Helfer, J.; Kim, O. M., "Psychoactive abuse potential of Robitussin-DM," *Am. J. Psychiatry*, **1990**, *147*, 672-673.

135. Degkwitz, R., "Dextromethorphan (Romilar) als Rauschmittel," *Nervenarzt*, **1964**, *35*, 412.

136. Murray, S.; Brewerton, T., "Abuse of over-the-counter dextromethorphan by teenagers," *Southern Medical Journal*, **1993**, *86*, 1151-1153.

137. McElwee, N. E.; Veltri, J. C., "Intentional abuse of dextromethorphan (DM) products: 1985 to 1988 statewide data," *Vet. Hum. Toxicol.*, **1990**, *32*, 355.

138. Rammer, L.; Holmgre, P.; Sandler, H., "Fatal intoxication by dextromethorphan: a report of two cases," *Forensic Sci. Int.*, **1988**, *37*, 233-236.

139. Wolfe, T. R.; Caravati, E. M., "Massive dextromethorphan ingestion and abuse," *Am. J. Emerg. Med.*, **1995**, *13*, 174-176.

140. Iaboni, R. P.; Aronowitz, J. S., "Dextromethorphan abuse in a dually diagnosed patient," *J. Nerv. Ment. Dis.*, **1995**, *183*, 341-342.

141. The respiratory depression some experience after mammoth doses seems to be reversible by naloxone: Schneider, S. M., et al., "Dextromethorphan poisoning reversed by naloxone," *Am. J. Emerg. Med.*, **1991**, *9*, 237-238. But the Utah doctors question the value of this (their patient did not benefit from naloxone) and note that in most cases where naloxone was used only a few symptoms of intoxication are reversed.

142. Koyuncuoglu, H.; "The combination of tizanidine markedly improves the treatment with dextromethorphan of heroin addicted outpatients," *Int. J. Clin. Pharmacol. Ther.*, **1995**, *33*, 13-19.

143. This from the author's informal survey of college classes. Most students had acquaintances who had experimented with DM; quite a few had tried it themselves. Unfortunately for any claim to scientific method, those who reported a euphoric effect also admitted on closer interrogation that they had simultaneously been drinking alcohol or smoking marijuana. Maybe the marijuana is necessary to overcome the nausea which comes from chugging down a cup or so of foul-tasting treacly syrup. A few experimenters admitted they couldn't stand it and "heaved."

144. Indeed, DM is used as a "marker substrate" for this enzyme; see, for example: Wu, D.; Otton, S. V., et al., "Inhibition of human cytochrome P450 2D6 (CYP2D6) by methadone," *Br. J. Clin. Pharmacol.*, **1993**, *35*, 30-34.

145. Franklin, P. H.; Murray, T. F., et al., "High affinity [3H]dextrorphan binding in rat brain is localized to a noncompetitive antagonist site of the activated N-methyl-D-aspartate receptor-cation channel," *Mol. Pharmacol.*, **1992**, *41*, 134-146.

146. (a) May, E. L.; Ager, J. H., *J. Org. Chem*, **1959**, *24*, 1432. (b) Archer, S.; Albertson, N. F., et al., *J. Med. Chem.*, **1964**, *7*, 123. See *OCDS*, **1**, 296-298.

147. For the synthesis of dezocine, see (a) *OCDS*, **4**, 59. (b) Freed, M. E.; Potoski, E. H., et al., *J. Med. Chem.*, **1973**, *16*, 595.

148. Showalter, C. V., "T's and blues: abuse of pentazocine and tripelennamine," *JAMA*, **1980**, *244*, 1224-1225.

149. Reed, D. A.; Schnoll, S. H., "Abuse of pentazocine-naloxone combination," *JAMA*, **1986**, *256*, 2562-2563.

150. Carter, H. S.; Watson, W. A., "IV pentazocine/methylphenidate abuse—the clinical toxicity of another T's and blues combination," *J. Toxicol. Clin. Toxicol.*, **1994**, *32*, 541-547.

151. (a) Jaffe, J. H.; Brill, L., "Cyclazocine, a long-acting narcotic antagonist: its voluntary acceptance as a treatment modality by narcotics abusers," *Int. J. Addict.*, **1966**, *1*, 99-123. (b) Resnick, R.; Fink, M., "Cyclazocine treatment of opiate dependance: a progress report," *Compr. Psychiatry*, **1971**, *12*, 491-502.

152. *ARMC*, **25**, p. 12.

153. *DE95*, p. 113.

154. *ARMC*, **27**, p. 326 (1992).

155. Jasinski, D. R.; Preston, K. L., "Assessment of dezocine for morphine-like subjective effects and miosis," *Clin. Pharmacol. Ther.*, **1985**, *38*, 544-548.

156. Sneader, W., *Drug Discovery: The Evolution of Modern Medicines*, Wiley: New York, 1985, p. 76.

157. Jaffe, J. H., "Drug addiction and drug abuse," in *G&G8*, 522-573.

158. Ziering, A.; Berger, L., et al., *J. Org. Chem.*, **1947**, 894-903.

159. *MI11*, entry # 307, p. 51.

160. Actually, this is unfair to the person, whoever it was, who made this underground drug. Whether using the so-called *Schotten-Baumann* procedure (esterification using propionyl chloride in the presence of added base to combine with the HCl that will be formed) would have worked is not all that clear. This may have been what chemist X was doing; he/she was probably using the acyl chloride at any rate, since that mode has problems according to the original Hoffmann-La Roche chemists in the one and only detailed description of the process in the English literature: "the piperidinols [i.e., the class of compounds like **2-67**] . . . can be readily acylated with acid anhydrides in pyridine or anhydrides catalyzed by a trace of sulfuric or perchloric acid. With acyl chlorides in pyridine less satisfactory results were obtained, partial dehydration [i.e. to **2-69**] usually occurring." The authors used only propionic anhydride with "a drop" of concentrated sulfuric acid in their experimental, and had no problems with dehydration: Ziering, A.; Berger, L., et al., *J. Org. Chem.*, **1947**, 894-903, cited material p. 894. Perhaps chemist X followed this procedure but used too much sulfuric acid. (Chemist X has never been found; the laboratory in which MPPP was synthesized have never been discovered. Langston and Ballard, in the article cited below, hypothesize that the person or persons may have been following Ziering's laboratory procedure, but there is no hard evidence for this.)

    On the other hand, F. Ivy Carroll, director for organic/medical chemistry at Research Triangle Institute, says that his group esterified alcohol **2-67** with propionyl chloride and produced MPPP (which he calls prodine), and "none of the toxic . . . MPTP was detected during the . . . synthesis." (Carroll, F. I.; Brine, G. A., "4-Phenylpiperidine analgesics, fentanyl and fentanyl analogues: methods of synthesis." In: *Clandestinely Produced Drugs, Analogues and Precursors: Problems and Solutions*, Klein, M.; Sapienza, F., et al., Eds., U.S. DEA: Washington, DC, 1989, pp. 67-90; cited material p. 73.)

161. College chemistry professors will be reminded of their students in early morning classes.

162. Langston, J. W.; Ballard, P., "Chronic Parkinsonism in humans due to a product of meperidine-analog synthesis," *Science*, **1983**, *219*, 979-980.

163. *Ibid.*

164. Richard B. Silverman (*The Organic Chemistry of Drug Design and Drug Action*, Academic: New York, 1992, p. 201) states that "MPPP decomposes upon heating or in the presence of acids with elimination of propionate to give . . . MPTP." This gives the impression that MPPP is quite heat labile, but it does not appear to be particularly so, since Ziering et al., (op cit) report it (as the hydrochloride, hence under moderately acid conditions) to have a stable melting point of 182-183°C; and quite a few of its analogues with the same propionoxy group have sharp mp as high as 220°C.

165. (a) Langston, J. W., *op. cit.* (b) Davis, G. C.; Williams, A. C., et al., *Psychiatry Res.*, **1979**, *1*, 249.

166. Langston, J. W.; Ballard, P. A., "Parkinson's disease in a chemist working with 1-methyl-4-phenyl-1,2,5,6-tetrahydropyridine," *NEJM*, **1983**, *309*, 310. MPTP is a useful synthetic intermediate that is sold by Aldrich Chemical Co. among others.

167. Langston, J. W.; Irwin, I., et al., "Pargyline prevents MPTP-induced Parkinsonism in primates," *Science*, **1983**, *225*, 1480-1482. The protection is afforded only by MAO B, and not by MAO A inhibitors; for further references, see Silverman, R. B., *The Organic Chemistry of Drug Design and Drug Action*, Academic: New York, 1992, p. 201-202.

168. Burns, R. S.; LeWitt, P. A., et al., "The clinical syndrome of striatal dopamine deficiency: Parkinsonism induced by 1-methyl-4-phenyl-1,2,5,6-tetrahydropyridine (MPTP)," *NEJM*, **1985**, *312*, 1418-1421.

169. Or it may be an endogenous process. It has recently been shown that harmine and norharman, which are found widely distributed in mammalian tissue (as well as in yage, a psychotropic snuff—see Chapter 6) can be *N*-methylated by enzymes in mammalian brain to produce β-carbolinium ions which bear a striking resemblance to MPP^+, and it is possible that this process is involved in idiopathic Parkinson's: (a) Fields, J. Z.; Albores, R.; Neafsey, E. J., et al., "Similar inhibition of mitochondrial respiration by 1-methyl-4-phenyl-pyridinium (MPP+) and by a unique N-methylated beta-carboline analogue, 2,9-dimethyl-norharman (2,9Me2NH)," *Ann. N. Y. Acad. Sci.*, **1992**, *648*, 272-4. (b) Matsubara, K.; Neafsey, E. J.; Collins, M.A., "Novel S-adenosylmethionine-dependent indole-N-methylation of beta-carbolines in brain particulate fractions," *J. Neurochem.*, **1992**, *59*, 511-518. (c) Matsubara, K.; Collins, M.A., et al., "Potential bioactivated neurotoxicants, N-methylated beta-carbolinium ions, are present in human brain," *Brain. Res.*, **1993**, *610*, 90-96.

There is some recent work which offers hope for victims of conventional and neurotoxin-induced Parkonson's: a molecule which occurs naturally in the brain, glial-cell-line–derived neurotrophic factor, seems to produce significant alleviation of the symptoms of Parkinson's in rhesus monkeys who had been exposed to MPTP. Gash, D. M., et al., "Functional recovery in parkinsonian monkeys treated with GDNF," *Nature*, **1996**, *380*, 252-255.

170. *G&G8*, 507.

171. *DE95*, 963.

172. Kram, T. C.; Cooper, D. A.; Allen, A. C., "Behind the identification of China white," *Anal. Chem.*, **1981**, *53*, 1379A-1386A.

173. Smialek, J. E.; Levine, B., et al., "A fentanyl epidemic in Maryland 1992," *J. Forensic Sci.*, **1994**, *39*, 159-64.

174. Zacny, J. P.; Lichtor, J. L, et al., "Subjective and behavioral responses to intravenous fentanyl in healthy volunteers," *Psychopharmacology* (Berlin), **1992**, *107*, 319-26.

175. Marquardt, K. A.; Tharratt, R. S., "Inhalation abuse of fentanyl patch," *J. Toxicol. Clin. Toxicol.*, **1994**, *32*, 75-78.

176. Kennedy, R., "Death highlights drug's lethal allure to doctors," *The New York Times*, 11 November 1995, 1, 29.

177. These analogues and others are described in Carroll, F. I.; Brine, G. A., "4-Phenylpiperidine analgesics, fentanyl and fentanyl analogues: methods of synthesis." In: *Clandestinely Produced Drugs, Analogues and Precursors: Problems and Solutions*, Klein, M.; Sapienza, F., et al., Eds., U.S. DEA: Washington, 1989, pp. 67-90. See also Morgan, J. P., "Controlled substance analogues: current clinical and social issues," *SA*, pp. 328-333.

178. *DE8*, p. 190f.

179. *G&G8*, p. 508.

180. Brizer, D. A.; Hartman, N., et al., "Effects of methadone plus neuroleptics in treatment-resistant chronic paranoid schizophrenia," *Am. J. Psychiatry*, **1985**, *142*, 1106-1107. Referenced in Beeder, A. B.; Millman, R. B., "Treatment of patients with psychopathology and substance abuse." In: *SA*, 675-690.

181. A study of this phenomenon can be found in: Levinson, I.; Galynker, I. I.; Rosenthal, R. N., "Methadone withdrawal psychosis," *J. Clin. Psychiatry*, **1995**, *56*, 73-76.

182. *G&G8*, pp. 509-510.

183. Loimer, N.; Presslich, O., et al., "Combined Naloxone/Methadone Preparations for Opiate Substitution Therapy," *Journal of Substance Abuse Treatment*, **1991**, *8*, 157.

184. Chiang, C. N., et al., "Clinical evaluation of naltrexone sustained-release preparation," *Drug Alcohol Depend.*, **1985**, *16*, 1-8.

185. Greenstein, R. A.; Arndt, I. C.; McClellan, A. T.; O'Brien, C. P.; Evans, B. "Naltrexone: a clinical perspective," *J. Clin. Psychiatry*, **1984**, *45*, 25-28.

186. Washton, A. M.; Pottash, A. C.; Gold, M. S., "Naltrexone in addicted business executives and physicians," *J. Clin. Psychiatry*, **1984**, *45*, 39-41.

187. Mendelson, J. H.; Ellingboe, J.; Kuehnle, J.; Mello, N. "Heroin and naltrexone effects on pituitary-gonadal hormones in man: tolerance and supersensitivity." In: Harris, L. S., Ed., *Problems of Drug Dependence*, **1979**. NIDA Research Monograph No. 27. Rockville, MD: Committee on Problems of Drug Dependence, **1979**, 302-308.

188. Burroughs, W. S., *Junky*, Penguin Books: New York, 1977, pp. 124-125. First published as *Junkie*, Ace Books: New York, 1953.

189. And the converse phenomenon appears to attend some addicts' withdrawal. Again, according to Burroughs, whose struggle with heroin addiction went on for decades, "spontaneous orgasms . . . [are] one of the few agreeable features of the withdrawal syndrome. And not limited to single orgasm, one can continue, with adolescent ardor, through three or four climaxes." William Burroughs to Jack Kerouac, Tangier, Morocco, 18 August 1954. In *The Letters of William S. Burroughs: 1945-1959*, Harris, O., Ed., Viking Penguin: New York, 1993, p. 224. See also: Burroughs, W. S., *Junky*, Penguin Books: New York, 1977, p. 103. (First published as *Junkie*, Ace Books: New York, 1953.)

190. Fabbri, A.; Jannini, E. A., et al., "Endorphins in Male Impotence: Evidence for Naltrexone Stimulation of Erectile Activity in Patient Therapy," *Psychoneuroendocrinology*, **1989**, *14*, 103.

191. Kaplan, J. L.; Marx, J. A., "Effectiveness and safety of intravenous nalmefene for emergency department patients with suspected narcotic overdose: a pilot study," *Ann. Emerg. Med.*, **1993**, *22*, 187-190.

192. Stone, N. N., "Nalmefene in the treatment of interstitial cystitis," *Urol. Clin. N. Am.*, **1994**, *21*, 101-106.

193. Glover, H., "A preliminary trial of nalmefene for the treatment of emotional numbing in combat veterans with post-traumatic stress disorder," *Isr. J. Psychiatry. Relat. Sci.*, **1993**, *30*, 255-263.

194. Rapaport, M. H.; Wolkowitz, O., et al., "Beneficial effects of nalmefene augmentation in neuroleptic-stabilized schizophrenic patients," *Neuropsychopharmacology*, **1993**, *9*, 111-115.

195. *DE92*, p. 7.

196. *MM15*, p. 1483.

197. *SA*, p. 202.

198. We do not mean to imply by the following discussion that because opioid drug usage poses a problem for many, there may not be still others for whom it is not a problem. *Abusus non tollat usum*, as the Romans said; or, in the words of a more recent United States President, "If it isn't broken, don't fix it."

199. Then there is another sort of problem which is fortunately very rare in occurrence: the dread scourge of *koro*. "This is the end; this is the end; free it, free it," a young man with this disease (it has thusfar only been seen in males) was recently observed to cry out "with a peculiar intonation." In the first week of his effort to quit heroin, he had been found on the floor of his family's bathroom "trembling with rolling eyeballs," and shouting the above words. Koro, for those who have not heard of this disease, is "an acute anxiety state with the perception of reduced penis length due to shrinkage from intra-abdominal traction and fear of impending death being the main psychopathology." The whole story with its happy outcome is reported in (and all above quotes taken from) the *Journal of Psychoactive Drugs* Short Communication *25*, Jul-Sep, 1993: Chowdhury, A. N.; Bagchi, D. J., "Koro in heroin withdrawal." Those who wish to examine this issue further are referred to a study by the same principal author in which revealing results were obtained by use of the "Draw-a-penis Test, a graphomotor projective test . . . administered on the same Koro patients three times over two years." See: Chowdhury, A. N., "Dysmorphic penis image perception: the root of Koro vulnerability: a longitudinal study," *Acta Psychiatr. Scand.*, **1989**, *80*, 518-520.

200. For a realistic description of the symptoms of opioid withdrawal too graphic to be cited here: Wallace, D. F., *Infinite Jest*, Little, Brown: Boston, 1996, pp. 299-306.

201. Cuthill, J. D.; Baroniada, V., et al., "Evaluation of clonidine suppression of opiate withdrawal reactions: a multidisciplinary approach," *Can. J. Psychiatry*, **1990**, *35*, 377.

202. Jackson, J., "Mystery high: Why is clonidine a hot new street drug?" *Baltimore City Paper*, 12 October 1994, (Mobtown Beat), p. 13.

203. Charney, D. C.; Heninger, G. R., et al., "The combined use of clonidine and naltrexone as a rapid, safe, and effective treatment of abrupt withdrawal from methadone," *Am. J. Psychiatry*, **1986**, *143*, 831.

204. From a study of detox protocols: Galloway, G.; Hayner, G., "Haight-Ashbury free clinics' drug detoxification protocols," *J. Psychoactive Drugs*, **1993**, *25*, 251-252.

205. *L&ID*, p. 83.

206. Burroughs, W. S., *Junky*, Penguin Books: New York, 1977, p. 117.

207. *A System of Medicine*, Clifford, Sir T. C.; Rolleston, H. D., Eds., Macmillan: London, Vol. II, part I, pp. 987-988. Cited in *L&ID*, p. 198.

208. I ran across this recently in an article in *High Times* (Derienzo, P., "The ibogaine factor," August, 1995, p. 23). It seemed extraordinarily unlikely, for a number of reasons: (1) Adolf did not spell his name with a *ph* but an *f*; (2) in any case, the original German trade name was Amidon, not Dolophine [see Sneader, W., *Drug Discovery*, Wiley: NY, 1958, p. 76]; (3) Dolophine was the trade mark used by Indianapolis's Eli Lilly, who are not known for their devotion to Adolf Hitler. Thoroughly exercised by all this, I contacted Eli Lilly. Fritz Frommeyer of their Media Relations department informed me that (1) Lilly no longer makes methadone, but sold their methadone manufacturing operations and the tradmark Dolophine to Roxane Laboratories of Columbus, OH, on 10 March 1995; (2) the folk etymology about the trade name has been bouncing around for many years, and is "an old wive's tale"—the correct origin of the word is from what Lewis Carroll would call a *portmanteau* operation obtained by the merging of *morphine* and *dolor*, the Latin word for pain; (3) the original trademark for Dolophine was not issued in Germany during the war but on 22 February 1949, long after Adolf's suicide in Berlin.

The article in *High Times* is characterized by numerous other efflorescences of paranoid ideation, such as the charge that methadone maintenance programs are run by a "methadone mafia . . . fighting tooth and nail against ibogaine treatment." [For those who do not know,

High Times is a somewhat brassy magazine celebrating the joys to be gotten from consuming marijuana products. It encourages what could be called an "in your face," Bronx-cheer approach to the government's attempts to eradicate this hardy plant. Citizens of the U.S. will recall, however, that similar attitudes in Bramin Boston towards tea taxes led to the founding of this great Republic, and I must rush to protest that my criticism of this one article ought not to be taken as a reflection on the quality of the magazine as a whole, any more than the origin of Dolophine should be taken as a reflection on the merits of methadone maintenance programs.]

209. Nogrady, T., *Medicinal Chemistry: A Biochemical Approach*, 2nd ed.; Oxford: New York, 1988, p. 322. But the author describes as "intriguing" the idea of developing an opiate antagonist which would act as an "immunizing agent . . . against persuading others to try an initially pleasurable sensation," despite the problems of how to (forcibly?) "deliver such agents to a population considered difficult from a sociopathic [sic] point of view."

210. Some articles on these programs: (a) Dole, V. P.; Nyswander, M. E., *JAMA*, **1976**, *235*, 2117. (b) Gossop, M., *Lancet*, **1978**, *1*, 812. (c) Newman, R. G.; Whitehill, W. G., *Lancet*, **1979**, *2*, 485.

211. Gerstein, D. R.; Harwood, H. J., "The effectiveness of treatment." In: *Treating Drug Problems*, Vol. 1; National Academy Press: Washington, DC, **1990**, pp.. 132-199.

212. Lowinson, J. H.; Marion, I. J., et al., "Methadone maintenance," *SA*, pp. 550-561. Cited material p. 554.

213. *MM15*, 1486.

214. Cooper, J. R., "Ineffective use of psychoactive drugs: methadone treatment is no exception," *JAMA*, **1992** *267*, 281-282.

215. D'Aunno, T.; Vaughn, T. E., "Variations in methadone treatment practices: results from a national study," *JAMA*, **1992**, *267*, 253.

216. There is the further consideration that more than one disease may be being medicated. We have noted anecdotal reports that some schizophrenics respond to methadone. An attempt to objectively validate the required maintenance dosage taking psychopathology into consideration can be found in: Maremmani, I; Zolesi, O., et al., "Methadone doses and psychopathological symptoms during methadone maintenance," *J. Psychoactive Drugs*, **1993**, *25*, 253-256.

217. *The New York Times* provided a lengthy series on this matter in the July 2, 3, and 4, 1995 issues; each part of the series starts on p. 1. As Sandra Ruiz Butter, Executive Director of Vocational Instruction, Project Community Services, a substance abuse treatment program in the South Bronx which provides either day treatment, methadone, or outpatient services, reminds us in a letter about this series to the *NYT* on Sunday 9 July 1995, p. 14: "our most expensive treatment program, and those of other residential communities, costs about $15,000 a year. The same period in a state prison costs $40,000—much more in New York City. . . . It seems that as a society we prefer the more costly prison, the loss of properly directed talent and productivity, the higher cost of family breakup and foster care to the possibility of success through treatment."

218. Neuss, N., "Indole alkaloids." In: Pelletier, S. W., *Chemistry of the Alkaloids*, Van Nostrand: New York, 1970, pp. 213-266.

219. Blakeslee, S., "A bizarre drug tested in the hope of helping drug addicts," *The New York Times*, 27 Oct 1993, C10-11.

220. Palumbo, P. A.; Winter, J. C., "Stimulus effects of ibogaine in rats trained with yohimbine, DOM, or LSD," *Pharmacology Biochemistry and Behavior*, **1992**, *43*, 1221-1226.

221. (a) Repke, D. B.; Artis, D. R., et al., "Abbreviated ibogaine congeners," *J. Org. Chem.*, **1994**, *59*, 2164-2171. (b) Sershen, H.; Hashim, A., et al., "The effect of ibogaine on kappa-opioid- and 5-HT3-induced changes in stimulation-evoked dopamine release in vitro from striatum of C57BL/6By mice," *Brain Res. Bull.*, **1995**, *36*, 587-591.

222. (a) Mash, D. C.; Staley, J. K., et al., "Properties of ibogaine and its principal metabolite (12-hydroxyibogamine) at the MK-801 binding site of the NMDA receptor complex," *Neuroscience Lett.*, **1995**, *192*, 53-56. (b) Mash, D. C.; Staley, J. K., et al., "Identification of a primary metabolite of ibogaine that targets serotonin transporters and elevates serotonin," *Life Sci.*, **1995**, *57*, 45-50.

223. A small sampling of more recent reports: (a) Dworkin, S. I.; Gleeson, S., et al., "Effects of ibogaine on responding maintained by food, cocaine and heroin reinforcement in rats," *Psychopharmacology (Berl)*, **1995**, *117*, 257-261. (b) Cappendijk, S. L.; Fekkes, D.; Dzoljic, M. R., "The inhibitory effect of norharman on morphine withdrawal syndrome in rats: comparison with ibogaine," *Behav. Brain Res.*, **1994**, *65*, 117-119. (c) Glick, S. D.; Kuehne, M. E., et al., "Effects of iboga alkaloids on morphine and cocaine self-administration in rats: relationship to tremorogenic effects and to effects on dopamine release in nucleus accumbens and striatum," *Brain Res.*, **1994**, *657*, 14-22.

224. The doses were 100 mg/kg given as an intraperitoneal injection, which would correspond to 7,000 mg of pure ibogaine if administered to a human being: O'Hearn, E.; Long, D.B.; Molliver, M. E., "Ibogaine induces glial activation in parasagittal zones of the cerebellum," *Neuroreport*, **1993**, 299-302. The usual psychedelic dose of ibogaine is 200-400 mg (about 4.3 mg/kg) taken orally. For comparison, a dose of 4 mg/kg ip is the LD_{50} for epinephrine (adrenalin), so one can assume that at doses >23 times greater adrenalin would be lethal. But no one suggests we should do without it.

3

Depressants: Alcohol, Benzodiazepines, Barbiturates

Be brave then, for your captain is brave and vows reformation. There shall be in England seven halfpenny loaves sold for a penny; the three-hooped pot shall have ten hoops, and I will make it felony to drink small beer.

—Shakespeare, *King Henry VI, Part II*[1]

"Is anything the matter with Mr. Snodgrass, Sir?" inquired Emily, with great anxiety.

"Nothing the matter, Ma'am. . . . Cricket dinner — glorious party — capital songs — old port — claret — good — very good — wine, Ma'am — wine."

"It wasn't the wine," murmured Mr. Snodgrass, in a broken voice. "It was the salmon."

— Dickens, *The Pickwick Papers*[2]

Introduction

The drugs discussed in this chapter are, like the opioids of the last chapter, depressants of the central nervous system (CNS). Like the opioids, alcohol and barbiturates can induce unconsciousness and even a potentially fatal coma if taken in sufficiently large amounts. Probably all of the drugs in this chapter act chiefly at the same locus in the brain, the γ-aminobutyric acid (GABA) inhibitory neurotransmitter (NT) system, and because of this, most exhibit the property of mutual synergism for one another. We will consider alcohol first because it is probably the most ancient drug of all, its use being coextensive with recorded human history. From there we consider two classes of widely used (and occasionally abused) synthetic drugs that have properties much like alcohol, the benzodiazepines and the barbiturates. We will close the chapter with a look at a few other synthetic drugs and a beverage made from an intriguing plant found in the Pacific islands, kava.

Alcohol

Dr. Alco-Jeckyl

Alcohol is perhaps paradigmatic of all the controversy that has ever arisen about the good and bad features of any drug. Alcohol in small quantities is a natural and innocuous ingredient of many foods, especially fruits and fruit juices. For example, the small amount of alcohol naturally present in orange juice contributes significantly to its flavor—as a volatile solvent, alcohol has the effect of "lightening" the heavier flavor components and allowing them to contribute to the aroma (most "taste" is actually smell). When orange juice is concentrated by evaporating the water, the alcohol is also lost, and this is one reason the reconstituted orange juice tastes flat.

Of course, stronger fermented beverages such as beer and wine have been known and used by the human family since prehistoric times. The praise of wine can be found in sources as old as the Psalms, which celebrate "wine that makes glad the heart of man," and Proverbs, which advise us to "give strong drink to those who are about to perish, and wine to those who are heavy of heart." Wine is used sacramentally in the Jewish feast of Passover, and an ancient ruling in the Mishnah[3] says that Jews are to drink wine while celebrating Purim until they are unable to distinguish between "Blessed is Mordecai" and "Cursed is Haman"—although with most Jews, this custom is now best honored in the breach.

In the Christian communion service the exhilarating and consoling effects of alcohol are meant to symbolize the sacramentally effected union with God. A prayer from the period of the *devotio moderna* in the late Middle Ages has the fervent communicant recite "sanguis Christi inebria me (blood of Christ, inebriate me)." In the Gospel attributed to John, Jesus's impromptu winemaking abilities at Cana are highly praised; and in the first letter to Timothy (ascribed to Paul), a little wine is said to be better for the stomach than much water.

At about the same time as these early Christian sources were being written, the Roman poet Horace was singing odes[4] in praise of partying with lots of his favorite wine, *caecubum*:

> *Now is the time for drinking, friends, the time*
> *to free your feet*
> *to beat*
> *the earth in dancing. . . .*
>
> *Your heir, more worthy of the wine you hoarded*
> *behind those scores of locks and doors,*
> *will pour it out to stain your splendid floors*
> *at feastings more lavish than pontiffs' banquets.*

More recent is Fitzgerald's translation of the poetry of Omar Khayyam, whose *Rubaiyat*,[5] a paean of praise for drinking as an escape from the uncertainties of life (and death), sounds all the more fervent when one recognizes its origin in a Muslim culture that condemned the use of alcohol:

A Book of Verses underneath the Bough,
A Jug of Wine, a Loaf of Bread—and Thou
Beside me singing in the Wilderness—
Oh, Wilderness were Paradise enow. . . .

You know, my Friends, how bravely in my house
For a new marriage I did make Carouse;
Divorced old barren Reason from my Bed,
And took the Daughter of the Vine to Spouse. . . .
And much as Wine has played the Infidel
And robbed me of my Robe of Honor—Well,
I often wonder what the Vintners buy
One half so precious as the stuff they sell.

Over the ages, the cultivation and appreciation of fine wines has even become a refined art form. The late and great connoisseur Frank Schoonmaker, who wrote rather soberly about wines in the prose of the 1940s, can nonetheless wax poetic when describing the truley greats: "Sauternes at their best are quite extraordinary: velvety and almost creamy despite their strength, remarkable for their fruit, their breed and their bouquet.... Chablis at its best is a dry, almost austere wine, pale straw in color (sometimes with a hint of green), remarkably clean on the palate, with a delicate, fleeting bouquet, and a characteristic flavor often described as 'flinty.'" In the more ebullient decade of the 1990s, *The Wine Spectator* describes a Napa Valley Cabernet Sauvignon as having a "complex, deep cherry, tobacco, and cassis nose, with a dry, big, woody, complex and powerful fruit, a good structure, and a long full finish."

In the Middle Ages the Europeans learned from Arabic alchemy the use of the alembic, and even stronger distilled liquors like brandy, whiskey, gin, rum, and vodka became available. Use of these seemed to result in increased drunkenness, particularly among the poor drawn in great numbers from the old rural areas of England to the rapidly growing cities. In the 1830s, Charles Dickens, then a free-lance journalist describing city scenes under the pen name of "Boz," observed how brilliantly illuminated, splendid gin "palaces" sprang up in precisely the most miserable slums of the city, and cautioned that the do-gooders of the middle-class temperance societies, organized largely to encourage the poor to refrain from alcohol, might be missing the point:

> Well-disposed gentlemen, and charitable ladies, would alike turn with coldness and disgust from a description of the drunken besotted men, and wretched broken-down miserable women, who form no inconsiderable portion of the frequenters of these haunts [the gin shops]; forgetting, in the pleasant consciousness of their own rectitude, the poverty of the one, and the temptation of the other. Gin-drinking is a great vice in England, but wretchedness and dirt are a greater; and until you improve the homes of the poor, or persuade a half-famished wretch not to seek relief in the temporary oblivion of his own misery, with the pittance which, divided among his family, would furnish a morsel of bread for each, gin-shops will increase in number and splendor. If Temperance Societies would suggest an antidote against hunger, filth, and foul air, or could

establish dispensaries for the gratuitous distribution of bottles of Lethe-water, gin-palaces would be numbered among the things that were.[6]

In 1919, the temperance movement led to the "Great Experiment" of Prohibition in the United States. From its inception, this 18th Amendment to the Constitution, known as the Volstead Act, was controversial. In time it seemed ruinous: because a sizable minority or even a majority of otherwise decent citizens were uninterested in following the law, it resulted in extensive corruption of public officials and enormous profits to the criminal underworld, which smuggled spirits to speak-easies. Here is an excerpt from a political speech purporting to debate the repeal of prohibition:

> You have asked me how I feel about whiskey; well, Brother, here's how I stand... If by whiskey, you mean the oil of conversation, the philosophic wine and ale that is consumed when good fellows get together, that puts a song in their hearts, laughter on their lips and the warm glow of contentment in their eyes; if you mean that sterling drink that puts the spring in an old man's steps on a frosty morning; if you mean that drink, the sale of which pours into our treasury untold millions of dollars which are used to provide tender care for our little crippled children . . . and to build our highways, hospitals, and schools—then, Brother, I am for it.[7]

Prohibition was repealed in 1933 by the 21st Amendment. However, the states were allowed the power to legislate the sale and use of alcohol in a manner that for other commodities would be an illegal restraint on interstate trade. To this day, the patchwork of often bizarre "blue laws" regulating alcoholic beverages varies from one community to the next.

Does Moderate Drinking Improve Health? An inverse association between alcohol consumption and the risk of coronary artery disease has been found consistently in several types of studies. However, it has been argued frequently that this association may be due to the use of a reference group of nondrinkers that could include heavy drinkers who deny that they drink at all, people who once drank heavily and now abstain (the "sick quitter" hypothesis), or those who abstain because of illness. The most careful study to date, the Health Professionals Follow-up Study (a prospective investigation of the relationship of diet to heart disease and cancer among 51,529 male dentists, veterinarians, pharmacists, and other health workers conducted jointly by the Harvard School of Public Health, the Harvard Medical School, and the Brigham and Women's Hospital) indicates that alcohol consumed *in moderation* functions as a powerfully effective life-prolonging drug, probably by raising beneficial high-density lipoprotein (HDL) cholesterol. The prospective nature of the study precludes recall bias, and there was 96% follow-up.[8]

The alcohol consumption patterns in this group are similar to those among the general U.S. population: 33% abstain from all alcohol and 91% drink less than 30 g/day. (The questionnaire included questions about average consumption of beer, white wine, red wine, and spirits, and the estimated alcohol content for standard portion sizes was: for beer 13.2 g, for wine 10.8 g, and for spirits 15.1 g.) There was a continual benefit from increasing amounts of alcohol up to 50 g per day; the number of respondents drinking more than this amount was too small to allow any conclusions to be drawn.

The authors of this study separately examined the association between coronary artery disease and alcohol from beer, wine, or spirits and found that the associations did not differ significantly, although the strongest association was found for men drinking spirits. Earlier, the same group had reported similar results in a prospective study of a smaller group of women.[9] A recent careful review of more than 20 studies concludes that it is the alcohol itself, and not other beverage components, which has a cardioprotective effect.[10]

Another benefit of drinking alcohol may be a diminished likelihood of suffering from the common cold. In a careful study where 391 subjects (and 26 controls) were quarantined in the Common Cold Unit of Britain's Medical Research Council and then deliberately exposed through nasal drops to one of five respiratory viruses, it was found that susceptibility to developing a cold was increased by smoking and decreased by moderate (up to four drinks a day) alcohol consumption. If you were both a smoker and a drinker, unfortunately, the benefits of drinking were entirely nullified by the tobacco.[11]

Alcohol consumption can also make one less liable to develop non-insulin-dependent diabetes mellitus (NIDDM). Two large prospective studies in Britain confirmed, as was long known, that obesity is a major risk factor for developing NIDDM, and found that smoking 25 or more cigarettes likewise increased the risk, but moderate drinking (16 to 42 drinks per week) was a protective factor.[12]

In another British study, it was found that, if total mortality from all factors was examined, rather than correlation with any specific disease, then there exists a U-shaped relationship to alcohol consumption: men who drank 8 to 14 drinks per week had the lowest mortality, 30% lower than nondrinkers; the mortality rate then rose as the drinking increased, and those in the highest category, more than 43 drinks per week, had the same mortality as nondrinkers.[13] This study was of a limited population: male doctors over the age of 50. But in a large prospective Danish study that followed over 7,000 men and 6,000 women for more than a decade, it was found that those reporting 1 to 6 drinks per week had the lowest overall risk for death; abstainers increased their risk for death by 1.37 and those who drank 70 or more drinks per week by 2.29. Of the 2,229 deaths observed in the population over the period, the number of *excess deaths* attributable to teetotalism was 159, while the number attributable to exceeding 6 drinks a week was 117.[14]

Of course, one can always mistrust the self-reporting of lifetime habits for which there is societal disapproval. Few people find it embarrassing to claim they are teetotalers, but most would be uncomfortable admitting they put away a six-pack every day before breakfast. The likelihood is that any statistics about alcohol consumption are skewed due to an underreporting of consumption. This could mean that the threshold level of alcohol intake whereupon the adverse effects of drinking begin to exceed the beneficial ones is, in the real world, even higher than what any study concludes.

Mr. Alco-Hyde

The picture is of course not all rosy; what has been for many a beneficial pleasure has proved for some an enslaving and destructive addiction. In a letter of Paul written much earlier than that to Timothy, he complains that for some of the Christians of Corinth their eucharistic celebrations had degenerated into occasions for drunkenness. And the same United States politician of the Prohibition era quoted above as "for whiskey, the oil of conversation" saw (and played) both sides of the issue:[15]

> If by whiskey, you mean the Devil's brew, the Poison scourge, the bloody monster that defies innocence, dethrones reason, creates misery and poverty, yea, literally takes the bread out of the mouths of babes; if you mean the Evil Drink that topples men and women from pinnacles of righteous, gracious living into the bottomless pit of despair, degradation, shame, helplessness and hopelessness—then certainly I am against it with all my power.

Many are those who have found that it was easier to use alcohol not at all than to use it moderately. In his later years, Samuel Johnson, the "Great Lexicographer" and preeminent figure of 18th-century English literature, drank nothing stronger than lemonade, although he had often enough in his youth cried out for "Poonch! Poonch!" (the pronunciation of "punch" in his native Lichfield). In explanation, and echoing Augustine, he said "abstinence is as easy to me, as temperance would be difficult."[16] (Augustine was speaking of sex, which to him had perhaps much the character of an addiction, when he said many years earlier, "to many, total abstinence is easier than perfect moderation.")[17]

The bard of A. E. Housman's *A Shropshire Lad*[18] provides us with some cautionary verse:

Oh many a peer of England brews
Livelier liquor than the Muse,
And malt does more than Milton can
To justify God's ways to man.
Ale, man, ale's the stuff to drink
For fellows whom it hurts to think:
Look into the pewter pot
To see the world as the world's not.
And faith, 'tis pleasant till 'tis past:
The mischief is that 'twill not last.
Oh I have been to Ludlow fair
And left my necktie God knows where,
And carried half way home, or near,
Pints and quarts of Ludlow beer:
Then the world seemed none so bad,
And I myself a sterling lad;
And down in lovely muck I've lain,
Happy till I woke again.
Then I saw the morning sky:
Heighho, the tale was all a lie;
The world, it was the old world yet,
I was I, my things were wet,
And nothing now remained to do
But begin the game anew.

And there is the cautionary life of the late Charles Bukowski, America's "drunk poet laureate." From his poem "Young in New Orleans":[19]

. . . sitting up in my bed
the lights out,
hearing the outside
sounds,
lifting my cheap
bottle of wine,

letting the warmth of
the grape
enter
me
as I heard the rats
moving about the
room,
I preferred them
to
humans.

There is some doubt that the "disease" of alcoholism always existed. In the Middle Ages, for instance, it seems that excessive drinking or drunkenness was not felt to be distinct from the problem of excessive eating: both were examples of *gule*, gluttony.[20] In our culture, alcoholism seems to exist as a distinct and recognizable entity; nonetheless, there exists the same problem in defining alcoholism as there is in defining drug addiction; adapting the definition from Chapter 1, we can say that alcoholism "is characterized by continuing compulsive use and overwhelming involvement with alcohol; additionally there is the risk of physical, social, or psychological harm to the individual and the need to stop alcohol use, whether the person understands and agrees or not."

Is Alcoholism a Disease? Of course, alcoholism is not a disease like measles, with a known infective agent: it is partly genetic, partly societal, and—like every addiction—partly psychological. The debate over terminology reflects society's slowly learned realization that it is counterproductive and inhumane to imprison alcohol abusers as criminals. On the contrary, with medical intervention and group support, in particular with programs like Alcoholics Anonymous (AA), many "recovering" alcoholics (the progressive present is used to emphasize the chronic, even accumulative, nature of the disease) become constructive members of society.

Closely related to the disease concept of alcoholism is the issue of whether an alcoholic can ever reassert control and become a moderate drinker. The majority opinion for some time has been no, and this is the position strongly endorsed by AA and other 12-step programs derived from it. The American Society of Addiction Medicine defines alcoholism as a disease, meaning that it is an "involuntary disability," because the alcoholic is ultimately unable to limit drinking once he or she begins to drink.[21]

Like all accepted beliefs in a revisionist world, this too has been challenged of late. Indeed, from a priori grounds it would seem impossible to deny that there exits *some* minimum amount of alcohol that even a recovering alcoholic could consume, given the requisite set and setting, without losing control, since there are measurable amounts of alcohol in many common foods that recovering alcoholics consume regularly without experiencing any problems.[22] In any case, the prevailing dogma has been challenged by therapists who claim that controlled drinking is possible for some problem drinkers—usually those who are younger, have less severe problems, and do not manifest a withdrawal syndrome when they stop drinking.[23] A self-help group called Moderation Management provides a program of support for problem drinkers who wish to become moderate controlled drinkers; the group recommends limiting women to three drinks a day and men to four. Similarly, in a Canadian study that examined data from 235 problem drinkers, it was found that men who limited themselves to 4 drinks a day and 16 a week, and women who limited themselves to 3 a day and 12 a week, provided the most accurate separation of those who experienced problematic incidents stemming from their alcohol use from those who did not.[24] But a recent study suggests that many problem drinkers are like Samuel Johnson: although relapse after 5 years of abstinence was rare, most of those who attempted controlled drinking, even those who succeeded for as long as 10 years, eventually relapsed.[25]

Is Alcoholism an Inherited Trait? Recently, Blum claimed to have found evidence[26] that the gene for the D_2 dopamine receptor was associated with an A1 allelic gene that was found in 69% of 35 brain samples of alcoholics and was absent in 80% of 35 nonalcoholics. Because the dopamine receptor has been linked strongly with the neural mechanisms for "reward" and behavior reinforcement, this may indicate a hereditary predisposition to alcoholism. Some later studies seem to indicate that in addition to alcoholism, this particular allele of the dopamine gene is associated with a number of other behavior disorders, such as Tourette's syndrome (Chapter 5), attention deficit hyperactivity disorder (Chapter 4), and autism.[27]

Other studies have long shown that close relatives of alcoholics have about four times the risk of themselves becoming alcoholic, and the risk is unchanged for children adopted away at birth and raised without knowledge of their biological parents' alcoholism. In research on twins, it has been shown that the probability of being an alcoholic is about 60% for identical twins and about 30% for fraternal/sororal twins. Some of these "behavioral genetic" results have been subjected to criticism: "I think it is by and large garbage," says Irving I. Gottesman of the University of Virginia, who is otherwise a strong defender of genetic models of human behavior.[28] On the other hand, three recent studies, two of them prospective, provide intriguing evidence that there are fundamental differences in the effect of alcohol on subjects with a familial history of alcoholism: in the first, a *lowered* response to alcohol at age 20 was an independent predictor of future alcoholism regardless of drinking practices at the time of the original study;[29] in the second, there was likewise a *lowered* encephalographic (EEG) response to a dose of alcohol among 19-year-olds who later developed alcohol dependence.[30] Both of these studies also confirmed the relationship between a familial history of alcoholism and future problem drinking. In the third study, an *enhanced* release of β-endorphin relative to controls was shown in subjects with a family history of alcholism.[31] "A significant fraction of vulnerability to alcoholism is inherited," concludes Enoch Gordis, Director of the National Institute on Alcohol Abuse and Alcoholism.[32]

Alcohol Morbidity

Blackouts; Wernicke's and Korsakoff's Syndromes. Heavy drinking can result in an alcoholic *blackout*, in which the imbiber is unable to remember part or all of what occurred while he or she was drinking. This problem is experienced by 30 to 40% of men in their late teens and early 20s, most of whom do not go on to develop more serious and pervasive problems.

In heavy drinkers, alcohol displaces food in the diet and increases the demand for B vitamins, which are needed to metabolize the alcohol; it may also impair absorption of vitamins from the intestines. The resultant deficiencies of the vitamin B class, chiefly thiamine, can cause severe nerve damage in the brain and CNS, leading to Wernicke's and Korsakoff's syndromes. These syndromes have developed among malnourished prisoners of war, where alcohol played no part; they are essentially the neurological components of the classical disease of vitamin-B_1 (thiamine, **3-1**, Figure 3.1) deficiency, *beriberi* (a word derived from the Singhalese for profound weakness). Korsakoff's syndrome is the psychic component of the disease; it involves profound amnesia and memory impairment, but a normal or unaffected IQ. Usually the level of recent memory loss is disproportionately high. Only about a fourth of such persons can achieve full recovery, even after total abstinence and thiamine supplementation. The memory of some patients is helped by administration of *clonidine* (Catapres, **3-2**), an adrenergic agonist that, as we saw in the previous chapter, is also used to ease the discomfort of heroin withdrawal. Many patients are left with a permanently damaged mental state characterized by large gaps in memory and an inability to sort out events in their proper temporal sequence—a condition loosely referred to as "alcoholic dementia." Wernicke's syndrome usually involves ocular disturbances such as seeing double (diplopia) and a wide-based, staggering gait (ataxia). The mortality rate can be as high as 17%.

Fetal Alcohol Syndrome. Fetal alcohol syndrome is a characteristic damage done to a developing fetus by alcohol consumption on the part of the mother. It consists of a group of mental and physical abnormalities such as mental retardation, hyperactivity, a small head with low-set ears, small eyes, a flat nose with upturned nostrils, and webbed fingers and toes. The exact amount of alcohol consumption required to produce fetal damage is unknown, and it is advisable for pregnant women to abstain completely.

Liver Damage. The three principal alcohol-induced liver diseases are (1) alcoholic fatty liver, (2) alcoholic hepatitis, and (3) alcoholic cirrhosis. They rarely occur in pure form, and features of each may overlap in a given patient. Alcoholic fatty liver occurs in most heavy drinkers but is reversible on cessation of alcohol consumption and is not thought to be an inevitable precursor of hepatitis or cirrhosis. The liver is enlarged (hepatomegaly), yellow, and distended with accumulation of fats, but symptoms are often minimal or entirely absent. Alcoholic *hepatitis* resembles viral hepatitis. Patients have anorexia, nausea, vomiting, jaundice, high fever, and malaise. Some patients die during acute attacks of alcoholic hepatitis, and even after complete abstinence, recovery can be slow. The quantity and duration of drinking necessary to cause alcoholic cirrhosis is not clear. The typical alcoholic patient with cirrhosis has had a daily consumption of a pint or more of whiskey, several quarts of wine, or an equivalent amount of beer for at least 10 years.

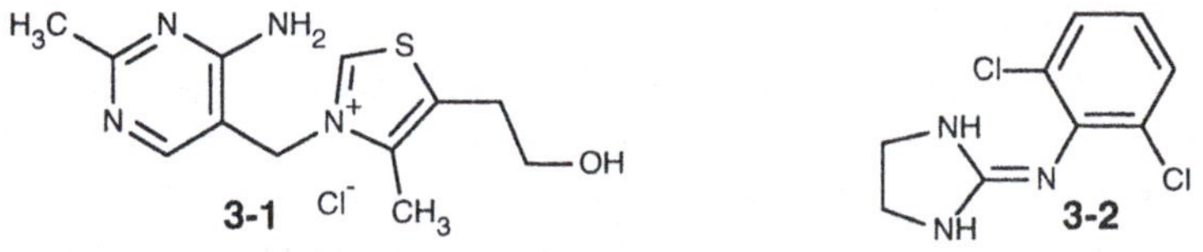

Figure 3.1 Vitamin B_1; clonidine.

However, only 10–15% of alcoholics develop cirrhosis. It is often a silent and insidiously progressive disease. The disease causes anorexia and a hormonal disbalance; consequently there is weight loss, reduction in muscle mass, weakness, and jaundice. In men there can be decreased body hair, gynecomastia, or testicular atrophy; in women there may be menstrual irregularities and virilization. Most patients with advanced cirrhosis die in a coma caused by bleeding from the esophagus or liver infection. Patients who have had a major complication from cirrhosis and continue to drink have a 5-year survival rate of less than 50%. In general, the prognosis for advanced liver disease is very poor; most victims usually die from massive hemorrhage.

Increased Cancer Risk. Cancer is the second leading cause of death in alcoholics, who have a rate of carcinoma 10 times higher than that expected in the general population. The sites with the greatest increased cancer incidence are the head and neck, esophagus, stomach, liver, and pancreas.

Acute Alcohol Toxicity

In addition to death or shortening of life from diseases caused by chronic heavy drinking, an acute massive intake of alcohol can cause death directly even on the part of a first-time user. Here are three recent cases:

> Wayne Parsons, a 19-year-old sophomore electrical engineering major with a B average at Virginia Tech, was the youngest of three brothers and two sisters. His parents and an older brother had warned him about the dangers of drinking. But when he went to a party with about 30 other people at a friend's apartment, he gulped a large quantity of beer, then in 15 minutes drank an oversized tumbler of about 32 oz. of tequila, friends told police.
>
> About an hour later, he collapsed. An emergency medical crew was called but could not revive him. He died in the hospital at six the next morning, his horrified parents at his side. The cause of death was alcohol poisoning.
>
> The parents of Alan Brodwater of Coeur d'Alene, Idaho, knew their son drank, sometimes excessively. But after several conversations about the dangers of drinking, they thought the worst was over. Alan was trying to improve his 3.0 GPA to 3.5 and used alcohol to ease the stress. At 1:10 A.M. on Nov. 1, 1991, he died after a bout of drinking that began at his apartment and continued at a Halloween party off campus.
>
> Larry K. Wooten, a 21-year-old history major with a 4.0 average at U of Florida, was watching a football game with his friends at a bar in Gainesville when he and an old friend challenged each other to see who could drink more. "It was something these boys always did," said his mother. "They competed through baseball, football, grades, girlfriends. Larry was going to beat Stevie's record." And he did: he drank 23 shots of liquor, one more than Steve, in an hour. Steve survived, but when friends took Larry to his apartment late that night, they noticed he had passed out and that his hands and feet were turning blue. They called an ambulance, but he died of asphyxiation on the way to the hospital.[33]

"Binge" drinking (defined as taking five or more drinks at a sitting at least once in the previous 2 weeks) is on the rise on college campuses; one study of 18,000 college students in 40 states claims that 50% of male students and 39% of women students met these criteria for binge drinking.[34] Rapid ingestion of large quantities of distilled spirits is particularly dangerous: alcohol accumulates in the stomach like a time bomb. Inevitably it will work its way into the bloodstream, and either the person will fall into a coma with respiratory failure, or his motor control will become so poor that he will vomit and breathe it in (the sanitized medical term for this is "aspiration"), thereby drowning. Many deaths in surgery are the result of the anesthetic, and alcohol in these immense doses is an anesthetic taken with none of the precautions of an intensive-care hospital setting.

The Role of Alcohol in Traffic Fatalities. People who drive while intoxicated do so repeatedly, claims a recent study by Dr. Robert Brewer;[35] and the risk of eventual death in an alcohol-related automobile crash is about 12 times as great for those who have been arrested while driving under the influence of this second-most common of drugs as it is for nondrinking drivers.[36] It is arguable that all mind-altering drugs affect judgment, but marijuana probably makes one overcautious (see Chapter 7), while alcohol typically makes people overconfident. Cocaine, caffeine, and tobacco tend to render their communicants more wakeful, even apprehensive; alcohol to the contrary has a pronounced hypnotic and anxiolytic effect. Thus habitually drunk drivers are relaxed and comfortable while snoozing at the wheel, their foot pressed ever more firmly and confidently on the accelerator—until they encounter "acute deceleration toxicity."

The Pharmacology of Alcohol

How Does Alcohol Work? Because alcohol can arguably be considered the oldest known drug, it is surprising to find that its mode of action is only poorly understood. At least three mechanisms have been proposed:

- The ethyl alcohol molecule is very small and quite soluble in both lipid and aqueous media. Hence it is easily carried in the bloodstream, which is primarily aqueous, and it easily crosses the blood-brain barrier into the brain, which is primarily lipid in nature. There it is thought to interact with *all* the synapses and affect *all* the NT–receptor interactions like the general anesthetics cyclopropane, diethyl ether, xenon, chloroform, and nitrous oxide, all of which are also small lipid-soluble molecules. Part of the effect of alcohol resembles that of the general anesthetics; indeed, before the discovery of chloroform and ether, whiskey or rum was used as a primitive anesthetic for operations like emergency battlefield amputations.
- Alcohol has been grouped into a class including local anesthetics (lidocaine, procaine, Chapter 4); phencyclidine, a dissociative anesthetic (PCP, Chapter 7), and chlorpromazine, an antipsychotic (Chapter 5), all of which modulate the nicotinic subclass of the acetylcholine receptors, but do not bind to the ACh sites. They are generally referred to as noncompetitive inhibitors, but there is much uncertainty about the number and location of the binding sites as well as their precise mechanism of action.
- The third mechanism is probably the most frequently proposed, and involves specific interaction with the GABA receptors in the CNS.

The GABA Receptor. It is thought that all the neurons in the brain are under inhibitory control by GABA (Chapter 1), with as many as 30% using GABA as their chief NT. We have grouped the benzodiazepines and the barbiturates together in this chapter because it is known that these drug classes owe their principle effect to an interaction at the GABA receptors. However, the case with alcohol is less clear. Some of the evidence linking the activity of alcohol to the GABA receptor is as follows:

- Alcohol, the benzodiazepines, and the barbiturates have a synergetic effect on one another—indeed, the interaction of alcohol and the barbiturates is particularly powerful and has led to many intended and unintended deaths by overdose.
- Many of the benzodiazepines and the barbiturates can be used to control epileptic seizures, and there is evidence that some epilepsies are caused by defective GABA neurons.
- Abrupt withdrawal from chronic barbiturate usage can result in epileptic seizures. Severe alcoholic delirium tremens can similarly be accompanied by epileptic seizures, and both of these withdrawal effects can be prevented by intravenous benzodiazepines.
- In animal studies, some of the effects of alcohol are blocked by flumazenil, a benzodiazepine antagonist.
- Large quantities of ethanol can open the GABA-coupled chloride channel in the rat brain even in the absence of the ordinary ligand, GABA.[37] Nonetheless, there is no specific receptor interaction in the sense of alcohol binding to a specific site on the GABA receptor.

Recently, the hypothesis has surfaced that although perturbation in neuronal lipid membranes by alcohol may affect all CNS neurons to some degree, two specific receptor-gated ion channels, GABA and NMDA are more sensitive to ethanol than other neurons in the CNS and are responsible for most of ethanol's effects on behavior and consciousness.[38] Other hypotheses suggest that the serotonin system may contribute to the craving for alcohol,[39] while the effectiveness of opioid antagonists in suppressing alcoholic recidivism shows that the endogenous opioid system is also somehow involved.

Metabolism of Alcohol. The major site of ethanol absorption is the small intestine. The rate of absorption increases in the absence of food, with dilution to about 20% by volume, and with carbonation (as in sparkling wines and beer). Between 2 and 10% of *ethanol (ethyl alcohol, "grain alcohol,"* **3-3**, Figure 3.2) is excreted unchanged through the lungs, urine, or sweat, but the greater part is metabolized to *acetaldehyde (ethanal,* **3-4**) in the liver via an enzyme, *alcohol dehydrogenase*. The resulting acetaldehyde is in turn metabolized by *aldehyde dehydrogenase* to *acetic acid (acetate,* **3-5**). The process results in about one drink, or 10 g ethanol, per hour being processed (12 oz. of beer, 4 oz. of unfortified wine, and a 1.5-oz. shot of 80-proof spirits each contains about 10 g of ethanol).

Methanol (methyl alcohol, "wood alcohol," **3-6**) is a common industrial solvent that causes relatively little intoxication compared to ethanol, but that is occasionally consumed accidentally or deliberately by desperate ethanol addicts. It is metabolized in the body by the same enzymes that metabolize ethanol, first to *formaldehyde (methanal,* **3-7**), then to *formic acid (methanoic acid, formate,* **3-8**). Unfortunately, both these metabolites are very toxic; the formic acid in particular tends to accumulate and is ultimately responsible for destruction of neural cells, and particularly the optic nerve.[40] Ingestion of methanol in significant quantities leads to blindness, paralysis, and death. Although death from

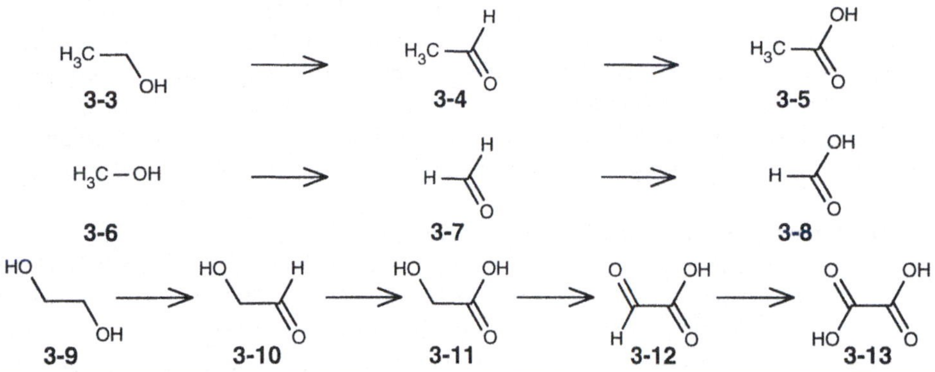

Figure 3.2 Biotransformations of ethanol, methanol, and ethylene glycol.

methanol has been reported after consuming as little as 30 mL, the usual fatal dose is about 200 mL.

When ethanol beverages are unavailable, the chronic alcoholic is tempted to use cheap methanol from a can of solvent that can be bought at the hardware store, or by ingesting ethyl alcohol or ethyl alcohol gels (such as Sterno) to which methanol has been added to make them poisonous ("denatured" alcohol). If brought to an emergency room promptly enough, a person in acute methanol intoxication can be saved from death—ironically by an IV drip of dilute ethanol. The alcohol dehydrogenase that converts methanol to the toxic formaldehyde binds selectively to ethanol if it is also present, and the more volatile and water-soluble methanol will eventually be excreted unchanged through the lungs and kidneys, provided the patient's blood ethanol concentration can be kept high. In other words, ethanol is an antidote to methanol poisoning. This is fortunate, because natural fermentation produces methanol: many wines, and especially the brandies or liqueurs distilled from them, contain quantities of methanol that might be dangerous if they did not also contain its antidote in still larger amounts.

Drinking *ethylene glycol,* **3-9**, a common component of automobile radiator antifreeze with a smooth, sweet taste, produces a CNS narcosis that is followed by coma, convulsions, and kidney damage. About 100 mL (2–3 oz.) can be fatal. Again, the toxicity is due to the metabolites rather than the ethylene glycol itself. The ethylene glycol is oxidized through several stages (by the same enzymes that transform ethanol) into *glycoaldehyde,* **3-10**; *glycolic acid,* **3-11;** *glyoxylic acid,* **3-12;** and, finally, to *oxalic acid,* **3-13**. It is the oxalic acid (although it occurs naturally in smaller quantities in foods such as rhubarb and other vegetables) that is particularly poisonous to the kidneys. Because ethylene glycol improves the flavor of cheap wine, providing a smooth finish, there have been instances of poisoning from less-than-conscientious vintners in Italy and Austria adding a to their products. However, fatalities only occurred among those who drank more than a liter at a quick sitting: for more moderate imbibers, the ethanol in the wine acted as an antidote. Tests with rats have shown that the LD_{50} for ethylene glycol is ordinarily 5.8 mL/kg, but it is only about half as lethal (LD_{50} = 10.5 mL/kg) if the rats are given an ethyl alcohol chaser 15 minutes after their glycol cocktail.

Even though legal intoxication requires a blood ethanol concentration of 80 to 100 mg/dL, some psychic and cognitive changes are noticeable at levels as low as 20 to 30 mg/dL—that is, after one or two drinks. Narcosis or deep sleep is induced in most people at twice the legal intoxication level, and death can occur with levels from 300 to 400 mg/dL. Ethanol, either alone or in combination with agents like the benzodiazepines, is plausibly said to be responsible for more toxic overdose deaths than any other agent.[41]

Factors Affecting Alcohol Metabolism: Sex, Race, and Drugs. Recent research indicates that the same amount of alcohol ingested by the average 130-lb male will result in a lower blood alcohol concentration than when ingested by the average 130-lb female. The reason? There is a significant concentration of alcohol dehydrogenase in the stomachs of most men that is lacking in the stomachs of most women; because of this, a significant amount of ingested alcohol is converted in the stomach to acetaldehyde and acetate without ever getting into the bloodstream.[42]

There are also genetic as well as sexual differences in the ability to metabolize ethanol: about 50% of Japanese, Korean, and other Asiatic groups have a significantly lower level of alcohol dehydrogenase compared to those whose ancestry is of African or European origin. In some individuals, alcohol intolerance is so high that it is the equivalent of endogenous Antabuse; consumption of quite modest amounts of an alcoholic beverage produces discomfiting facial reddening and burning.

Several prescription drugs affect the metabolism of ethanol; to a limited extent, so does aspirin. One of the many myths surrounding the use of alcohol is that one can avoid getting a hangover by taking an aspirin or two before drinking. But it has been shown that the blood-alcohol concentration increases when aspirin is taken before drinking, so that intoxication and most likely hangover also increases. The aspirin apparently inactivates gastric alcohol dehydrogenase.[43]

Alcohol and the Opioids. Over the years, there have been many hypotheses to the effect that addictions to mankind's oldest drugs, alcohol and opium, are somehow related. That this is at least true in some rats seems supported by a study in Finland. There is a strain of rats that is genetically alcohol preferring (known, perhaps whimsically, as the "AA" strain) to the extent that, given their pick of a 20-proof ethanol solution, water, or food, they will consistently choose the alcohol. The AA rats were compared to ordinary rats and to a third strain that was alcohol avoiding (ANA) as to their preference for drinking pure water versus water containing increasing concentrations of *etonitazene* (**2-11**, Figure 2.4), a powerful opioid. Only the AA rats liked etonitazene, and in fact consumed so much they became opioid intoxicated, arguably as a sort of substitute for alcohol intoxication.[44]

Many recovering alcoholics are convinced that the effects of alcohol that they experienced when they drank, some even with their very first drink, were qualitatively and quantitatively different—more intense and more pleasurable—than what ordinary people experience when they have what is for them an ordinary, ho-hum beverage. Could it be that in alcoholics there is a biochemical trait, perhaps something like the genetic trait in AA rats, whereby alcohol activates the individual's opioid system and thus produces a highly addictive interaction like heroin? A theory like this, the "opioid deficiency hypothesis," which proposes that alcoholics are deficient in their endogenous opioid system and that alcohol somehow compensates for this, was proposed by Blum and his colleagues.[45]

One intriguing way in which alcohol could compensate for an opioid deficiency was suggested by Cohen and Collins in 1970. These workers perfused cow adrenal glands with dilute solutions of acetaldehyde, the first metabolic product of the oxidation of alcohol. It condensed with norepinephrine or epinephrine in a classical Pictet-Spengler condensation to produce a variety of tetrahydroisoquinoline alkaloids, which Cohen and Collins hypothesized were involved in developing alcohol dependence and withdrawal symptoms. They theorized that acetaldehyde might also condense with dopamine in the CNS, producing structures similar to some natural alkaloids, including *anhalamine*, which occurs in the psychedelic peyote cactus **6-55**, Figure 6.10).[46] In a still more direct linkage of opioids with alcohol, Davis and Walsh reported earlier in the same year that when they incubated rat brainstem homogenate, dopamine, and alcohol or acetaldehyde, they could demonstrate the formation of *tetrahydropapaveroline* (**2-12**, Figure 2.5), an intermediate in the biosynthesis of morphine.[47] However, neither of these hypotheses was ever substantiated by any direct biochemical evidence from the CNS of human alcoholics.

Reid[48] proposed the opposite notion, an "opiate surfeit hypothesis," which attempts to explain the loss of control experienced by many alcoholics when they drink as due to the production of excessive amounts of endogenous opioids in response to alcohol intake—a sort of heroinlike rush on drinking where nonalcoholics experience only a mild buzz. This, too, has at least indirect support from animal studies: animals trained to consume alcohol will decrease their intake when given *naloxone*, an opioid antagonist, and increase their alcohol consumption when given morphine.[49]

Drugs Used in Alcohol Rehabilitation

Disulfiram. A somewhat controversial medication that has been used almost half a century in alcohol rehabilitation is *disulfiram (Antabuse,* **3-14**, Figure 3.3). "*Citrated*" (two parts citric acid to one part by weight of) *calcium cyanamide (Dipsan, Abstem, Temposil,* **3-15**) is used for the same purpose in Europe. These drugs act by inhibiting aldehyde dehydrogenase, and cause acetaldehyde to build up to toxic levels if alcohol is consumed. Disulfiram is a dangerous drug under these circumstances, and fatal reactions to alcohol ingestion have been reported. Short of death, the disulfiram-ethanol reaction produces tremor, hypertension, nausea, severe vomiting, and diarrhea. Disulfiram should never be used outside a setting of supportive counseling or psychotherapy, and perhaps not often then, since many alcoholics beginning a program of recovery have a few relapses before achieving lasting sobriety. Aldehyde dehydrogenase activity generally returns to normal within a week of taking Antabuse, but it may take 2 weeks or more.

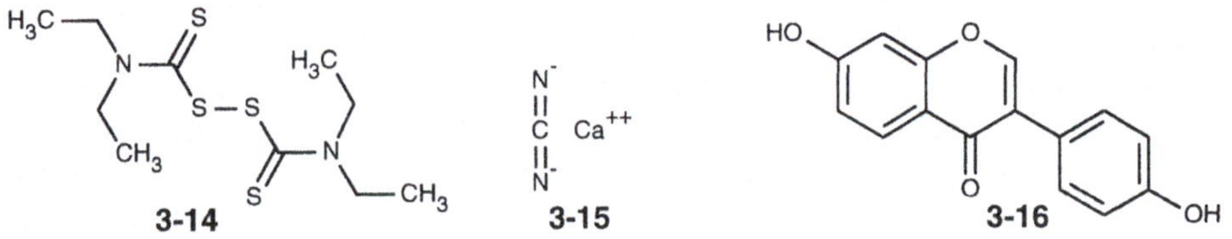

H_3C S N S—S H_3C CH_3 S N CH_3 **3-14**

N^- C N^- Ca^{++} **3-15**

HO O O OH **3-16**

Figure 3.3 Three drugs used in alcohol rehabilitation.

Disulfiram's interaction with alcohol was, of course, discovered by accident. It was the policy in the 1940s at a Danish drug company named Medicinalco to have employees test all drugs on themselves before marketing. An employee named Jens Hald was studying the effects of disulfiram in treating intestinal parasites, took the drug daily as required, and became violently ill when he had a snifter of brandy some days later.[50]

Opioid Antagonists in Alcohol Rehabilitation. A little earlier we discussed several theories attempting to relate alcohol consumption to some activation of the endogenous opioid system. Although none of these theories may be correct, the consistent effect shown by the opioid antagonists (Chapter 2) in suppressing alcohol intake in animal studies led finally to their being tried in humans attempting to abstain from alcohol. Although naltrexone is no instant cure, it was shown in a double-blind, placebo-controlled study conducted in 1992 at the University of Pennsylvania to be a significant help: whereas 55% of the placebo-treated subjects relapsed over a 12-week period, only 23% of those receiving naltrexone did. Those receiving naltrexone also reported lower craving for alcohol—although this was not the deciding difference between them and the placebo group. Rather, naltrexone exerted its most significant effect by decreasing the amount of alcohol consumed if and when relapse drinking occurred.[51] Whereas sampling alcohol provokes a crisis of binge drinking in the typical alcoholic, this phenomenon occurred with less frequency and intensity among those taking naltrexone than among those taking placebo. In a later follow-up study of those individuals who had relapsed, it was found that a greater proportion of those who were taking naltrexone than of those taking placebo reported experiencing a significantly diminished high from their drinking.[52]

The key to whether the reduced craving provided by naltrexone translates into fewer lapses in abstinence may be the sort of ongoing psychotherapy provided with the drug. In a study at Yale University, patients were randomized in a double-blind study to receive either naltrexone or placebo and one of two types of counseling: in the first, coping skills were taught that were intended to help them in case they experienced a relapse; in the second, the possibility of a relapse was never discussed and they were simply given support in their efforts to remain abstinent. Those who received the simple supportive therapy were much less likely to initiate drinking (unlike the group at Penn for whom diminished craving did not lessen the frequency of relapse incidents but only their severity); while those receiving the coping-skills therapy initiated drinking at a rate similar to those in the placebo groups but (as with those in the Penn study) were less likely than the placebo group to continue to a full-blown relapse.[53]

At least for some, naltrexone works where nothing else has. One Maryland man, an alcoholic for 30 years, was trying to become sober for the last 10. At least eight treatment programs and a trial of Antabuse all failed (he simply quit taking the Antabuse so he could continue drinking). Nothing kept him from drinking for longer than a few weeks until he entered a naltrexone study at the Baltimore Veterans Administration. For the first time, he reportedly felt free from the craving for alcohol and had gone for more than five months without drinking. "I just pray to God for that pill," he was quoted.[54]

The success of these studies led the FDA on December 30, 1994 (in the nick of time for New Year's) to approve naltrexone for the treatment of alcohol dependence. Du Pont-Merck is marketing naltrexone under the new name *ReVia* with a wholesale price of about $4.00 per one-a-day tablet.[55] Other drugs will likely follow: a preliminary study using *nalmefene* (Chapter 2), a newer opioid antagonist with a longer half-life and lower toxicity than naltrexone, has shown that it is as effective as naltrexone in preventing alcoholic relapse.[56]

What about people who are not alcoholics? A study at Brown University addressed this question in a double-blind crossover study of 19 nonalcoholics who were given, on two separate occasions, naltrexone or placebo followed by an intoxicating dose of ethanol. They were then evaluated for subjective and objective evidence of intoxication. Naltrexone did not affect either psychomotor performance skills or the metabolism of ethanol. But it did reduce the positive reinforcement effects of alcohol and intensify its sedative properties in what was perceived as an aversive manner (subjects felt themselves unpleasantly overintoxicated).[57] In all probability, therefore, the alteration of alcohol's high experienced by nonalcoholics is similar to that experienced by alcoholics, but there may be a difference in the degree or extent of this alteration.

Kudzu in Alcohol Rehabilitation. Kudzu? Readers from the southern United States will know what kudzu is: a nearly indestructible vine that spreads like a cancer through the Appalachian forests, smothering once stately trees. *Kudzu* is derived from the Japanese word for the vine, known to botany as *Pueraria lobata.* In its native East Asia, the kudzu plant is used for animal fodder and for the abundant starch that can be extracted from its roots. The roots of the plant have also been used in traditional Chinese folk medicine for more than a millennium; a Chinese pharmacopeia written about 600 A.D. claims that the roots will inhibit drunkenness.

Harvard Medical School's Dr. Wing-Ming Keung was intrigued by the continued use of kudzu for this purpose in modern-day Chinese medicine. After a visit to China to observe the use of the extract, he became convinced that the kudzu treatment helped recovering alcoholics experience a reduced craving for alcohol. He and Dr. Bert L. Vallee have subsequently performed a series of tests on the Syrian golden hamster, a species that will consume large quantities of alcohol in preference to water. Kudzu root as a crude extract exerted a significant "antidipsotropic" effect on these hamsters, suppressing ethanol intake by more than 50% without affecting body weight or intake of food or water.[58] It turns out that the active ingredient in kudzu root is a fairly simple compound, *diadzein* (4′,7-dihydroxyisoflavone, **3-16**).

Other experiments show that daidzein is an inhibitor of human aldehyde dehydrogenase,[59] but the capacity of diadzein to suppress alcohol intake must involve more than this, because doses smaller than needed to block aldehyde dehydrogenase are sufficient to suppress alcohol intake with no effect on alcohol metabolism.[60] In further studies by workers at Indiana University School of Medicine, this time using fasted rats, *daidzin* (which is simply the glucoside of diadzein and occurs with it in kudzu root) was shown to delay and decrease the peak blood alcohol concentration as well as to shorten the sleep time induced by ethanol.[61] The extrapolation of these effects from rats to humans gives one, as it is said, furiously to think—as does the potential "drink and drive" benzodiazepine partial inverse agonist RO 15-4513, which we will encounter in the next section of this chapter. Only time will tell whether these substances will find any useful application in human medicine.

Benzodiazepines

History

The chemical ring system consisting of a benzene ring fused to a seven-atom ring, two atoms of which are nitrogen, is called a *benzodiazepine*. This ring system, and the immense class of benzodiazepine drugs (BZDs) that contain it, was originally discovered, as were so many useful drugs, by a series of accidents intersecting with careful observation. In the mid-1950s, Leo Sternbach of Hoffmann-LaRoche's research laboratories in Nutley, NJ, was impressed with the success of *chlorpromazine* (**3-17**, Figure 3.4), the first antipsychotic drug, which had just been introduced by Rhône-Poulenc. The structure of chlorpromazine reminded Sternbach of a series of tricyclic compounds he had made as a postdoctoral fellow in Cracow some 20 years previously, which contained the ring system of **3-18**, a "benzheptoxdiazine." Sternbach's idea was to append a dimethylamino group like chlorpromazine's onto his ring system by reacting **3-18** with dimethylamine, which he presumed would form **3-19**.

The results of about 40 variations on this theme were all negative. Sternbach thereupon made a more careful study of the substance that he had thought had structure **3-18**. It turned out not to have this benzheptoxdiazine structure after all; rather, it had structure **3-24** (Figure 3.5). The synthesis of this substance began with the Friedel-Craft reaction between two simple materials, chloroaniline and benzoyl chloride, forming **3-20**, whose oxime, **3-21**, produces **3-22** on reaction with chloroacetyl chloride. When **3-22** cyclized it formed not **3-23**, but **3-24**. (Apparently, the nitrogen of the oxime was more nucleophilic than the oxygen.) In any case, all but one single compound made by this route were tested and none showed any sedative effects, and so the project was abandoned. Almost 2 years later, Earl Reeder, a tidy fellow researcher with a sense for closure, decided to clean up the lab bench and noticed this last compound, which he sent in for testing. It turned out to have quite significant sedative and muscle-relaxant properties and a much lower toxicity than chlorpromazine.[62]

This last compound was the product formed in the reaction of methyl amine with **3-24** (shown in Figure 3.6 as **3-25**), and Sternbach and Reeder presumed that its structure was **3-28** (Figure 3.6). But when they studied the compound more closely they found this was not the case. Instead it had the completely unexpected structure **3-27**.[63] In another unexpected reaction, the methylamine had not displaced the chlorine to form **3-28**, as anticipated, but had attacked the carbon between the two nitrogens, opening the ring as

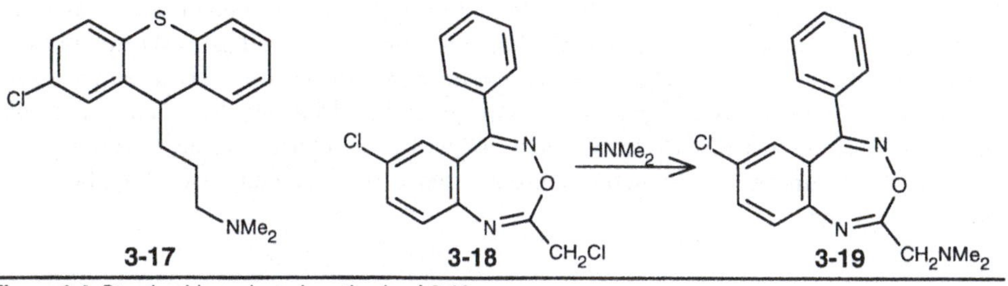

S
Cl
NMe_2
3-17
Cl
N
O
N
3-18
CH_2Cl
$HNMe_2$
Cl
N
O
N
3-19
CH_2NMe_2

Figure 3.4 Sternbach's projected synthesis of **3-19**.

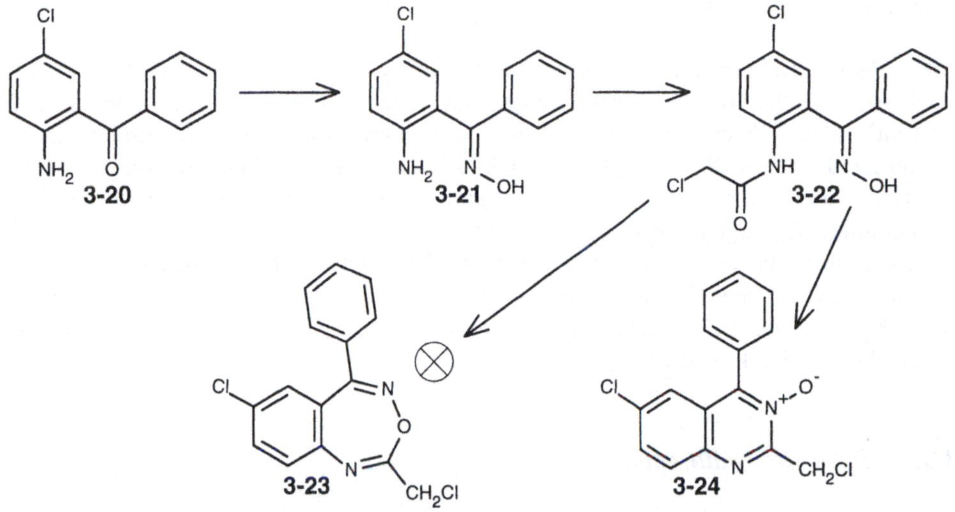

Figure 3.5 The actual outcome of Sternbach's synthesis.

in **3-26** and allowing it to close again as a seven-membered benzodiazepine ring, **3-30**, which was named *chlordiazepoxide*. When marketed as *Librium* in 1961, it became the first of many runaway best-sellers. Studying the chemistry of chlordiazepoxide further, they found that in two easy steps, a methyl group could be tagged on to the amide nitrogen and the oxygen of the *N*-oxide removed, providing what is probably the best known of all the BZDs and one of the most widely used drugs of all time, *diazepam (Valium*, **3-29**).

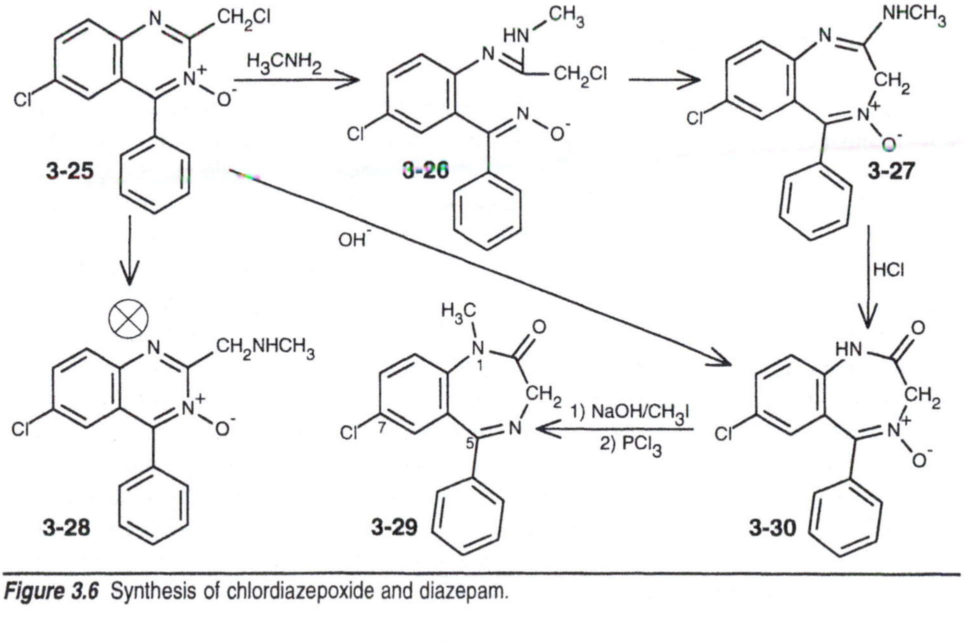

Figure 3.6 Synthesis of chlordiazepoxide and diazepam.

Benzodiazepines: Structure-Activity Relationships

Since the original synthesis of Librium and Valium in 1960, more than 3,000 other BZDs have been synthesized; hundreds of these have been tested for activity, and there are currently about 35 in use throughout the world. With this many structural variations synthesized and tested, the SAR of the BZDs is fairly well understood: For significant activity, there must be a small electron-withdrawing group at position 7, like the chlorine of Valium, **3-29**; large groups or electron-donating groups weaken potency, as do substituents anywhere else on this fused benzene ring. There must also be a phenyl ring, the so-called "pendant" phenyl, substituted at position 5. Antagonist properties develop when, as with *flumazenil,* the pendant ring is replaced with a keto function and a methyl group is attached to the 4 position.

Uses of the Benzodiazepines

BZDs have a number of uses, but they are most commonly prescribed as *hypnotics* or *soporifics* (drugs that induce sleep), or as *anxiolytics* (drugs that diminish anxiety). The most significant factor that affects the usefulness of one particular BZD over another for these two purposes (and one of the few effects significantly changed by structural modification of the basic BZD structure) is how long the effects of the particular drug last. Most people don't want to sleep much more than 8 hours a day, and a sleeping pill that induced a sound 30-hour nap, however restful, would not please most consumers. In turn, a major factor determining how long the effects of a given BZD last is the half-life of any active metabolites produced when it is ingested. Before we get into all this, we should make a brief excursion into that neglected third of most people's lives, sleep.

Sleep, an Altered State of Consciousness

It is doubtless a reflection on how vast is our ignorance of the functioning of the brain that we have no certain notion what physiological function is achieved by the third of our lives we spend in the very different form of consciousness we call sleep. Sleep is not a coma; we are not unconscious, and can be awakened. Many of us can awaken ourselves with a chronological precision that, were we capable of it when awake, would deal a fatal blow to the wristwatch industry. Few of us could sleep overnight on a narrow ledge thousands of feet up a mountainside. But this is the sort of thing mountaineers do when they do their thing; and only a few early years out of the crib, ordinary college students (at least when sober) can be trusted to fall sound asleep on a narrow bunk bed seven feet over a cement dormitory floor and not roll off in the night with quite likely fatal consequences. On the other hand, evolution seems not to have provided us with the ability to steer down the highway at 65 mph while sound asleep, and this aspect of our daily state of altered consciousness is thought by some to cause as many as half of all highway fatalities.[64] Dr. William C. Dement, director of the Stanford University Sleep Research Center and one of the pioneering researchers who discovered rapid-eye-movement sleep, feels that sleepy drivers should be treated with the censure presently reserved for drunk drivers. "We don't allow drunk driving," says Dr. Dement. "But if your family has been wiped out by someone who fell asleep, the damage is exactly the same."[65]

What is sleep, and why do we do it? If, in some Alice-in-Wonderland world, nobody slept but you, you would be considered to have an idiopathic, fortunately self-limiting but severely debilitating disease state. But everybody does it, so it must not be a disease. In fact, it is considered a disease (literally, a dis-ease) when you *can't* sleep. Indeed, insomnia is one of the most common ailments of mankind.

Abnormal Sleep: Parasomnias. Parasomnias are sleep disorders consisting of various unusual phenomena during sleep—talking, grinding the teeth (*bruxism*), walking around (*somnambulism*), urinating (*nocturnal enuresis*), night terrors (*pavor nocturnus*), etc. They are very common during early childhood, and relatively rare in adulthood. There may very likely be genetic factors or epilepsies involved with several of these disorders when they continue into maturity. Somnambulism can have some very dramatic consequences, including a few murders[66] committed during sleepwalking: one report of a 14-year-old boy who awakened at 2 o'clock in the morning and stabbed his 5-year-old cousin is accompanied by the admonition that "sleepwalking is not an hysterical condition . . . [and] can be accompanied by violent injury to the self or others."[67] People who walk in their sleep have no awareness of what they are doing and no recollection afterward. Some cases of "episodic nocturnal wanderings" which include unintelligible speech, screaming, and violent behavior have been shown to be associated with encephalographic evidence of epilepsy and to respond to anticonvulsant drugs.[68] In at least one case, an adult with a 30-year history of somnambulism and night terrors was successfully treated with *paroxetine*, one of the newer selective serotonin reuptake inhibitors usually used as an antidepressant (Chapter 5).[69] All this is fascinating and just goes to show how little we know and how much there is to be discovered about a phenomenon so common as sleep. It is quite likely that medications will play a part in providing help for these unusual conditions as understanding of them develops. Many, however, may remain so rare as to fall into the category of "orphan diseases." In any case, the one sleep problem for which BZDs and other hypnotics discussed in this chapter have been used with considerable success is the relatively normal disorder of normal sleep, insomnia—for which the one infallible cure, in the words of W. C. Fields, is to "go home and get a good night's sleep."

Normal Sleep in Absentia: Insomnia and the Benzodiazepines. Many physicians secretly side with W. C. Fields. It is claimed that few clinical disorders are more "casually and carelessly treated"[70] than insomnia. After all, no one will die of it. Yet the perceived lack of refreshing sleep plagues many people, and the business of providing these people with cures (usually only partially successful) profits almost as many more. Insomniacs have always been supplied with an abundance of free advice as well. Seventeenth-century author Richard Burton cites an ancient authority who tells us that profound sleep may be promptly induced by anointing "the soles of the feet with the fat of a dormouse, [and] the teeth with ear-wax of a dog."[71]

In recent times, sleep has been studied much more thoroughly, and electroencephalographic analysis has shown that there are several phases of sleep and a general sleep pattern, called sleep *architecture*, characterizing normal sleep. The two most important phases of sleep are probably the rapid-eye-movement (REM) sleep, which occurs most frequently during the last half of the night, during which dreaming takes place, and which is a relatively shallow phase of sleep as measured by metabolic rate; and slow-wave sleep, occurring in the first half of the night, and which, in terms of metabolic rate and the secretion of growth hormone, is the deepest phase. When REM sleep is deliberately and excessively curtailed, it causes irritability and a heightened sense of anxiety. Most hypnotic drugs decrease the proportion of REM sleep time (although BZDs do this much less than older classes of soporifics like the barbiturates), and when these drugs are discontinued, there is a measurable rebound in REM sleep. But how people feel about their sleep is not always well-correlated either with the general architecture of their sleep or with the REM sleep time that it can be shown they have experienced.[72] Thus, tolerance to such effects as shortened sleep latency (how long it takes to fall asleep) and lengthening of overall sleep time develops rapidly when people use BZDs for as along as a week. Nonetheless, in double-blind, placebo-controlled crossover administration of typical BZD hypnotics like *lormetazepam* or *nitrazepam* for as long as 24 weeks, according to the subjects themselves, the quality of their sleep was significantly improved when they used BZDs rather than placebo, and they did not perceive any diminution in the benefit derived from the drug over the course of the entire 24-week period, although they were acutely aware of the quality of their sleep deteriorating when the drug was switched with placebo at the 25th week.[73]

Instituting Vertigo in Daily Life. Of course, there are benefits to be derived from the creative use of almost any adversity. The late Romanian anti-philosopher and novelist E. M. Cioran, known for his elegant French prose style, was commended by the noted critic Edmund White as one who "has contemplated suicide for decades, esteems extremists, fanatics and eccentrics of all sorts and has instituted vertigo into his daily life." His first book, *On the Heights of Despair*, won the Prize of the Royal Academy; he later rejected two even more prestigious French literary awards, saying that "deep down, I have always sided with the Devil; unable to equal him in power, I have tried to be worthy of him, at least, in insolence, acrimony, arbitrariness and caprice." Whence the source of this world view? Insomnia. "It is much worse than sitting in prison. I went out of the house at about midnight or later and roamed through the alleys. And there were only a few lunatics and me, all alone in the entire city, in which absolute silence reigned. . . . My vision of things is the result of this years-long wakefulness. I saw that philosophy had no power to make my life more bearable. Thus I lost my belief in philosophy."[74] Had he but been provided with the fat of a dormouse, the wax of a dog's ear!

Endogenous Sleep Factors. As early as 1913, Pieron transfused cerebrospinal fluid from sleep-deprived dogs to the system of normal dogs and induced 6–8 hours of sleep; as a control, transfusion of fluid from normal, rested dogs to other normal dogs did not induce sleep. These experiments have been repeated and validated, and a delta sleep-inducing peptide (DSIP) has been isolated that consists of nine amino acids.[75] DSIP may interact with a number of other endogenous sleep factors, synergistically and antagonistically, to modulate the patterns of normal sleep.[76] And it may play a role in response to stress, perhaps by stimulating the release of met-enkephalin (Chapter 1).[77] In any case, the fact that DSIP and other endogenous sleep factors exist leads to the speculation that at least some insomnia may be due to an impaired ability to synthesize these factors in the CNS.

If the mechanism for their syntheses were understood, it might be possible to devise a drug that would enhance their production.

Cerebrodiene (Oleamide), Erucamide. More recently, a team of scientists at the Scripps Research Institute in San Diego has reported the isolation of a substance accumulating in the spinal fluid of exhausted cats.[78] Identification of the extremely minute amount of this substance was not easy, and at first it was thought to be an 18-carbon diene with a primary amine and alcohol functionality, a structure suggestively related to the widely distributed brain component sphingosine. The group provisionally named it "cerebrodiene."[79] But when the compound was finally unambiguously identified, it turned out to be an even simpler structure, *Z-9,10-octadecenoamide*; i.e. it is simply the amide of *oleic acid*, the main constitutent (as the triglyceride) of olive oil.[80] Synthesis of this compound was, of course, not at all difficult, and when it was injected into the peritoneum of well-rested rats, they showed all the signs of going into an entirely natural sleep, with a lowering of body temperature and REM cycling. One of the researchers, Steven J. Henriksen, says it "creates a natural deep sleep, but one from which the test animals could be aroused. They were not drugged."[81] The compound is probably one of a family of related structures found in the cerebrospinal fluid of cats, rats, humans, and probably all mammals; at least one other member of the family was isolated and identified as *Z-13-docosenoamide*, i.e. the amide of *erucic acid*, which is "Lorenzo's oil," the 22-carbon monounsaturated acid found in some forms of canola oil. Both of these structures bear some similarities to that of *anandamide*, the endogenous agonist at the cannabinoid receptors (Chapter 7), and to that of the newest class of soporifics, *zolpidem* (below).

Whatever the physiological function of sleep, we all experience its psychological necessity. In the words of the insomniac Macbeth: "Sleep . . . knits up the ravell'd sleave of care, / The death of each day's life, sore labour's bath, / Balm of hurt minds, great nature's second course, / Chief nourisher in life's feast."[82] Sleep does seem to knit up the mind overwrought and unravelled by care. Asleep, we are freed from care; and this is only appropriate since the non-sleeping hours of the day are mostly spent in what Heidegger would describe as the existential preoccupation of care, worry, or *Sorge*.[83] And this brings us, probably not coincidentally, to the other use of BZDs, as drugs that relieve the wearing chafe of anxiety.

Benzodiazepines as Anxiolytics

To be weighted with cares and responsibilities is an essential aspect of the genuinely human life; it would be irresponsible and subhuman to be totally, carelessly carefree. Exactly why this dimension of life so profoundly reveals the human must be left to the philosophers. But there can be too much of an otherwise good thing, and what is reasonable concern in some turns into malignant, tormenting anxiety in others. Probably there is a biogenetic component to this; in any case, for many people with this condition, the BZDs, presumably by altering the chemistry of the brain, have been found to ease the torment of the heart.

Speaking in the terminology of clinical psychology, BZDs are useful in relieving what is referred to as generalized or situational anxiety, which in turn is to be distinguished from psychosis or clinical depression, for which the BZDs are not helpful (but for which the antipsychotics and antidepressants are indicated, as we will see in Chapter 5). Of course, anxiety can be mixed with depression, and diagnosis may not always be easy. It has been said that anxiety has been treatable since the discovery of alcohol, and the BZDs have been called "alcohol in pill form." Both of these remarks reflect the similarities in use and abuse of the BZDs and alcohol, and concern *(Sorge?)* has been voiced in recent years that these drugs are overprescribed, especially to certain portions of the population, such as women and the elderly. However, the counterclaim from careful analyses of prescription practices is that anxiolytics are actually employed quite moderately, or even—like medical practice with regard to opioids—underutilized. The fact that more women than men use BZDs may reflect a male's societal reluctance to admit to needing help and his greater likelihood to flee to the neighborhood bar to relieve his anxieties. "At a tavern," exults the Great Lexicographer, "there is general freedom from anxiety. You are sure you are welcome: and the more noise you make, the more trouble you give, the more good things you call for, the welcomer you are. . . . There is nothing which has yet been contrived by man, by which so much happiness is produced as by a good tavern or inn."[84] (Of course, women could not enter taverns in 1776 so readily as they can grace the palmy brie-courts of today.) A further reason why women account for more BZD prescriptions than the hirsute sex is because they live longer and because difficulty in sleeping, for which BZDs are also prescribed, becomes much more common as we advance in age.[85]

Benzodiazepine Metabolism and Duration of Action

We return now to a discussion of the relative duration of action of the various BZDs in current use. It is obvious that how long a given BZD remains active after it is taken will greatly affect its utility as a sleeping aid, when more brief action is desired, or as an anxiolytic, when a more constant effect is required. In most cases, the overall duration of action of the BZDs depends significantly on the biotransformations the drug undergoes once ingested.

As can be seen from Table 3.1, many of the BZDs in common use have metabolites that themselves are active agonists at the BZD receptor and have quite significant half-lives. The half-lives given here are rounded averages for healthy young male adults; it has been shown that these times vary considerably among individuals and usually are very greatly lengthened in elderly patients. There have been cases reported where metabolism is so slow in older persons that what seemed to be a safe daily dosage (or what was a safe dosage when the patient began taking the drug on a daily basis many years before at a much younger age) has accumulated active metabolites until the person suffered toxic effects. In the case of BZDs, these effects (inattentiveness, somnolence, memory loss) easily can be mistaken for signs of senility.

Metabolic Transformations of the BZDs. The biotransformations indicated in Table 3.1 are generally quite similar and can be exemplified by the metabolism of chlordiazepoxide, **3-31** (Figure 3.7). The first step in the metabolism of most BZDs is the removal of any alkyl group attached to either the N-1 atom or, in the case of chlordiazepoxide, the methyl group of the methylamino substituent, forming *desmethylchlordiazepoxide*, **3-32**. The oxide and amino groups are then successively plucked off forming *demoxepam*, **3-33**, and *desmethyldiazepam* (also known as *nordazepam*), **3-35**.

Table 3.1 Benzodiazepines with active metabolites[86]

BENZODIAZEPINE	ACTIVE AGENT(S)	HALF-LIFE (HOURS)
Quazepam (Doral)	Quazepam Oxoquazepam Desalkylflurazepam	40 40 75
Diazepam (Valium)	Diazepam Desmethyldiazepam (Nordazepam) Oxazepam	45 75 7
Flurazepam (Dalmane)	Desalkylflurazepam 3-Hydroxydesalkylflurazepam	75 15
Halazepam (Paxipam)	Halazepam Desmethyldiazepam (Nordazepam) Oxazepam	15 75 7
Prazepam (Centrax)	Desmethyldiazepam (Nordazepam) Oxazepam	75 7
Clorazepate (Tranxene)	Desmethyldiazepam (Nordazepam) Oxazepam	75 7
Chlordiazepoxide (Librium)	Chlordiazepoxide Desmethylchlordiazepoxide Demoxepam Desmethyldiazepam (Nordazepam) Oxazepam	10 40 > 20 75 7

A glance at Table 3.1 will show that **3-35** is the common intermediate in the metabolism of diazepam, halazepam, and prazepam as well. It is also formed by decarbox-

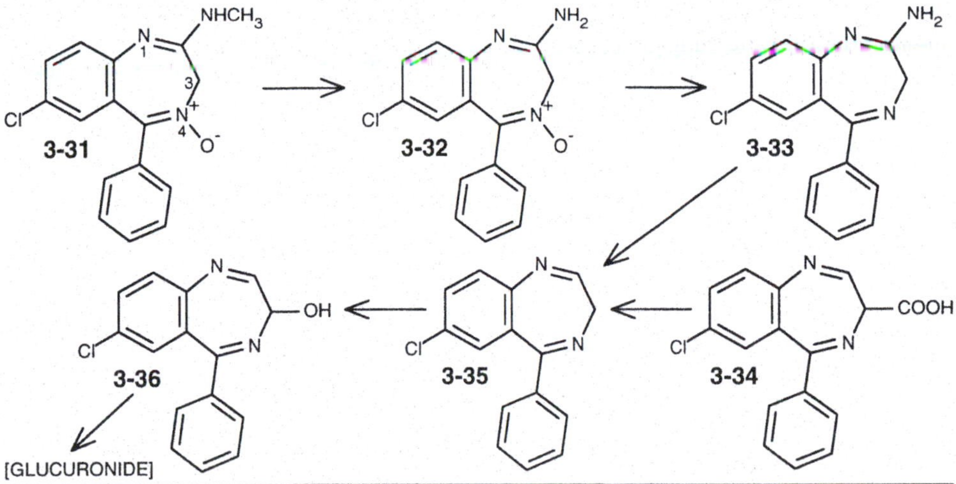

Figure 3.7 Biometabolic transformations of the BZDs.

ylation of clorazepate, **3-34**. In fact, clorazepate is really a prodrug, since this reaction occurs upon contact with the acid in the stomach, so that what is actually absorbed into the system is not clorazepate but nordazepam. The final metabolite of all of these drugs, oxazepam, **3-36**, is eventually excreted as a glucuronide after conjugation to glucuronic acid by its OH handle.[87]

The net effect of all this is that a drug like chlordiazepoxide, when administered to a given patient, will result in the perhaps simultaneous presence of four other active BZD entities in addition to itself. Since the half-life (or rate constant) of each of the transformations shown in Figure 3.7 can vary widely from individual to individual, the potential for misdosage or overdosage would seem to be considerable. Nonetheless, most often no such problems are encountered: a person desiring anxiolytic treatment is given a starting small dose of one of the longer-acting agents, and the dosage is slowly increased over a period of a few weeks until some satisfactory pharmacokinetic steady-state equilibrium is reached.

But the possibility of variation of the BZD structure is considerable (as can be seen from Figures 3.8 and 3.9) and a wide variety of other BZD drugs have been synthesized and marketed that are rapidly hydroxylated and conjugated to glucuronic acid directly after a half-life of a few hours. Two of these, oxazepam and nordazepam, are metabolites in the pathway we have just discussed. The most common of the other shorter-acting BZDs are given with their half-lives in Table 3.2.

It should be said that an "elimination half-life" is generally much longer than the duration of the perceptible effects of these drugs, and those which appear to be most successful in providing a good night's sleep with minimal side effects are the drugs with half-lives of 10-15 hours.

Problems with the Benzodiazepines. The controversy about BZD abuse liability is compounded when they are taken by persons who suffer from a sort of chronic anxiety syndrome that is not attributable to any crisis in their lives but is more or less an intractable feature of their personality. The anxiety they experience when they stop taking BZDs is not so much a withdrawal syndrome as the recurrence of their problem, and it does not seem suprising that they feel the need to use BZD on a daily basis. Again, we are up against cultural value judgments: most red-blooded Americans feel that people should not have to take drugs because they are anxious, even if the anxiety is debilitatingly painful.

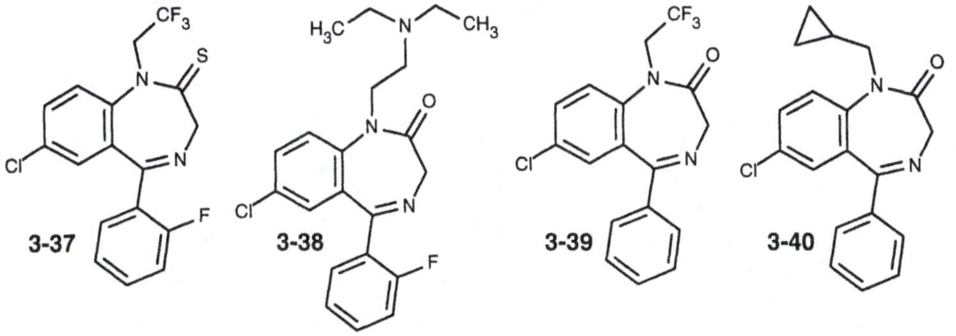

3-37 **3-38** **3-39** **3-40**

Figure 3.8 Quazepam, flurazepam, halazepam, prazepam.

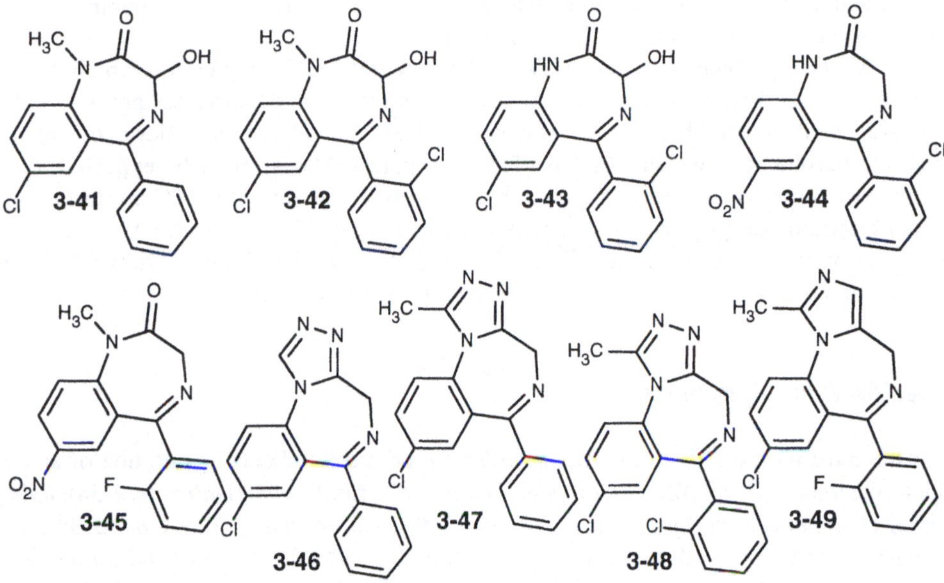

Figure 3.9 Temazepam, lormetazepam, lorazepam, clonazepam, flunitrazepam, estazolam, alprazolam, triazolam, midazolam.

Either they are "really" drug addicts or they are just "worrywarts." The proper cure for their problem is simply to stop worrying, you see.[88]

Table 3.2 Benzodiazepines with short half-lives[89]

Benzodiazepine	Half-life (hours)
Clonazepam (Klonopin)	25
Lorazepam (Ativan)	15
Estazolam (ProSom)	15
Temazepam (Restoril)	13
Alprazolam (Xanax)	10
Flunitrazepam (Rohypnol)	10
Oxazepam (Serax)	7
Triazolam (Halcion)	3
Midazolam (Versed)	1

On the other hand, there is a certain "buzz" from some of the short-acting BZDs that some people do seem to enjoy, but this seems to be a problem mostly with those few who will try anything for its psychic effect. In any case, in the United States, all BZDs are Schedule IV drugs, and in some states they are Schedule II. In other countries, some short-acting BZDs are sold OTC for insomnia.

Another problem with the BZDs is the synergetic CNS depression which can be produced when they are used with alcohol; although this combination has proved fatal if the BZD is taken in deliberate massive overdose, BZDs are many times safer in this regard than the barbiturates, which they have largely supplanted. As might be expected, there is likewise a synergetic sedation caused by taking BZDs along with OTC antihistamines or antihistamine sleeping aids, and there is some antagonism of sedative effects when BZDs are taken with caffeine or theophylline. Additionally, a variety of other drugs (Tagamet, oral contraceptives) prolong the elimination half-life of the BZDs.

Specific Benzodiazepines

For the most part, the uses of particular BZD put primarily reflects its duration of action. Thus, the longer-acting BZDs, *quazepam (Doral,* **3-37**), *flurazepam (Dalmane,* **3-38**), *halazepam (Paxipam,* **3-39**), *prazepam (Centrax,* **3-40**), *chlorazepate (Tranxene,* **3-34**), and *clonazepam (Klonopin,* **3-44**) (Tables 3.1, 3.2; Figures 3.7, 3.8, 3.9) , are used for the same purposes as diazepam, the prototypical BZD: generalized anxiety disorders, alcohol withdrawal, and insomnia, particularly insomnia associated with anxiety. *Chlordiazepoxide (Librium,* **3-31**) is thought to be more effective in relieving anxiety than diazepam while exhibiting less pronounced anticonvulsant and muscle relaxant activity.[90] *Lorazepam (Ativan,* **3-43**), which has a half-life of intermediate duration, is used orally in panic disorder and parenterally to suppress status epilepticus and the nausea of cancer chemotherapy procedures.[91] Most of the shorter-acting BZDs of Table 3.2, *estazolam (ProSom,* **3-46**), *temazepam (Restoril,* **3-41**), *oxazepam (Serax,* **3-36**), and *triazolam (Halcion,* **3-48**) are used almost exclusively as hypnotics. Some of the BZDs have more particular histories, uses, or effects, and we will consider these in the paragraphs that follow.

Diazepam. Diazepam (Valium, **3-29**) is the most widely used BZD. It is prescribed primarily as an anxiolytic, but it is also used as a sleeping aid, as an antiemetic, as an intravenous anticonvulsant, and as a muscle relaxant (after athletic injuries, for spasms in neurological disorders, and to relieve urinary retention after anesthesia). It traditionally has been used prior to endoscopy or surgery because it relieves anxiety and apprehension on the part of patients and diminishes their recall of the procedure. This phenomenon is termed *anterograde* amnesia (because it induces a forgetfulness for events following its administration) as opposed to the more familiar notion of *retrograde* amnesia (the forgetting of events prior to an injury typical of some concussions). Because lorazepam and particularly *midazolam* (below) are even more effective in this regard, they are supplanting diazepam for this use. Diazepam is also widely employed to alleviate the abstinence syndrome during alcoholic withdrawal and to suppress delirium tremens.

Alprazolam. Alprazolam (Xanax, **3-47**) seems to be almost unique in its utility for the treatment of agoraphobia and panic disorder; unlike most BZDs, it probably also has some antidepressant activity. However, withdrawal reactions are potentially severe, ranging from mild dysphoria and insomnia to a major syndrome that may include abdominal and muscle cramps, vomiting, sweating, tremors, and convulsions. For patients receiving large doses of alprazolam for long periods, an 8-week tapering period with a gradual dosage reduction has been recommended.

Flunitrazepam. Flunitrazepam (Rohypnol, "rophies, roofies, roach, rope," **3-44**) is a sleeping pill manufactured by Hoffmann-La Roche in Columbia and sold on prescription in Europe and South America, but not in the United States. Large quantities appear to be readily divertable from Columbia or from Mexico, where some pharmacies sell it OTC; and it is smuggled into Florida, Louisiana, and Texas, where it has sounded alarms. Jim Shedd, spokesman for the DEA in Miami, has called it the "Quaalude of the 90s." He says that several batches of 4,000-6,000 Rohypnol tablets have been confiscated in Florida, where they sell on the street for about $5 each. "They take a pill and they drink a beer and they're high for an entire day," said police Sgt. John Johnston of Coral Gables. "You feel like [sic] you could do anything," says Cristian, an 18-year-old Columbian. "You could have thousands of problems and it doesn't matter." Cristian was up to 10 pills a day, and is now in rehab in Bogota. Some Columbian youths prefer Rohypnol to cocaine, which is even easier to obtain there, because "roofies" (as they are called in Florida) provide a longer, more soothing high.[92]

By late 1995 , there were several rape cases in Broward County, FL, where men allegedly slipped the drug to women in their drinks and then sexually assaulted them. The internet was abuzz with similar stories, and concerned for its reputation, Hoffman-La Roche began an epidemiological study in Texas to examine why Rohypnol was being abused.[93] Indeed, the actual pharmacology of the drug would appear to be virtually identical to many of the shorter-acting BZDs legitimatly prescribed in the United States. But in March, 1996, Treasury Secretary Robert E. Rubin banned the importation of Rohypnol, claiming that it was "a growing threat to teen-agers and young adults and has no legitimate therapeutic use."[94]

In a survey of nearly 1,000 heroin addicts undergoing detoxification in Barcelona, about 80% claimed a history of BZD usage starting after their heroin habit began. Asked to express their preferences, 68% of the addicts "voted for"flunitrazepam, with clorazepate and diazepam tied for a distant (13%) second place.[95] In England, it has become popular to snort flunitrazepam; and in a placebo-controlled study there using 20 healthy volunteers asked to snort the drug, all reported sedation but no unpleasant side effects. They all "liked the drug effects," and their subjective ratings of the dosage they were given correlated accurately with the levels of the drug in their plasma and their fondness for the effects.[96]

Triazolam. Evaluation of triazolam (Halcion, **3-48**) for usefulness as a soporific in a sleep laboratory showed that with regard to both sleep induction and maintenance it was superior to flurazepam; however, after 2 weeks of use, all sleep efficacy parameters were close to what they had been initially, indicating significant tolerance had developed. Upon withdrawal, there was a considerable worsening of sleep function. These characteristics may be typical of most BZDs used for long-term hypnotic effect. But, as we have seen, patients may nonetheless subjectively feel that their sleep is improved despite the laboratory findings.

There have been frequent complaints from the European market as well as the United States[97] that while Halcion induces sleep more rapidly and leaves one more alert the following day, it also can cause amnesia, anxiety, and idiosyncratically hostile outbursts (perhaps by a process similar to which excess alcohol induces in some people a "mean drunk" syndrome). A San Francisco novelist named Cindy Ehrlich wrote a 1988 article for a California magazine claiming Halcion made her "convinced that the world was on the brink of nuclear war or invasion from space." While taking Halcion, Mrs. Ilo Grundberg killed her 83-year-old mother and at the murder trial, two physicians testified that triazolam had contributed to her violent behavior. She filed a $21 million suit against Upjohn in 1989; the case was settled out of court in 1991.[98] Testifying for Mrs. Grundberg was Scottish physician and sleep researcher Dr. Ian Oswald, who with his wife, Dr. Kirstine Adam, has crusaded against Halcion.

William Styron, in his book *Darkness Visible: A Memoir of Madness,* ascribes his suicidal depression to use of Halcion.[99] And at least two doctors have described their own dramatic memory lapses upon one-time use of Halcion: Dr. Joseph Mendels, director of the Philadelphia Medical Institute, said he took the drug only one time the night before an important meeting. "The next thing I knew it was early afternoon the following day. I found myself at the train station in Washington. I apparently got up, shaved and got dressed and went to the meeting. But I have no knowledge of doing any of these things," he says.[100] A similar story is told by Dr. Mark Silverman of Mercy Hospital in Pittsburgh: "One Sunday afternoon in June 1987 I found myself sitting at my parents' home at 1 p.m., with no recollection of how I had arrived there. I was unable to remember anything that had happened after I went to bed the night before at my own house 3 miles away. I had not consumed any alcohol, and I was a healthy 30-year-old nonsmoker not taking any medication." He finally recalled having had insomnia the night before and taking one 0.125-mg tablet of Halcion; his family told him that he had eaten breakfast, driven to his parent's house, and acted in every way as he usually did, but he had almost no recollection of any of his actions.[101]

In April 1991, a study of next-day memory impairment with use of triazolam versus temazepam and placebo was reported in *The Lancet.* In this double-blind parallel-group study, it was found that "impairment of delayed recall was significantly and several times greater" with triazolam than with temazepam or placebo, and the memory impairment and amnesia tended to increase with continued use. The authors concluded that "cognitive impairments associated with triazolam probably represent a spectrum of organic brain dysfunction, with memory impairment/amnesia and confusion being the commonest, and milder manifestations and hallucinations and delusions the more severe and less common, features." Nonetheless, the subject remains controversial, and other studies have shown that triazolam is no more likely to induce memory lapses, daytime anxiety, or rebound insomnia than any of the other BZDs.[102] It should also be pointed out that people experiencing insomnia occasionally have spontaneous episodes of amnesia like those described above even though they are taking no drug whatsoever; obviously, a person who had experienced such a memory lapse and had also taken some medication would quite naturally ascribe the memory lapse to the medication.[103]

In no small part because of the efforts of Drs. Oswald and Adams, the sale of Halcion was suspended in Britain in October 1991, the government citing potentially dangerous side effects including memory loss and depression. The British Government accused Upjohn of misrepresenting these potential side effects. Finland and Argentina joined in suspending licensing of the drug.[104] In November 1991, the Upjohn company announced that it would provide stronger written warnings and new packaging for the more than 90 countries in which it is still prescribed. Over the years since the first problems with dosage arose in the Netherlands in 1980, Upjohn has lowered the recommended dosage: from 1.0 to 0.5 to 0.25 and finally to 0.125 mg; and after numerous appeals in the United Kingdom, in 1993 Upjohn won the approval of the Medicines Commission for the 0.125-mg dosage.[105] Upjohn also sued the BBC for allegedly libelous statements in its critical *Panorama* documentary program "The Halcion Nightmare"; they likewise sued Dr. Oswald for his statements to the *New York Times*. The company won significant damages but was also rebuked by the High Court judge for being "at least reckless . . ." and for their "serious errors and omissions."[106]

In May 1996, a task force set up by the United States Food and Drug Administration recommended that the Justice Department assess whether the Upjohn Company (now known, after a merger, as Pharmacia and Upjohn) committed any crimes by failing to report serious side effects caused by Halcion.[107] Because of the continued controversy over the drug (as well as the expiration of patent protection), worldwide sales have slumped from a high of $237 million in 1991 to about $100 million in 1995.[108]

Synthesis of Triazolam. The last four structures in Figure 3.9 contain an extra heterocyclic ring fused to the BZD structure, and it might be thought that this would significantly change their pharmacologic profile. In actuality, the chief effect is to make these BZDs capable of very rapid metabolism and elimination. The way this extra ring is tacked on to the structure is fairly simple and can be exemplified by the synthesis of triazolam, shown in Figure 3.10. The intermediate **3-50** can be synthesized easily in a manner analogous to the synthesis of diazepam; treatment with P_2S_5 converts this lactam to the thiolactam, **3-51**, which is condensed with acetyl hydrazide to give triazolam, **3-52**.

Midazolam. Midazolam (Versed, **3-49**), is available in the United States for IV injection, its aqueous solubility making it particularly suited for this. It is used for induction of anesthesia because of its short half-life and quick action. After IV injection, a subject will become unconscious in about 80 seconds and recover in about 5 minutes.[109]

Another desirable feature of midazolam is its reliable induction of anterograde amnesia. The drug is used alone to provide a conscious ("basal") sedation for unpleasant diagnostic procedures such as endoscopies; patients become cooperative with the medical team during the procedure and conveniently incapable of recalling afterward what was done to them. Midazolam may also find use in the treatment of status epilepticus.

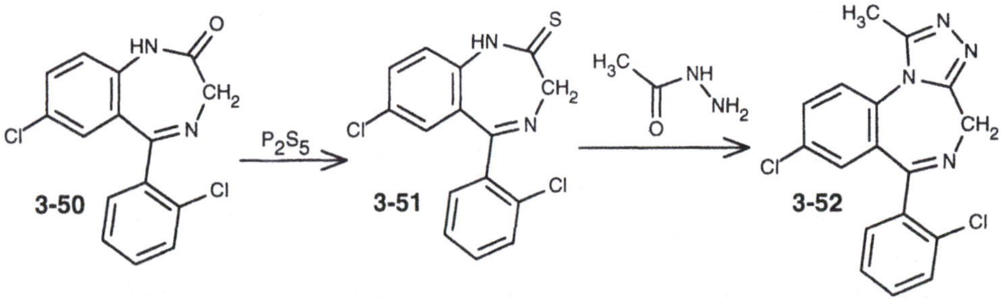

Figure 3.10 Synthesis of triazolam.

The Benzodiazepine Site on the GABA Receptor Complex

Extensive research on the GABA receptor system has been motivated in no small part by the realization that the BZDs produce their effects by binding to GABAergic receptors. There are two major types of GABA receptor, $GABA_A$ and $GABA_B$; by far the most important in the mammalian brain are the $GABA_A$ receptors, which are responsible for most of the inhibitory synapses in the CNS. It is this class of receptors at which the BZDs and barbiturates (as well as some steroids, general anesthetics, and probably alcohol) are active. There are five subunits constituting the GABA receptor complex, symmetrically arranged around a central ion channel selective for the chloride ion. There are two sites on the complex at which GABA binds; when both are occupied, the ion pore opens to allow more chloride ions into the cell, causing depolarization (inhibition of neuron firing). There are three other sites on the same complex; a barbiturate site, a steroid site, and the BZD site. The BZD site itself can be occupied in three (probably overlapping) ways by BZD inverse agonists, BZD antagonists, or BZD agonists (which includes all of the BZDs we have considered thus far). Binding of a BZD agonist such as diazepam to the BZD receptor will not cause the ion channel to open by itself unless GABA is also bound to the complex, but it will induce the channel to open more frequently for a given occupancy level of GABA than would otherwise be the case. Agonist binding at the BZD probably affects the GABA binding by causing an "allosteric modification" (a distortion in conformational stereochemistry) of the GABA site that makes the GABA binding more effective.[110] Conversely, binding of inverse agonists reduces the effectiveness of GABA binding, while binding of antagonists simply blocks access to the site, rendering both agonists and inverse agonists ineffectual.

Since the BZD site is so specific and ubiquitous in the CNS the question naturally arises: what was its function there all those millennia before Sternbach stumbled on the BZD structure? There is some evidence for an endogenous BZD agonist found in the body fluids of patients with hepatic encephalopathy;[111] however, the source of some of this (probably chlorinated) substance may be the food chain. (BZDs are probably not as "synthetic" as might be thought; many of them, including diazepam itself, seem to occur naturally in small amounts in plant and animal sources.)[112] There is no dispute, however, about the existence of endogenous *inverse* agonists called GABARINS (GABA receptor inhibitors) at the BZD site. One has been purified and its structure determined: a peptide termed DBI (diazepam binding modulator), which seems to be elevated in the spinal fluid of severely depressed persons.[113] As this and other structures have been discovered that do not have the BZD structure but are active at the BZD site, it has seemed appropriate to refer to this site in a more neutral way. They are now often called *omega* receptors, type I, type II, etc.

Non-Benzodiazepine Benzodiazepine-Receptor Agonists

There are several newer drugs that while not BZDs, act on the $GABA_A$ receptor complex through subtypes of the BZD site. *Zolpidem* (Stilnox, Ambien, **3-53**, Figure 3.11) binds preferentially at a BZD receptor site called Ω-1. Unlike traditional BZDs such as Valium, it is not a muscle relaxant, anxiolytic, or anticonvulsant; but it seems to induce deep sleep

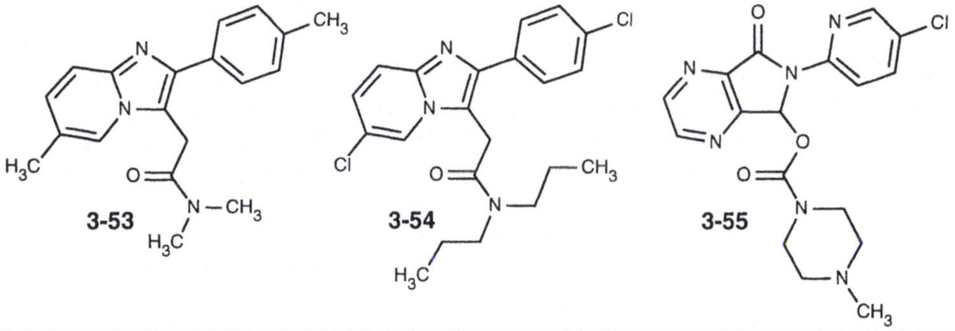

Figure 3.11 Some newer "non-benzodiazepine" hypnotics.

for a relatively short time, with little distortion of sleep architecture. In a number of double-blind, placebo-controlled trials, it was found to be effective in improving sleep for as along as 4 weeks. It has been approved by the FDA for use in the United States and is, like the BZDs, a Schedule IV drug.[114] The remaining two drugs are used outside the United States. *Alpidem* (Ananxyl, **3-54**) has both anxiolytic and anticonvulsant properties, and is also selective for the Ω-1 modulatory site of the $GABA_A$ receptor.[115] *Zopiclone* (Amoban, **3-55**) binds to yet another, "type III" receptor site that BZDs do not bind to, and seems to induce a conformational change at the BZD site; it is a strong anxiolytic and hypnotic that creates neither hangover nor rebound insomnia[116] on withdrawal.[117] Despite all these seeming improvements, a large study (funded, however, by Upjohn, the makers of triazolam, and not double-blinded) seemed to show that there were no differences between zopiclone, zolpidem, midazolam, brotizolam, temazepam, lormetazepam, and loprazolam compared with triazolam.[118] As for its abuse liability, it seems that baboons at least find it very similar to the frequently abused barbiturate pentobarbital and self-inject it in preference to triazolam.[119]

BZD Inverse Agonists, Partial Inverse Agonists, Partial Agonists, Antagonists

As previously mentioned, there are drugs (most of which have a β-carboline structure, like *DMCM*, **3-56**, Figure 3.12)[120] that are *inverse agonists* and have effects opposite to those

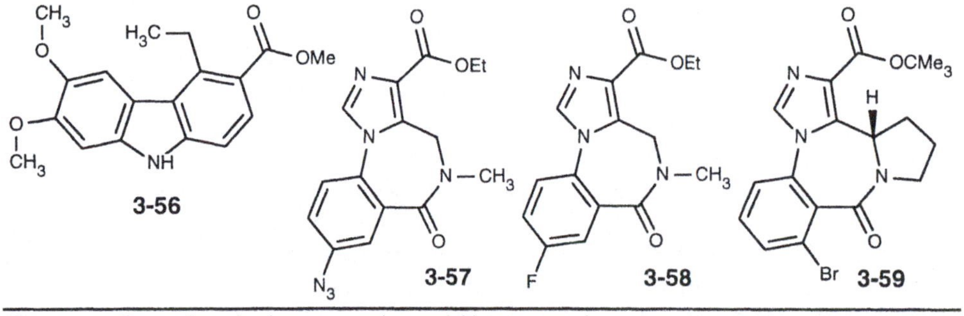

Figure 3.12 Benzodiazepine inverse agonist, partial inverse agonist, antagonist, partial agonist.

of the BZDs. Substances like DMCM induce anxiety and epileptic seizures in experimental animals. There is also evidence in animal models that at least one β-carboline (methyl β-carboline-3-carboxylate, β-CCM) provokes an opposite effect to the anterograde amnesia caused by BZDs: this β-carboline actually enhances animal performance in three different tasks used to probe memory and learning.[121] (Whether this effect, which would seem advantageous for those of us who exhibit a history of endogenous anterograde amnesia, will ever find use outside the rat-and-mouse kingdom remains as yet quite uncertain.)

Removal of the pendant benzene ring found in all the BZD agonists (along with several other minor modifications) produces structures with inverse agonist or antagonist properties. One of the most intensely studied of these is *Ro 15-4513*, **3-57**, which is described as a BZD partial inverse agonist. It seems to have a quite significant effect in blocking the ability of ethanol to disrupt motor coordination and release punished behavior. These terms come from animal studies: their correlatives in human behavior would presumably be those effects of alcohol that make driving hazardous and cause people to act with little guidance from their "conscience" or other higher mental centers. Needless to say, the development of a "drink and drive" drug based on these effects would be fraught with consequences unthinkable. As Rall states, use of this drug by human drinkers would "thoroughly confuse the medicolegal relationship between concentrations of ethanol in the blood and degree of impaired driving performance."[122]

A variation on the structure of RO 15-4513 provides *flumazenil* (Anexate, Lanexat, **3-58**), a BZD antagonist that has been marketed as a prescription drug by Hoffmann-LaRoche: flumazenil is useful as a fast-acting antidote in the treatment of BZD overdose and in reversing the CNS sedative effects of BZDs during anesthesia. Initial studies of this compound with laboratory animals seemed to indicate that, like RO 15-4513, it not only reversed the effects of BZD but also some of the sedative and soporific effects caused by hepatic failure and by alcohol.

Bretazenil (RO16-6028, **3-59**) which is a partial agonist, in a double-blind randomized crossover study conducted in 28 male volunteers produced less psychomotor and memory impairment than diazepam or alprazolam and, while it was clearly more euphorigenic than placebo, this effect had a "plateau" response—that is, it was *not* dose dependent, as was the case with diazepam and alprazolam. Hence, it is likely to be less abuse liable.[123]

Azaspirone Anxiolytics. *Buspirone* (BuSpar, **3-60**, Figure 3.13) is the first antianxiety drug of non-BZD structure to be developed for clinical use in the United States since the BZDs were introduced in 1958. It is as effective as diazepam or other BZDs, but produces less sedation, lethargy, and depression than do the BZDs.

Buspirone does not act on the GABA receptor. Its anxiolytic effects are probably due to its partial agonist activity at one or more classes of serotonin receptors, although it also shows some weak activity at dopamine receptors. Unlike the BZDs, buspirone is ineffective as a hypnotic, anticonvulsant, muscle relaxant, or anti-panic medication, and it does not synergize the CNS depressant actions of alcohol or the barbiturates. It has been particularly recommended for the management of anxiety in the elderly, where the CNS depressant actions of the BZDs can be more problematic. Unlike the BZDs, but like the antidepressants (whose effects are also attributable to action at the serotonin receptors; see Chapter 5), buspirone must be given for 1 or 2 weeks before the antianxiety effect is

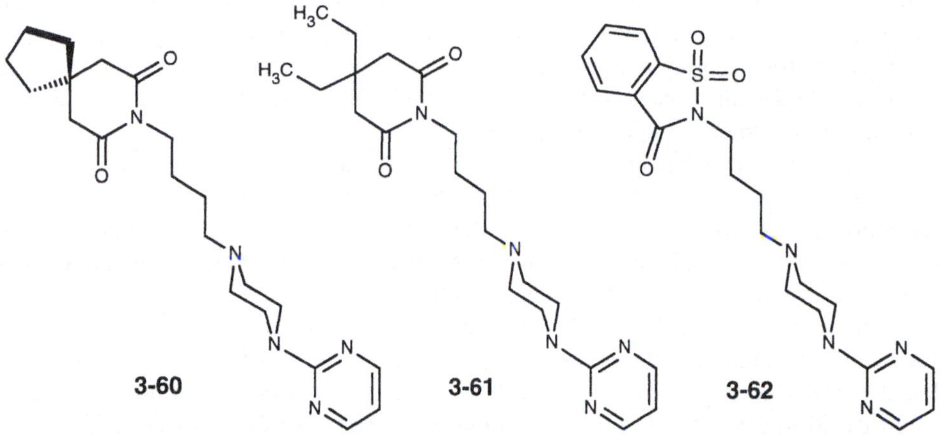

3-60 **3-61** **3-62**

Figure 3.13 Three azaspirone anxiolytics.

noticed. The dependence liability of buspirone seems to be negligible; indeed, in large doses it probably is dysphoric. Nor does it appear to affect motor skills such as driving a car.

The observant reader may have noticed that the structure of buspirone is yet a further variation on the haloperidol-fentanyl theme discussed in Chapter 2; indeed, buspirone was originally developed as a potential antipsychotic agent like haloperidol. It is impressive that the exploration of this path has produced antidiarrhetics, opioids, antipsychotics, and now anxiolytics. Two other drugs with similar structure, *gepirone*, **3-61**, and *ipsapirone*, **3-62**, are in late-stage clinical trials and may soon be introduced in the United States.[124] (Buspirone, gepirone, and ipsapirone are collectively denominated the "azaspirone" anxiolytics, but purists may wish to quibble that only buspirone has a true "spiro" structure—which refers to the joining of a five- and six-atom ring by one atom, as in the top left corner of **3-60**.)

Barbiturates

Discovery

The discovery of the medically useful properties of the barbiturates in the early 1900s marked a milestone in the progress of medical pharmacology. At the time of their discovery, the only effective drug for inducing sleep (other than the opioids or alcohol) was chloral hydrate—the "hydrate" of trichloroacetaldehyde, i.e., trichloroethane-1,1-diol, or $Cl_3CCH(OH)_2$—a harsh and foul-tasting compound that first had been used in 1869. It is still used in some hospital settings despite its high toxicity and tendency to irritate the stomach; must be credited with being cheap, effective, and fast-acting—whence its reputation as the "Mickey Finn" knock-out drops of the 1930s gangster films. Barbiturates soon became best-sellers as soporifics and remained so for a half-century until they were displaced by the BZDs. More significant than their use as sleeping aids was the discovery that the longer-acting barbiturates like phenobarbital could suppress seizures in many epileptics. In this they were the first miracle drugs of their day, since they made a normal life possible for many epileptics. Again, it was more than 50 years before antiseizure medications other than phenobarbital became available, and phenobarbital still finds use today in the treatment of epilepsy.

The barbiturates are all derivatives of a simple six-atom cyclic structure, *barbituric acid* (**3-65**, Figure 3.14), which is easily made by condensing *urea*, **3-63**, and *malonic ester*, **3-64** $\mathbf{R_1=R_2=H}$). Barbituric acid itself, however, has no sedative or hypnotic properties. Nor was it originally made by this simple process, but in a complex series of studies carried out by Johann Friedrich Wilhelm Adolf von Baeyer in 1863, at the very dawn of synthetic organic chemistry. Little was understood at this time of the nature of organic chemical structures, and the 40 or so pages of Baeyer's original article describe only lengthy purification procedures and tedious elemental analyses resulting in uncertain empirical formulas. Baeyer, Liebig, and the other chemists of the day could only name the substances they isolated by allusion to their properties or the process of their discovery. Thus, Baeyer discusses three chemically related acids in his paper: *Violursäure, Dilitursäure,* and *Barbitursäure.* Strange as it may seem, the origins of these words are probably the characteristic violet color of the first acid and its salts, the deletion of this color when it is transformed into dilituric acid, and the fact that barbituric acid provides the "key" (*Schlüsselbart* or *Bart* in German, but *barba* in Latin) to the nature of the other two acids: namely, that violuric acid is a nitroso-substituted barbituric acid and that dilituric acid is the oxidation product of violuric acid. It was not until 1904 that the work of Emil Fischer raised barbituric acid from utter obscurity to become the parent of a series of famous and profitable drugs. In 1905, at the age of 70, Baeyer received the Nobel prize for chemistry, and there are several oral traditions claiming that, when asked why he named structure **3-65** barbituric acid, he said it was because he had been enamored of a lass named Barbara. But he was quite likely joshing.[125]

In any case, if von Baeyer first made barbituric acid in 1863, it was von Mering, Emil Fischer, and Fischer's nephew Alfred Dilthey[126] who developed an effective process for synthesizing the barbiturates; and the first of these, *barbital* (barbitone, Veronal, **3-66**) was promptly marketed as a sleeping agent in 1904 by F. Bayer (no relation to A. Baeyer) and Company of Elberfield, Germany. This was the same company, originally a dyestuffs manufacturer, whose pharmaceutical fame and fortune were already world-renowned from

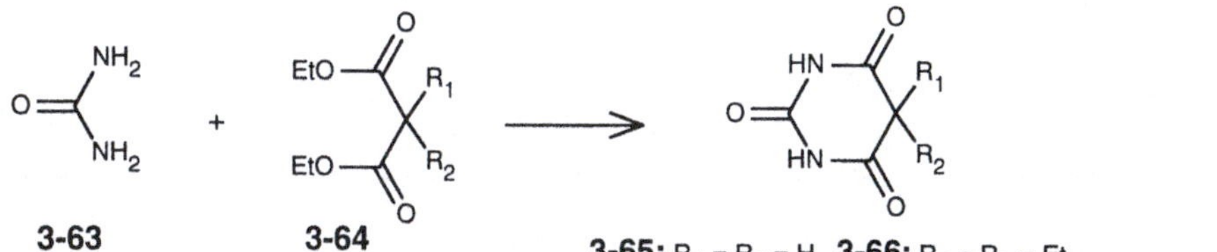

3-63 **3-64**

3-65: $R_1 = R_2 = H$ **3-66:** $R_1 = R_2 = Et$
3:67: $R_1 = Et; R_2 = Ph$
3:68: $R_1 = Et; R_2 = \text{-}CH_2CH_2CHMe_2$
3:69: $R_1 = Et; R_2 = \text{-}CHMeCH_2CH_2CH_3$
3:70: $R_1 = \text{-}CH_2CH{=}CH_2; R_2 = \text{-}CHMeCH_2CH_2CH_3$

Figure 3.14 Some more important barbiturates.

their introduction only a few years previously of Aspirin and Heroin (which in those days were Bayer-patented trade names). Fischer continued to make variations on the barbiturate theme, and in 1911 he developed an ethyl phenyl variant, *phenobarbital* (phenobarbitone, Luminal, **3-67**) that was superior in many ways to the original barbital. This turned out to be more than just a better sleeping pill; because it had a much longer half-life than barbital, it was soon recognized (by an observant physician named Hauptmann, as we shall narrate below) that it had the desirable property of suppressing epileptic seizures.

The Epilepsies

We pause briefly to specify more accurately a few of the major types of epilepsy. *Epilepsies* are chronic seizures due to disturbances in the brain caused by congenital defects or by trauma in the CNS (e.g., from fever, tumors, drug/alcohol poisoning). About 6 of every 1,000 people are afflicted with some form of epilepsy, of which there are numerous varieties. Only the four more common types of epilepsy will be considered here.

- *Primary generalized tonic-clonic seizures (grand mal)* is the classic "falling sickness" attributed to historical figures such as Julius Caesar, Napoleon, and Dostoyevsky. There is sudden loss of consciousness, a *tonic* (constant) contraction of the muscles, a cry caused by forced expulsion of air from the lungs: the person falls to the ground (often with injury) with all muscles rigid. There follows a *clonic* phase (alternate contraction and relaxation of muscles) in which tongue-biting and so forth may occur. A period of relaxed muscles and unconsciousness is followed by slow return to consciousness with amnesia. Several days may be required for full recovery. *Status epilepticus* is a state of constant, unremitting grand mal seizures that can be fatal if not arrested.
- *Primary generalized absence seizures (petit mal)* is in its pure form the sudden loss of ongoing conscious awareness without any convulsive muscular activity or loss of postural control (falling). The seizures may be unnoticeable because they are so brief and may occur 50 to 100 times daily; they are often discovered in children only when they cause learning difficulties.
- *Simple partial seizures* may consist of recurrent contractions of the muscles of one part of the body without loss of consciousness. A localized part of the brain is involved, and depending on the functions it controls, symptoms may range from hallucinations to an overwhelming impression that present sensations have already happened in the past (*déjà vu*).
- *Complex partial seizures* may start with a simple partial seizure but advance to loss of conscious contact with the environment accompanied by cessation of activity or by some repetitive *automatism*: lip smacking, picking at one's clothes, and so forth. Occasionally the person may perform highly skilled activities like driving a car or playing complicated music. (Sleepwalking is usually not due to this sort of seizure, although it may resemble it.) Amnesia of the event follows the seizure and full consciousness only gradually returns. In general, partial seizures are more difficult to control than generalized seizures and often require a combination of medications.

The discovery of the antiseizure properties of phenobarbital was a classic case of serendipity. A German physician named Hauptmann was given living quarters over the ward of epileptic patients for whom he was responsible. When he was unable to sleep because of the noise they made from their seizures, he used the then-new drug to put them to sleep. To his surprise, the incidence of seizures was decreased during their waking hours as well as at night.[127] Phenobarbital is still one of the most widely used antiepileptic drugs. It is effective in generalized tonic-clonic and simple partial seizures. Complex partial seizures do not respond as well, and absence seizures may even be exacerbated. It is often used as the initial drug for the treatment of epilepsy in young children, but less sedating drugs such as carbamazepine, phenytoin or valproic acid (below) are increasingly preferred. As with all antiepileptic drugs, there is the possibility of impaired attention, concentration, memory, and motor coordination. A good degree of tolerance to these side effects develops in most people who use the drug regularly, a tolerance that fortunately does not extend to its antiseizure activity.

The Barbiturate High

Barbiturates, BZDs, and alcohol are strikingly similar in their syndromes of dependence, withdrawal, and chronic intoxication. Indeed, it is known that there is a specific barbiturate site distinct from the BZD site on the GABA receptor complex that enables barbiturates, like BZDs, to enhance the activity of endogenous GABA. A mutual but incomplete cross-tolerance exists between these three classes of drugs.

Addiction to barbiturates, and recreational use of them, has declined following the introduction of the BZDs in the 1950s. Those who do use barbiturates for their mind-altering properties usually prefer the rapid-onset drugs such as the following, which produce a distinct and intense rush: *Amobarbital* ("blue heavens," "blue dolls," "blues," **3-68**); *Pentobarbital* ("yellows," "yellow jackets," "yellow bullets," "nebbies," **3-69**); *Secobarbital* ("reds," "red devils," "seccies," "F-40's," "mexican reds," **3-70**). *Tuinal* ("rainbows," "tuies," "double trouble") is a 50/50 mixture of secobarbital and pentobarbital. All these drugs are Schedule II, have similar short-acting effects, and are available as oral prescription drugs for use either as sedatives or hypnotics. The last two are also used for premedication as an adjunct to anesthesia. Abrupt withdrawal from extensive regular use of these barbiturates can produce a severe, frightening, and potentially life-threatening crisis that in its mildest manifestation resembles delirium tremens and in its most severe form status epilepticus. Indeed, the withdrawal syndrome subsequent to long-term high-dosage use of barbiturates is probably the most dangerous withdrawal phenomenon known, and should always be undertaken in a hospital setting; once symptoms have started, they are difficult to reverse. In 75% of patients taking 800 mg/day or more of the short-acting barbiturates, convulsions occur during the second or third day of withdrawal; the convulsions may progress to status epilepticus and death. To avoid this, a person dependent on barbiturates should be given an equivalent substitute dose of a long-acting barbiturate such as phenobarbital and the dose then tapered slowly in a hospital setting over 10 or more days. Complete reestablishment of CNS stability requires about 30 days.

Other Drugs with Barbituratelike Activity

Methaqualone. Methaqualone (Quaalude, Sopor, "ludes," "sopes," "soapers," "Q's," **3-71**, Figure 3.15) was originally synthesized by a group of chemists in India in an effort to find a better drug against malaria (a continuing problem to this day). Methaqualone was not active against malaria, but it had a strong sedating effect on laboratory animals. It was introduced as a hypnotic-sedative in the early 1960s and soon became both widely prescribed and widely abused. It was very similar in its pharmacological activity to the short-acting barbiturates, but with a recreational plus: a prolonged period of mild euphoria resembling the "buzz" from alcohol. By the early 1980s it was withdrawn from most markets. However, it is relatively easy to synthesize, and numerous clandestine labs in the United States and Hong Kong are able to provide ample supplies to their domestic black market as well as to the developing countries of Africa, where it is a popular recreational drug. Two other obsolete sedative-hypnotics with similar pharmacology are *Glutethimide* (Doriden, "goofballs," "goofers," **3-72**), and *Ethchlorvynol* (Placidyl, "green weenies," **3-73**). Both of these are still available in the United States and can be prescribed for insomnia. However, "situations that warrant selection of glutethimide [or] ethchlorvynol over a BZD are extremely rare."[128] They are still occasionally encountered as street drugs.

Other Nonbarbiturate Antiepileptic Drugs

Most of the drugs below, while of great medical importance, do not produce the sort of high that the short-acting barbiturates do and are not addictive or abused. But they are listed briefly here to summarize developments in antiepileptic drugs since the 19th century and to allow the reader to cursorily view the interplay of structure and activity.[129] Of greater relevance to the concerns of this book, several of these antiepileptic drugs have been found to be useful in treating various psychiatric conditions, particularly mood disorders, and even in ameliorating the craving associated with the withdrawal from addictive drugs. It can be seen from inspection of Figures 3.16 and 3.17 that primidone, phenytoin, and ethosuximide are variations on the alternating CO/NH structure found in the barbiturates; while valproic acid, gabapentin, and felbamate are modifications of GABA.

Primidone. The structure of primidone (Mysoline, **3-74**, Figure 3.16) is closely related to that of phenobarbital: in fact, it is converted in the body to two active metabolites,

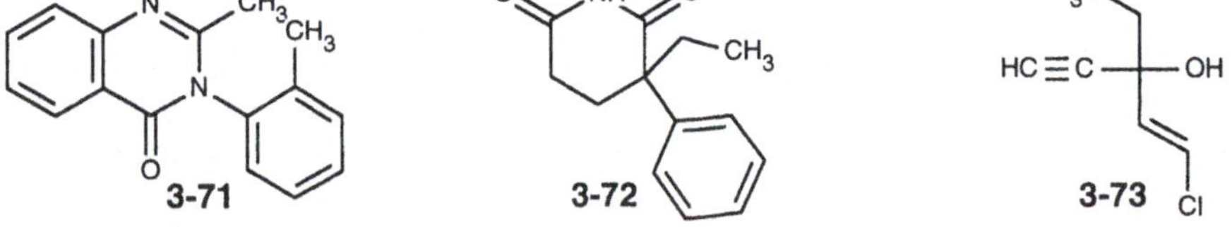

Figure 3.15 Three barbituratelike sedative-hypnotics.

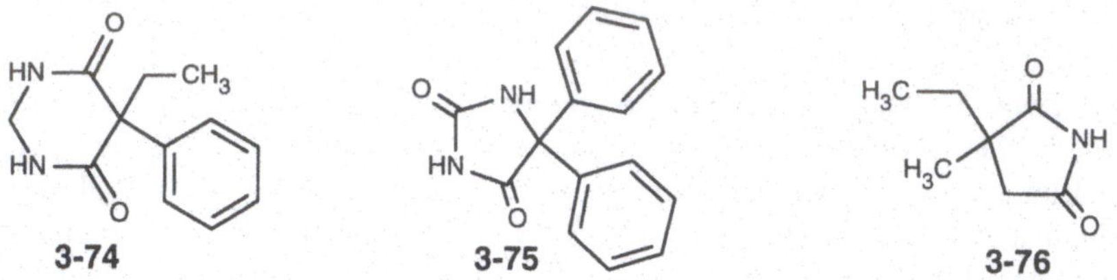

3-74 3-75 3-76

Figure 3.16 Primidone, phenytoin, ethosuximide.

phenobarbital and phenylethylmalonamide. It is used principally in generalized tonic-clonic seizures and in complex partial seizures; it is ineffective in absence seizures. The side effects, as might be expected, are similar to phenobarbital: sedation that diminishes with continued use.

Phenytoin. The first serious study of electroencephalographic data was begun by Tracey Putnam at Boston City Hospital in the 1930s. He was soon able to initiate epileptiform seizures in experimental animals by applying a minimum electric current to the brain. Putnam then requested a series of phenyl-substituted compounds from the Parke, Davis research laboratories and tested them to find which if any raised the seizure threshold; one of these, phenytoin (Dilantin, **3-75**) was found to do this better than phenobarbital and with significantly less sedation. When tested in humans during the next decade it was able to prevent seizures in many people for whom phenobarbital had proved ineffective. By the 1950s, phenytoin had became the drug of choice for tonic-clonic seizures, since it produced less intellectual impairment and little sedation relative to phenobarbital. It has also been used in simple partial and complex partial seizures, but it is ineffective in absence seizures. There are a number of unfortunate adverse reactions from continual dosing with phenytoin, including hirsutism and coarsening of facial features; when given to children, gingival hyperplasia (gum tissue overgrowing the teeth) is common. Still more tragic is its potential when given during pregnancy to cause *fetal hydantoin syndrome*, characterized by a cleft lip or palate and/or heart defects. Since, unfortunately, other epileptic drugs and unmedicated maternal seizures themselves can also cause fetal injury, the decision whether to have children is always a difficult one on the part of epileptic women.

Ethosuximide. Ethosuximide (Zarontin, **3-76**) is considered the drug of choice for absence seizures unaccompanied by other types of seizures. While valproic acid is as effective as ethosuximide, it can occasionally cause liver damage, and so it is usually reserved for those cases where ethosuximide is ineffective. The side effects of ethosuximide include nausea, drowsiness, and urticaria (rash). There have been rare reports of more serious problems, including systemic lupus erythematosa or aplastic anemia.

Valproic acid. Valproic acid (Depakene, Depakote, **3-77**, Figure 3.17) is more effective in generalized tonic-clonic seizures than in partial seizures, and it is also helpful in the treatment of absence seizures and in photosensitive seizures (a type of seizure disorder that is triggered by exposure to light or strobelike pulsations of light). Recently, it was found to be as effective as lithium in controlling manic episodes in hospitalized pa-

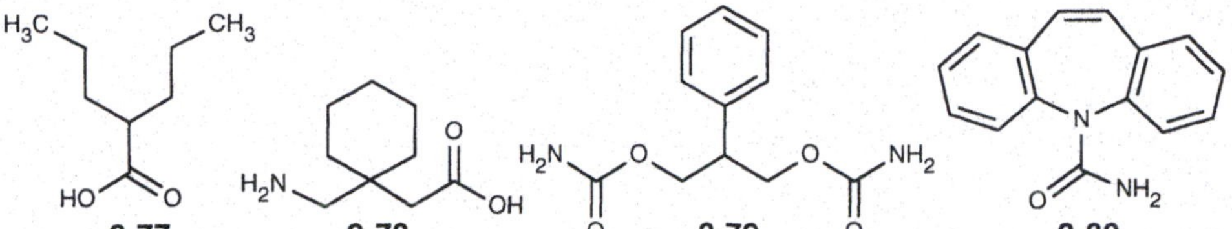

Figure 3.17 Newer antiseizure drugs.

tients[130] (see Chapter 5). There are also reports that it is helpful either alone or in combination with lithium as a prophylactic to suppress the recurrence of mania in the treatment of bipolar affective disorder (manic-depression).

Gabapentin. Gabapentin (Neurontin, **3-78**) is a newer antiepileptic with a structure based on that of GABA. The cyclohexane ring locked onto the GABA structure makes the molecule considerably more lipophilic than its parent; it easily crosses the blood-brain barrier. It has been found effective in treating partial seizures, usually as an adjunct to other antiepileptic medication when the first drug has only partial success. Despite the structure and effectiveness of the drug, it seems that the GABAergic system is not involved in its activity. It may interact with the transporter system of some of the excitatory amino acid neurotransmitters.[131]

Felbamate. Felbamate (Felbatol, **3-79**) was approved by the FDA in 1993. Felbamate has a dicarbamate structure that, like gabapentin, bears some resemblance to GABA,[132] and it acts at the GABA complex to enhance inhibitory activity.[133] At the same time, it inhibits NMDA responses, it has been shown to have some neuroprotective activity in stroke and hypoxia.[134] The drug showed great promise in treating hitherto poorly controlled partial seizures and Lennox-Gastaut Syndrome, a childhood epilepsy unresponsive to conventional drugs.[135] Adult users of the drug welcomed it because it enhanced mental alertness and concentration while inducing significant weight loss. But in early August 1994, the FDA and the manufacturer recommended suspending use of the drug after two patients experienced fatal complications from aplastic anemia.[136]

Carbamazepine. Carbamazepine (Tegretol, **3-80**) is a tricyclic compound is structurally related to antidepressants such as imipramine, desipramine, and clomipramine (see Chapter 5). It has potent antiepileptic properties in partial seizures and is increasingly preferred to phenobarbital in pediatric patients because it has less effect on alertness and behavior; indeed, it seems to elevate mood in depressed epileptic patients. Side effects can include drowsiness, nausea, gastritis, or photosensitivity.

Perhaps because of its structural relationship to the antidepressants, carbamazepine has at least one other, quite surprising, experimental use: to control compulsive gambling. Dr. Ralph Ryback of the Psychiatric Institute reports carbamazepine helps compulsive gamblers "become calmer so they can use the support systems of Gamblers Anonymous more effectively."[137] The most famous patient to use this drug is former Baltimore Colts NFL star Art Schlichter, who was suspended from the NFL in 1983 after losing more than $400,000 to his gambling habit. Schlichter says "gambling was a rush. It took my mind off of what I thought were tremendous problems." Carbamazepine "helped me adjust to regular life, and I'm not battling and I'm not struggling off the field." This is one of several anecdotal reports of the value of carbamazepine in treating impulsive and aggressive behavior in nonpsychotic individuals.[138]

Two GABAergic Botanicals: Fly Agaric, Kava

Fly Agaric

"Fly agaric" is the folk term for the mushroom *A. muscaria*, whose toxic properties were exploited in bygone days by chopping it up and putting it in a saucer of milk to attract flies ("agaric" is an old term for mushroom).[139] The mushrooms contain two active ingredients, *muscimol* (**3-81**, Figure 3.18) and *ibotenic acid* (**3-82**). Muscimol and ibotenic acid (in different proportions depending on the vagaries of fungal life) are likewise the active constituents of the related *A. pantherina*. Ingestion of either of these produces a state probably best described as delirium. Within an hour of eating these mushrooms, symptoms develop such as restlessness, ataxia (staggering and uncoordinated muscle movements), and hallucinations.[140] The word "hallucination" is used here in the strict medical sense where judgment as to the reality or unreality of one's perceptions is lost, unlike the phenomenon accompanying the use of psychedelics like LSD or "magic" mushrooms. *Amanita* are not the "magic" mushrooms widely available on college campuses and for which spore prints are advertised in the pages of *High Times*. "Magic" mushrooms are truly psychedelic, with effects identical to LSD, and are various subspecies of *Psilocybe* (Chapter 6). Nor are *A. muscaria* (which have bright red or orange caps with white dots, like the Walt Disney cartoons)[141] or *A. pantherina* (which is brown with white dots) the same as the *A. phalloides*, the white "death cap/cup" or "death angel" mushroom. The *A. phalloides* mushrooms are responsible for most fatal cases of mushroom poisonings among amateur wild-mushroom fanciers (for example, a 1994 case in Long Island where one elderly woman died and three others were hospitalized after cooking up a mess of mushrooms from their front lawns).[142] *A. phalloides* contains an extraordinarily potent toxin, *amanitin*,[143] which binds irreversibly to RNA-polymerase II in the liver and kidneys, blocking transcription from DNA to messenger RNA, and causing cell death. Nor, finally, are these *Amanita* mushrooms to be confused with *Inocybe* and *Clitocybe* mushrooms, which are toxic—though not fatally—because they contain large quantities of *muscarine*, **3-83**. Muscarine acts at a subset of cholinergic

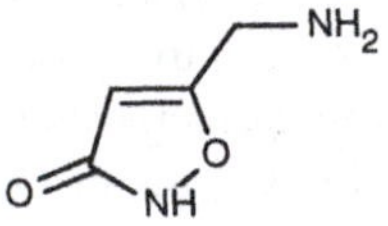

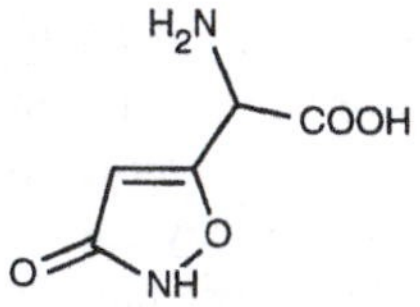

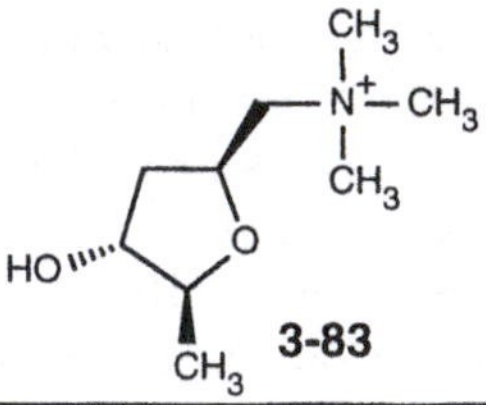

Figure 3.18 Alkaloids of *Amanita*.

receptors (named for its specific action) to produce symptoms such as salivation, nausea, diarrhea, hypotension, and shock; but these symptoms are fortunately easily abolished by intramuscular injections of atropine.[144] There are trace amounts (about 0.003%)[145] of muscarine in the *A. muscaria* and *A. pantherina* mushrooms, and indeed muscarine was originally isolated from *A. muscaria*, hence its name. But there is not enough to produce significantly toxic symptoms.[146]

Muscimol itself is known to be a CNS depressant by virtue of its specific interaction as a GABA agonist at the $GABA_A$ receptors, and since this is (probably) the more predominant agent responsible for the effects of *A. muscaria*, the use of this mushroom is included in this chapter. (Note that both muscimol and ibotenic acid embody a significant part of the GABA structure.) People acting under the influence of these mushrooms (which are usually brewed into a tea or dried and smoked) seem to do some strange, even dangerous things in a delirious state. Andrew Weil provides a first-hand account of a man whose first trial of *A. pantherina* produced a "dreamy" state and eyes-closed imagery. A second trial, using two medium-sized mushrooms, resulted in a trip to the hospital with a bruised and cut face after the man kept compulsively crawling out on a limb (literally) over a pond in the Oregon woods, and deliberately falling off, over and over again, because he "couldn't tell whether it had happened or was going to happen."[147] Occasionally young children eat these brightly colored mushrooms with the cute spots. In nine toddlers who were reported to have eaten *A. muscaria* or *A. pantherina* in Seattle in 1992, symptoms occurred within an hour or so of eating the mushrooms, and were described as "CNS depression, ataxia, waxing and waning obtundation, hallucinations, intermittent hysteria or hyperkinetic behavior."[148] There was some mild seizure activity in four of the children, which was controlled by administering anticonvulsants. All of the children recovered rapidly.

If none of this sounds like fun to you, it has not proved appealing to many others either, and the use of this species of mushroom is increasingly rare in Western culture. But *A. muscaria* has a very ancient history of use by Siberian shamans (and their reindeer),[149] and there is a considerable body of literature in anthropological circles debating the hypothesis of R. Gordon Wasson, who played a key role in rediscovering the use of *Psilocybe* mushrooms by Central American natives (see Chapter 6), that *A. muscaria* was the mysterious *Soma*, the sacred plant of the Aryans when they migrated into the Indus valley in the second millennium B.C., a plant revered and celebrated in the *Rig Veda.* Wasson also hypothesizes that the mysterious "fruit" of the "tree of the knowledge of Good and Evil" in the Book of Genesis was actually the little red-capped mushrooms growing beneath it.[150]

The Pacific Drug: Kava

Kava (also known as "kava-kava," "kawa," and "ava-ava") is a shrub of the pepper family, botanically known as *Piper methysticum*, which is used to make a beverage that is consumed by most of the Pacific Ocean native communities from New Guinea to Hawaii.[151] The dried rhizome and roots of the plant are ground, grated, and/or masticated (by women or children who themselves usually do not partake of the final beverage) and then steeped in water which, after being strained, is drunk. In most Pacific

Island societies that use kava, it is traditionally a drink of adult males only.[152] (There are extracts of the dried root sold in health food stores[153] and by mail order in the United States; in France, the National Health Scheme recognizes a kava extract manufactured by Merrell Dow called Kaviase.)[154] The effect of drinking a half pint or so of the final brew made from about 15 g of root is said to be a pleasant sense of paralysis of the lower limbs and musculature. "A euphoric state develops during which the mind remains clear; the drinker is tranquil and friendly, and refuses to be annoyed; and finally, if the dose is strong enough, sleep ensues. Several hours of dreamless sleep follow: the drinker awakes feeling marvelously well, and no hangover occurs."[155] There appears to be conflicting evidence as to whether these effects occur via the GABA system.[156]

Kava extract is not fermented, nor is it consumed with alcohol; the psychic properties of the brew are attributable to a unique nonnitrogenous series of lactones, the kavalactones, of which the most important are the five shown in Figure 3.19. (There are at least ten others, all of quite similar structure.)[157] These were studied by H. L. Meyer of the Freiburg University Institute of Pharmacology, and their separate properties determined, in a series of papers published during the late 1950s and the 1960s: *Analgesia.* The two most potent kavalactones, *dihydrokawain* (DHK, **3-85**), and *dihydromethysticin* (DHM, **3-88**), have an equal analgesic potency about twice that of aspirin on a mg/kg basis and about 2% that of morphine. *Local anesthesia.* The women and children assigned to chew kava in its preparation experience a numbing of their oral cavity; according to Van Veen,[158] these anesthetic effects are equivalent to and as long-lasting as those from cocaine, and are primarily due to the effect of *kawain*, **3-84**. There are also significant anticonvulsive and muscle-relaxing effects attributable to DHK and DHM.

Controversy has arisen in Australia concerning excessive use of kava by aboriginal people in the Northern Territory, where it is a recent introduction; there are reports that it causes more extensive liver damage when taken in large quantities than does alcohol.[159] A peculiar scaly eruption of the skin attributable to kava consumption, a "reversible ichthyosiform kava dermopathy," has also been recognized since the crewmen of Captain James Cook's exploratory expeditions in the Pacific; it is possibly traceable to some interference with cholesterol metabolism.[160]

Nonetheless, kava might prove to be a good substitute for alcohol, since it seems to have many of its beneficial properties (a mild euphoriant and anxiolytic) and fewer of

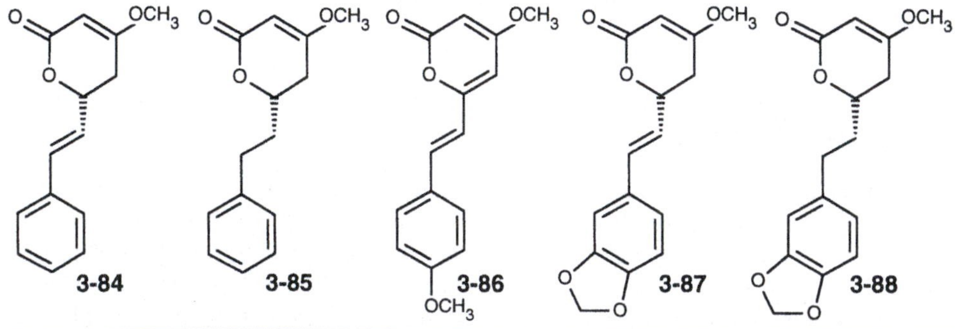

Figure 3.19 The major kavalactones.

its baneful ones. At least this is suggested by the work of a German group in 1993 that tested kava extracts in a placebo-controlled double-blind study to see whether the kava had any synergistic effect on alcohol-induced safety-related performance. The kava did not aggravate the impairment induced by alcohol; rather, it caused a significant reversal of alcohol's detrimental effects.[161] And the last chapter of one thorough study of kava is entitled "Kava: A World Drug?" and ends on a hopeful note:

> We expect that the religious, economic, and political functions and meanings of kava will continue to evolve, both within the Pacific and beyond [attention is drawn by the authors to a photo of Pope John Paul II drinking kava with the Fijian prime minister during a visit to Fiji in 1986]. This mélange of expanding symbolic meanings and social functions is of course grounded ultimately in the chemical properties of kavalactones and their physiological effects on the human body. We look for future adoptions and elaborations of kava use, rather than abandonment. Kava's traditional cultural meanings and social functions are now overlaid with new uses in the contemporary Pacific: kava as symbol of Christian atonement; kava as icon of the new state; . . . kava as ethnic Valium or alcohol; . . . kava as the shared pick-me-up of urban Pacific kava bars. The story of kava is far from ended.[162]

Our own story is far from ended with the conclusion of the CNS depressants. We turn in the next chapter to a set of drugs that have at least superficially an opposite effect to those we have studied here. We move from "downers" to "uppers," to the CNS stimulants, which include two of the most widely used drugs (caffeine and nicotine), as well as some of the most addictive and controversial (cocaine and amphetamines).

References and Notes

The following abbreviations are used for frequent sources in these References/Notes:

ARMC: *Annual Reports in Medicinal Chemistry*; Academic Press: New York.

BBN: *The Biochemical Basis of Neuropharmacology*, Cooper, J. R.; Bloom, F. E.; Roth, R. H., 6th ed., Oxford: New York, 1991.

DE95: *Drug Evaluations: Annual 1995*; American Medical Association Division of Drugs and Toxicology, Department of Drugs; Bennett, D. R., Ed.; AMA: Chicago, 1995.

DWPS: *Drugs in Western Pacific Societies: Relations of Substance*, Association for Social Anthropology in Oceania Monograph No. 11, Lindstrom, L., Ed.; University Press of America: Lanham, MD.

G&G8: *Goodman and Gilman's The Pharmacological Basis of Therapeutics*, 8th ed., Gilman, A. G.; Rall, T. W.; Nies, A. S.; Taylor, P., Eds; Pergamon: New York, 1990.

G&G9: *Goodman and Gilman's The Pharmacological Basis of Therapeutics*, 9th ed., Hardman, J. G.; Limbird, L. E., et al., Eds.; McGraw-Hill: New York, 1996.

MI11: *The Merck Index*, 11th ed.; Budavari, S.; O'Neil, M. J., et al., Eds.; Merck: Rahway, NJ, 1989. References are given by monograph number.

MI12: *The Merck Index*, 12th ed.; Budavari, S.; O'Neil, M. J., et al., Eds.; Merck: Rahway, NJ, 1996. References are given by monograph number.

OCDS: *Organic Chemistry of Drug Synthesis,* Lednicer, D.; Mitscher, L. A.; Georg, G. I.; Wiley: New York, 1977, 1980, 1984, 1990, 1995. The **bold** number following OCDS indicates the volume number (1-5); this is followed by the page number.

SA: *Substance Abuse: A Comprehensive Textbook,* 2nd ed.; Lowinson, J. H.; Ruiz, P.; Millman, R. B.; Langrod, J. G., Eds.; Williams & Wilkins: Baltimore, 1992.

1. Shakespeare, W., *King Henry VI, Part II*, IV, ii, 73.

2. Dickens, C., *The Pickwick Papers*, ch. 8.

3. (*Megilla 7b*) Cf. Moore, C. A., *The Anchor Bible: Esther*, Doubleday: Garden City, NY, 1971, p. xxxiii.

4. *Nunc est bibendum, nunc pede libero pulsanda tellus. . . . absumet heres Caecuba dignior servata centum clavibus et mero tinguet pavimentum superbo pontificum potiore cenis.* Horace, *Odes*, I. xxxvii. 1-2; II. xiv. 45-48.

5. Fitzgerald, E., *The Rubaiyat of Omar Khayyam*, ed. 4, xii; ed. 2, lvii, ciii.

6. Dickens, C., *Sketches by Boz*, "Scenes: xxii: Gin Shops."

7. The origin of this classic *if-by-whiskey* speech, a parody of a politician taking both sides of an issue, is Judge Noah S. (Soggy) Sweat Jr. of Corinth, Miss. (The nickname "Soggy" is based not on his last name but on "Sorghum Top," referring to the way his hair resembled the tassel that grows on top of sugar cane.) In 1952, as wets and drys were debating local prohibition of booze, Judge Sweat copyrighted his speech. All this according to William Safire's column *On Language* from the NYT Magazine for January 12, 1992, p. 12.

8. "Prospective study of alcohol consumption and risk of coronary disease in men," *The Lancet*, **1991**, 338, 464-468. For an earlier opinion in considerable disagreement, see "Alcohol and the Cardiovascular System," *JAMA*, **1990**, *264*, 377-381.

9. "A prospective study of moderate alcohol consumption and the risk of coronary disease and stroke in women," *NEJM*, **1988**, *319*, 267-273.

10. Rimm, E. B., "Review of moderate alcohol consumption and reduced risk of coronary heart disease: Is the effect due to beer, wine, or spirits?" *Brit. Med. J.*, **1996**, *312*, 736-741.

11. Cohen, S.; Tyrrell, D. A.; et al., "Smoking, alcohol consumption, and susceptibility to the common cold," *Am. J. Public Health,* **1993**, *83*, 1277-1283.

12. (a) Rimm, E. B., et al., "Prospective study of cigarette smoking, alcohol use, and the risk of diabetes in men," *BMJ*, **1995**, *310*, 555-559. (b) Perry, I. J., et al., "Prospective study of risk factors for development of non-insulin dependent diabetes in middle aged British men," *BMJ*, **1995**, *310*, 560-564.

13. Doll, R., et al., "Mortality in relation to consumption of alcohol: 13 years' observations on male British doctors," *BMJ*, **1994**, *309*, 911-918.

14. (a) Grønbœk, M., et al., "Influence of age, body mass index, and smoking on alcohol intake and mortality," *BMJ*, **1994**, *308*, 302-306. For a recent review of the effects of alcohol on the liver and the heart, see (b) Achord, J. L., "Alcohol and the liver," *Scientific American Science and Medicine*, **1995**, *2*, 16-27. (c) Klatsky, A. L., "Cardiovascular effects of alcohol," *Scientific American Science and Medicine*, **1995**, *2*, 28-37.

15. See Note 7.

16. *Johnsonian Miscellanies*, Hill, G. B., Ed., 1897, Vol. II, "Anecdotes by Hannah More," p. 197. But it does not seem that Johnson was ever what we would now call a problem drinker. He boasts that "I did not leave off wine, because I could not bear it; I have drunk three bottles of port without being the worse for it" (Boswell, *Life of Johnson*, aetat 69, 7 April 1778, Modern Library, p. 776). It may have been a sort of conservation of energy. "Few people," he opines elsewhere to Boswell, "[have] intellectual resources sufficient to forgo the pleasures of wine. They could not otherwise contrive how to fill the interval between lunch and dinner" (cited in *The Oxford Companion to Wine*, J. Robinson, Ed., Oxford U Press: Oxford, 1994, p. 364.) Like good wine, our taste improves with age.

While pouring port for Boswell and Goldsmith, himself drinking water, he observed: "The lad does not care for the child's rattle, and the old man does not care for the young man's whore" (*Life of Johnson*, aetat 57, 1776, Modern Library, p. 310.)

17. *"Multi quidem facilius se abstinent ut non utantur, quam temperent ut bene utantur,"* Augustine, *On the Good of Marriage*, xxi.

18. Housman, A. E., *A Shropshire Lad*, lxii.

19. Bukowski, C., "Young in New Orleans," *The Last Night of the Earth Poems*, Black Sparrow Press: Santa Rosa, CA, p. 354.

20. Warner, J., *Contemporary Drug Problems*, **1993**, *19*, 409-429. Reviewed in *J. Psychoactive Drugs*, **1993**. *25*, 278.

21. ASAM, *ASAM News*, **1990**, *5*, 1, 9.

22. This sounds like a quibble, since the amount of alcohol in an orange is very small. But set and setting mean a great deal in this as in every addiction: some decades ago, it was common for recovering alcoholic Roman Catholic priests to celebrate a daily private Mass using a very small thimbleful of wine, and this did not seem to cause problems. (Later, the more progressive post-Vatican II church allowed priests with this problem to use unfermented grape juice, just like proper Methodists.)

23. (a) A review article with leading references: Levy, M. S., "The disease controversy and psychotherapy with alcoholics," *J. Psychoactive Drugs*, **1992**, *24*, 251-256. For a strongly dissenting opinion: Maltzman, I., "Why alcoholism is a disease," *J. Psychoactive Drugs*, **1994**, *26*, 13-31.

24. Sanchez-Craig, M., et al., "Empirically based guidelines for moderate drinking: 1-year results from three studies with problem drinkers," *Am. J. Public Health*, **1995**, *85*, 823-828.

25. Vaillant, G. E., "A long-term follow-up of male alcohol abuse," *Arch. Gen. Psychiatry*, **1996**, *53*, 243-249.

26. (a) Blum, K.; Noble, E. P., et al., "Allelic association of human dopamine D-2 receptor gene in alcoholism," *JAMA* **1990**, *263*, 2055-2060. (b) Arinami, T.; Itokawa, M., et al., "Association between severity of alcoholism and the A1 allele of the dopamine D2 receptor gene TaqI A RFLP in Japanese," **1993**, *33*, 108-114.

27. Comings, D. E.; Comings, B. G., et al., "The dopamine D-2 receptor locus as a modifying gene in neuropsychiatric disorders," *JAMA* **1991**, *266*, 1793-1800. But see also Gelernter, J.; O'Malley, S., et al., "No association between an allele at the D_2 dopamine receptor gene (*DRD2*) and alcoholism," *JAMA*, **1991**, *266*, 1801-1807. What this conflicting data probably means is that "the A1 allele of *DRD2* appears to be associated in the general population with alcoholism but not tightly linked to alcoholism in pedigrees [and] modifies the expression of other genes that have a major and direct influence on susceptibility to alcoholism." Cloninger, C. R., "D_2 dopamine receptor gene is associated but not linked with alcoholism," *JAMA* **1991**, *266*, 1833-1834. And an editorial by Shirley Y. Hill ("Is there a genetic basis of alcoholism?" *Biol. Psychiatry*, **1992**, *32*, 955-957), reviews the evidence and concludes that "there appears to be no specific genetic transmission of alcoholism in the strict Mendelian sense, but there may well be a genetic diathesis [heritable predisposition] for its development. However, the complexity of the endpoint phenotype (alcoholism) may obscure significant latent traits that are under genetic control."

28. Other behavioral areas in which a genetic link has been claimed and denied include crime, manic depression, schizophrenia, intelligence, and homosexuality. See: (a) Horgan, J., "Trends in behavioral genetics: eugenics revisited," *Scientific American*, **1993**, June, pp. 122-

131. (b) Horgan, J., "D_2 or not D_2: a barroom brawl over an 'alcoholism gene,'" *Scientific American*, **1993**, April, pp. 29-32.

29. Schuckit, M. A.; Smith, T. L., "An 8-year follow-up of 450 sons of alcoholic and control subjects," *Arch. Gen. Psychiatry*, **1996**, *53*, 202-210.

30. Volavka, J.; Czobor, P., et al., "The electroencephalogram after alcohol administration in high-risk men and the development of alcohol use disorders 10 years later," *Arch. Gen. Psychiatry*, **1996**, *53*, 258-263.

31. Gianoulakis, C. G.; Krishnan, B., et al., "Enhanced sensitivity of pituitary β-endorphin to ethanol in subjects at high risk of alcoholism," *Arch. Gen. Psychiatry*, **1996**, *53*, 250-257.

32. Gordis, E., "Alcohol research: At the cutting edge," *Arch. Gen. Psychiatry*, **1996**, *53*, 199-201.

33. *The New York Times*, 31 December 1991.

34. Wechsler, H; et al., "Health and behavioral consequences of binge drinking in college," *JAMA*, **1994**, *272*, 1672-1677. Unfortunately, the definition of "binge" drinking seems to be constructed so as to engender a self-fulfilling prophecy of doom. College students are polled as to whether they have "taken five drinks at a sitting." Students who have had five beers from the beginning of the TGIF party at 3:00 in the afternoon until its end at 2:00 a.m. Saturday thereby define themselves as "binge drinkers," when their actual alcohol intake has been quite modest and left them at all times legally, morally, and pharmacologically cold sober. There is plenty of excessive and dangerous drinking on college campuses, but credibility is lost by this sort of exaggeration.

35. Brewer, R. D.; Morris, P. D., "The risk of dying in alcohol-related automobile crashes among habitual drunk drivers," *NEJM*, **1994**, *331*, 513-517.

36. Angell, M.; Kassirer, J. P., "Alcohol and other drugs—toward a more rational and consistent policy," *NEJM*, **1994**, *331*, 537-539.

37. *J. Pharmacol. Exp. Ther.*, **1991**, *256*, 922.

38. "Receptor-gated ion channels may be selective CNS targets for ethanol," **1991**, *TIPS*, 1.

39. Krystal, J. H.; Webb, E., et al., "Specificity of ethanollike effects elicited by serotonergic and noradrenergic mechanisms," *Arch. Gen. Psychiatry*, **1994**, *51*, 898-911.

40. Klaassen, C. D., *G&G8*, pp. 1624f.

41. "Alcohol and Alcoholism," Marc A. Schuckit, in *Harrison's Principles of Internal Medicine*, Braunwald et al. Eds., Chap 365, p. 2107.

42. (a) Frezza, M.; di Padova, C., "High blood alcohol levels in women: the role of decreased gastric alcohol dehydrogenase activity and first-pass metabolism," *NEJM*, **1990**, *322*, 95-99. (b) Schenker, S.; Speeg, K. V., "The risk of alcohol intake in men and women: all may not be equal," [editorial] *ibid.*, 127-129.

43. "Aspirin increases blood alcohol concentrations in humans after ingestion of ethanol," *JAMA*, **1990**, *264*, 2406-2408.

44. Hyyatia, P.; Sinclair, J. D., "Oral etonitazene and cocaine consumption by AA, ANA and Wistar rats," *Psychopharmacology*, **1993**, *111*, 409-14.

45. Blum, K.; Briggs, A. H., et al., "A common denominator theory of alcohol and opiate dependence: review of similarities and differences." In: Rigter, H.; Crabbe, J. C., Eds., *Alcohol Tolerance and Dependence,* Elsevier: New York, 1980, 371-391.

46. Cohen, G.; Collins, M., "Alkaloids from catecholamines in adrenal tissue: possible role in alcoholism," *Science*, **1970**, *167*, 1749-1751.

47. Davis, V. E.; Walsh, M. J., "Alcohol, amines, and alkaloids: a possible biochemical basis for alcohol addiction," *Science*, **1970**, *167*, 1005-1007. For leading references to this and other theories of alcoholism, see the excellent review article, in *SA*: Tabakoff, B.; Hoffman, P. L., "Alcohol: neurobiology," pp. 152-185.

48. Reid, L. D.; Delconte, J. D., et al., "Tests of opioid deficiency hypotheses of alcoholism," *Alcohol*, **1991**, *8*, 247-257.

49. Reid, L. D.; Hunter, G. A., "Morphine and naloxone modulate intake of ethanol," *Alcohol*, **1984**, *1*, 33-37.

50. This according to John Emsley, science writer-in-residence at the Imperial College of Science, London, writing to correct earlier mistaken theories: Reese, K. M., "Newscripts: Disulfiram as alcohol deterrent—version II," *C&E News*, 4 July 1994, p. 96.

51. Volpicelli, J. R.; Alterman, A. I., et al., "Naltrexone in the treatment of alcohol dependence," *Arch. Gen. Psychiatry*, **1992**, *49*, 876-880.

52. Volpicelli, J. R.; Watson, N. T., et al., "Effect of naltrexone on alcohol high in alcoholics," *Am. J. Psychiatry*, **1995**, *152*, 613-615.

53. O'Malley, S. S.; Jaffe, A. J., et al., "Naltrexone and coping skills therapy for alcohol dependence," *Arch. Gen. Psychiatry*, **1992**, *49*, 881-887.

54. Bor, J., "Drug for alcoholism receives qualified praise from experts: alcoholics, doctors assess naltrexone," *The Baltimore Sun*, 22 January 1995, 1B, 3B.

55. Leary, W. E., "Drug for heroin addiction is being marketed for treatment of alcoholism," *The New York Times*, 18 January 1995.

56. Mason, B. J.; Ritvo, E. C., et al., "A double-blind, placebo-controlled pilot study to evaluate the efficacy and safety of oral nalmefene HCl for alcohol dependence," *Alcohol Clin. Exp. Res.*, **1994**, *18*, 1162-1167.

57. Swift, R. M.; Whelihan, W., et al., "Naltrexone-induced alterations in human ethanol intoxication," *Am. J. Psychiatry*, **1994**, *151*, 1463-1467.

58. Keung, W. M.; Vallee, B. L., "Daidzin and daidzein suppress free-choice ethanol intake by Syrian golden hamsters," *Proc. Natl. Acad. Sci. U.S.A.*, **1993**, *21*, 10008-10012.

59. Keung, W. M.; Vallee, B. L., "Daidzin: a potent, selective inhibitor of human mitochondrial aldehyde dehydrogenase," *Proc. Natl. Acad. Sci. U.S.A.*, **1993**, *90*, 1247-1251.

60. Keung, W. M.; Vallee, B. L., "Therapeutic lessons from traditional Oriental medicine to contemporary Occidental pharmacology," *EXS*, **1994**, *71*, 371-381.

61. Xie, C. I.; Lin, R. C., et al., "Daidzin, an antioxidant isoflavonoid, decreases blood alcohol levels and shortens sleep time induced by ethanol intoxication," *Alcohol Clin. Exp. Res.*, **1994**, *18*, 1443-1447.

62. Sneader, W., *Drug Discovery: The Evolution of Modern Medicines*, Wiley: New York, 1985, pp. 183f.

63. Sternbach, L. H.; Reeder, E., *J. Org. Chem.*, **1961**, *26*, 4488-4497; *J. Org. Chem.*, **1961**, *26*, 4936-4941.

64. Brody, J. E., "Personal health: avoiding the hazards of drowsy driving," *The New York Times*, 21 December 1994, p. C10.

65. Wald, M. L, "Wrecks on land, air, and sea: The costs of sleeping on the job," *The New York Times*, 19 November 1995, iv, 4.

66. (a) Broughton, R.; Billings, R., et al., "Homicidal somnambulism: A case report," *Sleep*, **1994**, *17*, 253-264. (b) Ovuga, E. B., "Murder during sleep-walking," **1992**, *East Afr. Med. J.*, *69*. 533-534.

67. Oswald, I., "On serious violence during sleep-walking," *Br. J. Psychiatry*, **1985**, *147*, 688-691.

68. Montagna, P., "Nocturnal paroxysmal dystonia and nocturnal wandering," **1992**, *Neurology*, *42*, 61-67.

69. Lillywhite, A. R.; Wilson, S. J.; Nutt, D. J., "Successful treatment of night terrors and somnambulism with paroxetine," *Br. J. Psychiatry*, **1994**, *164*, 551-554.

70. Rall, T. W., "Benzodiazepines." In: *G&G8*, p. 369.

71. *Plantam pedis inungere pinguedine gliris . . . et quod vix credi potest, dentes inunctos ex sorditie aurium canis.* Cardan, *De rerum varietat.* cited by Richard Burton, *The Anatomy of Melancholy: What it is, with all the kinds, causes, symptomes, prognostickes and severall cures of it,* Pt. 2, Sec. 5, Subsec. 6, "Correctors of accidents to procure sleep." The student of Latin will note that this authority concedes himself that this remedy "vix credi potest," i.e., is "hardly to be believed." But, absent a controlled study, who can say? In justice to Burton, this is one of his remedies of last resort; one of his first is to use a few grains of Paracelsus's laudanum, which would surely help.

72. In German-speaking countries, it is traditional to ask one another on first meeting in the morning, whether one has slept well: *"Haben Sie gut geschlaffen?"* But this is only a little removed from the seeming idiocy of asking someone *are you asleep?* How can anyone know whether they have slept well? They were asleep the whole time.

73. Oswald, I., *Br. Med. J.*, **1982**, *284*, 860-864.

74. All citations from Cioran's obituary: Pace, E., "E. M. Cioran, 84, novelist and philosopher of despair," *The New York Times*, 22 June 1995.

75. Trp-Ala-Gly-Gly-Asp-Ala-Ser-Gly-Glu. For references, see *MI11*, #3445, p. 544.

76. Inoue, S.; Kimura-Takeuchi, M; Honda, K., "Co-circulating sleep substances interactingly modulate sleep and wakefulness in rats," *Endocrinol. Exp.*, **1990**, *24*, 69-76. DSIP has also been found be stored in neurosecretory vesicles with luteinizing-hormone releasing factor (LH-RH, aka gonadotropin-releasing hormone, GnRH): (a) Vallet, P. G.; Charnay, Y.; Bouras, C., "Distribution and colocalization of delta sleep-inducing peptide and luteinizing hormone-releasing hormone in the aged human brain: an immunohistochemical study," *J. Chem. Neuroanat.*, **1990**, *3*, 207-214. (b) Charnay, Y.; Leger, L., et al., "Immunohistochemical mapping of delta sleep-inducing peptide in the cat brain and hypophysis. Relationships with the LHRH system and corticotropes," *J. Chem. Neuroanat.* , **1990**, *3*, 397-412. (c) Vallet, P. G.; Charnay, Y., et al., "Colocalization of delta sleep inducing peptide and luteinizing hormone releasing hormone in neurosecretory vesicles in rat median eminence," *Neuroendocrinology*, **1991**, *53*, 103-106. (d) Pu, L. P.; Charnay, Y., et al., "Light and electron microscopic immunocytochemical evidence that delta sleep-inducing peptide and gonadotropin-releasing hormone are co-expressed in the same nerve structures in the guinea pig median eminence," *Neuroendocrinology*, **1991**, *53*, 332-338.

77. Nakamura, A.; Sakai, K., et al., "Characterization of delta-sleep-inducing peptide-evoked release of Met-enkephalin from brain synaptosomes in rats," *J. Neurochem.*, **1991**, *57*, 1013-1018.

78. Those who have closely observed this species may question how an animal that sleeps virtually 24 hours a day could become "exhausted." The answer is, they were put on a moving treadmill for 18-22 hours.

79. Lerner, R. A.; Siuzdak, G., et al., "Cerebrodiene: a brain lipid isolated from sleep-deprived cats," *Proc. Natl. Acad. Sci. U.S.A.*, **1994**, *91*, 9505-9508.

80. Cravatt, B. F.; Prospero-Garcia, O., et al., "Chemical characterization of a family of brain lipids that induce sleep," *Science*, **1995**, *268*, 1506-1509.

81. *The New York Times*, 13 June 1995, p. C5.

82. *Macbeth* II, i, 36-38. Often do the sages warn us that the price of riches and fame is usually insomnia: "Uneasy lies the head that wears a crown," says King Henry IV (Part II, III, i. 30). And Horace warns that nothing can bring the sleep of the carefree shepherd to those over whose head black care (*atra Cura*) hangs like a suspended sword: *destrictus ensis cui super impia / cervice pendet, non Siculae dapes / dulcem elaborabunt saporem, / non avium citharaeque cantus / somnum reducent. somnus agrestium / lenis virorum non humiles domus / fastidit umbrosamque ripam, / non zephyris agitata Tempe . . . cur valle permutem Sabina / divitias operosiores?* (Carminum Liber III, i.)

83. In one of the most famous meditations in modern philosophy, there is a passage in Heidegger's *Sein und Zeit* where he recalls the ancient Latin fable of humankind (*homo*) being molded from the clay of a riverbed by the goddess Care (*cura*). After Jove, Care, and Earth (*Tellus*) squabble over who gets to name the new creature, Saturn judges between them: *Tu Jovis quia spiritum dedisti, in morte spiritum, tuque Tellus, quia dedisti corpus, corpus recipito, Cura enim quia prima finxit, teneat quamdiu vixerit. Sed quae nunc de nomine eius vobis controversia est, homo vocetur, quia videtur esse factus ex humo.* "Since you, Jupiter, have given its spirit, you shall receive that spirit at its death; and since you, Earth, have given its body, you shall receive its body. But since Care first shaped this creature, she shall possess it as long as it lives. And as for the dispute among you about its name, let it be called '*homo*', for it is made out of *humus* (earth)." Heidegger, M., *Being and Time*, J. Macquarrie; E. Robinson, Trans., San Francisco: Harper and Row, 1962, p. 242 (German original of 1926, p. 198).

84. Boswell, J., *Life of Dr. Johnson*, aetat 67, Thursday 21 March 1776 (Everyman's Library [Knopf]: New York, 1992, p. 613).

85. Perhaps this in turn is a matter of perception. As we advance in years, even when we do not—but especially when we do—advance alike in wisdom, we become the more liable to boredom and find ourselves wishing we were asleep much more of the time than this can be physically managed. How much drug—or, still more lamentable, TV—addiction among the elderly may be simply the result of sheer, unremitting *boredom?*

86. Adapted from *DE95*, p. 228, 231 and *G&G9*, p. 369.

87. These glucuronidated hydroxylated metabolites (conjugates) are probably not truly inert, but they play a very minor pharmacological role in normal individuals since they are so rapidly excreted by the kidneys. However, in patients with renal failure, the conjugates accumulate and seem to be responsible for a comatose state that can be reversed by flumazenil: Bauer, T. M.; Ritz, R., "Prolonged sedation due to accumulation of conjugated metabolites of midazolam," *Lancet*, **1995**, *346*, 145-147.

88. On the other hand, the advertising of the drug companies directed to physicians during the 1960s and 1970s can certainly seem questionable in retrospect. For a critical view of this phenomenon and the overuse of BZDs: Bargmann, E.; Wolfe, S. M., et al., *Stopping Valium: and Ativan, Centrax, Dalmane, Librium, Paxipam, Restoril, Serax, Tranxene, Xanax,*

Warner: New York, 1982.

89. Adapted from *DE95*, p. 228, 231, *G&G9*, p. 369; data on flunitrazepam from Korolkovas, A., *Essentials of Medicinal Chemistry*, 2nd ed., Wiley: New York, 1988, p. 209.

90. *DE95*, p. 239.

91. *DE95*, p. 243.

92. "Strong sedative cheap high on street: Columbia traffickers pushing Rohypnol," *The Baltimore Sun*. If one Rohypnol reduces beer consumption to one bottle per day amongst the street folk, it should be distributed free by the FDA to encourage this healthful trend.

93. Navarro, M., "A new abused drug in Florida is prescribed abroad," *The New York Times*, 9 December 1995, 6.

94. "U.S. bans a drug used abroad for insomnia," *The New York Times*, 4 March 1996.

95. San, L.; Tato, J., et al., "Flunitrazepam consumption among heroin addicts admitted for in-patient detoxification," *Drug Alcohol Depend.*, **1993**, *32*, 281-286.

96. Bond, A.; Seijas, D., et al., "Systemic absorption and abuse liability of snorted flunitrazepam," *Addiction*, **1994**, *89*, 821-830.

97. Kolata, Gina, "Maker of sleeping pill hid data on side effects, researchers say," *The New York Times*, 20 January 1992, p. A1.

98. Cowley, G.; Springen, K; Iarovici, D.; Hager, M., "Sweet dreams or nightmare?" *Newsweek* 19 August 1991, pp. 44-51.

99. Kolata, Gina, "Finding a bad night's sleep with Halcion," *The New York Times*, 20 January 1992, p. B7.

100. Kolata, Gina, *op. cit.*

101. Silverman, M. S., Letter to *NEJM*, **1991**, *325*, 1742.

102. For a lively debate and a plethora of leading references, see the Correspondence of Drs. Silverman, Kales, Ayd, Thompson, Robinson, Kurt and responses of Drs. Greenblatt and Gillin in: *NEJM*, **1991**, *325*, 1742-1745.

103. Bixler, E. O.; Kales, A., et al., "Next-day memory impairment with triazolam use," *The Lancet* **1991**, *337*, 827-31.

104. Brahams, D., "Triazolam suspended," *The Lancet*, **1991**, *338*, 938.

105. Brahams, D., "Triazolam licensing in UK," *The Lancet*, **1993**, *341*, 1587.

106. Like Jarndyce and Jarndyce in *Bleak House*, "the massive trial (involving some 36,000 sheets of paper) before May *J*, a High Court judge (sitting without a jury) produced damages awards totalling only £210,000 while the costs run to millions." Brahams, D., "Upjohn (Halcion) libel actions," *The Lancet*, **1994**, *343*, 1422.

107. Eichenwald, K., "A justice department review is sought on a sleeping pill's side effects," *The New York Times,* 1 June 1996, p. 10. But the task force, a committee of experts from outside the FDA, advised that the drug could remain on the market because it appeared to be safe if the recommended dosage limits were adhered to.

108. (a) Brahams, Diana, *op. cit.* (b) Eichenwald, K., *op. cit.*

109. *DE95*, p. 184.

110. How this takes place is beginning to be understood on a molecular level. The BZD site and the GABA site are both partly located on the α1 protein subunit of the complex: Smith, G. B.; Olsen, R. W., "Functional domains of $GABA_A$ receptors," *TIPS*, **1995**, *16*, 162-168.

111. *The Lancet*, **1990**, *336*, 81.

112. Wildmann, J., *Pharmacol. Res.*, **1989**, *21*, 673.

113. *BBN*, pp. 158-159.

114. *DE95*, pp. 228, 247-248.

115. *ARMC* **27**, 322.

116. Lader, M., "Rebound insomnia and newer hypnotics," *Psychopharmacology (Berl)*, **1992**, *108*, 248-255.

117. (a) Trifiletti, R. R.; Snyder, S. H., "Anxiolytic cyclopyrrolones zopiclone and suriclone bind to a novel site linked allosterically to BZD receptors," *Mol. Pharmacol.***1984**, *26*, 458-469. (b) Hoehns, J. D.; Perry, P. J., "Zolpidem: a nonbenzodiazepine hypnotic for treatment of insomnia," *Clinical Pharmacy*, **1993**, *12*, 814-828.

118. Jonas, J. M.; Coleman, B. S., et al., "Comparative clinical profiles of triazolam versus other shorter-acting hypnotics," *Clin. Psychiatry*, **1992**, *53, Supp:* 19-31.

119. Griffiths, R. R.; Sannerud, C. A., et al., "Zolpidem behavioral pharmacology in baboons: self-injection, discrimination, tolerance and withdrawal," *J. Pharmacol. Exp. Ther.*, **1992**, *260*, 1199-1208.

120. (a) Nogrady, T., *Medicinal Chemistry: A Biochemical Approach*, 2nd ed., Oxford: New York, 1988, p. 233. DMCM is methyl 4-ethyl-6,7-dimethoxy-β-carboline-3-carboxylate and one of several β-carbolines that act as inverse agonists: (b) Triggle, D. J.; Langs, D. A., "Ligand-gated and voltage-gated ion channels," *ARMC* **1990**, *25*, 225-234. (c) Browne, L. J.; Shaw, K. J., "New anxiolytics," *ARMC*, **1991**. *26*, 1-10.

121. Venault, P.; Chapouthier, G., et al., "Benzodiazepine impairs and β-carboline enhances performance in learning and memory tasks," *Nature,* **1986**, *321*, 864-866.

122. *G&G8*, p. 375.

123. Busto, U.; Kaplan, H. L., et al., "Pharmacologic effects and abuse liability of bretazenil, diazepam, and alprazolam in humans," *Clin. Pharmacol. Ther.* **1994**, *55*, 451-463.

124. Kaplan, H. I.; Sadock, B. J., *Pocket Handbook of Psychiatric Drug Treatment*, Williams & Wilkins: Baltimore, 1993, p. 239 f.

125. By then Baeyer had relocated southward from Prussian Berlin to Bavarian Munich with a lengthy stopover in Francophilic Strassbourg. He had probably picked up a sense of humor. But the 1863 article of the 28-year-old Baeyer from volume 127 of the *Annalen der Chemie und Pharmacie*, "Mittheilungen aus dem organischen Laboratorium des Gewerbeinstitutes in Berlin: Untersuchungen über die Harnsäuregruppe" (the second part of two, pp. 199-236), betrays no evidence of whimsy. The name for barbituric acid is introduced on page 209 in a context that, as Fieser and Fieser say, "so stresses his [von Baeyer's] conception of this substance as the key compound in the series of related ureides as to suggest the derivation from the German word *Schlüsselbart* . . . plus uric acid" (Fieser, L. F.; Fieser, M., *Organic Chemistry*, D. C. Heath: Boston, 1944). Other charming but unlikely stories are that Barbara was the name of the waitress at the *Stammtisch* in his favorite *Lokal*, where he went to wet his whistle after synthesizing new compounds. One Hans Schindler of NYC claims that Nobel Laureate (1930) Hans Fischer, in his general organic chemistry lecture at the *Technische Hochschule*, Munich (but long after Baeyer's death), "stated quite definitely" that Baeyer named the acid after Saint Barbara, the patron saint of artillerymen (*C&EN*, Feb 26, 1990, p. 168). Schindler writes, "I do not recall the explanation for this choice, and my lecture notes of, alas, 59 years ago are lost." He mentions two possibilities. One is that

Baeyer's father was a general well known for geodetic work. A second possibility is that Baeyer wanted to celebrate the important role of the artillery in Prussia's victory in the war with Denmark in 1864, the year he synthesized barbituric acid (*C&EN*, March 19, 1990, p. 94). All these theories have been rehashed many times. For those eager to obsess on the matter, Kauffman, G. B., "Adolf von Baeyer and the naming of barbituric acid," *J. Chem. Ed.*, **1980**, *57*, 222-223 provides the most serious documentation of the Lady Barbara legends. The chemistry is explained to the English-speaking reader by Carter, M. K., "The history of barbituric acid," *J. Chem. Ed.*, **1951**, *28*, 524-526. A slew of other references (along with the story of a molecule, "barbaralone," which actually was named after a real person, Dr. Barbara M. Ferrier) can be found in the delightful and informative work on chemical namings: Nickon, A.; Silversmith, E. F., *Organic Chemistry: The Name Game: Modern Coined Terms and Their Origins*, Pergamon: New York, 1987, pp. 133-134.

126. Fischer, E.; Dilthey, A., *Ann.*, **1904**, *335*, 334. The discovery of the hypnotic properties of the barbiturates was, like so much of the history of medicinal chemistry, semi-serendipitous: von Mering hypothesized (for the wrong reason) that a diethylbarbituric acid would be a hypnotic from analogy to a soporific compound, sulphonal, which had just been discovered by Bayer & Co. See Sneader, W., *Drug Discovery: The Evolution of Modern Medicines*, Wiley: New York, 1985, pp. 28-30.

127. Sneader, W., *Drug Discovery . . .* , p. 30.

128. *Drug Evaluations: Annual 1994,* American Medical Association Division of Drugs and Toxicology, Department of Drugs: Bennett, D. R., Ed., AMA: Chicago, 1994.

129. For a review of recent developments in antiepileptic medications: Upton, N., "Mechanisms of action of new antiepileptic drugs: rational design and serendipitous findings," *TIPS*, **1994**, *15*, 456-463.

130. Bowden, C. L.; Brugger, A. M.; Calbrese, J. R., "Efficacy of Divalproex vs lithium and placebo in the treatment of mania," *JAMA*, **1994**, *271*, 918-924. Divalproex is a patented formulation of Abbott Laboratories consisting of a 1:1 molar mixture of sodium valproate and valproic acid. Abbott provided partial funding for the study, and several of the many contributors to the study own stock in Abbott. It is unlikely that Divalproex has effects significantly different from generic valproic acid or generic sodium valproate.

131. (a) *ARMC*, **29**, 338. (b) Handforth, A.; Treiman, D. M., *Epilepsia*, **1993**, *34*, Suppl. 6, 109. (c) Pierce, M. W.; Anhut, H.; Sauermann, W., *Epilepsia*, **1993**, *34*, Suppl. 2, 181.

132. The structure of Felbamate (2-phenyl-1,3-propanediol dicarbamate) is even more closely related to an older antianxiety medication, meprobamate (Miltown, 2-ethyl-2-methyl-1,3-propanediol dicarbamate); but the pharmacology is significantly different.

133. *DE95*, p. 381-382.

134. (a) *ARMC*, **29**, 337. (b) White, H. S.; Wolf, H. H., et al., *Epilepsia*, **1992**, *33*, 564. (c) McCabe, R. T.; Wasterlaink C. G., et al., *Pharmacol. Exp. Ther.*, **1993**, *264*, 1248.

135. Palmer, K. J.; McTavish, D., "Felbamate: A review of its pharmacodynamic and pharmacokinetic properties, and therapeutic efficacy in epilepsy," *Drugs*, **1993**, *45*, 1041-1065.

136. *G&G9*, p. 481.

137. Armen Keteyian (28 July 1991), ABC TV News.

138. Arana, G. W.; Hyman, S. E., *Handbook of Psychiatric Drug Therapy*, 2nd ed., Boston: Little, Brown, 1991, p. 114.

139. Weil, A.; Rosen, W., *From Chocolate to Morphine: Everything You Need to Know About Mind-Altering Drugs,* 2nd ed., Houghton Mifflin: New York, pp. 134-135.

140. *G&G8*, 129.

141. The brighter the color, the more powerful the effect: Weil, A.; Rosen, W., *From Chocolate. . .* , pp. 134-5.

142. Feinfeld, D. A.; Mofenson, H. C., et al., "Poisoning by amatoxin-containing mushrooms in susburban New York—report of four cases," *J. Toxicol. Clin. Toxicol.*, **1994**, *32*, 715-21. Two of the four ate a different variety of mushroom, *Lepiota chlorophyllum*, which however contains the same toxin.

143. Amanitin is an interesting glycopeptide, $C_{39}H_{54}N_{10}O_{13}S$, containing an unusual 2-thio-6-hydroxytryptamine moiety; for its structure and that of related toxins cf *MI11*, entry 378, p. 373.

144. *G&G8*, 129.

145. *G&G8*, 129.

146. A book that will satisfy the questions of mycologist, physician, pharmacologist, and amateur chef alike in mushroom matters: Benjamin, D. R., *Mushrooms: Poisons and Panaceas*, W. H. Freeman: New York, 1995.

147. Weil, A.; Rosen, W., *From Chocolate . . .* , p. 208.

148. Benjamin, D. R., "Mushroom poisoning in infants and children: the *Amanita pantherina/muscaria* group," *J. Toxicol. Clin. Toxicol.*, **1992**, *30*, 13-22.

149. The yellow snow created by the Siberian shamans who have consumed this mushroom is particularly prized by their domesticated reindeer, who will rush to consume it. Urine from *Amanita*-intoxicated shamans is likewise consumed by fellow human cultists, and Wasson ties this in with the Hindu practice of drinking the urine of a guru possessed of particularly inspiring teaching. (There is also an ancient Hindu custom, sometimes called *self-urine therapy*, favored by figures as eminent as Nehru and Gandhi, of sipping one's own urine upon arising in the morning.) For a refreshing overview of some of these intriguing customs see: Wasson, S. H., "The divine mushroom of immortality." In: *Flesh of the Gods: The Ritual Use of Hallucinogens*, Furst, P. T., Ed., Praeger: New York, 1972, pp. 185-200.

150. Wasson, G.; Kramrisch, S.; Ott, J.; Ruck, C. A. P., *Persephone's Quest: Entheogens and the Origins of Religion*, Yale University Press: New Haven, CT, 1986, p. 25.

151. An oversimplification; for a detailed demographic survey of the use of kava: Marshall, M., "An overview of drugs in oceana." In: *DWPS*, pp. 13-49, specifically pp. 21-23.

152. For peculiar uses of kava in somewhat indelicate rites of male-bonding: Knauft, B. M., "Managing sex and anger: Tobacco and kava use among the Gebusi of Papua New Guinea," *DWPS*, pp. 73-98.

153. I sampled the contents of one of the small bottles of kava tincture sold at a local health food store and found the flavor intensely bitter and tannic; if this is the flavor of the active ingredients, beverages containing it are unlikely to compete in the United States market. Nor could I detect any psychic effects other than an intense craving for mouthwash. Global World Media, which markets a controversial "herbal ecstasy" (see Chapter 4) also markets Nexus (in imitation of the psychedelic drug described in Chapter 6), the chief ingredient of which is *P. methysticum*.

154. Lebot, V.; Merlin, M.; Lindstrom, L., *Kava: The Pacific Drug*, Yale University Press: New Haven, CT, 1992, p. 196.

155. Tabrah, F. L.; Eveleth, B. M., "Evaluation of the effectivenes of ancient Hawaiian medicine," *Hawaii Medical Journal*, **1966**, *25*, 223-230.

156. (a) Jussofie, A.; Schmiz, A., et al., "Kavapyrone enriched extract from *Piper methysticum* as modulator of the GABA binding site in different regions of rat brain," *Psychopharmacology*, (Berl.) **1994**, *116*, 469-74. (b) Davies, L. P.; Drew, C. A., et al., "Kava pyrones and resin: studies on GABAA, GABAB and benzodiazepine binding sites in rodent brain," *Pharmacol. Toxicol.*, **1992**, *71*, 120-126.

157. Lebot, V., et al., *op. cit.,* pp. 67-72.

158. Van Veen, A. G., "Isolation and constitution of the narcotic substance from kawa kawa (*Piper methysticum*)," *Rec. Trav. Chim. Pays-Bas,* **1939**, *58*, 521-527.

159. Mathews, J. B.; Riley, M. D., et al., "Effects of the heavy usage of kava on physical health: summary of a pilot survey in an aboriginal community," *Med. J. Australia*, **1988**, *148*, 548-555.

160. Norton, S. A.; Ruze, P., "Kava dermopathy," *J. Am. Acad. Dermatol.*, **1994**, *31*, 89-97.

161. Herberg, K. W., [Effect of Kava-Special Extract WS 1490 combined with ethyl alcohol on safety-relevant performance parameters] (Ger.), *Blutalkohol*, **1993**, *30*, 96-105.

162. Lebot, V., et al., *op. cit.,* pp. 210-211.

4

Stimulants: Nicotine, Caffeine, Cocaine, Amphetamines

Cigars are sheer poetry to me . . . something almost sacred. . . . When I smoke now, my whole childhood comes back to mind—all my ideals as they were, clear and pure.

—Jean Sibelius[1]

"Then," says the trooper, not yet lighting his pipe . . . . "the two got mixed up . . . with a flinty old rascal . . . hard, indifferent, taking everything so evenly—it made flesh and blood tingle, I do assure you."

"My advice to you," returns Mrs. Bagnet, "is to light your pipe, and tingle that way. It's wholesomer and comfortabler, and better for the health altogether."

—Dickens[2]

Tea, though ridiculed by those who are naturally of coarse nerves, or are become so from wine-drinking, and are not susceptible of influence from so refined a stimulant, will always be the favorite beverage of the intellectual.

—Thomas De Quincey[3]

I know I ain't gonna get hooked. My friends have been snorting 15 years, and they ain't hooked.

—Richard Pryor[4]

Introduction

There was a considerable similarity among the drugs of the last chapter, the central nervous system (CNS) depressants, not only in terms of their effects (often mutually synergistic) but also in terms of their mode of action: nearly all are known to act at the ubiquitous γ-aminobutyric acid (GABA) inhibitory synapse.

The picture is somewhat more opaque for the diverse drugs which are the subject of the present chapter. In a general sense, all are CNS stimulants; but they affect the synapses of several distinct neurotransmitter (NT) stimulatory systems. Nicotine, betel, and datura are cholinergics, binding to acetylcholine receptors; caffeine and the other xanthine drugs in tea, coffee, and cocoa bind to adenosine synapses; and cocaine, the amphetamines, and stimulant herbs like ephedra and kat affect catecholamine systems, primarily dopamine and norepinephrine. And nicotine, when it is used in its most common mode, as a component of smoked tobacco, seems to act subjectively either as a stimulant or as a depressant, being to the same person at different times in the same day a drug which enhances concentration and attention or relieves stress, tension, and anxiety.

Tobacco: 'A Very Fine Drug'

Tobacco: A Brief History

The encounters between the European and Amerindian cultures from Columbus's day on brought a fearsome toll of death upon the native populations, either through direct slaughter or indirectly through such Old World diseases as smallpox. But in grim reciprocity, probably an equal or greater return of death and disease through syphilis and tobacco were brought back to Europe from America by Columbus's crew.

Since then, several antibiotics have been found which can cure syphilis and inhibit its spread, but the same cannot be said for the addiction to tobacco, which has so conquered every culture and civilization that it can be found literally everywhere, even in the most remote and isolated village of Africa or Siberia. This is not to say that its introduction was not resisted: use of tobacco was prohibited and severely punished at one time or another by the English, the French, and the Japanese; by a czar, a pope, and a sultan. All finally capitulated, and it is an amazing indication of the subtle but irresistible power of this fragrant weed that "no country that has ever learned to use tobacco has given up the practice."[5]

It is not tobacco itself which is so addictive, but the alkaloid *nicotine* which it contains, without which tobacco would be no more likely to be smoked than basil or bay leaves. Nicotine owes its name to Jean Nicot, a cultivated man of letters and a minister of Francis II of France. In 1559 he visited the Royal Pharmacy in Lisbon and was given a strange plant recently brought from Florida. He gave some of the plant's leaves to Catherine de Medicis, the mother of Francis II, who became an enthusiastic user of the powdered tobacco snuff, and soon it was so chic that everyone at court had a fancy box of tobacco. Eventually, Nicot began importing large shipments of tobacco to Paris, where its use became fashionable and where the plant was first referred to as "nicotiana." But there was soon religious opposition to the new drug:

> The first enemy was Scotland's James VI (soon to become James I of England), who fought Nicot's plant as well as papism throughout his life. . . . Pope Innocent X even went so far as to excommunicate tobacco-users.
>
> Tobacco divided the Sorbonne and became yet another bone of contention between the Jansenists (who were for it) and the Jesuits (who were decidedly against it). The Jesuits finally

> conceded that tobacco might not be forbidden fruit in itself, but only when used for the satisfaction of depraved desires; that is, only those *intentionally* defying God's command by sniffing or smoking would be excommunicated! Outside France things were even worse. Amurat IV condemned smokers to death; the Czar ordered that their noses be cut off; the Shah Sifi simply had them impaled. In Switzerland the Senate of Berne had "smoking" inserted with "stealing" and "killing" in the ten commandments.[6]

Then the French found a diabolically practical solution: tax the stuff at two francs per hundredpound and net the state about a million francs a year.[7] But now the government was addicted; addicted to that most potent of all substances, money. This same pattern of acceptance-prohibition-legalization cum sin tax has since been mirrored many times for other ineradicable attractions such as gambling and alcohol, and may yet come to pass for other drugs.

Shamanic Uses of Tobacco

Throughout the world, shamans must prove themselves and validate their vocation, usually by undergoing a symbolic death and resurrection, or in some manner crossing the threshold between this world and the spirit world. "Familiarity with death as a gateway to life on a different plane of existence is what sets the shaman apart from the average person."[8] In South America, numerous aboriginal tribes employ hallucinogens in their shamanic rituals; almost all of them also employ tobacco extensively. Since nicotine is very toxic and rapidly metabolized, if a shaman ingests a large enough dose he will undergo what appears to be a near-death experience—and may physiologically correspond to near-death as well. The shamans in their training learn how to quickly dose themselves with enormous quantities of nicotine-rich tobaccos: depending on tribal custom, tobacco concoctions or extracts are chewed, licked, drunk, eaten, applied to the eyes or skin, inserted subcutaneously, taken as an enema, or smoked. Often several routes are employed simultaneously, and other psychotropic substances such as hallucinogens and alcohol are taken as well to achieve a variety of psychic and religious effects. In sufficiently large doses, nicotine will cause nausea, vomiting, prostration, tremors, convulsions, seizures, and near-catatonic paralysis. When a novice shaman falls unconscious, his mouth may be forced open, a funnel inserted, and a cup of tobacco juice force-fed him until he vomits blood. This is how a man of the Tapirapé tribe described his "journey to the House of Thunder:"

> I smoked much and then I smoked again. I sang, I saw one large sun and it came toward me and disappeared. I saw many small suns. They came and they left. I saw Thunder. It was small and came (to the house) in a small canoe. It was Thunder's child (a *topu*). It wore a small headdress of parrot feathers. It had a small lip plug. I reached to pull out the lip plug but it left. [The shaman did not conquer the *topu* and he was shot down by an arrow while he was unconscious] . . . all was dark. I saw many suns. I travelled singing as I walked. I spent three days walking. I climbed a large mountain on the other side of the Araguaya. There it is that the sun comes up. I saw Kanawana. He was big and his body was covered with much hair. He had many red parrot feathers. There were many *topu* and many souls of shamans. I did not talk but came back. [If he had touched Thunder's rattle he would have stayed (died); but he was properly treated with tobacco smoke and returned to consciousness.][9]

The Active Constituents of Tobacco

The chief alkaloid in the ordinary tobaccos used for cigarettes, cigars, and so on is *nicotine* (**4-1**, Figure 4.1). Two other alkaloids, nornicotine, **4-2**, and anabasine, **4-3**, seem to have essentially the same physiological properties as nicotine. Of some 50 species of *Nicotiana*, nicotine is the chief and essentially single alkaloid in 14; it occurs in about equal concentration with nornicotine in 22; it is absent in about 12 species in which nornicotine is the principle alkaloid; and in the remaining species it is a minor component, with nornicotine or anabasine the principal alkaloid.[10]

The two principal cultigens—that is, the species used in commercial tobacco—are *N. tabacum* and *N. rustica*. In these, nicotine is the major alkaloid, but the content in *N. tabacum* varies considerably, from 0.6 to 9.0%. In *N. rustica*, it can be as high as 18%. The alkaloid is found in flowers, stems, and roots, but 64% occurs in the leaves.[11]

Pharmacology of Nicotine

Dosage, Effects, Duration of Action. In the form of an infusion of tobacco leaves in water, nicotine has been used as an insecticide since 1746. It is still sold commercially as Black Leaf 40. The toxic properties derive from not only the nicotine, but also the two previously described toxic alkaloids, nornicotine and anabasine. In humans, nicotine is a rapidly acting poison that affects the autonomic ganglia first as a stimulant and then as a depressant, resulting in paralysis and functional failure of vital organs. It is intensely toxic to mammals upon inhalation or absorption through the skin. The free base is more rapidly absorbed than the salts: the average lethal dose is 60 mg; 4 mg can produce serious symptoms. The symptoms are similar to those from acetylcholinesterase inhibitors: salivation, vomiting, muscular weakness, clonic convulsions, and fibrillation. The amount of nicotine in an ordinary cigar, if extracted and injected, would be enough to kill two adult humans. Fortunately, cigars are smoked, which takes some time, and the body's rapid metabolic degradation of nicotine protects the smoker.

This ability of nicotine to be absorbed through skin and mucous membranes has made possible two nicotine-delivery systems—patches and nasal sprays—which are helping addicts (smokers) cling to the comforting solace of their addiction while abstaining from the carcinogenic components of smoked tobacco.[12]

Health Hazards from Long-Term Use of Tobacco. Smoking tobacco in the form of cigarettes, cigars, or pipes leads to coronary artery disease, chronic obstructive pulmonary disease (emphysema), and cancers of the mouth, throat, and lung. Smoking further

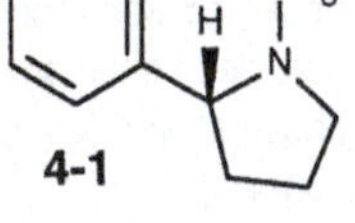

4-1 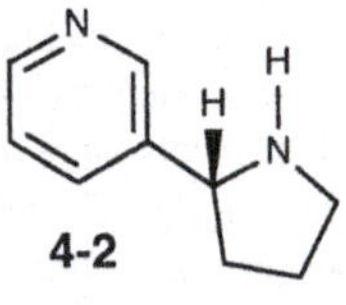 **4-2** 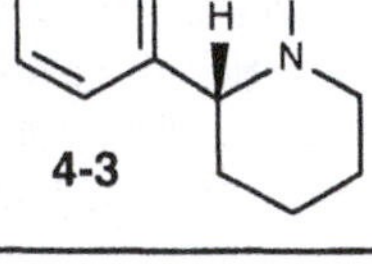 **4-3**

Figure 4.1 The *Nicotiana* alkaloids.

contributes to the development of cancer of the bladder, pancreas, and kidney. Lung cancer, according to the National Cancer Institute, kills more smokers than heart disease; and has now replaced breast cancer as the most frequent cause of cancer deaths among American women. According to former Surgeon General C. Everett Koop, "cigarette smoking is associated with more death and illness than drugs, alcohol, automobile accidents and AIDS *combined*."[13] Specifically, cigarette smoking and smokeless tobacco use are responsible for more than 400,000 deaths per year, whereas heroin and cocaine combined produce only about 6,000.[14] Even when these figures are adjusted for the reality that far more people use tobacco than cocaine, tobacco is still much more lethal: there are about 85 deaths from tobacco per 10,000 weekly users and about 30 deaths from cocaine per 10,000 weekly users, which makes cocaine roughly equal to alcohol with 21 deaths per 10,000 weekly users. Marijuana, as has been said before, has never been known to cause a single fatality in all its millions of users over several millenia.[15]

However, the enormous monetary cost reputedly inflicted on the smoke-free portion of society by wicked smokers seems actually more a matter of "smoke and mirrors." According to Richard Kluger, smokers die so much sooner than the rest of us—thereby collecting less Social Security, Medicare, and private pension payments—that they more than pay for their medical bills. And he points out that they pay more taxes: in Massachusetts, smokers paid $237 million in the one year of 1994 under the state's 51-cents-a-pack tax, well above the $200 million in public health-care costs that Massachusetts Attorney General Harshbarger is suing cigarette makers to recover.[16] From a purely monetary point of view, the state seems to have little reason for complaint.

Addictivity of Tobacco Products. "To cease smoking is the easiest thing I ever did. I ought to know, because I've done it a thousand times," said Mark Twain.[17] Perhaps Oscar Wilde described best the ambivalent essence of tobacco's appeal: "A cigarette is the perfect type of a perfect pleasure. It is exquisite, and it leaves one unsatisfied. What more can one want?"[18] Many people drink alcohol only on weekends, still more use marijuana only occasionally, and there are "chippers" who use cocaine or amphetamines less than daily. But this pattern of habituation is almost never found with tobacco products—when they are used, they are used daily, most often many times a day.

As we saw in Chapter 1, nicotine has been ranked by one group of experts as more addictive than any other common drug of abuse, including amphetamines, cocaine, barbiturates, alcohol, or heroin. Physical dependence on nicotine is manifested by the rapid onset of a withdrawal syndrome after cessation of smoking. Symptoms of withdrawal are nausea, headache, constipation or diarrhea, and increased appetite. In one study, 21% of former smokers reported craving cigarettes at least intermittently 5 to 9 years after cessation.[19]

Toxic Substances in Tobacco Smoke. The higher incidence of atherosclerosis (buildup of plaque in the coronary arteries) associated with tobacco smoking may be due to the effects of nicotine, inasmuch as some studies seem to indicate that it lowers HDL. But this effect may be due to other factors in smoked tobacco; a recent a study of the health effects of smokeless tobacco (ST) among baseball players showed that there were no significant differences between ST users and nonusers in systolic or diastolic blood

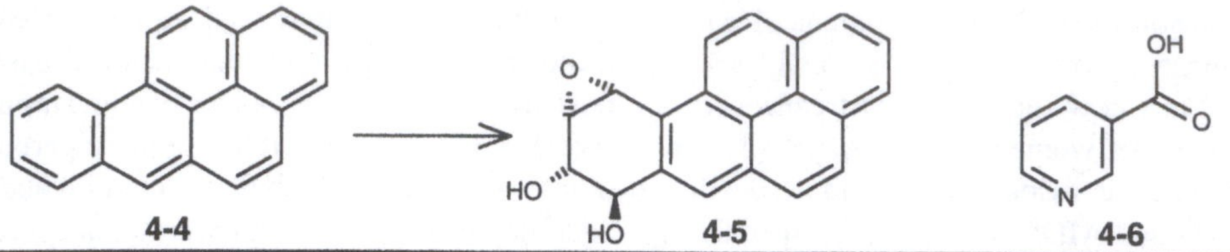

4-4 4-5 4-6

Figure 4.2 Biotransformation of Benzo[*a*]pyrene; nicotinic acid.

pressure, in pulse rate, or in total or HDL cholesterol levels.[20] However, the carcinogenicity of tobacco is not due to the nicotine but to other substances in the smoke of burning tobacco (and marijuana) leaves. Chief among these is *benzo[a]pyrene* (**4-4**, Figure 4.2), which is converted by the ubiquitous P-450 enzyme system to its *trans*-7,8-diol-9,10-epoxide, **4-5**, one of the most potent carcinogens known. (Nicotine itself is, somewhat ironically, oxidized—not, however, in the process of smoking—to form *nicotinic acid*, or *niacin*, **4-6**, which is one of the B vitamins.) As early as 1958, researchers at Brown and Williamson's parent corporation, British American Tobacco Company (Batco), had determined that "cigarette smoke contains about 5 μg of 3,4-benzo[*a*]pyrene per 500 g of cigarette, and that 90% of 3,4-benzo[*a*]pyrene was formed during combustion." The leader of the Batco research group, Sir Charles Ellis, while conceding that "smoking is a habit of addiction," went on to eulogize nicotine as a "remarkable, beneficent drug that both helps the body to resist external stress and also can as a result show a pronounced tranquilizing effect. . . . Nicotine is not only a very fine drug, but the technique of administration by smoking has considerable psychological advantages."[21]

Medical Uses of Nicotine and Tobacco

Physicians caring for victims of chronic ulcerative colitis had often remarked that it seemed those who smoked had less severe attacks. In olden days, smoking was recommended to some sufferers. With the recent advent of the nicotine patch, an opportunity arose to test whether nicotine really helped this condition. A controlled study of 72 nonsmokers who suffered from ulcerative colitis confirmed what had always seemed circumstantially the case: after 6 weeks, those who used a real nicotine patch instead of a placebo patch had twice the rate of complete remission of symptoms. They also reported lower stool frequency, less abdominal pain, and less fecal urgency.[22]

In view of the fact that nicotine itself does relatively little harm; that some find themselves unable to function well without it, for whatever reason; that objective tests indicate that people addicted to nicotine do indeed have improved functional performance when given nicotine—in view of all this, it would seem a good policy to find a way in which people who want nicotine could get it with the least possible harm. But past attempts of cigarette manufacturers to provide a smokeless cigarette, one that would be free of carcinogens yet satisfy the craving for nicotine, have encountered FDA disapproval as "nicotine delivery devices." In a similar vein, the FDA recently attempted to show that the tobacco-product manufacturers, by using high-nicotine strains of tobacco, are actually selling not cigarettes but nicotine delivery devices. However, because double-blind studies have shown that people smoking cigarettes with lower levels of nicotine compensate by smoking more cigarettes, the FDA's effort to put limits on the nicotine content of cigarettes could well backfire, exposing smokers to ever more carcinogens as they attempt to satisfy their craving for nicotine.

In 1996, R. J. Reynolds began test marketing a nearly smokeless cigarette called Eclipse. This product heats tobacco by drawing a stream of hot air from a burning charcoal tip through the cigarette. Thus nicotine is vaporized and "delivered" to the smoker, but over 90% of the tars and carcinogens found in burning tobacco are eliminated. Eclipse is described by Jed Rose, one of the inventors of the nicotine skin patch, as "an important alternative that has been underexplored."[23]

Cigars and Kreteks

'A Good Cigar.' "A woman is only a woman," mused Rudyard Kipling, "but a good cigar is a smoke."[24] There is something more serious about a cigar, compared to a cigarette. A cigar is a commitment. Recently, there has been a resurgence in cigar smoking, at least among the rich and famous: Demi Moore, Whoopi Goldberg, Jack Nicholson, David Letterman—and, of course, George Burns, whose undying love for Gracie and cigars sustained him to his centenary year. Yearly consumption of cigars in the United States has risen from a low of 2.1 billion in 1993 to 2.5 billion in 1995.[25] It is a resurgence perhaps fueled by the unspoken belief that cigars are less hazardous than cigarettes. Indeed, most cigar smokers do not inhale the smoke, and the overall incidence of lung cancer is lower for cigar smokers than for cigarette smokers; nonetheless, it is still three times as high among cigar smokers as among nonsmokers. As might be expected, those who do inhale cigar smoke are subject to the same risk of lung cancer as those who inhale cigarette smoke. And the tenfold increase in other cancers among cigar smokers is the same as that experienced by cigarette smokers.[26]

Clove Cigarettes. The clove cigarette was developed in Indonesia in the early 20th century; called "kreteks," they are composed of up to 40% cloves and the rest dark tobaccos. The *eugenol* (**4-7**, Figure 4.3) in cloves acts as a local anesthetic to make the harsh local tobaccos more tolerable—a purpose also served by the menthol in some conventional cigarettes. Kreteks became popular among some teenage groups in California during the mid 1980s; a media furor erupted when lung injuries occurred among users of kreteks, and some localities banned their sale. The likelihood is that the eugenol anesthesia allowed inadvertent aspiration of some stomach contents.

Betel and Datura

Betel. Betel chewing is a habit of almost one-tenth the world's population; although almost unknown in the West, it is extremely common throughout southern Asia. Betel chewers use a combination of two different plants, and the name betel is used for both:

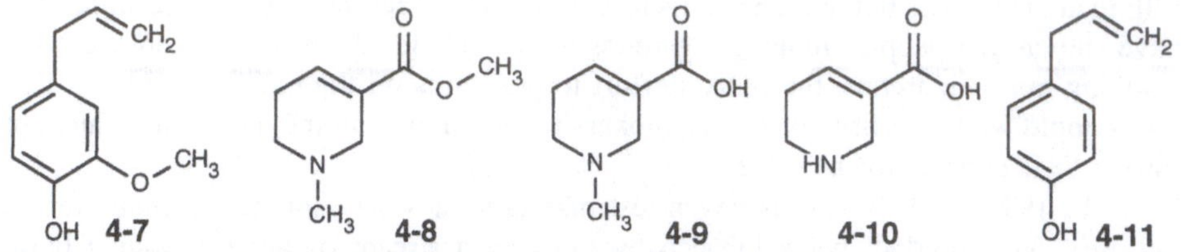

4-7 4-8 4-9 4-10 4-11

Figure 4.3 Eugenol, arecoline, arecaidine, guvacine, chavicol.

areca (betel nuts, pinang, penang), the seed of the betel palm *Areca catechu*; and *pan* (betel leaf), the leaf of the betel pepper, *Piper betle*. The seed of the first plant is wrapped in the leaf of the second for chewing. The fruit of the areca is about the size of a small egg; just before it is ripe, it is harvested, husked, boiled, cut into pieces, and dried. The pieces are wrapped in a leaf of the betel pepper along with a pellet of shell lime that lowers the pH and thereby releases the alkaloid content of the areca. (The natives of South America chew coca leaves with lime for the same reason.) The flavor of the areca is said to be like mild cheese and that of the betel leaf like pepper; the combination perhaps tastes like jalepeño-flavored cheddar. Chewing betel causes intense salivation, and the spittle is colored a deep brick red. The lips and gums of the betalist are thereby stained the same color.

Recently, the government of Myanmar (formerly Burma) outlawed the sale of betel in the capital city, Yangon (formerly Rangoon)—not because of its pharmacology, but because the copious red expectoration from its frequent users ruins the appearance of the city streets. This hurts tourism. The ban has caused great distress, much as if smoking were prohibited in the United States. "I am sorry that the foreigners do not like to see the red stains in the street, but this is a habit I learned as a boy, just as my father did, and his father," laments one hapless Yangoner.[27] Another describes the effects of betel as "a little like the feeling of alcohol. Betel gives me the feeling that everything is good in this life. Maybe the government does not want me to feel this way anymore."

The areca part of the betel combination contains the alkaloids *arecoline* (**4-8**), *arecaidine* (**4-9**), *guvacine* (**4-10**), and some related alkaloids (all of which are structural cousins of nicotinic acid with two of the three double bonds of the pyridine ring hydrogenated). These substances seem to stimulate the nicotinic cholinergic receptors and may actually have some of the subtle but powerful addictive properties of nicotine.[28] The pepper leaf contains *chavicol* (**4-11**, a close relative of eugenol, **4-7**) and other aromatic and flavor constituents. Unfortunately, there is extensive evidence that betel mastication increases the incidence of oral carcinoma (although, as with the nicotine of tobacco, it is likely that the areca alkaloids are not themselves carcinogenic).[29]

Datura. The common flowering plant *Datura stramonium* ("jimsonweed," "devil's apple") contains two powerful anticholinergic drugs: *atropine* (**4-12**, Figure 4.4), and *scopolamine* (**4-13**). The psychic effect of munching the flowers, seeds, or leaves of this and related plants such as deadly nightshade, thorn apple, or the Florida trumpet vine, is best described as a state of delirium like that associated with a very high fever. Consuming very large amounts can lead to coma or death. A similar state is caused by the *amanita*

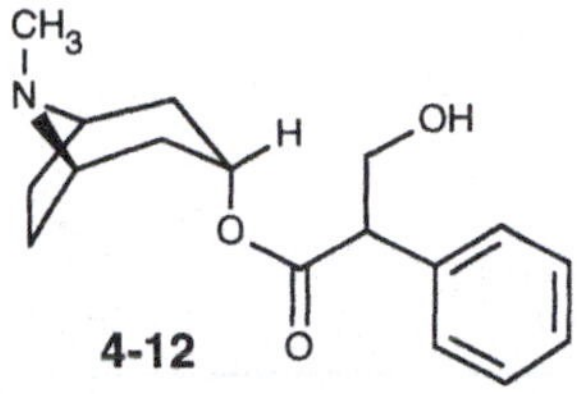

4-12 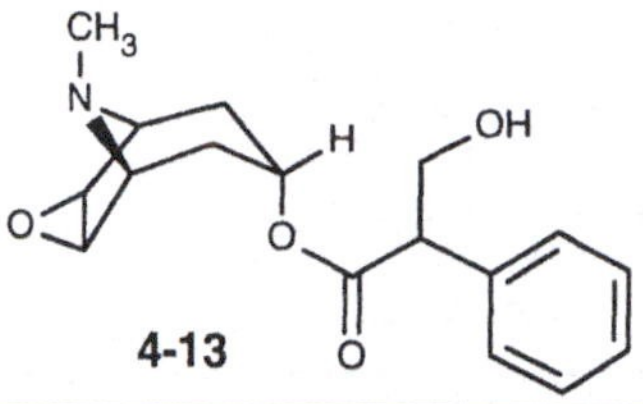 4-13

Figure 4.4 Toxic principles of *Datura.*

mushrooms mentioned in the previous chapter. In Kentucky, a review of 29 cases of people who consumed as much as a half-cup of jimsonweed seeds concluded that conservative treatment with sedatives like Valium was adequate.[30] But a review of 24 cases in Texas, New York, and California reported 2 fatalties from drinking jimsonweed tea.[31]

Caffeine and the Xanthines

Caffeine, Theophylline, Theobromine

These are three alkaloids of similar structure (all contain the xanthine ring system with one or more methyl substituents) found in several popular beverages. All are mild stimulants, and use of these beverages induces physical dependence with dysphoric withdrawal symptoms such as irritability, nervousness, restlessness, and headache. It has been said that this group of drugs, more than any other, provides an example of society's ability to cultivate and control drug use with some degree of moderation, making legal restraints unnecessary.

Of these three alkaloids, theophylline, **4-15** (Figure 4.5), is the most potent stimulant. It is used in the treatment of asthma because of its bronchodilating property, and it and ephedrine are the active ingredients of the common OTC medication Primatene, used for asthma and hay fever. Although small amounts of theophylline occur in tea, the principal stimulant effects of tea, coffee, and cola drinks are due to the larger quantities of caffeine, **4-14**, that they contain. Caffeine is only slightly less potent a stimulant than theophylline, while theobromine, **4-16**—the principal alkaloid of the cacao bean from which chocolate is made and a minor alkaloid in cola nuts and in tea—is the least potent stimulant of the three and may even be inactive in humans.[32] All three alkaloids act by blocking the ubiquitous adenosine receptor sites in the CNS—adenosine being an endogenous NT with a generally depressant action in the brain, heart, and kidneys. All produce diuresis, although not to a clinically useful degree. All produce vasoconstriction of cerebral arteries, and caffeine in combination with ergot (Cafergot) in a prescription formulation is used to relieve migraine headaches. Caffeine and theophylline are a standard and effective medication in treating the syndrome of primary apnea of newborn babies. (Oddly enough, neonates are not only more sensitive generally than adults to caffeine; they also metabolize theophylline to caffeine, an anomalous metabolic pathway not found in adults.)[33] Caffeine is also found with aspirin or acetaminophen in several OTC analgesics (Anacin, Excedrin, Midol, Vanquish).

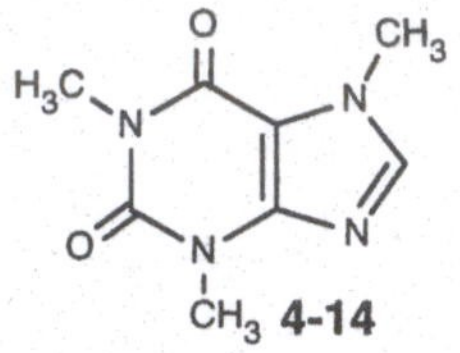

4-14 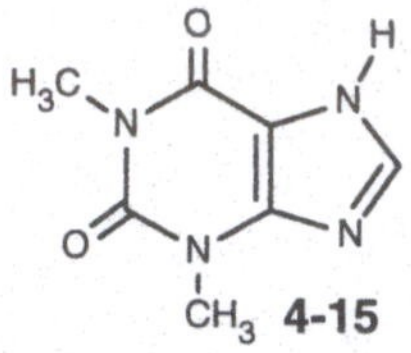 4-15 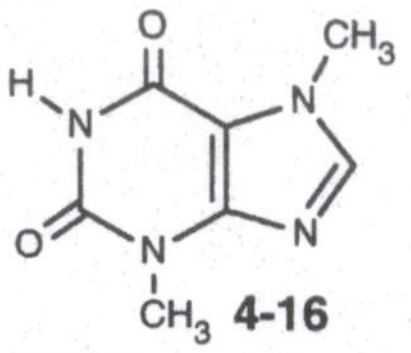 4-16

Figure 4.5 The xanthine alkaloids.

The amount of caffeine in a cup of drip-brewed coffee is about 100-150 mg; percolated or instant coffee contains somewhat less, 70-125 mg; a cup of tea contains about 50 mg of caffeine, as do most 12-oz bottled colas; and a cup of cocoa contains only about 25 mg. It is also found in the maté tea and guarana paste used in South America. The estimated fatal dose of caffeine in humans is 10 g; this corresponds to the rapid consumption of about 100 cups of coffee or 100 No-Doz tablets.

Many value caffeine's ability to stimulate mental alertness, overcome fatigue, and enhance endurance in physically exhausting work. But it has been considered off and on to be a dangerous drug. According to Dr. T. D. Crothers, author of the 1902 text *Morphinism and Narcomanias from Other Drugs* and editor of the *Journal of Inebriety*, caffeinism was similar to morphinism and alcoholism. He cites a case of "coffee psychosis" on the part of "a prominent general in a noted battle in the Civil War; after drinking several cups of coffee he appeared on the front of the line, exposing himself with great recklessness, shouting and waving his hat as if in a delirium, giving orders and swearing in the most extraordinary manner. He was supposed to be intoxicated. Afterward it was found that he had used nothing but coffee."[34]

There is no doubt that, by the classical definition, caffeine is an addictive drug. It induces not only tolerance but physical dependence. Regular users deprived of their daily caffeine are subject to mental sluggishness, depression, inability to think clearly, and a dull, generalized headache by the noon hour. All these symptoms are promptly relieved by caffeine.[35] Like nicotine and heroin, and unlike alcohol and cocaine, even moderate users of caffeine usually must have their fix on a daily basis. Sometimes the passion for caffeine approaches the dedication of a heroin or cocaine user. Dr. Roland Griffiths, one of the psychiatrists at the Johns Hopkins School of Medicine who conducted a study of caffeine dependence in 1994, described a pregnant woman in the study who had avoided caffeine during her pregnancy. As soon as she went into labor she asked her husband to stock up on caffeinated soda so she would have it as soon as she returned from the hospital. When he refused, she went to a supermarket, although she was in labor, and bought the sodas herself.[36]

Others are repelled by the effects of caffeine. The somewhat bizarre protagonist of Mark Helprin's recent novel, *Memoir from Antproof Case,* regards coffee as a Mephistophelian evil, "a filthy corruption brewed from a bean that poisons its own tree, . . . turns your inner self into a happy sparkling clockwork, hypnotizes you with artificial joy, and takes from you the sadness and deliberation that are the anchors of love."[37]

There are instances of individuals becoming addicted to caffeine tablets, several brands of which (No-Doz, Vivarin) are sold OTC as stimulants, to the point of mental confusion, disorientation, and violence. Indeed, caffeine is the only drug used by humans that, when administered in large quantities to rats, will cause them to physically attack themselves or one another. In one 1967 study, caffeine-crazed rats were seen to bite themselves and chew off their feet; some continued this frenetic self-mutilation until they died of hemorrhagic shock.[38] Cocaine will not do this to rats, nor will LSD or heroin. Marijuana *pacifies* rats. As Brecher comments in the classic *Consumers Union Report on Licit and Illicit Drugs:* "If the drug producing this effect in rats were marijuana, or LSD, or amphetamine, the report would no doubt have made headlines throughout the country. One of the distorting effects of categorizing drugs as 'good,' 'bad,' and 'non-drugs' is to protect the 'nondrugs' such as caffeine from warranted criticism while subjecting the illicit drugs to widely publicized attacks—regardless of the relevance of the data to the human condition."[39]

On the positive side, extensive studies have failed to support any mutagenic or teratogenic effects from caffeine in humans; there seems to be no relationship between moderate coffee drinking and increased heart attacks or cardiac arrhythmias.[40] Indeed, coffee consumption may be life-saving for some: two large prospective studies have shown a strong inverse relation between coffee consumption and suicide among women. Compared with those who did not drink coffee, the relative risk for suicide was reduced by 66% in women who drank two to three cups per day; other caffeinated beverages seemed to show the same protective effect.[41]

As most of us know from our acquaintances, some people seem to cheerfully chug down quarts of caffeinated colas on top of coffee and sleep like the proverbial logs; others cannot touch anything caffeinated after the noon hour without tossing and turning sleepless through the endless night; some cannot tolerate caffeine at all. There may be a physiological basis for these variations:

> Probably the major source of individual differences is related to the remarkable extremes in the rates of caffeine metabolism. For some the half-life is only two hours, whereas others may show a half-life for caffeine of grater than fifteen hours. The longer half-lives are consistent with some individual's observation that even a cup of coffee at noon or early afternoon may result in difficulty in falling asleep at night. On a population basis, the more rapid metabolizers are those who are the heaviest consumers; the slower metabolizers tend to consume much less.[42]

There appear to be degrees of psychological reaction to caffeine on the part of people with generalized anxiety disorder (GAD) or panic disorder (PD) compared with the ordinary population. Caffeine in 500-mg doses can induce panic attacks in normal subjects; but in a double-blind, triple crossover, placebo-controlled study, it was found that those with PD are even more sensitive to the drug, and those with GAD are extraordinarily sensitive.[43]

Coca and Cocaine

Early Uses of Cocaine

History. The Spanish colonists noted how the Indians of what are now Bolivia, Chile, Columbia, and Peru were able to allay fatigue by chewing the leaves of the coca shrub

Erythroxylon coca, and in 1569 the earliest written account of the plant occurred in a medical treatise published in Seville. However, it was ignored by European science and medicine until the middle of the 19th century, when Albert Niemann at Göttingen University first obtained pure crystals of cocaine in 1860. For the next 25 years, the alkaloid was used as a mild stimulant much like caffeine in tea, and was added to numerous proprietary beverages and wines, the most famous of which were Vin Mariani and Coca-Cola.

Vin Mariani. Angelo Mariani was a Parisian druggist of Corsican extraction whose eponymous beverage was already immensely popular by the early 1880s. Many were the rich and famous who praised his cocaine-laced wine, which contained a quite ordinary 11% alcohol but into every ounce of which, according to the French Pharmaceutical Codex, was admixed 0.10 grains, or 6.5 mg of cocaine. The recommended daily intake was a "claret glass" thrice daily after each meal, which comes out to almost 100 mg/day (assuming a claret glass holds about 5 oz).

Mariani was an enthusiastic salesman of his patent medicine, and he went to great lengths to collect and advertise the endorsements of his product. The indefatigable Leo XIII (pope from 1878-1903), who "for sheer productivity surpassed all his predecessors in modern times,"[44] presented Mariani with a gold medal and cited him as a benefactor of humanity (after all, he had been "for years . . . supported in his ascetic retirement by a preparation of Mariani's coca"); composers Gounod, Fauré, and Massenet honored him in their music.[45] If he had only used Vin Mariani earlier in life, said Frédéric-Auguste Bartholdi, he would have engineered the Statue of Liberty a few hundred meters higher.[46] Thomas Edison, Sarah Bernhardt, William McKinley, Emile Zola, and Lillian Russell likewise praised the beverage.[47]

An American imitation of Vin Mariani, "Metcalf's Coca Wine" was advertised as "A Pleasant Tonic and Invigorator [which] Public Speakers, Singers, and Actors have found . . . to be a valuable tonic to the vocal chords. . . . Athletes, Pedestrians, and Base Ball Players have found by practical experience that a steady course of coca taken both before and after any trial of strength or endurance will impart energy to every movement, and prevent fatigue. . . . Elderly people have found it a reliable aphrodisiac superior to any other drug.[48]

The medical profession was just as enthusiastic. American neurologist Dr. James Leonard Corning, who first reported in 1855 a method of inducing anesthesia by spinal block (using, of course, cocaine)[49] gave the beverage his highest commendation: "Of Vin Mariani I need hardly speak as the medical profession is already aware of its virtues. Of all the tonic preparations ever introduced . . . this is undoubtedly the most potent for good in the treatment of exhaustive and irritative conditions of the central nervous system."[50]

The last few years of the life of Ulysses S. Grant were marked by a feverish writing of his superb *Personal Memoirs* (which he wrote to provide a living for his family after losing all his money in fraudulent investment schemes) and the torments of cancer eating away at his throat and mouth. The cancer was probably a result of his lifetime enthusiasm for cigars and alcohol (he was thrown out of the army in 1854 because of continual drunkenness, and only regained his commission at the outbreak of the Civil War). Grant's pain was in no small measure relieved (and, one wonders, his writing accelerated?) by frequent compassionate applications of cocaine, both locally and internally, provided gratis by Mariani. The general died in 1885, only a few days after the manuscript for his memoirs was completed, his family well provided for. And Mariani, too. Sales of Vin Mariani boomed in the United States, and soon there were native imitations: there was Parke, Davis's nationally marketed Coca Cordial; there was Metcalf's Coca Wine; and then there were some more obscure local products, among them Atlanta pharmacist "Doc" Pemberton's French Wine of Coca, probably only remembered now because it provided part of the inspiration for Coca-Cola.

Coca-Cola. By 1885, the demand for Pemberton's French Wine of Coca was exceeding capacity, and Pemberton needed capital. In 1886, he was joined by a local advertising company run by Frank Robinson, and together they incorporated the Pemberton Chemical Company. Robinson also helped to stir the creative pot. Atlanta passed a local prohibition ordinance in 1886 and, while the medical uses of Wine of Coca made it exempt, it also limited its sales. Robinson and Pemberton saw the need for a new, nonalcoholic cold drink which would provide the pep of coffee or tea (or Vin Mariani) during the steamy summer days in Atlanta. Pemberton concocted a new beverage, a syrup which would be mixed at the local drugstores with cold soda water. Since it contained extracts of both the coca leaf and the African kola bean, Robinson suggested the name Coca-Cola.

There was probably much less cocaine in the original Coca-Cola than there had been in French Wine of Coca; instead there was a lot of caffeine (about four times the amount presently used). Originally the caffeine came from the kola bean, but Pemberton found its bitterness so intense that it could be used only in very small quantities, and so he began to add synthetic caffeine purchased from Merck in Germany . Other ingredients were lots of sugar, caramel coloring, lime juice, citric and phosphoric acids, vanilla, orange, lemon, nutmeg, coriander, neroli, and Chinese cinnamon.[51]

In a series of somewhat complicated transactions initiated by Pemberton's terminal sickness and indebtedness, by 1888 Coca-Cola had come under the control and direction of Asa Candler, a prosperous Atlanta druggist, who soon built it into one of the most successful corporations in history. But the first thing he did was to change the formula. The country was beginning to be concerned about the possibilities of "cocainism," and even the minute amount of kola extract in the formula still left a bitter taste. Candler decided to cut back the proportions of coca and kola in the concoction.

But the demon was still there. Two years later, Candler asked the Georgia Pharmaceutical Association to analyze his syrup for traces of cocaine; they easily found it. A fine Christian man who himself neither smoke nor drank, Candler proceeded to work up a process of percolating, macerating, steaming, thumping and whacking the cola leaves until it would seem that there could be little more left in them than cellulose. He was convinced that there was no longer any cocaine in Coca-Cola. But the techniques of the analytical chemists continued to improve, and in 1902, in a controversy over whether Coca-Cola should be subjected to a new federal tax on "medicines," the Georgia State Board of Pharmacy determined that there was indeed cocaine in Coca-Cola: a 6-oz bottle by their calculations contained just under 1 mg.[52] This was declared too tiny an amount to qualify Coca-Cola as a drug, but it didn't satisfy Asa Candler. In 1903, he went to the Dr. Louis Schaefer's Alkaloid Works in Maywood, NJ, the chief supplier of cocaine in the United States, and contracted for Schaefer's most thorough extraction process to be used on his coca leaves. It was a fine partnership: Schaefer could sell Candler what he would have otherwise thrown out. The Schaefer process consisted of powdering the leaves, making them alkaline with bicarbonate, extracting them exhaustively with toluene, and then subjecting them to steam distillation.[53]

But Candler's problems were not over: in 1906, the Pure Food and Drugs Act was passed by Congress, and the official in charge of its enforcement, Dr. Harvey Wiley, set out to prove that the "Coca-Cola habit" was harmful to one's health. In a federal suit against Coca-Cola in 1911, Wiley was unable to prove that there was cocaine or alcohol in the brew, as he first claimed, and his efforts to prove the beverage was "adulterated" with an addictive drug, caffeine, likewise failed. The suit was thrown out of court, but the government appealed. Finally, in 1918, the matter was dropped in a private settlement wherein Coke agreed to cut the amount of caffeine by half. But Wiley, out of office after 1912, never gave up. As late as 1922 he claimed in *Good Housekeeping* that a child who drank three or four Cokes a day "would probably ruin his health for life."[54]

Early Users of Cocaine

Sherlock Holmes. The great fictional detective is most memorably shown using cocaine in the opening scene of "The Sign of the Four," written by Arthur Conan Doyle, M.D.,[55] in 1890:

> Sherlock Holmes took his bottle from the corner of the mantelpiece, and his hypodermic syringe from its neat morocco case. With his long, white, nervous fingers he adjusted the delicate needle, and rolled back his left shirt-cuff. For some little time his eyes rested thoughtfully upon the sinewy forearm and wrist, all dotted and scarred with innumerable puncture-marks. Finally, he thrust the sharp point home, pressed down the tiny piston, and sank back into the velvet-lined arm-chair with a long sigh of satisfaction.

Sherlock's Boswellian companion and medico has been observing this practice thrice daily for many months, and finally is fed up with exasperation. "Which is it today," asks Dr. Watson, "morphine or cocaine?" "Cocaine," replies Sherlock; "a 7% solution. Would you care to try it?" The good doctor indignantly declines the offer, but the great detective rationalizes: "I suppose that its influence is physically a bad one. I find it, however, so transcendently stimulating and clarifying to the mind that its secondary action is a matter of small moment." Watson delivers a stern admonishment: "Your brain may, as you say, be roused and excited, but it is a pathological and morbid process, which involves increased tissue-change, and may at last leave a permanent weakness."[56]

The story ends with Watson interrogating Sherlock: "You have done all the work in this business. I get a wife out of it, Jones gets the credit; pray what remains for you?" "For me," replies the Supreme Sleuth, "there still remains the cocaine-bottle." And with this, doubtless to Watson's chagrin, "he stretched his long, white hand up for it." Granted, some (those unenlightened enough to suppose that Holmes never lived) think this is fiction. But it was written by a physician, and the claim on the part of Holmes that the physical liabilities of cocaine are easily outweighed on a cost-benefit analysis by its property of being "transcendently stimulating and clarifying to the mind" can still be heard today.

Sigmund Freud. Dr. Doyle could have learned of the beneficial properties of cocaine through reading an 1864 article by Dr. Sigmund Freud of Vienna, although it is also quite possible that his interest was aroused by an independent reading of the same works that inspired Freud. In 1883, a German doctor, Theodor Aschenbrandt, wrote of his experiments using cocaine for a variety of indications during the maneuvers of the Bavarian Artillery. The soldiers so treated (sometimes by surreptitiously slipping cocaine into their coffee, hence providing a sort of single-blinded control), were marvelously restored from various states of debilitating exhaustion to full participation in the most strenuous activities of the regiment. Freud also read an 1880 article in the *Detroit Therapeutic Gazette* describing Dr. W. H. Bentley's claims of success in curing morphine addiction by administering addicts an extract of cocaine, thus counteracting the depressant opioid with a powerful stimulant. Freud was intrigued and started experimenting with the drug on himself. In 1884, he published his first paper on cocaine, "Über Coca," and a translation of this paper appeared in the December 1884 *St. Louis Medical and Surgical Journal.*

From the letters of Freud, it is clear that he was in the habit of using small amounts of cocaine himself to relieve his occasional despondency, energize periods of fatigue, and embolden him in challenging social situations. From Paris, where he was just beginning a brief but critical period of apprenticeship under the renowned neurologist Jean Martin Charcot (where Freud's studies of hypnosis and hysteria were to inaugurate the theories of psychoanalysis), Freud writes to his fiancée, Martha Bernays, of his first invitation to a soirée at the Charcots':

> We spent all afternoon preparing for the evening. . . . Ricchetti [another young doctor studying under Charcot] was terribly nervous, I quite calm with the help of a small dose of cocaine, although his success was assured and I had reasons to fear making a blunder. We were the first after-dinner guests and. . . we had to wait for the others to come from the dining room. . . . But then they came and we were under fire: M. and Madame Charcot . . . M. Strauss, an assistant of Pasteur and well known for his work on cholera . . . M. Giles de la Tourette, former assistant to Charcot. . . . And now you will be anxious to know how I fared in this distinguished company. Very well. I approached Lépine, whose work I knew, and had a long conversation with him; then I talked to Strauss and Giles de la Tourette, and accepted a cup of coffee from Mme Charcot; later on I drank beer, smoked like a chimney, and felt very much at ease without the slightest mishap occurring. [Freud goes on to describe a successful bon mot he makes, comparing his momentary lag in apprehending spoken French to a case of tabes, a disease first characterized by Charcot.] These were my achievements (or rather the achievements of cocaine), which left me very satisfied.[57]

There is no indication whatever that Freud ever suffered anything that could be called a toxic reaction to cocaine or an addictive relationship to it. But his efforts to help his friend Dr. Ernst von Fleischl went sadly awry. Fleischl had contracted an infection while conducting research in pathological anatomy; his right thumb eventually required amputation, and he suffered excruciating nerve pain thereafter. He became dependent on increasing amounts of morphine, and Freud, a close friend, attempted to help him by prescribing cocaine, with nearly fatal results. Fleischl eventually injected so much cocaine that he went into toxic psychosis with a vivid experience of formication, or "cocaine bugs" (the sensation of ants running over and under the skin). Freud began to have some doubts about the harmlessness of cocaine and in 1887 he published an article titled "Remarks on craving for and fear of cocaine."[58] Freud had also begun to receive some criticism for his advocacy of cocaine from other famous scientists of the day. A. Erlenmeyer called cocaine "the third scourge of mankind." And Louis Lewin stated in 1885 what is now known to be the case, that "according to all available evidence, coca is no substitute for morphine and a morphine addiction cannot be cured by its use." Although Freud later avoided comment on his earlier papers endorsing the use of cocaine, he continued to take cocaine as late as 1895, when he writes about a dream he had under its influence in his *Interpretation of Dreams*.[59]

William Stewart Halsted. Meanwhile, another Vienna friend of the young Freud, ophthalmologist Karl Koller, had found that cocaine anesthetized the conjunctiva of the eye and for the first time allowed practical surgery (e.g., cataract removal) on the eye.[60] This stimulated Dr. William Stewart Halsted, then a leading surgeon in New York City and later to become the first head of surgery of the Johns Hopkins University School of Medicine, to try cocaine for himself and to attempt to inject it into nerves as a blocking agent and local anesthetic. However, by 1885, he and perhaps three of his closest associates were addicted to the drug to the extent that he stopped attending meetings of the New York Medical Society and found difficulties in teaching and practice. Late in 1885, he published an article ("Practical comments on the use and abuse of cocaine, suggested by its invariably successful employment in more than a thousand minor surgical operations")[61] commenting on an earlier paper in *The New York Medical Journal* in which a Dr. Henry J. Garrigues had complained that cocaine did not work, as claimed by Halsted, when used for local anesthesia. Halsted pointed out rather trenchantly that Dr. Garrigues was not using cocaine correctly: he was using too dilute a solution, injecting it too deeply, and waiting so long after the injection (7 minutes) to begin the operation that the anesthetic effects would have already worn off. Most of this material is perfectly clear. But the opening paragraph of Halsted's article is an incoherent goulash which is generally thought to have been affected by the racing thought processes of the typical late-stage cocaine addict. The incoherence of the paper's opening lines could only have been permitted by the editors because of the great respect the reputation of Halsted's name conveyed:

> Neither indifference as to which of how many possibilities may best explain, nor yet at a loss to comprehend, why surgeons have, and that so many, quite without discredit, could have exhibited scarcely any interest in what, as a local anesthetic, had been supposed, if not declared, by most so very sure to prove, especially to them, attractive, still I do not think that this circumstance, or some sense of obligation to rescue fragmentary reputation for surgeons rather than the belief that an opportunity existed for assisting others to an appreciable extent, induced me, several months ago, to write on the subject in hand the greater part of a somewhat comprehensive paper, which poor health disinclined me to complete.[62]

Halsted withdrew from public life and committed himself to a sanatorium in an effort to rid himself of his addiction. His friends and admirers felt that the experience had benefited him; according to a National Academy of Sciences memoir:

> But, in thus experimenting upon themselves and students, they did not realize that cocaine is a demoralizing habit-forming drug and neither Dr. Halsted nor his co-worker, Hall, escaped, but through superhuman strength and determination Halsted at last overcame it. After an interval of more than a year he came back to a more thoughtful leisurely life with time for reflection and contemplation of his surgical problems, a life in the end far more fruitful than could ever have been the strenuous rush of his existence in New York, if he had kept on at that pace. After all, in his case it was probably no misfortune but rather the reverse.[63]

Another of his biographers agrees:

> Doctor Halsted and the others who injected their own nerves in the preliminary experiments with cocaine unfortunately but understandably fell victims to the "shadow side" of the drug. That Halsted had the perseverance and character to control and eventually overcome this addiction and to return to a splendid productive life is one of his crowning glories. It required about a year in a hospital in Providence, Rhode Island, for Halsted to control his craving for the drug. Welch [first Professor of Pathology of Hopkins] . . . had great faith in Halsted's future, even after this unfortunate cocaine episode. In December, 1886, Doctor Welch invited Halsted to come to Baltimore. . . . This enforced withdrawal from the pressures of his busy life in New York gave Halsted leisure to think and to develop powers within himself that otherwise might have lain dormant. Viewed from this angle Halsted's accidental cocaine addiction was a blessing.[64]

Unbeknownst to these admirers, Halsted had succeeded in overcoming his cocaine addiction only—in a reversal of Freud's treatment of Fleischl—by taking morphine (see Chapter 2). He continued to inject morphine for the rest of his life, but was able nonetheless to pursue a brilliant career as the most respected surgeon in the United States.

William Hammond. William Alexander Hammond was named Surgeon General of the United States Army in 1862, in the midst of the Civil War, when he was only 34 years old. He was later professor of neurology at Bellevue Hospital. In 1887 he published a report on a series of experiments he performed on himself to test the toxicity of increasing levels of cocaine. He first injected a grain (65 mg) of cocaine hydrochloride under the skin of his forearm one evening at eight o'clock. He experienced a "pleasant thrill" and his pulse rate rose from 82 to 94. Retiring at midnight, he was unable to sleep until dawn. So the next night he started early, injecting twice as much at seven o'clock. He felt a similar "exhilaration" and a great desire to write, which he did with freedom and ease, expecting all the while it might prove later to be nonsense he was writing. By midnight, his pulse had risen to 112; he was unable to sleep and tossed and turned "thinking of the most preposterous subjects." His symptoms persisted until noon the next day, when he felt normal again after drinking "two or three cups of strong coffee." And he was "agreeably disappointed" to discover that his writings of the previous night were perfectly coherent. From this he proceeded to inject 4 grains (260 mg), noticing this time that his heart alternatively beat so rapidly he was unable to count his pulse, and then would occasionally skip a beat or slow suddenly to 60 beats a minute. And he "wrote page after page, throwing the sheets on the floor without stopping to gather them together," believing at the time that they were quite profound, but finding the next day they were "a series of high-flown sentences altogether different from my usual style," and concerned with "matters in which I was not the least interested."

Undaunted by these ominous signs, on subsequent evenings (separated by a few days to catch up on his sleep), he increased the dose to 6, 8, 10, 12, and finally an astonishing 18 grains (1,170 mg). Within 5 minutes of the injection of this enormous dose, his pulse had risen to 140, and within 30 minutes he felt that his mind was passing beyond his control and he lost any awareness of his behavior. At nine o'clock the next morning he found himself in bed "with a splitting headache and a good deal of cardiac and respiratory disturbance." On going downstairs, he found his library "strewn with encyclopedias, dictionaries and other books of reference, and one or two chairs overturned." He concludes that he had come very near the fatal dose.[65]

As late as 1901, cocaine continued to be strongly recommended by eminent physicians, Mortimer writing that "whether the condition combated be muscular tire, nerve exhaustion from worry, or a physical incapacity due to chemical changes in the blood, the action of coca is depurative [purifying]."[66]

Cocaine: Freebase and Hydrochloride Salt

Terms of Endearment. Now, at the end of the 20th century, cocaine is most familiar as a drug of abuse, and the extensive use of cocaine in contemporary culture is suggested by the numerous slang expressions which exist for its two forms. As the freebase, **4-17** (Figure 4.6), it is known as "base," "rock," "crack," "pasta," "hubba," "bazooko," "petillos," "slab," "crazy 8ths," "moonrock," and "lace joints"; as the hydrochloride salt, **4-18**, "coke," "blow," "toot," "snow," "flake," "girl," and "lady." The effect of either is fairly short, and dedicated users may indulge in binges, using the drug every 20 minutes or so. In the early 1980s, sophisticated coke users did their own freebasing: the hydrochloride was dissolved in water, the water made alkaline with household ammonia or lye, and the lipophilic freebase extracted into ether. The ether was then evaporated to give the smokable free base, or—in the case of incautious cigarette smokers—the volatile ether solution was accidently ignited like a Molotov cocktail.[67]

As freebasing became more popular, inventive drug dealers in Detroit developed a simpler procedure: high-grade cocaine hydrochloride was simply dissolved in water with baking soda and boiled down to a solid residue. When the mixture was smoked, the baking soda released carbon dioxide and made a crackling noise. Hence the origin of the term "crack." According to a 1996 best-selling novel, some experienced users "cook" their own crack:

> From the purse she removes . . . two little thick glycine bags each holding four grams of

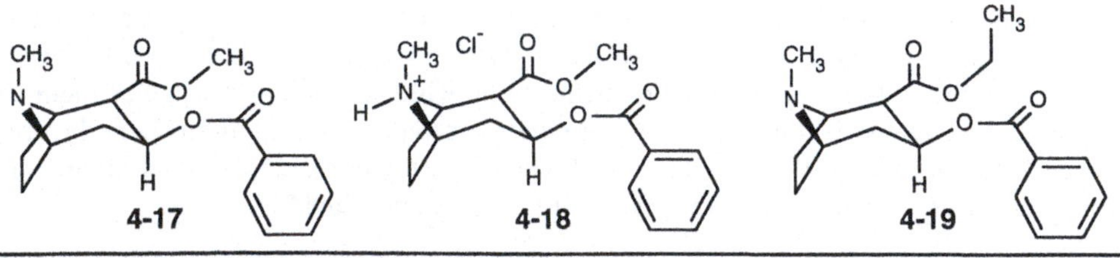

Figure 4.6 Cocaine freebase and hydrochloride; cocaethylene.

> pharmaceutical-grade cocaine. . . . She taps half a glycine's worth into the cigar tube and adds half again as much baking soda, . . . [and] fills the rest of the tube to the top with water. She holds the tube up straight and gently taps on its side with a blunt unpainted nail, watching the water slowly darken the powders beneath it. She produces a double rose of flame in the mirror that illuminates the right side of her face as she holds the tube over the matches' flame and waits for the stuff to begin to bubble. . . .
>
> The improbable thing . . . is that when the soda and water and cocaine are mixed right and heated right and stirred just right as the mix cools down, then when the stuff's too stiff to stir and is finally ready to come on out it comes out slick as s— from a goat . . . its snout round from the glass tube's bottom.[68]

In Europe, specifically the Netherlands, crack is said not to be used. However, while crack as defined above and found in the United States is not sold to users, cocaine hydrochloride in quite pure form is. And it has become common for users to "cook" the cocaine salt with ammonia, boiling away the water and then smoking the residue, which is called "base." This is, of course, essentially the same substance, perhaps even purer, than American crack. And it has developed a reputation for causing compulsive use escalating to psychoses.[69]

In the popular imagination, crack is believed to be much more addictive than snorted cocaine, and more likely to induce violence. The hard clinical evidence shows only a slight difference between these modes: the blood levels from intravenous cocaine use are the highest of any mode of ingestion; smoked freebase (crack) produces slightly less intense a spike (with some loss of material from burning); and snorting only a little less than smoking.[70] Many "chippers" have tried one or all of these methods once or several times and yet have not become "enslaved." Some (e.g., television talk-show host Oprah Winfrey)[71] have later gone on to become models of success and good behavior.

But there seems no doubt that the subjective "kick" from injected cocaine is perceived as more satisfying than that which comes from snorting the drug. For this there is the testimony of William S. Burroughs:

> The full exhilaration of cocaine can only be realized by an intravenous injection. The pleasurable effects do not last more than five or ten minutes. If the drug is injected in the skin, rapid elimination vitiates the effects. This goes doubly for sniffing.[72]
>
> Before you can clean the needle the pleasure dims. Ten minutes later you take another shot. No visceral pleasure, no satisfaction of need, no increase of enjoyment, no sense of well-being, no alteration or widening of perspective, C[ocaine] is electricity through the brain stimulating pleasure connections that can only be known with C. . . . Once C channels are stimulated, there is urgent desire to restimulate and fear of falling from C high. The urgency lasts as long as C channels are activated—an hour or so. Then you forget it because it corresponds to no pleasure ordinarily experienced.[73]

Too Much of a Good Thing: Problems with the Lady. Cocaine resembles alcohol in that, at least for some people, it strongly evokes the phenomenon of "acute tolerance." That is, once the drug is taken in a sufficiently high first dose, it tends to be binged in an endless and always unsuccessful effort to repeat the initial burst of euphoria:

> You just keep doing it to try to get that rush, it's never quite right. I mean the first hit is the best anyway. After that whatever you do all night long is just like shit, it's like . . . that's not IT

. . . that's not what I wanted. Let's try it again and get it right this time. And again, No, that's not it either, and on and on it goes.[74]

But with chronic escalating use of cocaine, toxic effects eventually develop such as tachycardia, hypertension, formication ("cocaine bugs"), visual hallucinations, and paranoid psychosis. Sudden death can result from cardiovascular collapse, even among young persons in otherwise excellent physical health. As of December 1989, at least 58 cases of acute myocardial infarction related to intranasal cocaine abuse had been reported in the literature.[75] Considering the high usage of cocaine during the late 1980s, one could argue that this side effect, while unfortunately lethal, is still relatively rare. But it is troubling that the actual interaction of cocaine with the body that causes cardiac arrest is not well enough understood to be able to suggest a "safe" way to use cocaine and minimize cardiac risk.

The toxic effects of cocaine on the heart almost certainly come from its inducing a temporary arrhythmia (something that other local anesthetics can also cause when administered systemically). But the propensity of an individual to develop cardiac arrhythmia when using cocaine seems to some degree idiosyncratic: some people obviously can use large amounts with only the psychological problems of addiction or stimulant psychosis to deal with; others can't. The case of Len Bias (a University of Maryland basketball star who died in 1986 of cocaine intoxication) is typical—so far as anyone knows, Bias's heart was in fine shape absent cocaine. It is indisputable that cocaine in very large amounts can induce intense spasms in the coronary arteries, causing either a short-term episode of angina or a myocardial infarction.[76] But the problem is when lower amounts provoke angiospasm.[77]

Another problem may arise in the unusual pharmacokinetic interaction of high levels of alcohol and cocaine, which is a common enough combination in the North American recreational drug scene. The human liver, that busy little laboratory of chemical synthesis, actually combines these two drugs in significant amounts when they are taken simultaneously in recreational doses: the synthetic product, *cocaethylene* (**4-19**, cocaine in which the methyl ester has been replaced with an ethyl ester), can be detected in the urine of study subjects given a few shots of vodka followed by a 100-mg snort of cocaine.[78] Neuropharmacologist Deborah Mash at the University of Miami School of Medicine claims that combining alcohol use and cocaine can cause death from overdose at cocaine blood levels of only a tenth of those known to be fatal when cocaine is used alone, and hypothesized that the increased toxicity was due to cocaethylene.[79] But in a single-blind crossover study of cocaethylene and cocaine in human volunteers, cocaethylene produced a lessened high and a lower effect on heart rate than cocaine.[80] It is more likely that the synergistic cardiotoxicity comes from the combined direct effect of each drug: in a study on dogs given IV ethanol followed by cocaine, the authors concluded that "cocaine combined with ethanol produces a significant synergistic depression of ventricular contraction and relaxation that substantially exceeds the arithmetic sum of the depressive effects of either cocaine or ethanol alone."[81]

Probably the safest way to take cocaine is the way the native Peruvians do; slowly chewing the leaves in a cud, or sipping a coca-leaf tea. The tea thus made is said to be "a mild stimulant [that] has none of the addictive or intoxicating effects of cocaine."[82] It also reportedly eases the headaches, nausea, and weakness of altitude sickness, which most visitors experience in the 12,000-ft Andean cities of Bolivia and Peru. Upon arrival for a state visit to El Alto, Bolivia, Queen Sofia of Spain publicly drank coca leaf tea at the airport, and Bolivian President Jaime Paz Zamora dismayed drug enforcement agencies throughout the world when, in 1992, he proposed balancing the budget by exporting legal "herbal" tea compounded of coca.[83] The proposal seems to have been dropped.[84]

Mode of Action of Cocaine. Although there is always a certain epistemological gap in relating behavior to neuronal systems, it seems almost certain that the euphorigenic properties of cocaine are attributable to its effects on the dopaminergic (DA) system. DA neurons *look* like the sort of neurons that might affect something so all-encompassing as "mood": both the axons and dendrites of DA neurons are more "arborized"than most, having as many as 100,000 synapses for each neuron, and there are many interconnections between dendrons as well as the usual axon-dendron connections. These morphologic features suggest a modulatory or tonic function.

The principal psychic effects of cocaine appear to be the result of its ability to block the dopamine receptor site at the dopamine reuptake shuttle in the presynapse (most significantly in the area of the brain known as the nucleus accumbens); the ultimate effect of this blockade is to increase the concentration of dopamine in the synapses. (It may be that similar activity accounts for the evolutionary pressures driving the humble coca plant to produce the alkaloid in the first place: cocaine is an even more powerful inhibitor of the reuptake of the insect-specific NT octopamine; by causing the insect to overdose on its own octopamine, it thus functions in the plant as a natural pesticide.)[85]

Cocaine Metabolism. The effects of cocaine are notoriously short-lived; this phenomenon is due to the rapid cleavage of the two ester functions in the cocaine molecule, either of which is subject to hydrolysis from the common esterases in biological systems or by simple nonenzymatic hydrolysis. In human beings, it appears that 80-90% of cocaine is metabolized by ester hydrolysis; some cocaine is also degraded by *N*-demethylation to norcocaine followed by oxidation to *N*-hydroxynorcocaine. Urinary excretion in humans accounts for almost 90% of ingested cocaine, with as little as 1% unchanged cocaine.[86] Urine screens for cocaine use usually test for the presence of benzoylecgonine, the product of cleavage of the methyl ester function, rather than for ecgonine methyl ester, the product of cleavage of the benzoyl group. As with the opioids or marijuana, ordinary tests (e.g. for employment) employ a preliminary urinalysis screening using thin-layer chromatography (TLC) or enzyme immunoassay (EIA). TLC is inexpensive and can test for up to 40 drugs in one sample; but a number of substances in the urine can interfere and lead to false positives. EIA and the related radioimmunoassay (RIA) rely on antibody binding to ecgonine methyl ester; in either procedure, cross-reactivity can occur whereby chemicals with some similarity—in the case of cocaine this includes several antibiotics—can also bind to the antibody, leading again to false positives. Either of these methods must be confirmed by gas chromatography-mass spectroscopy (GC-MS) which is the most reliable—but most expensive—method presently available.

Cocaine continues to be used legitimately in medicine for ear, nose, and throat anesthesia (e.g., for bronchoscopy) and less often for eye (corneal) anesthesia, where better agents have supplanted it. It is used in nasal surgery and as a vasoconstrictor before examination. Because patients legitimately exposed to cocaine might be subject to urine testing for illegal use, a group of otolaryngologists at the University of Texas, San Antonio, tested a group of patients who were given cocaine in the usual preoperative manner: 4 ml of a 4% solution on cotton swabs. They tested positive for benzoylecgonine, the most common screening test for cocaine use, 24 hours later, but all were negative in 72 hours.[87]

Syntheses of Cocaine

No synthesis of cocaine produces material in sufficient yield to be at all cost-competitive with the natural supply. The first synthesis was that of Willstätter in 1923;[88] the later modifications by Robinson were remarkably close to the presumed biosynthetic pathway. The most elegant synthesis to date is that of Tufariello, which is based on an intramolecular cycloaddition of nitrone **4-20** (Figure 4.7), to form **4-21**, from which *dl*-cocaine, **4-22**, can be formed in good yield.[89]

Local Anesthetics Modeled on the Cocaine Structure

Procaine. One of the first local anesthetics developed to replace cocaine was procaine (Novocaine, **4-25**, Figure 4.8); it was one of several compounds synthesized between 1890 and 1905 by Albert Einhorn, professor of chemistry at the University of Munich, in an effort to make a nonaddicting molecule that would still have the anesthetic properties of cocaine. Einhorn believed (mistakenly) that the addictive properties of the cocaine molecule came from its six-membered heterocyclic ring containing the nitrogen atom, *piperidine* (**4-23**). He believed this because the piperidine ring is also found in the alkaloid *coniine* (2-propylpiperidine, **4-24**), which is the toxic principle in hemlock responsible for the death of Socrates. So he made a series of molecules having most of the remaining pieces of the cocaine structure without the piperidine ring. Novocaine was one of these, and it produced a local anesthesia not unlike cocaine. Novocaine was used for many years in dentistry, but it has now been replaced by similar, superior agents such as lidocaine.

It appears that the psychoactivity of procaine has gone largely unrecognized. In a recent study, high doses of procaine were injected IV in 32 volunteers. The effects differed among several subgroups. While all experienced some level of auditory halluci-

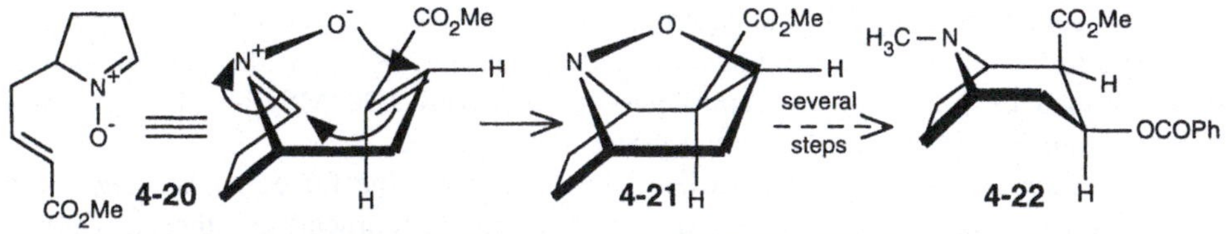

4-20 4-21 4-22

Figure 4.7 Tufariello's synthesis of cocaine.

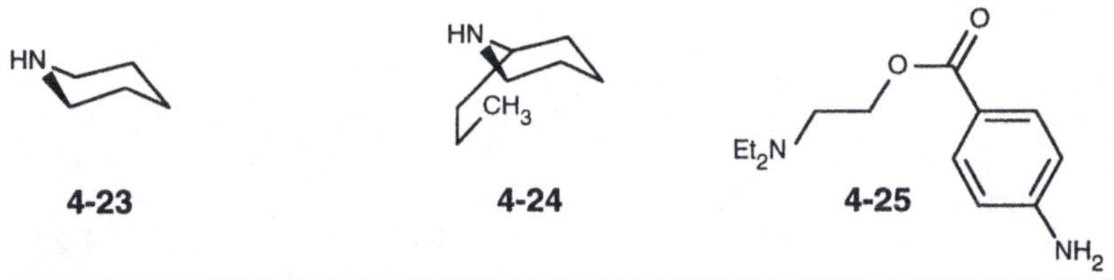

4-23 4-24 4-25

Figure 4.8 Piperidine, coniine, procaine.

nations, nine experienced intense unformed visual hallucinations; the remaining subjects had no hallucinations or only mild visual effects. Nine subjects experienced intense fear meeting the criteria for clinical panic attacks, and nine other subjects experienced intense euphoria. There was no overlap between those subjects whose response was affective (fear or euphoria) and those whose response was visual. Subjects with ongoing depression experienced transient alleviation of their symptoms.[90] Procaine and other local anesthetics are widely used to "cut" (dilute) cocaine, and they are sometimes sold as "synthetic" cocaine. Depending on which subset of responders the buyers belong to, they may be getting more (or less, or something quite different) than they bargained for.

Amphetamines and the Psychostimulants

This section will consider the *psychostimulants*, of which the paradigm is amphetamine. The great majority of these, like amphetamine itself, incorporate an underlying 1-phenyl-2-aminopropane skeleton; a subset of less powerful stimulants—including the active principles of two ethnobotanicals, ephedra and qat, which have been in use for millenia—are closely related in structure but have an additional oxygen substituent on the carbon bearing the phenyl substituent.

Amphetamines

Amphetamine. Amphetamine (Benzedrine, "crosstops," "black beauties," "whites," "bennies," "cartwheels," "crank," **4-26**, Figure 4.9) is considered the prototype CNS stimulant. It is a relatively simple structure, and its chemical synthesis dates back to 1897. But it was not introduced into clinical medicine (by Smith Kline and French) until the late 1920s, when Gordon Alles investigated its use as a substitute for ephedrine, which at that time had become expensive due to restricted availability of the Chinese plant from which it was extracted.[91]

The subjective effects produced by the most powerful amphetamines—amphetamine itself, dextroamphetamine, and methylamphetamine—and the pattern, among heavy abusers, of binge use and final psychotic toxicity, are essentially identical to that found with cocaine. There are also, as with cocaine, few physical withdrawal symptoms, although psychological craving can be intense. Withdrawal is followed by a state of mental depression and physical fatigue qualitatively similar to the withdrawal from

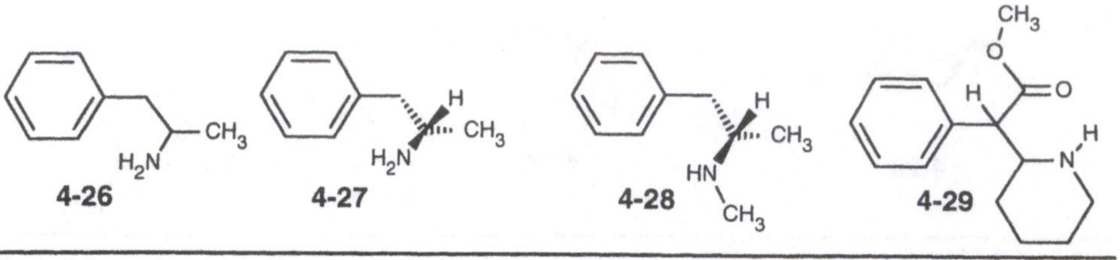

4-26 4-27 4-28 4-29

Figure 4.9 Amphetamine, dextroamphetamine, methamphetamine, methylphenidate.

cocaine. The differences between amphetamines and cocaine lie chiefly in the length of time the effect of amphetamines lasts; depending on the particular amphetamine, this can be as long as 24 hours, whereas with cocaine it is notoriously brief. Unlike cocaine, amphetamine induces very significant tolerance, which develops slowly but eventually allows ingestion or injection of amounts several hundred times greater than the original dose. The tolerance is uneven, however: sleeplessness usually persists, and psychotoxic effects such as hallucinations and delusions begin to develop with increasing doses. Chronic users have reported injecting as much as 15,000 mg in 24 hours without fatality, while novices can be killed by rapid injection of 120 mg.

A paranoid psychosis involving feelings of omnipotence, delusions of persecution, and auditory or visual hallucinations can result from long-term use of high doses of amphetamines. As a matter of fact, if any drug can be called a psychotomimetic, it would have to be high-dose amphetamines, which create a "psychosis" that even experienced psychiatrists find indistinguishable from the endogenous variety—except that amphetamine psychosis eventually dissipates when the drug is withdrawn. Experienced users of amphetamines learn to expect psychotic episodes and can often compensate by isolating themselves until they recover; recovery from even prolonged amphetamine psychosis is usual within a few days or weeks of withdrawal. Solomon Snyder has pointed out other similarities between the home-grown and the amphetamine-induced psychotic states: antipsychotic medications like the phenothiazines or butyrophenones (see Chapter 5) are effective in ameliorating the psychosis, and the hallucinations in both conditions are usually auditory (unlike the visual distortions in perception experienced by users of psychedelics). On the other hand, the flattened, obtunded affect characteristic of schizophrenia is not found with those experiencing an amphetamine crisis.[92]

Paranoid psychosis is understandably associated with violence, and it seems reasonable to assume that amphetamine use predisposes to violent or more aggressive behavior. But the evidence in both animals and humans is mixed: in some studies amphetamines increase aggressive behavior, in some they reduce it.[93] Higher doses for longer periods probably predispose to violence (perhaps simply from the irritability which accompanies loss of sleep), but social circumstances and learned expectations seem to play as significant a role as pharmacology: compare the quite different responses to amphetamine use on the part of William S. Burroughs and Graham Greene described below.

Amphetamine as a free base is a volatile oil, and this was put to good effect in its original use, absorbed on cotton wadding with menthol in a tube marketed in 1932 by

Smith Kline and French as the Benzedrine Inhaler for the relief of nasal congestion. It soon found other uses, as narrated in *Junky* by William Burroughs:

> "Benzedrine is a good kick," she said. "Three strips of the paper or about ten tablets. Or take two strips of benny and two goof balls [nembutal]. They get down there and have a fight. It's a good drive. . . ." Mary cracked a Benzedrine tube expertly, extracting the folded paper, and handed me three strips. "Roll it up into a pill and wash it down with coffee."
>
> The paper gave off a sickening odor of menthol. Several people sitting nearby sniffed and smiled. I nearly gagged on the wad of paper, but finally got it down. . . . I began talking very fast. . . . I was full of expansive, benevolent feelings, and suddenly wanted to call on people I hadn't seen in months or even years, people I did not like and who did not like me.[94]

Amphetamine was used in the past as an appetite suppressant, but development of tolerance to this anorectic effect is rapid and more effective agents have been developed. It is still used to treat narcolepsy—a rare syndrome consisting of uncontrolled episodes of excessive REM sleep accompanied by sudden loss of muscle tone and hallucinations—and as an alternative to methylphenidate in the treatment of children with attention-deficit hyperactivity disorder (ADHD). Amphetamine was widely available as an OTC drug in the 1930s, and it was used by the military of both sides, particularly pilots, during World War II.

In 1938, as a struggling writer with a wife and two children to support, Graham Greene describes how amphetamine enabled him to write a money-making spy novel, *The Confidential Agent* in only 6 weeks:

> I fell back for the first and last time in my life on Benzedrine. For six weeks I started each day with a tablet, and renewed the dose at midday. Each day I sat down to work with no idea of what turn the plot might take and each morning I wrote, with the automatism of a planchette, two thousand words instead of my usual stint of five hundred words. In the afternoons *The Power and the Glory* proceeded towards its end at the same leaden pace, unaffected by the sprightly young thing who was so quickly overtaking it. . . .
>
> Six weeks of a Benzedrine breakfast diet left my nerves in shreds and my wife suffered the result. At five o'clock I would return home with a shaking hand, a depression which fell with the regularity of a tropical rain, ready to find offense in anything and to give offense for no cause. For long after the six weeks were over, I had to continue with smaller and smaller doses to break the habit. Sometimes looking back I think that those Benzedrine weeks were more responsible than the separation of war and my own infidelities for breaking our marriage.[95]

The precise mode of action of amphetamine probably involves the close similarity of its structure and the endogenous catecholamine NTs. There are four major neural effects: (1) inhibition of reuptake of several NT amines such as norepinephrine, epinephrine, and dopamine; (2) enhancement of release of catecholamines; (3) direct α-adrenergic receptor stimulation; (4) inhibition of monoamine oxidase in higher concentrations. Controversy exists as to whether the dopamine or norepinephrine systems are the most affected.

Dextroamphetamine. The only difference between amphetamine and dextroamphetamine (Dexedrine, "dexies," "Christmas trees," "beans," **4-27**) is that the latter compound is the (exclusively) dextrorotatory (optical, chiral) isomer, whereas what is marketed under the name of amphetamine is a 50/50 racemic mixture of both levo and dextro isomers.

Dextroamphetamine is a more potent appetite suppressant than amphetamine, and its effect on the cardiovascular system is slightly less pronounced. It is used as an alternative to methylphenidate for attention deficit disorder in children and for narcolepsy.

Methamphetamine. Methamphetamine (Methedrine, Desoxyn, "crank," **4-28**) is essentially equivalent to dextroamphetamine in its effect on the CNS and in suppressing appetite. As a street drug, it has been injected as the water-soluble hydrochloride salt ("crystal meth"); more recent custom favors smoking it in its freebased form ("ice," "glass," "batu").[96] It provides a 24-hour high versus the 4-hour effect from smoking crack cocaine. The press and some health officials sounded alarms over methamphetamine in the late 1980s. According to Dr. Jon Jackson of the University of Hawaii at Manoa, use of smokable crystal mushroomed during the summer of 1988 and all but replaced crack as the drug of choice.[97] The drug was in wide use among prostitutes in Hawaii, and Jackson conceded that "one benefit of the shift may be the elimination of needles, thus reducing the risk of infection with HIV." However, he stated that chronic users came to emergency rooms in acute psychosis with auditory hallucinations and extreme paranoia. "Often, these patients demonstrate destructive behavior. . . . Unlike the acute paranoid disorders seen in users of cocaine, these psychotic symptoms do not resolve during the next few hours, and they may persist for days, weeks, or even longer."[98]

As of 1993, the anticipated epidemic use of methamphetamine had not developed. Two social scientists interviewed users of ice in the San Francisco area and discovered that many found the ice high was too long and too intense; several had stories of severe paranoia and hallucinations. However, authors of the study conclude that the primary factor influencing the form of this or any drug use is likely to be social trends rather than drug pharmacology.[99]

Methamphetamine can be synthesized from ephedrine or pseudoephedrine, the active ingredients in several OTC nasal decongestants like Sudafed; it is claimed that mixing the red phosphorus of an emergency flare and the iodine of common antiseptic tinctures can provide the hydrogen iodide needed to effect the reduction. In late 1995, there were reports that Mexican drug cartels had obtained tons of ephedrine to synthesize methamphetamine, and that high-purity, cheap "meth" from this source had already begun to appear in quantity on the streets of Los Angeles.[100] In 1996, there were reports of clandestine laboratories synthesizing methamphetamine and of several deaths from its use in central Iowa.[101]

Methylphenidate. Methylphenidate (Ritalin, "pellets," **4-29**) is well known, for its use in the treatment of children with attention-deficit hyperactivity disorder (ADHD).[102] This is a syndrome which usually emerges in early childhood and consists of excessive restlessness, impulsiveness, and difficulty in sustaining attention.[103] There is considerable controversy concerning the diagnosis of this disorder and the use of methylphenidate to treat it. Production of Ritalin has increased by nearly 500% in the last 5 years. Is this because "the treatment is catching up with the illness," which is what Dr. Joseph Biederman, chief of pediatric psychopharmacology at the Massachusetts General Hospital believes? Biederman thinks that as many as 10% of the nation's children could benefit from the drug. (About 5% of American boys and 2% of girls are estimated to take the drug as of 1996.)[104] Others feel that the upsurge in Ritalin use is "a cause for alarm," in the words of Dr. James Swanson, professor of pediatrics at the University of California at Irvine. It is his conviction that use of the drug is unwarranted in about a third of the children now taking it.[105]

According to a commentary in the *Archives of General Psychiatry,* the use of methylphenidate in the treatment of childhood ADHD proved to be "unqualified" and "amply documented" success.[106] Symptoms of the disorder for many children are not only improved, they are eliminated. Here is testimony from one of those helped, Annie I. Anton of Atlanta, writing to *The New York Times*:

> At 11 years old I was diagnosed with attention deficit disorder and dyslexia. I have taken Ritalin for 18 years. Thanks to the drug, the support of my parents, and hard work, and in spite of having always been told by my teachers that I would never go to college because I was "too social and lazy," I graduated from college in 1990. In 1992 I earned my master's degree in computer science.
>
> Dr. Breggin [in a letter of 20 May 1996] criticizes the drugging of "our best and brightest." Yet I am now 29 years old, and with the help of this medication I have been able to set my goals higher than I dreamed possible. And I have achieved them. This year I will obtain my Ph.D.
>
> There are young adults with attention deficit disorder who are leading successful lives as contributing members of our communities thanks to this medication. Perhaps Dr. Breggin would do well to identify a few of Ritalin's success stories. Maybe then the fears of some parents will be alleviated.[107]

There is even more controversy about the use of methylphenidate to treat adults with ADHD. But a 1995 double-blind, crossover study by a group from the Massachusetts General Hospital and Harvard Medical School found a quite marked therapeutic response (78% versus 4%) for methylphenidate treatment of adults with ADHD symptoms who had suffered ADHD as children.[108] And there are probably many adults who have been taking psychostimulants daily for decades; this class of drugs was once prescribed quite freely for symptoms such as low blood pressure. William F. Buckley, Jr., has been taking methylphenidate daily for more than 30 years—with no deleterious consequences and with apparent benefit—because, in his words, after fainting his "first and last time," his doctor said that his blood pressure "was so low that I should either take a quarter pound of chocolate in mid afternoon, or a Ritalin. Big deal! I doubt, by the way, that a doctor would nowadays say that because some people are affected adversely by Ritalin."[109] There are cyclical trends in prescribing this class of drugs, responding to periodic alarms expressed in the media.[110] Presently, there is some movement back to using methylphenidate for depressed adults who have not responded well to standard antidepressants, and for apathetic and withdrawn elderly patients.[111]

Methylphenidate is probably less likely than some other psychostimulants to provide the sort of quick high which may encourage abuse. It is otherwise not believed to be essentially different in its psychopharmacological effects from other psychostimulants. And even with methylphenidate, there are isolated reports of patients, their friends, or their siblings grinding up and snorting or injecting Ritalin capsules[112] in pursuit of a more interesting rush.[113] In a recent study using positron emission tomography to compare the pharmacokinetics and distribution in the human brain of methylphenidate and cocaine, it was indeed concluded that the drugs acted at the same parts of the brain, competing for identical binding sites. Both provided a "high" corresponding to rapid uptake in the striatum, but methylphenidate cleared very slowly from the brain, and cocaine cleared rapidly.[114] The authors point out that methylphenidate is taken orally, not injected IV, as it was in this study, and hypothesize that this and the slow clearance from the brain serve as limiting factors in promoting its frequent self-administration.[115]

Pemoline. Pemoline (Cylert, "popcorn coke," **4-30**, Figure 4.10) is another drug used in the management of children with attention deficit disorder, although some studies have indicated that it is inferior to methylphenidate. Although no potential for abuse was found in studies in primates, psychotic symptoms have been reported in adults who misused the drug.

Phenmetrazine. Phenmetrazine (Preludin, "pink hearts," **4-31**) is comparable to dextroamphetamine in suppressing the appetite and also equivalent in its abuse potential. One study has shown the drug to be significantly helpful, however, in alleviating or abolishing premenstrual syndrome (PMS). Another essentially similar drug is *phendimetrazine* (Phenazine, Bacarate, Prelu-2, **4-32**).

Fenfluramine. Fenfluramine (Pondimin, Dexfenfluramine, **4-33**, Figure 4.11) is a newer anorectic, and it seems to mark a genuine step forward in the effort to create an appetite suppressant with minimal abuse potential. Although a phenethylamine, it can be given late in the day because it actually has a sedative rather than a stimulant effect. It improves glucose tolerance and so is quite helpful for patients with diabetes (who also frequently suffer from obesity).[116] In addition, it has a slight hypotensive effect and so can be used by persons with high blood pressure. In one study, fenfluramine given for one year induced an average weight loss of 25 lb within the first 5 to 7 months. Despite these promising characteristics, it seems to be euphorigenic for some individuals and has become a recreational drug in South Africa. It can cause physical dependence after long use, and sudden withdrawal can cause severe depression, while overdosage can cause convulsions, coma, and death.

Ricaurte's group at Johns Hopkins University, which studied the neurotoxic effects of MDMA ("Ecstasy") in animals and humans (see Chapter 6) has shown that fenfluramine has effects on serotonin neurons similar to those caused by Ecstasy. Moreover, most Ecstasy users use this drug only occasionally, but the indication for using fenfluramine, weight loss, is such as to require its daily use for months or years. In a 1994 article, the authors conclude that "concern over possible dexfenfluramine neurotoxicity in humans is warranted, and that physicians and patients alike need to be aware of dexfen-

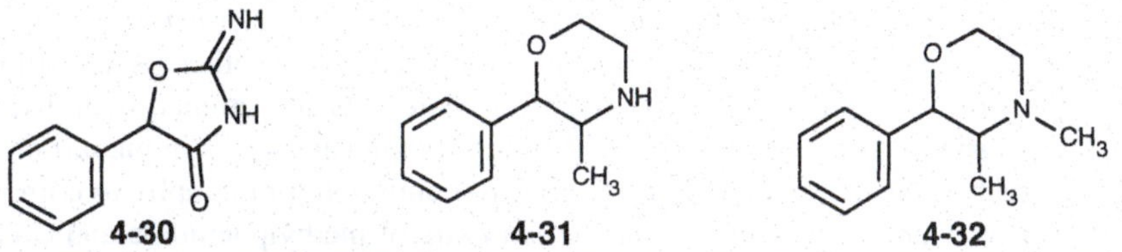

4-30 4-31 4-32

Figure 4.10 Pemoline, phenmetrazine, phendimetrazine.

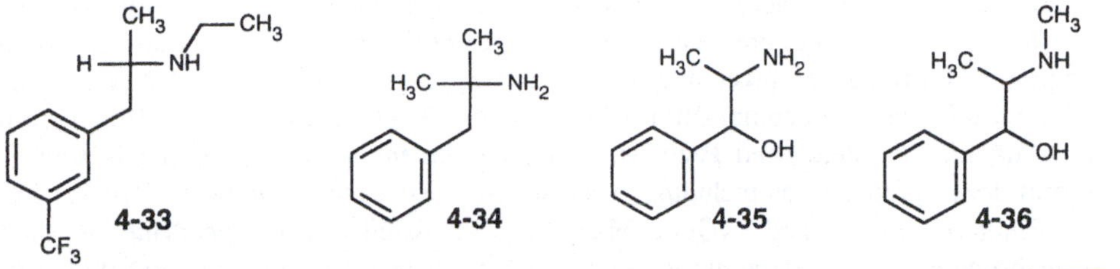

Figure 4.11 Fenfluramine, phentermine, phenylpropanolamine, ephedrine.

fluramine's toxic potential towards brain serotonin neurons."[117] Despite these and other concerns, dexfenfluramine (Redux) was approved by the FDA in early 1996 for the treatment of obesity.[118]

Phentermine. The abuse potential of phentermine (Adipex-P, Fastin, Obermine, Phentrol, "robin's eggs," "black and whites," **4-34**) seems genuinely low, perhaps because it rarely produces euphoria, but has a tendency to cause insomnia. A 1992 report[119] from a 4-year federally financed study shows that a combination of fenfluramine and phentermine significantly helped patients lose weight and maintain the weight loss. The study was conducted with a variety of double-blind controls. The two drugs were used in combination because their mode of action is believed to be different—fenfluramine increasing serotonin in the brain, and phentermine (like the amphetamines) increasing dopamine and norepinephrine. Also, the sedative effects of fenfluramine counter the stimulant effects of phentermine.

Phenylpropanolamine. Phenylpropanolamine (Acutrim, Dexatrim, **4-35**) is a nonprescription drug that has been sold OTC for more than half a century, both for appetite control and as a nasal decongestant. Although it has a structure like amphetamine, the hydroxy group reduces its lipid solubility so that it does not cross the blood-brain barrier as readily as the amphetamines; it seems to have little or no significant CNS stimulant effects. In controlled clinical trials, it improves weight loss by about 4 lb over placebo. Side effects include tachycardia, sleeplessness, and headache, but their incidence is low. Phenylpropanolamine is structurally related to ephedrine (**4-36**); it is racemic norephedrine, a mixture of equal amounts of both enantiomers of ephedrine without the *N*-methyl group.

Ephedrine. Ephedrine (**4-36**) is the active principle found in the Chinese medicinal herb, mahuang[120] (*Ephedra sinica* Stapf) and in other species such as *E. vulgaris* and *E. equisetina* Bunge. The last contains nearly 1% ephedrine and is the best natural source of the alkaloid. The herb has been used in Chinese folk medicine for more than 5,000 years, and was introduced to Western medicine in the 1920s by Ku Kuei Chen, a pharmacologist at Peking Union Medical College who had recently completed his doctorate at the University of Wisconsin.[121] There are two chiral centers in **4-36**, and four possible isomers: two of these occur naturally in ephedra—*l*-ephedrine, which is (1*R*,2*S*)-(–)-α-(1-methylaminoethyl)benzyl alcohol (*erythro* in the older terminology), and *d*-pseudoephedrine, which is the (1*S*,2*R*)-(+)-α-(1-methylaminoethyl)benzyl alcohol (*threo*)—but all four isomers are readily and inexpensively available from synthetic routes, and all isomers as well as the racemic mixtures exhibit similar physiological activity. They relieve the symptoms of asthma, hay fever, and colds by acting as bronchodilators and decongestants, and are used alone or in combination in numerous OTC preparations: Ephedrine is found in Bronkaid, Primatene, and Pazo hemorrhoid ointment; pseudoephedrine is found in several dozen preparations including Allerest, Comtrex, Contac, Dimacol, Dristan, Sinarest, Sinutab, Sudafed, and NyQuil. Many truck stops sell vials of ephedrine, nominally for its bronchodilating effect, but in reality for long-distance truckers who use it as a mild stimulant like caffeine. (Unlike caffeine, it is neither a diuretic nor taken with large quantities of water and therefore does not necessitate frequent pit-stops.)

"Herbal Ecstasy." In 1995 and 1996, a variety of so-called "herbal ecstasy" products began to be advertised in publications like *High Times* and sold at "raves." These are touted as being "all natural," "herbal," and "legal" (which they certainly are, but then so are numerous lethal plants from the average garden) and, by implication, safe. Here are the ingredients in one brand, Ecstacy [sic], manufactured by Global World Media Corporation, which is said to be "100% natural vegetarian": *Tibetan ma-huang, wild Brazilian guarana, Chinese black ginseng, German wild ginko* [sic] *biloba, African raw cola nut, Russian gotu-kola, Indonesian wild fo-ti-tieng, Chinese green tea extract, rou gui (rare form of Chinese nutmeg)*, all "synergistically blended to insure visionary vibrations." There are ten 750-mg tablets in each cleverly designed pyramidal package. Buyers are warned not to use the product if they have high blood pressure, diabetes, heart or thyroid disease, or if they are taking a MAO inhibitor, and not to exceed the recommended dosage of five tablets. The label further states that it was "developed by many of the same doctors who created the chemical version," and that it "contains no chemicals or other impurities."[122]

It is difficult to judge how dangerous such a concoction is. The toxicity of taking 3,750 mg of such a mixture would depend on how concentrated the extract of each of these plants is, what the active ingredients of each of the herbs are and what percentage of the gross weight of plant material these active ingredients comprise, and what synergetic effects might or might accompany their simultaneous use. There is also the issue of quality control in so loosely formulated a mix; could a particular "batch" have an atypically large amount of one or other ingredient? In any case, the predominant physiologically active ingredients of this particular brand (others are quite similar) would seem to be ephedrine (from the ephedra) and caffeine (from the kola, guarana, and tea). Very large amounts of these drugs, particularly ephedrine, could be predicted to cause significant hypertension and even cardiac arrhythmia.[123] In some individuals, this can lead to cardiac arrest, and that is what seems to have happened in the tragic case of 20-year-old Peter Schlendorf, a resident of Northport, Long Island, NY. During spring break in Panama City, FL, on 6 March 1996, Peter took eight tablets of Ultimate Xphoria (the package's recommended dose is four). The local medical examiner reported that he died from the "synergistic effect of ephedrine, pseudoephedrine, phenylpropanolamine, and caffeine," the active ingredients in Ultimate Xphoria.[124]

On April 16, 1996, Nassau County (NY) officials banned the sale of all herbal stimulant pills containing ephedrine;[125] the FDA administered a warning saying that it had received reports of 15 deaths from ephedrine products.[126]

4-Methylaminorex. This designer drug has the catchy street name of "U4Euh." Its structure, **4-38** (Figure 4.12), is similar to that of the prescription drug pemoline and is a modification of *aminorex*, **4-37**, an appetite suppressant which was marketed in Europe in the 1960s under the trade name of Menocil, but later recalled when it was found to cause chronic pulmonary vascular obstruction due to arteriopathy in some individuals. (Arteriopathy is a side effect characterized by shortness of breath and exercise intolerance and can lead to lengthy hospitalization or death; whether the 4-methyl variant of aminorex has this problem is unknown.)[127] The drug seems to produce effects similar to those of cocaine or amphetamine, releasing stores of dopamine;[128] it maintained rates of self-injecting comparable to or greater than cocaine in rhesus monkeys.[129] Like cocaine, it can induce convulsions, and there has been one death in the United States attributed to overdose from U4Euh.[130]

Other Psychostimulants

Qat, 'Flower of Paradise.' Qat[131] is a weedy evergreen shrub, *catha edulis,* that grows in Kenya, Yemen, and Ethiopia. It is chewed like tobacco—or perhaps more closely like the coca leaf of the high Andes—in Somalia, Ethiopia, and other East African countries, and functions both as mild stimulant and a social ritual like afternoon tea. It was feared that this "narcotic," as some news reports characterized it, would return with the troops sent by the United States in 1992 to Somalia as part of a United Nations force; in 1994, it began to make an appearance in the Netherlands. Qat contains several stimulant substances (cathinone, cathine, cathidine, cathedulin), some of which are closely related in structure to amphetamine. The principal ingredient is *cathinone*, **4-39**, with some lesser amounts of *cathine* (norpseudoephedrine, **4-40**); the overall physiological effect lies somewhere perhaps between espresso and amphetamine. In the early 1990s, it was still sold openly in many New York City grocery stores catering to East African immigrants, although it was technically a forbidden drug. "It's just like drinking strong coffee," one East African cafe manager told a reporter from *The New York Times*.[132] Abdullatif El-Taieb, manager of a travel agency in Brooklyn, said he did not use qat because it was too time-consuming. But he said that many of his friends like it, and it is a way of life he knows well. "You chew it slowly and rotate it in your mouth," he said. "You smell the aroma and you drink water and the water tastes sweet. Sometimes it makes you happy. Poets will just sit down and start chewing qat and they write poems. Good poems."[133] Kenyans interviewed in Nairobi speak highly of the plant: "Miraa [qat] brings people

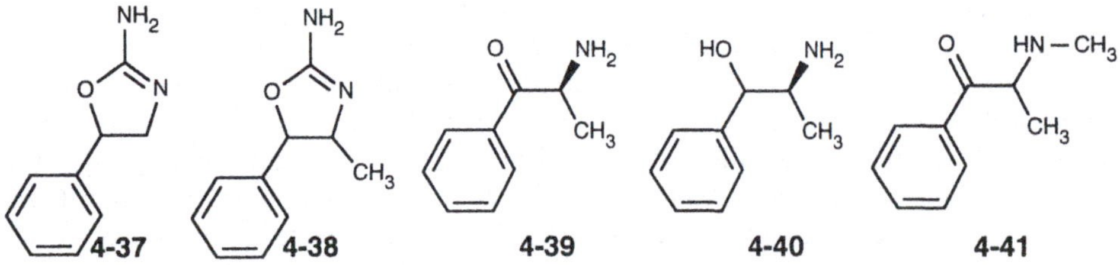

4-37 **4-38** **4-39** **4-40** **4-41**

Figure 4.12 Aminorex, 4-methylamorex, cathinone, cathine, methcathinone.

together. The crime rate here is very low." Another concludes "I think miraa is not more dangerous than a bottle of whisky."[134]

In "Khat," Theodore Dreiser describes a small village whose social life (and social ostracisms) revolve around the afternoon ritual of khat. The story begins by quoting an old Arabian song in honor of khat: "O, thou blessed that contains no demon, but a fairy! When I follow thee thou takest me into regions overlooking Paradise. My sorrows are as nothing. My rags are become as robes of silk. My feet are shod, not worn and bleeding. I lift up my head —O Flower of Paradise!" Halfway through the story, and halfway through the story's one day in the village of Hodeidah, the author describes the surcease this herb brings the villagers:

> And by now it was that time in the afternoon when the effect on the happy possessor of khat throughout all Arabia was only too plainly to be seen. The Arab servant who in the morning had been surly and taciturn under the blazing sun was now, with a wad of the vivifying leaves in his cheek, doing his various errands and duties with a smile and a light foot. The bale which the ordinary coolie of the waterfront could not lift in the morning was now but a featherweight on his back. The coffee merchant who in the morning was acrid in manner and sharp at a bargain, now received your orders gratefully and with a pleasantry, and even a bid for conversation in his eye. . . . Everywhere the evasive apathetic atmosphere of the morning had given way to the valor of sentient life. Chewing the life-giving weed, all were sure that they could perform prodigies of energy and strength, that life was a delicious thing, the days and years of their troubles as nothing.[135]

While Western observers of the use of qat in Yemen have traditionally judged it quite negatively from a distinctly ethnocentric viewpoint, it appears to contribute positively to social interaction as it is used in this small Muslim country that lies just south of Saudi Arabia. "Quat consumption is a complex social and cultural phenomenon which is primarily sustained by social not biological factors," writes anthropologist Shelagh Weir.[136] When dry spells make supplies unobtainable, or when a native Yemini travels abroad for an extended period, there are no physical withdrawal symptoms; and the psychological loss corresponds pretty well to the social and sensual deprivation experienced by the sons and daughters of Great Britain straying into a region where they are deprived of their afternoon tea.[137]

Social Use of Kat in Yemen. A typical qat party in Yemen will last perhaps 4 hours, from early afternoon to dusk. Participants gather to smoke tobacco from large communal water pipes, to chew the fragrantly tender leaves of qat, and to sip chilled scented water. Although some Yemenis say they are completely unaffected by qat, and only use it to be polite, the more usual effect occurs after about an hour, when most people attain a state called *kayf*, once prized by Shi´ah men of religion to enhance their mystical communion with Allah. There is a heightened sense of optimism, alertness, and mental attention to the conversation, which becomes louder and more vivacious. But after about 3 hours, the noise level diminishes, and a mood of calm and reflection called "Solomon's Hour" replaces the hitherto vigorous discussion.[138] The day is done then, and darkness falls from the wings of night (much as a feather is wafted downward from an eagle in his flight). With no folding their tents like the Arabs (but dispensed of the cares of the day), in the stillness that follows the nightfall all then silently steal away—after a courteous "Salaam Alaykum."

Despite the apparent harmlessness of this culturally structured use of qat, the active ingredient, cathinone, when synthesized or extracted as a pure substance, is capable of eliciting responses in laboratory animals that are very similar to cocaine.[139] In large enough quantities, it is as reinforcing as cocaine or amphetamines.

Methcathinone. Methcathinone ("cat," **4-41**) is an underground drug synthesized by treating ephedrine with any of a variety of commonly available oxidizing agents. The resulting compound, methcathinone, which is cathinone with an extra methyl group on the nitrogen atom, is capable of producing a amphetaminelike rush if injected IV; it can also be snorted. Street cost is said to be about $100 a gram, with typical doses ranging in some reports from 500 to 1,000 mg[140] (other reports give 80 to 250 mg).[141] "Powerful, addictive new drug 'cat' sweeps Michigan like an epidemic," read a press report of 1993.[142] Users were said to describe the effects as a "flying euphoria with eyesight becoming brighter and clearer, accompanied by color enhancement. . . . followed by a five-to-eight-hour period of feeling tough [and] invincible, with [an] increased libido and desire to be physical."[143] Although there were more unpleasant side effects such as headaches and tachycardia, the drug was cheaper and more available than cocaine. The epidemic seems to have died out. There are a few studies of intravenous self-injection of methcathinone by baboons; in sufficiently high doses (1 mg/kg/injection), it seems to be as reinforcing as cocaine.[144]

Methcathinone was actually synthesized and tested by Parke Davis in the 1950s and found to be essentially similar to amphetamine but with slightly greater toxicity; it was patented, but never marketed.[145] Cat first appeared in its designer form as a recreational stimulant in Leningrad in 1982, and usage in Russia (where it is known as "ephedrone," "jeff," or "mulka")[146] is said to be almost as extensive as cocaine. As with cocaine and the amphetamines, there are reports of death from cardiac arrhythmias.

Drugs Used to Treat Amphetamine and Cocaine Addiction. It has been reported that the mixed agonist/antagonist opioid buprenorphine, which was discussed in Chapter 2, suppresses self-administration of cocaine by rhesus monkeys and thus may be an effective therapy for cocaine abuse. A Harvard group reports that some 30 days of buprenorphine treatment decreased cocaine self-administration more than 90% below baseline levels but did not significantly reduce the monkeys' appetite, indicating that the drug's effect on cocaine use did not stem from any generalized suppression of behavior. However, how the drug works is not yet understood. Because human trials of buprenorphine had earlier shown that it curbs heroin use without causing serious withdrawal symptoms, the drug is also a potential therapeutic for dual cocaine and heroin abuse. But the researchers point out that clinical evaluation of buprenorphine treatment will require double-blind trials involving independent indices of compliance with the treatment regimen (for example, buprenorphine blood levels) and objective measures of drug use (frequent drug urine screens).[147]

Dopamine Receptor Partial Agonists. In contrast to the daily maintenance pattern of opioid addiction, a common form of quite destructive abuse of cocaine and the amphetamines is characterized by intensive and compulsive binges. This pattern has been duplicated to some extent in laboratory animals allowed unrestricted access to these drugs. Withdrawal, or attempts at withdrawal, are characterized by intense craving and anhedonia. As was mentioned in Chapter 1, it is believed that the reinforcing center for many addictions has a common locus in the D_2 dopamine receptors of the basal forebrain.

Drugs with full or partial agonist activity at the D_2 receptor that have been used with some success in animal and human subjects to ameliorate withdrawal symptoms from psychostimulants are *bromocriptine* (Parlodel, **4-42**, Figure 4.13), *lisuride* (Dopergin, **4-43**), *terguride* (Dironyl, **4-44**), *SDZ208911* (**4-45**), *SDZ208912* (**4-46**), and *preclamol* (**4-47**). The rationale for the use of partial agonists is that the effect of this class of drugs depends on the level of occupancy of the receptor by full agonists. They can act as agonists in conditions of low receptor occupancy (as is the case in withdrawal from psychostimulants, when the craving is presumed to come from abnormally low levels of endogenous dopamine) but as antagonists in the presence of high levels of agonist (as during use of cocaine or amphetamines). A drug with these properties at the dopamine centers in the forebrain would, at least in theory, have the desirable effect of blocking the effects of cocaine or amphetamines if they were used *and* of blocking craving if these drugs were not used.[148]

Combination Regimens. An alternative treatment for drug abuse and alcoholism which seems to show promise is to simultaneously administer an agonist and antagonist. Thus the combined effects of nicotine and *mecamylamine* (Inversine, Mevasine), a nicotinic receptor blocking agent, have been claimed to decrease craving for tobacco and facilitate smoking cessation better than nicotine patches alone.[149] And phentermine combined with fenfluramine, two amphetaminelike drugs considered in this chapter, which

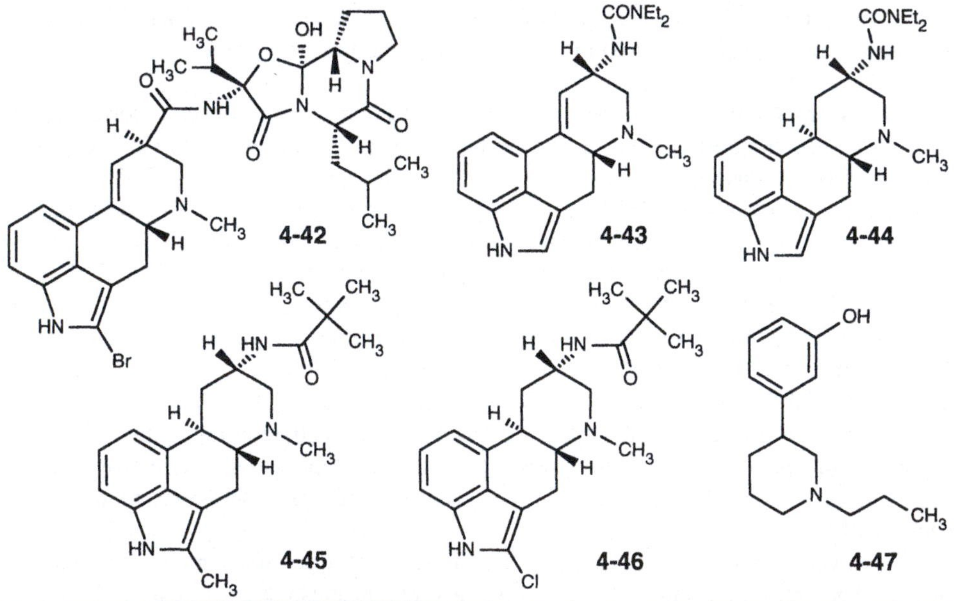

Figure 4.13 Dopamine receptor partial agonists.

have been shown to provide synergetic efficacy in the treatment of obesity, are now being invoked as a promising treatment for cocaine and stimulant addiction.[150]

The sedative drugs of Chapter 3 and the stimulant drugs of this chapter will seem like rather blunt and crude instruments compared to the drugs to be considered in the following chapters, all of which affect consciousness in its higher functions. In the next chapter we will consider the antipsychotics and antidepressants, drugs that can bring sanity to madness and hope to despair. And some will probably think that the drugs we discuss in the final two chapters (the psychedelics and cannabinoids) only bring the madness back again. But the picture is by no means as simple as that, as we shall see.

References and Notes

The following abbreviations are used for frequent sources in these References/Notes:

Ashley: Ashley, R., *Cocaine: Its History, Uses and Effects*, St. Martin's Press: New York, 1975.

CP: *Cocaine Papers by Sigmund Freud*, Robert Byck, Ed., Stonehill: New York, 1975.

CPPC: *Cocaine: Pharmacology, Physiology, and Clinical Strategies*, Lakoski, J. M; Galloway, M. P.; White, F. J., Eds., CRC Press: Boca Raton, FL, 1992.

DE95: *Drug Evaluations: Annual 1995*; American Medical Association Division of Drugs and Toxicology, Department of Drugs; Bennett, D. R., Ed.; AMA: Chicago, 1995.

G&G8: *Goodman and Gilman's The Pharmacological Basis of Therapeutics*, 8th ed., Gilman, A. G.; Rall, T. W.; Nies, A. S.; Taylor, P., Eds.; Pergamon: New York, 1990.

G&G9: *Goodman and Gilman's The Pharmacological Basis of Therapeutics*, 9th ed., Hardman, J. G.; Limbird, L. E., et al., Eds.; McGraw-Hill: New York, 1996.

Gold: Gold, M. S., *Cocaine* [Vol. 3 of *Drugs of Abuse: A Comprehensive Series for Clinicians*], Plenum Medical Book Company: New York, 1993.

HOC: Mortimer, W. G., *History of Coca: "The Divine Plant" of the Incas*. The original edition was published in 1901, and is largely unavailable. Republished in 1974 by the Fitz High Ludlow Memorial Library through And/Or Press, San Francisco.

L&ID: *Licit and Illicit Drugs: The Consumers Union Report on Narcotics, Stimulants, Depressants, Inhalants, Hallucinogens, and Marijuana—Including Caffeine, Nicotine and Alcohol*, Brecher, E. W. and the Editors of Consumer Reports, Eds., Little, Brown: Boston, 1972.

MI11: *The Merck Index*, 11th ed.; Budavari, S.; O'Neil, M. J., et al., Eds.; Merck: Rahway, NJ, 1989. References are given by monograph number.

MI12: *The Merck Index*, 12th ed.; Budavari, S.; O'Neil, M. J., et al., Eds.; Merck: Rahway, NJ, 1996. References are given by monograph number.

NA: *Nicotine Addiction: Principles and Management*, Orleans, C. T.; Slade, J.; Eds., Oxford University Press: Oxford, 1993.

NR: *Nicotine Replacement: A Critical Evaluation*, Pomerleau, O. F.; Pomerleau, C. S., Eds., Pharmaceutical Products Press (Haworth): New York, 1992.

SA: *Substance Abuse: A Comprehensive Textbook,* 2nd ed.; Lowinson, J. H.; Ruiz, P.; Millman, R. B.; Langrod, J. G., Eds.; Williams & Wilkins: Baltimore, 1992.

SF: Allen, F., *Secret Formula: How Brilliant Marketing and Relentless Salesmanship Made Coca-Cola the Best-Known Product in the World*, Harper: New York, 1994.

TSSA: Wilbert, J., *Tobacco and Shamanism in South America*, Yale University Press: New Haven, CT, 1987.

1. Quoted in Blum, D., "At a rustic retreat, Sibelius explored his many selves," *The New York Times*, 24 July 1994, 34.
2. Dickens, C., *Bleak House*, Chapter XLIX, "Dutiful Friendship" (Mr. George arrives).
3. De Quincey, T., *Confessions of an English Opium-Eater*, Ticknor, Reed, & Fields: Boston, 1853 [reprint of the 1822 edition], p. 98.
4. Quoted in Plummer, W.; Feeney, E. X., "Nowhere to hide," *People*, *43*, 29 May 1995, p. 79.
5. *L&ID*, p. 213. In this excellent book edited by the late Edward M. Brecher, the wise counsel of which has sadly never been heeded, can be found the extreme sanctions meted out on tobacco users by the Sultan Murad IV (execution by beheading, quartering, crushing the hands and feet), by the Romanoff czar Michael Feodorovitch (slitting the nostrils, bastinado, the knout); by the Shôgun (confiscation of all property)—all to no avail. Within a century in every case the legislators were themselves avid users of tobacco; in that most traditional of societies, Japan, tobacco had even become part of the solemn tea ceremony.
6. Sorel, N. C., *Word People*, American Heritage Press: New York, 1970, pp. 173-176. This is perhaps not the most serious history in the world, nor intended to be such. Some of it is nonchalantly vague—the "Father Superior of Malta" (?!), the "Czar." Sorel also informs us that Nicot as a precocious lad of 29 had already commenced work on the first dictionary of the French language; after the fuss died down about his wonderful "powder" he was able to finish it, although it was not printed until 1600, several years after his death.
7. When urged to prohibit tobacco for the good of his citizenry, Napoleon III of France (1803-1873) replied: "This vice smoking brings in 100,000,000 francs in taxes every year. I will certainly forbid it at once—as soon as you can name a virtue that brings in as much revenue." Cited in *Gale's Quotations: Who Said What?* CD-ROM, Gale Research: Detroit, MI, 1995.

8. *TSSA*, p. 156.
9. Wagley, C., "World view of the Tapirape Indians," *J. Am. Folk-Lore*, **1940**, *53 (210)*, 258. Cited in *TSSA*, p. 159.
10. *TSSA*, p. 134-136.
11. Other genera than the nicotiana contain nicotine, albeit in much smaller amounts. There is some nicotine present in marijuana and in the common zinnia; if proponents of tobacco prohibition have their way, addicts will find that their only legal recourse may be to smoke zinnia leaves.
12. "Nasal nicotine spraying in smoking cessation," *Lancet,* **340**, 324-329 (1992).
13. Koop, C. E., *Koop: The Memoirs of America's Family Doctor,* Random House: New York, 1991, p. 163.
14. Jarvik, M. E.; Schneider, N. G., "Nicotine," *SA*, p. 334-356.
15. Statistics taken from Husak, D. N., *Drugs and Rights,* Cambridge University Press: New York, 1992 [Cambridge Studies in Philosophy and Public Policy, MacLean, D., Ed.], p. 95.
16. Kluger, R., "A peace plan for the cigarette wars," *The New York Times Magazine*, 7 April 1996, 28-50. Thus, in purely monetary terms, cigarette smoking produces an overall savings to society and a transfer of money from the insurance industry to the health-care industry.
17. Cited in *Gale's Quotations: Who Said What?* CD-ROM, Gale Research: Detroit, MI, 1995.
18. Wilde, O., *The Picture of Dorian Gray*, 1891, Chapter 6.
19. According to NIDA's Dr. Jack Henningield (interviewed on the NBC Evening News 3 June 1996) animal studies originally conducted by Philip Morris show that the acetaldehyde in cigarette smoke powerfully reinforces the addictivity of nicotine; there are indications that additional sugars may have been added to the Marlboro formulation in order to increase the concentration of acetaldehyde in the smoke.
20. Ernster, V. L.; Grady, D. G., "Smokeless tobacco use and health effects among baseball players," *JAMA*, **1990**, *264*, 218-224.
21. Quoted from notes dated July 1962 that were stolen from the Brown & Williamson Tobacco Company and published in the *New York Times* 16 June 1994. Sir Charles's remarks ring somewhat dated, and younger readers will need to be told that in the early 1960s, "tranquilizers" represented a new and very chi-chi approach to the sorrows of life. Nonetheless, Sir Charles hits on the heart of it all: those deep drags of pure Virginia, soggy with nicotine! Even as they scoured esophagus and lungs, did they not sooth and console the Soul? Even as we sucked fire and ate smoke, did we not feel an impregnable assurance that we were, windowless and airless monads we, eternal inviolable seed of the gods? Think the bleak night which saw the breakup of our first—we feared our only—Love: after, we sat alone on the porch steps and chain-smoked a pack or two of Luckies into the blackness of that sultry July midnight. And . . . and we survived! The next day's dawn was met with a wrenching and rasping—but a courageous and confident—*Cough.*
22. Pullan, R. D., et al., "Transdermal nicotine for active ulcerative colitis," *N. Eng. J. Med.*, **1994**, *330*, 811-815.
23. Feder, B. J., "A safer smoke or just another smokescreen?" *The New York Times*, 12 April 1996, D1, D17. Meanwhile, the FDA has approved the OTC sale of nicotine chewing gum and, more recently, Johnson & Johnson's one-dose nicotine patch, Nicotrol: *The New York Times*, 4 July 1996, A15.

24. Kipling, R., "The Betrothed." In *The Oxford Dictionary of Quotations*, 2nd ed., Oxford: New York, 1966, p. 294.

25. Brody, J. E., "Personal Health: Smokescreen of glamour hides dangers of cigars," *The New York Times*, 29 May 1996, C9.

26. *Ibid.*

27. Shenon, P., "Yangon Journal: Burmese generals ask less spit, more polish," *The New York Times,* 14 June 1995. He would probably be no happier here.

28. Burton-Bradley, B. G., "Arecaidinism: Betel chewing in transcultural perspective," *Can. J. Psychiatry*, **1979**, *24*, 481-488.

29. (a) Mangla, B., "Saccharin to sweeten bitter facts about betel nut," *Lancet*, **1993**, *342*, 293. (b) Mangla, B., "India: betel nut warning," *Lancet*, **1993**, *341*, 818-819. (c) Thomas, S. J., "Slaked lime and betel nut cancer in Papua New Guinea," *Lancet*, **1992**, *340*, 577-578.

30. Rodgers, G. C., Jr.; Von Kanel, R. L., "Conservative treatment of jimsonweed ingestion," *Vet. Hum. Toxicol.*, **1993**, *35*, 32-33.

31. "Jimson weed poisoning—Texas, New York, and California," *Morbidity and Mortality Report,* **1995**, *44,* 41-44. However, this should be viewed in perspective. Jimsonweed is not among the list of the 20 plants most frequently associated with poisoning. The top 5 are philodendron, pepper, dumb cane, poinsettia, and holly.

32. There is no bromine in theobromine, which is named from the Greek for "food of the gods" (βρωμα, food); the element bromine derives from the Greek for "stench" (βρωμος). Theophylline is named by analogy to theobromine (φυλλον, leaf).

33. *DE95*, pp. 213-214.

34. Crothers, T. D., *Morphinism and Narcomanias from Other Drugs*, W. B. Saunders: Philadelphia, 1902, pp. 303-304. Cited in *L&ID*, p. 197.

35. Hughes, J. R.; Higgins, S. T., et al., "Caffeine self-administration, withdrawal, and adverse effects among coffee drinkers," *Arch. Gen. Psychiatry*, **1991**, *48*, 611-617.

36. Blakeslee, S., "Yes, people are right. Caffeine is addictive," *The New York Times*, 5 October 1994.

37. Quoted by Bernstein, R., "Beware the first sip of ol' demon coffee," Books of the Times, *The New York Times* 20 April 1995.

38. Peters, J. M., "Caffeine-induced hemorrhagic automutilation," *Arch. Int. Pharmacodynamics*, **1967**, *169*, 141.

39. *L&ID*, p. 205.

40. *G&G9*, p. 675.

41. Kawachi, I., et al., "A prospective study of coffee drinking and suicide in women," *Arch. Intern. Med.*, **1996**, 521-525.

42. Bonson, K. R.; Winter, J. C.; Smith, C. M., "Drug-induced psychiatric symptoms." In: *Textbook of Pharmacology*, Smith, C. M.; Reynard, A. M., Eds., Saunders: Philadelphia, 1992, pp. 380-397. Cited material p. 392.

43. Bruce, M.; Scott, N., "Anxiogenic effects of caffeine in patients with anxiety disorders," *Arch. Gen. Psychiatry*, **1992**, *49*, 867-869.

44. *The Columbia Encyclopedia*, 3rd ed., Bridgwater, W. ; Kurtz, S., Eds., Columbia University Press: New York, 1201.

45. "Famous musical composers, such as Gounod, Fauré, Ambrose Thomas, Massenet, and many others have sung their hosannas in unique bars of manuscript melody." So, at any rate, says Mortimer in *HOC*, p. 180. This is as specific as Mortimer gets, and whether he is referring to bars of melody composed for the famous *edition de luxe* of Mariani—gratefully received by Queen Victoria—or to actual published compositions of these musicians, is not clear.

46. Cited in *SF*, p. 23.

47. *SF*, p. 23.

48. Quoted in *Ashley*, p. 47.

49. *The Columbia Encyclopedia*, 3rd ed., Bridgwater, W.; Kurtz, S., Eds., Columbia University Press: New York, 492.

50. Mariani, A., *Coca and Its Therapeutic Application*, J. N. Jaros: New York, 1896. Cited in Mortimer, W. G., *HOC*, p. 462 and in Hortense Koller Becker, "Carl Koller and Cocaine," *Psychoanal. Q.*, **1963**, *32*, 322. (The latter, reprinted as "Coca Koller," forms Chapter 21 of *CP*, pp. 261-320; cited material on p. 275.)

51. *SF*, pp. 24-28.

52. I am indebted for all this information on the history of this most American of beverages to Allen's *SF*; in this instance, pp. 34-46. Allen cites Dr. George Payne of the Georgia Pharmacy board as calculating there was "1/400th of a grain" per oz; 1 grain is 65 mg, so this works out to 0.975 mg/6 oz bottle.

53. It is tempting to speculate whether modern GC-MS techniques could still detect some residue of cocaine in today's "Classic" beverage. An obvious solution, it might seem, would be to drop the coca leaves and kola beans altogether. But Candler was convinced that he would then lose the copyright to the name, Coca-Cola, which certainly seemed to imply that the beverage contained these two ingredients. (His fears were not entirely specious; Dr. Wiley simultaneously accused Candler of marketing a medicine because Coke *had* cocaine and of fraudulent advertising because, despite its name, it *hadn't*). While providing some marketing allure, Candler's insistence on keeping some amount of coca leaves in the drink also provided a lot of headaches. There was a period when the U.S. Government forbade exportation of Coke syrup since it contained a narcotic; there was an embarrassing incident when a dutiful employee in the company's Peruvian subsidiary, which was charged with extracting the cocaine from coca leaves, decided to maximize profits for the company by selling a few kilos of pure cocaine to a drug marketer in Paris. All these juicy details can be read at greater length in *SF*.

54. *SF*, p. 90.

55. "The Sign of the Four" was the second Sherlock story to appear (after "A Study in Scarlet"), being published in the February 1890 issue of *Lippincott's* magazine. For an inexhaustible source of Holmes commentary and trivia, see Baring-Gould, W., *The Annotated Sherlock Holmes*, Clarkson Potter: New York, 1967.

56. Devoted admirers of the Doughty Deerstalker will be relieved to know that he is never elsewhere observed to have used morphine, although there are several other instances where he indulges in cocaine. According to Dr. David F. Musto, he eventually saw the error of his ways, and renounced cocaine as well (but never his pipe): Musto, David F., "Sherlock Holmes and Sigmund Freud: A study in cocaine," *JAMA*, **1968**, *204*, 125-130.

57. Letter to Martha Bernays of 20 January 1886, trans. James and Tania Stern, cited in *CP*, pp. 162-63. Freud goes on in this letter to tell of his attending an autopsy of the victim of a celebrated murder, performed the next day by Prof. Brouardel at the Paris Morgue.

58. Freud, S., "Beiträge über die Anwendung des Cocaïn. Zweite Serie. I. Bemerkungen über Cocaïnsucht und Cocaïnfurcht mit Beziehung auf einem Vortrag W. A. Hammond's," *Wien. Med. Wochensch.,* **1887**, pp. 929-932.

59. The relevant passages can all be found with commentary in *CP*, pp. 205-236. In a passage from the "Dream of Irma's Injection," Freud says "I was making frequent use of cocaine at that time to reduce some troublesome nasal swellings, and I had heard a few days earlier that one of my women patients who had followed my example had developed an extensive necrosis of the nasal mucous membrane."

60. Koller was a close friend of Freud, and Freud inscribed a reprint of his 1885 paper, *Beitrag zur Kenntniss der Cocawirkung,* with the phrase: *seinem lieben Freunde Coca-Koller* (to his dear friend Coca-Koller). It is hard to believe that Freud was aware of the Atlanta-based beverage in 1885; it had only been named the year before. Perhaps the inscription and reprint were sent to Koller much later, but this seems unlikely. Perhaps this is an instance of what Freud's later disciple Jung would call synchronicity. . . In any case, the story of Drs. Koller and Freud in their youth, and an engrossing revelation of that fabled *gute, alte Wien*, complete with bloody duels and anti-Semitism, is told by Koller's daughter in: Becker, Hortense Koller, "Karl Koller and Cocaine," originally published in *The Psychoanalytic Quarterly*, **1963**, *32*, 309-343; reprinted in *CP*, pp. 261-319.

61. Halsted, W., "Practical comments on the use and abuse of cocaine, suggested by its invariably successful employment in more than a thousand minor surgical operations," *N. Y. Med. J.*, **1885**, *42*, 294-295.

62. *Ibid.*

63. MacCallum, W. G., "Biographical Memoir of William Stewart Halsted, 1852-1922," *National Academy of Sciences of the United States of America, Biographical Memoirs, Volume XVII, Seventh Memoir*, National Academy of Sciences: Washington, DC, 1935, p. 153.

64. Crowe, S. J., *Halsted of Johns Hopkins: The Man and His Men*, Charles C. Thomas: Springfield, IL, 1957, pp. 29-30.

65. Hammond, W. A., *Trans. Med. Soc. V.*, **1887**, pp. 212-226. Cited in *CP*, pp. 177-193.

66. *HOC*, p. 462.

67. Richard Pryor spent 6 weeks in a burn unit in 1980 after what he then said was an accidental fire while free-basing. But since then he has stated that, "coked out of his mind, he doused himself with a bottle of cognac and put a lighter to it" (Plummer, W.; Feeney, E. X., "Nowhere to hide," *People*, **1995**, *43*, 29 May, p. 78).

68. Wallace, D. F., *Infinite Jest*, Little, Brown: Boston, 1996, pp. 236-238.

69. Dr. Hans van Mastrigt, officer for Addiction Policy for the Rotterdam Municipal Health Service, brought me one midnight in June 1994 to a shelter for prostitutes located in the midst of the one area of Rotterdam which is zoned for prostitution. (The women—and men, for nearly half are transvestites—are offered the opportunity for counseling, medical examinations, and needle exchange in a program with government funding run by the private Bulldog organization.) There we talked with some of the Bulldog staff. A prostitute from England, overhearing our conversation, interrupted to tell me how she had returned to using heroin after a few months of whiplash highs and lows on base. She sternly admonished me not to try base myself but to stick to heroin, preferably smoked.

70. The criminal penalties for possession and/or sale of "rock" or crack are much higher than for "crystal" or powder cocaine HCl. This reflects drug hysteria or racism in the view of most experts.

71. Oprah Winfrey confessed using crack back in the 1970s, before she was famous: "I had a perfect round little Afro. I went to church every Sunday. . . and I did drugs." She was led on to do so, like so many, not by a "pusher" but by a man "I was so in love with I would have done anything for. . . ." *People*, 30 Jan 1995, p. 55.

72. Burroughs, W. S., "Letter from a master addict to dangerous drugs." Appendix to Burroughs, W. S., *Naked Lunch*, Grove Weidenfeld: New York, 1959, pp. 215-232. [This letter was originally printed at the request of a British doctor in *The British Journal of Addiction*, *53* (2).] It is a fascinating document from an intelligent and gifted writer whose drug experiences must rank among the most extensive of all time.

73. William Burroughs to Allen Ginsberg, Tangier, Morocco, 7 April 1954. In *The Letters of William S. Burroughs: 1945-1959*, Harris, O., Ed., Viking Penguin: New York, 1993, p. 201.

74. A recovering coker quoted in: Shaffer, H. J.; Jones, S. B., *Quitting Cocaine: The Struggle against Impulse,* Lexington/Heath: Lexington, MA, 1989, p. 114.

75. "Cocaine and vasospasm," *N. Engl. J. Med.*, **1989**, *321*, 1604-1606.

76. *The Merck Manual of Diagnosis and Therapy*, 16th ed., Berkow, R.; Fletcher, A. J., Eds., Merck: Rahway, NJ, 1992, p. 507.

77. The difficulty of cocaine-induced cardiac arrhythmias can be resolved, according to William S. Burroughs, by promptly injecting morphine—an expedient, however, with which not everyone will feel comfortable. He writes: "There is no tolerance with C[ocaine], and not much margin between a regular and a toxic dose. Several times I got too much and everything went black and my heart began turning over. Luckily I always had plenty of M[orphine] on hand, and a shot of M fixed me right up." Burroughs, W. S., *Junky*, Penguin Books: New York, 1977, p. 124. (First published as *Junkie*, Ace Books: New York, 1953.)

78. de la Torre, R.; Farre, M., et al., "The relevance of urinary cocaethylene following the simultaneous administration of alcohol and cocaine," *J. Analyt. Toxicol.*, **1991**, *15*, 223.

79. Cited in *C&EN*, 12 November 1990.

80. Perezreyes, M.; Jeffcoat, A. R., et al., "Comparison in humans of the potency and pharmacokinetics of intravenously injected cocaethylene and cocaine," *Psychopharmacology*, **1994**, *116*, 428-432.

81. Henning, R. J.; Wilson, L. D.; Glauser, J. M., "Cocaine plus ethanol is more cardiotoxic than cocaine or ethanol alone," *Crit. Care Med.*, **1994**, *22*, 1896-1906.

82. Nash, N. C., "A cup of coca tea, anyone?" *The New York Times*, 17 June 1992. The tea is routinely offered visitors to the high Andean cities of Peru; the author knows several Jesuits who have drunk it and found it very mild. One, who lived for several years in a remote village in Northern Peru, said he had tried to replace his regular coffee with coca tea. But after some months he resumed coffee drinking; he found the coca tea to be far too weak a substitute.

83. Nash, N. C., *op. cit.* In the same article, Bolivia's Minister of Planning, Samuel Doria Medina, claims that "it takes about 150 kilos of coca leaves to make one kilo of cocaine that a peasant can sell in Bolivia to traffickers for about $2,000. If you make mate de coca out of the same amount of leaves, the export price would be more like $15,000." This assumes

demand for the tea would equal demand for the pure alkaloid. For some realistic considerations from a professional economist on the economics of coca growing, see Passell, P., "Economic Scene: Coca dreams, cocaine reality," *The New York Times*, 14 August 1991, D2.

84. Yvonne Freund of New York City writes in comment on the article by Nash cited above: "As anyone who has traveled in the high Andes knows, drinking coca tea, which tastes like most herbal teas and does not become addictive, is great at relieving altitude sickness. We traveled into the new National Park between Chile and Bolivia, at 18,000 feet altitude, and if it hadn't been for the coca leaf tea we had brought along, many of us would have become very ill." But Herbert S. Okun, a member of the United Nations International Narcotics Control Board, showing an exquisite sensibility for the nuances of set and setting, strongly disagrees, writing: "Despite claims to the contrary, coca leaf is hardly innocent. In view of broad socioeconomic, geographic and cultural differences, drinking coca tea in lower Manhattan could have very different effects from drinking it in the higher Andes." Both letters appear in *The New York Times*, 29 June 1992.

85. "Organic farming with cocaine," *Science*, **1993**, *262*, 651.

86. Shuster, L., "Pharmacokinetics, metabolism, and disposition of cocaine." In: *Cocaine: Pharmacology, Physiology, and Clinical Strategies*, Lakoski, Galloway, White, Eds., CRC Press: Boca Raton, FL, 1992, pp. 1-14.

87. Reichman, O. S.; Otto, R. A., "Effect of intranasal cocaine on the urine drug screen for benzoylecgonine," *Otolaryngology—Head and Neck Surgery*, **1992**, *106*, 223-225.

88. Willstätter, R.; Wolfes, O., et al., *Ann. Chem.*, **1923**, *434*, 111.

89. A synthesis which more nearly approaches practical production of gram quantities of cocaine is a 1987 process developed by a group at the Research Triangle Institute: Lewin, A. H.; Naseree, T., et al., *J. Heterocycl. Chem., 1987*, 24, 19-21.

90. Ketter, T. A.; Andreason, P. J., et al., "Anterior paralimbic mediation of procaine-induced emotional and psychosensory experiences," *Arch. Gen. Psychiatry*, **1996**, *53*, 59-69.

91. Sneader, W., *Drug Discovery: The Evolution of Modern Medicines*, Wiley: New York, 1985, pp. 100-101.

92. Snyder, S. H., "Catecholamines in the brain as mediators of amphetamine psychosis," *Arch. Gen. Psychiatry*, **1972**, *27*, 169-179.

93. For references and a balanced discussion, see: King, G. R.; Ellinwood, Jr., E. H., "Amphetamines and other stimulants," *SA*, pp. 247-270.

94. Burroughs, W. S., *Junky*, Penguin Books: New York, 1977, pp. 14-15. It is interesting to compare this sense of expanded benevolence with the "entactogenic" effect attributed to methylenedioxymethamphetamine (MDMA, ecstasy, Chapter 6).

95. Greene, G., *Ways of Escape*, Simon & Schuster: New York, 1980, pp. 92-93. Greene admits to a "manic-depressive temperament," and it would be interesting to contrast *The Confidential Agent* to *The Power and the Glory* as the works of respectively the manic and the depressant author. Later in life, he says he endured a period of depression so severe, even while simultaneously "happy in love," that he asked a psychiatrist friend for electroshock treatment—he was refused. Lacking the courage for suicide or the Russian roulette he played earlier, he would nonetheless travel to dangerous parts of the world as a sort of alternative (ibid, p. 146). Parallels come to mind of Hemingway's fascination with war and bull-fighting and his final suicide; see Chapter 5 for psychiatrist Kay Jamison's theory that writers are more likely to suffer from mood disorders.

96. "Batu" is the Filipino word for "rock," and is the usual term for freebase methamphetamine among the large Filipino population in Hawaii. See Smith, D. E.; Seymour, R. B., et al., "Smokable drugs," *J. Psychoactive Drugs,* **1992**, *24*, 91-98.

97. Jackson, J., [letter] *N. Engl.J. Med.*, **1989**, *321*, 907.

98. *Ibid.*

99. Lauderback, D.; Waldorf, D., "Whatever happened to ice? The latest drug scare," *J. Drug Issues*, **1993**, *23*, 597-613.

100. "In San Diego, an old drug comes back," *The New York Times*, 20 February 1996.

101. Johnson, D., "Good people go bad in Iowa, and a drug is being blamed," *The New York Times*, 22 February 1996, A1, A19. The chemistry reported in this article is confused, to say the least. Example: "Years ago, the authorities said, a typical street dose of methamphetamine consisted of perhaps 20 percent of ephedrine, the ingredient that delivers the kick. . . . Now the drug contains over 90 percent of the active ingredient." Since ephedrine and methamphetamine are distinct substances, one of them cannot be the active ingredient of the other. The journalist makes no effort to evaluate the claim that methamphetamine is "the most malignant, addictive drug known to mankind," or that the death of a student from meningitis was caused by methamphetamine use, which had "broken down his immune system." There has been criticism from within the press' own ranks of how drug stories are often "sensational, colorful, gruesome, alarmist, with a veneer of social responsibility." Shaffer, H. J.; Jones, S. B., *Quitting Cocaine: The Struggle Against Impulse,* Lexington Books: Lexington, MA, 1989, p. 81.

102. An earlier designation of this syndrome was "attention deficit disorder" (ADD).

103. *G&G9*, p. 224.

104. All quotations and statistics in this paragraph from: Kolata, G., "Boom in Ritalin sales raises ethical issues," *The New York Times*, 15 May 1996, C6.

105. *Ibid.*

106. Klein, R. G.; Wender, P., "The role of methylphenidate in psychiatry," *Arch. Gen. Psychiatry*, **1995**, *52*, 429-433.

107. *The New York Times*, 24 May 1996.

108. Spencer, T.; Wilens, T., et al., "A double-blind, crossover comparison of methylphenidate and placebo in adults with childhood-onset attention-deficit hyperactivity disorder," *Arch. Gen. Psychiatry*, **1995**, *52*, 434-443.

109. Letter to Peter McWilliams, cited in McWilliams, P., *Ain't Nobody's Business If You Do: The Absurdity of Consensual Crimes in a Free Society,* Prelude Press: Los Angeles, 1993, p. 57.

110. Safer, D. J.; Krager, J. M., "Effect of a media blitz and a threatened lawsuit on stimulant treatment," *JAMA*, **1992**, *268*, 1004-1007.

111. Thase, M. E.; Rush, A. J., "Treatment-resistant depression." In Bloom, F. E.; Kupfer, D. J., Eds., *Psychopharmacology: Fourth Generation of Progress,* Raven Press: New York, 1995, pp. 1081-1097.

112. Parran, T. V., Jr.; Jasinski, D. R., "Intravenous methylphenidate abuse. Prototype for prescription drug abuse," *Arch. Intern. Med.*, **1991**, *151*, 781-783.

113. Jaffe, S. L., "Intranasal abuse of prescribed methylphenidate by an alcohol and drug abusing adolescent with ADHD," *J. Am. Acad. Child Adolesc. Psychiatry,* **1991**, *30*, 773-775.

114. Volkow, N. D.; Ding, Y.-S., et al., "Is methylphenidate like cocaine?" *Arch. Gen. Psychiatry*, **1995**, *52*, 456-463.

115. *Ibid.*

116. Arora, R.; Dryden, S., et al., "Acute dexfenfluramine administration normalizes glucose tolerance in rats with insulin-deficient diabetes," *Eur. J. Clin. Invest.*, **1994**, *24*, 182-187.

117. McCann, U.; Hatzidimitriou, G.; Ridenour, A., et al., "Dexfenfluramine and serotonin neurotoxicity: further preclinical evidence that clinical caution is indicated," *J. Pharmacol. Exp. Ther.*, **1994**, *269*, 792-798.

118. Hilts, P. J., "New diet pills raise old safety questions," *The New York Times*, 21 February 1996.

119. a) Weintraub, M., "Long-term weight control study: The National Heart, Lung, and Blood Institute funded multimodal intervention study," *Clin. Pharmacol. Ther.*, **1992**, *51*, 581-585. (b) Weintraub, M., "Long-term weight control study: Conclusions," *Clin. Pharmacol. Ther.*, **1992**, *51*, 642-646.

120. Other common English transliterations are *Ma Huang* and *Ma-huang*; the meaning in Mandarin is *má*, "hemp" + *huáng*, "yellow."

121. Sneader, W., *Drug Discovery: The Evolution of Modern Medicines*, Wiley: New York, 1985, p. 100.

122. These citations are from an ad in *High Times* (November 1995, p. 7) and from an (empty) package given me by a student, who reported that the effects on her were barely perceptible, but that some of her friends liked it. The same company, Global World Media, also markets a "herbal" Nexus (in imitation of the designer-drug psychedelic described in Chapter 6) whose principal ingredient is *Piper methysticum*, the plant used to make kava (Chapter 3).

123. *G&G9*, 221.

124. Cowley, G., "Herbal warning," *Newsweek*, 6 May 1996, 61-64, 67-68. Quoted material on pp. 61-63.

125. Lambert, B., "Nassau to ban sale of herbal stimulant linked to a death," *The New York Times*, 17 April 1996, B1, B4. According to the same source, other municipalities and states are framing their own regulations, some of which are peculiar: in Iowa a bill is pending which would outlaw synthetic ephedrine as a controlled substance—but not ephedrine extracted from the mahuang plant. Herbal Ecstacy and similar products were banned in New York as of May 1996: Kraus, C., "Pataki outlaws herbal stimulant used by youths," *The New York Times,* 24 May 1996, B1, B6.

126. "FDA says use of 'legal highs' is hazardous," *The New York Times,* 11 April 1996.

127. Bunker, C. F.; Johnson, M., et al., "Neurochemical effects of an acute treatment with 4-methylaminorex: a new stimulant of abuse," *Eur. J. Pharmacol.*, **1990**, *180*, 103-111.

128. Glennon, R. A.; Misenheimer, B., "Stimulus properties of a new designer drug: 4-methylaminorex ("U4Euh")," *Pharm. Biochem. Behav.*, **1990**, *35*, 517-521.

129. Mansbach, R. S.; Sannerud, C. A., et al., "Intravenous self-administration of 4-methylaminorex in primates," *Drug Alcohol Depend.*, **1990**, *26*, 137-144.

130. Davis, F. T.; Brewster, M. E., "A fatality involving U4Euh, a cyclic derivative of phenylpropanolamine," *J. Forensic Sci.*, **1988**, *33*, 549-553.

131. *Qat* is the Arabic term for *Catha edulis* and is used throughout Arabic-speaking areas. The q is meant to represent the initial Arabic letter *qaf.* Two other transliterations, *khat* and *ghat* are not as correct, and it is unfortunate that *khat* has been adopted as the official transliteration of the World Health Organization and the International Council on Alcohol and Addictions. See Weir, Shelagh, *Qat in Yemen: Consumption and Social Change*, Dorchester, Dorset: Trustees of the British Museum, Dorset Press, 1985, p. 10. In Kenya, the drug is also known as *miraa*, in Ethiopia as *tschat.*

132. *The New York Times*, 14 December 1992.

133. *Ibid.*

134. Lorch, D., "Despite war, famine and pestilence, the khat trade thrives in East Africa," *The New York Times*, 14 December 1994, A8.

135. Dreiser, T., "Khat." In: Dreiser, T., *Chains: Lesser Novels and Stories*, Fertig: New York, 1987, pp. 156-180. Cited material on pp. 172-173.

136. Weir, Shelagh, *Qat in Yemen: Consumption and Social Change*, Trustees of the British Museum, Dorset Press: Dorchester, England, 1985, pp. 9, 38.

137. *Ibid.* For a more negative view of the impact of qat consumption in these countries, see the article by Peter Kalix, a Swiss pharmacologist: Kalix, P., "Khat, an amphetamine-like stimulant," *J. Psychoactive Drugs*, **1994**, *26*, 69-74.

138. Kennedy, J. G.; Teague, J.; Fairbanks, L., "Qat use in North Yemen and the problem of addiction: a study in medical anthropology," *Cult. Med. Psychiatry*, **1980**, *4*, 311-344.

139. Woolverton, W. L.; Johanson, C. E., "Preference in rhesus monkeys given a choice between cocaine and *d,l*-cathinone," *J. Exp. Anal. Behav.*, **1984**, *41*, 35-43.

140. Goldstone, M. S., "'Cat': methcathinone—a new drug of abuse," *JAMA*, **1993**, *269*, 2508.

141. Emerson, T. S.; Cisek, J. E., "Methcathinone: a Russian designer amphetamine infiltrates the rural Midwest," *Ann. Emerg. Med.*, **1993**, *22*, 1897/129-1903/135.

142. Schaefer, J., "Powerful, addictive new drug 'cat' sweeps Michigan like an epidemic," *The Baltimore Sun*, 28 March 1993.

143. Emerson, T. S.; Cisek, J. E., "Methcathinone. . . "

144. Kaminski, B. J.; Griffiths, R. R., "Intravenous self-injection of methcathinone in the baboon," *Pharmacol. Biochem. Behav.*, **1994**, *47*, 981-983.

145. Emerson, T. S.; Cisek, J. E., "Methcathinone. . . ."

146. Zhingel, Y. K.; Dovensky, B. S., et al., "Ephedrone: 2-methylamino-1-phenylpropan-1-one (Jeff)," *J. Forensic. Sci.* 1991 *36*, 915-920.

147. Mello, N. K., *Science*, **245**, 859 (1989). Reported in C&EN, 28 August 1989, p. 24.

148. (a) Pulvirenti, L.; Koob, G. F., "Dopamine receptor agonists, partial agonists and psychostimulant addiction," *Trends Pharm. Sci.*, **1994**, *15*, 374-379. See also: (b) Meyer, R. E., "New pharmacotherapies for cocaine dependence . . . revisited," *Arch. Gen. Psychiatry*, **1992**, *49*, 900-904.

149. Rose, J. E.; Behm, F. M., et al., "Mecamylamine combined with nicotine skin patch facilitates smoking cessation beyond nicotine patch treatment alone," *Clin. Pharmacol. Ther.* **1994**, *56*, 86-99.

150. Brauer, L. H.; Johanson, C.-E., et al., "Evaluation of phentermine and fenfluramine, alone and in combination, in normal, healthy volunteers," *Neuropsychopharmacology*, **1996**, *14*, 233-241.

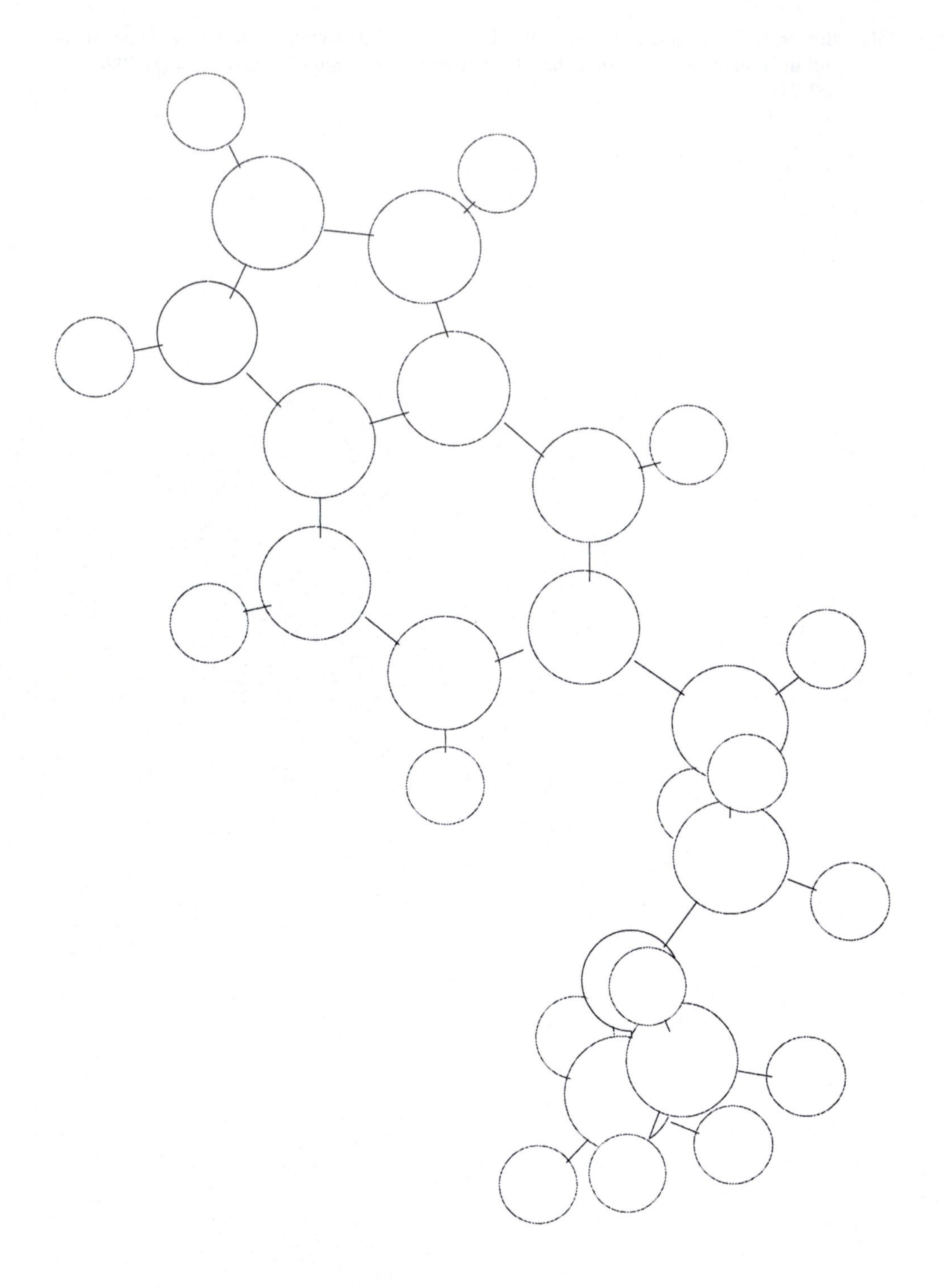

5
Antipsychotics and Antidepressants

If a madman were to come into this room with a stick in his hand, no doubt we should pity the state of his mind; but our primary consideration would be to take care of ourselves. *We should knock him down first, and pity him afterwards.*
—Samuel Johnson[1]

O then, if in my lagging lines you miss
The roll, the rise, the carol, the creation,
My winter world, that scarcely breathes that bliss
Now, yields you, with some sighs, our explanation.
—G. M. Hopkins[2]

Introduction

Few drugs more profoundly affect the mind than the antipsychotic and antidepressant drugs. We saw in the previous chapter that a psychotic state can be induced in otherwise normal individuals who use very high doses of cocaine or the amphetamines, drugs which cause intense activity at the dopaminergic (DA) neurons. It should come then as no surprise that antipsychotic drugs can reverse these drug-induced psychoses, and that the older antipsychotic drugs block DA receptors.

But newer, more effective, and less toxic antipsychotics act at serotonin synapses as well, the same class of neurotransmitters (NT) affected by the antidepressants now so widely employed in our culture. It is becoming increasingly evident that serotonin plays a critical role in many of the mental functions we think of as essential in defining human personality. Newer, less toxic antidepressants, fluoxetine (Prozac) in particular are increasingly being used as "mood brighteners" in an effort to fine-tune the personality of individuals who are by no means psychotic or even seriously neurotic. Such uses will likely increase as more selective agents at the various CNS receptor sites are discovered.

We will see in the chapter following this that some of the psychedelics have been extensively used in psychotherapeutic settings in order to provide a subtle change of mood and emotional tone resembling the more long-term effects of the antidepressants; it is probably significant in this regard that both these agents act at related subtypes of the serotonin receptor system.

The Antipsychotics

The antipsychotic drugs mitigate or abolish the symptoms of active psychoses (the psychosis becomes inactive or dormant, but it is not cured by these drugs and will usually recur if use of the drug is discontinued). These drugs are also called *major tranquilizers* (as opposed to the minor tranquilizers or anxiolytics, which we have discussed in the previous chapter); they are also frequently referred to as *neuroleptics* (from the Greek meaning "affecting the nerves").

Definitions

Some brief definitions of the various psychiatric states treated by these drugs may be helpful:

Psychosis. A mental disorder so severe that it involves loss of contact with reality, particularly in the sense that the afflicted person cannot maintain an intelligible interaction with others. In contrast, many less severe psychiatric syndromes, formerly referred to as *neuroses* (e.g., disproportionate anxiety, obsessively recurring thoughts, hysteria) may impede but do not abolish communication. Two major types of psychoses frequently found together are schizophrenia and paranoia.

Schizophrenia. A group of poorly defined, chronic psychotic disorders; the symptoms usually become evident during adolescence, and an older term reflecting this phenomenon is *dementia praecox* (Latin for early-onset madness). Schizophrenia is a chronic and lifetime disease, but one that has acute episodes of manifest psychosis interrupted by more or less normal functioning. During the acute, florid episodes, schizophrenics are typically subject to auditory hallucinations (usually hearing voices, often perceived as hostile) delusions concerning their own identity (Jesus Christ, Elvis Presley) or role (candidate for president, consultant to the pope, the center of an international conspiracy). There are frequently *ideas of reference*: belief that the individual's thoughts control or are being controlled by outside forces or persons. Coherent thought is impossible or fleeting during florid episodes; there is considerable bizarre ideation, frequently with a religious or sexual content.

The symptoms of schizophrenia are often classified into *positive* (those that do not occur with normal individuals—hearing voices and delusions of identity) and *negative* (those that represent deficits in normal behavior—obtunded or absent affect, poor social interactions). It is often said that the antipsychotic drugs treat the positive symptoms more successfully than the negative, but this is perhaps in part simply a reflection of the dramatic and conspicuous nature of the positive symptoms, in the presence of which the negative symptoms are essentially invisible. Even among those who respond well to the antipsychotic drugs, some residual positive as well as negative symptoms are occasionally in evidence.

A recent effort to improve the negative symptomatology of schizophrenia seems promising. Patients are given megadoses (up to 60 grams daily) of glycine, saturating the bloodstream so that significant amounts of this polar amino acid cross the blood-brain barrier (BBB). Patients seemed to show improvement in initiating interactions with others and general social awareness. The operative hypothesis is that the glycine interacts with the *N*-methyl-D-aspartate (NMDA) receptors in the brain; the challenge is to find a prodrug that crosses the BBB and is transformed in the brain to glycine or some similar NMDA agonist.[3]

Many schizophrenias probably reflect some underlying genetic predisposition that can be activated or aggravated by stress; there are definite familial correlations, and data from the extensive Danish health archives has allowed Dr. Seymour Kety of the U.S. National Institute of Mental Health to show that the genetic legacy of an adopted child is a far more powerful predictor of schizophrenia than the psychological dynamics of the family in which the child grew up.[4] In another recent study, it was shown that nonschizophrenic siblings of schizophrenic patients displayed deficits in several tests of spatial abilities and sensory-motor functions that were intermediate between those of the schizophrenic patients and normal controls.[5]

Severe, acute, active schizophrenia is often treated by intramuscular injection of an antipsychotic drug; long-term management or prophylaxis of recurrent acute episodes is achieved by oral antipsychotics. Unfortunately, about a third of all schizophrenics respond either poorly or not at all to antipsychotic drugs.

Folk wisdom has it that insanity and genius are often found together, and there may be some truth to this in the case of manic-depressive disorder. Although schizophrenia is no more common among the gifted than the average, there are some particularly prominent individuals who experienced periods of memorable madness. Sir Isaac Newton is considered to have been psychotic for several years; it is uncertain whether this was due to mercury intoxication from his intense involvement with alchemical experiments. A modern parallel case is that of John Forbes Nash, Jr., recipient of the 1994 Nobel Prize in economics, who for som 30 or more years after his brilliant contributions to mathematics in the 1950s was devastated by paranoid schizophrenia:

> His life, once so full of brightness and promise, became hellish. There were repeated commitments to psychiatric hospitals. Failed treatments. Fearful delusions. A period of wandering around Europe. Stretches in Roanoke, Va., where Mr. Nash's mother and sister lived. Finally, a return to Princeton, where he had once been the rising star. There he became the Phantom of Fine Hall, a mute figure who scribbled strange equations on blackboards in the mathematics building and searched anxiously for secret messages in numbers.[6]

After years of repeated hospitalization and therapy, a "miraculous remission" occurred without benefit of any drug or treatment, the disease lifting as mysteriously as it originally came. At this writing, Nash was working quietly at Princeton.

Paranoia. Schizophrenia is frequently accompanied by paranoia, which is characterized by persistent delusions of persecution. The syndrome can also occur following high doses of some drugs—alcohol, amphetamines, cocaine, or phencyclidine in particular—but use of antipsychotic drugs in these instances often worsens the symptoms.

Side Effects

The chief liability of most antipsychotic drugs is the debilitating side effects they cause, and most of the newer antipsychotics are described as "atypical" precisely in that they do not have these side effects. The side effects are peculiar and unique to this class of drugs, and it will be helpful to describe and define them at the outset. Unlike antibiotics, antipsychotics do not cure the underlying disease, so they must be taken daily for life. Under these circumstances, problems eventually emerge that might never become apparent with other classes of drugs used only on a short-term basis. The side effects are unpleasant and discomforting; a patient who has been restored to more or less sound judgment can perceive that the drug causes these problems, and the paradoxical consequence is often a decision on the part of the patient to stop taking them. Frequently, patients on medication can function well and are able to be employed and support themselves; when they discontinue the antipsychotic medication, even if they do not lapse into acute psychosis they can become so eccentric in their behavior that they are fired. At the same time, they are not (at least in a legal sense) a danger to themselves or others and cannot be forced to resume their medication. Nor can they be persuaded, because various paranoid or hallucinatory elements have entered into their arguments against taking the drugs: e.g., "the Devil wants me to take drugs; the voice of God tells me I shouldn't."

The most debilitating side effect of the antipsychotic drugs, *tardive dyskinesia,* occurs with about the same frequency with all of the older antipsychotic drugs despite the considerable variation of structure between the different classes. And the most significant advance in the treatment of psychoses is the development of the so-called atypical antipsychotics, which are atypical precisely in that they do not cause tardive dyskinesia. It and the other serious side effects of the traditional antipsychotics are all considered *extrapyramidal syndromes* because they are effects of damage to or inhibition of the extrapyramidal system in the brain. This system includes the corpus striatum and the substantia nigra (damage to which is known to cause parkinsonism) and controls many common motor activities. The extrapyramidal reactions caused by most antipsychotic drugs are described briefly below:

- *Dystonia.* Most frequently noted after parenteral administration in young male patients, it is characterized by abnormal, long-sustained posturing and grimacing of the neck, jaw, and eyes; often with spasms of the neck or back and protrusion of the tongue.
- *Akathisia.* The literal Greek meaning of this word is "without sitting." Patients often feel uncontrollably restless and show this by constantly pacing back and forth like a caged animal. This behavior is one of the several dysphoric symptoms that doubtless contribute to the significant patient noncompliance associated with the use of these drugs.
- *Parkinsonism.* Rigidity, a shuffling gait, a mask-like facial expression, hypersalivation, tremor, and "pill-rolling" of the fingers are some of the symptoms associated

with Parkinson's disease, which is known to be caused by failure of dopaminergic neurons in the substantia nigra (see Chapters 1 and 2). Its occurrence as a side effect of antipsychotic medication is a result of blockade of these neurons by the drug.

- *"Rabbit Syndrome."* The name derives from the resemblance to a rabbit's characteristic sniffing and nose-wiggling. It is a fine, rapid tremor of the cheeks and tongue that sometimes responds to the administration of common antihistamines such as *diphenhydramine* (Benadryl).[7]
- *Tardive Dyskinesia.* The American Psychiatric Association reported that 20% of patients given antipsychotic drugs for a 5 years will develop this condition,[8] which is most specifically characterized by single muscle jerks or tics of the face, jaw, tongue, trunk, and extremities. Like most dyskinesias, it disappears during sleep. It is often not reversible, even after discontinuation of the antipsychotic medication. On the other hand, there is some recent evidence that in many cases, "oral-facial dyskinesias in patients with intellectual impairment and negative symptoms may actually represent spontaneous movement disorders associated with hebephrenic or deficit forms of schizophrenia."[9]
- *Neuroleptic Malignant Syndrome (NMS).* As the most serious reaction to the antipsychotic drugs, NMS can cause kidney failure and death. The most dramatic symptoms are a "lead pipe" body rigidity and a fever as high as 106° F. NMS is more likely when the patient is exposed to heat stress and dehydration. Some of the symptoms can be decreased by administering *bromocriptine* (Parlodel), a brominated ergot alkaloid with mixed dopamine agonist/antagonist activity usually employed in the treatment of Parkinson's disease.[10] Another dopamine agonist used in the treatment of NMS and Parkinson's is *amantadine* (Symadine, Symmetrel), a drug perhaps more commonly encountered in the prophylaxis and treatment of influenza A virus infection.[11] Finally, *dantrolene* (Dantrium), a powerful skeletal relaxant with significant side effects of its own, is often used with these dopamine agonists to relax the muscle rigidity associated with NMS. Dantrolene is thought to act by inhibiting the release of second-messenger calcium from the sarcoplastic reticulum.[12]

Some Beneficial Side Effects. It has been noted for many years that the schizophrenic population has an unusually low incidence of cancer; many have speculated that this reflects some sort of unusual body-mind interaction. However, it is probably a physiological effect of antipsychotic medication, most clearly linked to chlorpromazine, which has been shown to slow the growth of experimentally induced tumors in animals. The human results come from the unusually complete data on psychiatric patients available in Denmark, where the unparalleled combination of a universal social welfare system and obsessive record-keeping has resulted in a gold mine of demographic correlations. Male schizophrenics show about half the rate of prostate cancer as the general population; female schizophrenics have a halved rate of death from cervical cancer. Schizophrenics treated with chlorpromazine had a lower incidence of cancer overall than those not so treated.[13]

The Phenothiazines

History and Discovery. The three-ring unit containing a sulfur and nitrogen atom in chlorpromazine (and in all but two of the other antipsychotic structures pictured in Figure 5.1) is referred to as a *phenothiazine* system ("thi" is the infix for sulfur, "aza" for nitrogen.) Phenothiazine itself, without any substituents, is a chemical that has been known since the 1880s, when it was synthesized by fusing sulfur with diphenylamine. It was used for decades as a treatment for intestinal parasites (anthelmintic). In 1947, Paul Charpentier's group at Rhone-Polenc, in an effort to create an improved antihistamine by varying the structure of *diphenhydramine* (**5-1**, Figure 5.1), (the active ingredient in Benadryl, Dramamine, and Sominex, synthesized promethazine, a phenothiazine derivative with a dimethylaminopropyl group attached to the nitrogen of the central ring.[14] Like most antihistamines, the drug had pronounced central sedating effects, and French Navy surgeon Henri Laborit used it in Tunisia in 1949 to inhibit the development of shock in surgery. In an effort to enhance these effects, Charpentier added a chlorine substituent to one of the rings forming *chlorpromazine* (Thorazine, **5-2**), which was tested by Laborit in 1950 and found to be even more effective than promethazine in inhibiting shock. (Phenothiazine and butyrophenone neuroleptics are still used to potentiate surgical anesthesia.) Laborit noticed a pronounced antianxiety effect of these drugs when given prior to surgery and asked his psychiatric colleagues at the Val de Grâce hospital to try the drug on schizophrenics.[15] By 1952, the drug was marketed as an antipsychotic and a new era in psychiatric medicine had begun. But the phenothiazines still betray their parentage; most have many antihistamine and soporific properties, and share with the classical antihistamine diphenhydramine the side effect of being antiemetics.

SAR of the Phenothiazine Neuroleptics. The structures of some of the more common phenothiazines are shown in Figure 5.1. About a dozen other similar structures might have been included; the constant variations on a theme represent efforts to modify the many undesirable side effects of these structures without diminishing activity. Although this has never been accomplished with this class of drugs, an understanding of the structure-activity relationships (SAR) was developed and a new class of drugs, the tricyclic antidepressants, was discovered in the process. The constant features of the nine drugs shown have been found to be essential for antipsychotic activity: there must be three atoms between the central ring and the nitrogen atom on the side chain (if there are only two atoms, the molecule is a powerful sedative and antihistamine but has no antipsychotic activity); an electron-withdrawing group must occupy position 2 of one of the rings for full activity. The central ring does not need to have a nitrogen infix; *chlorprothixene*, **5-5**, and *thiothixene,* **5-9**, which have a *thioxanthene* ring system, are both active (interestingly, only as their *cis* isomers—i.e., the orientation of the side chain must be to the same side of the molecule as the benzene ring bearing the substituent). Antipsychotics that, like thiothixene, incorporate a piperazine ring into their side chain (e.g., *fluphenazine,* **5-7**, and *trifluoperazine*, **5-8**) are generally more potent than those with a single nitrogen in the chain. The piperazine phenothiazines also have less sedative effect than other phenothiazines; autonomic effects such as orthostatic hypotension (fainting on assuming an upright posture) are also less frequent. High-potency piperazines like fluphenazine (Permitil, Prolixin) are

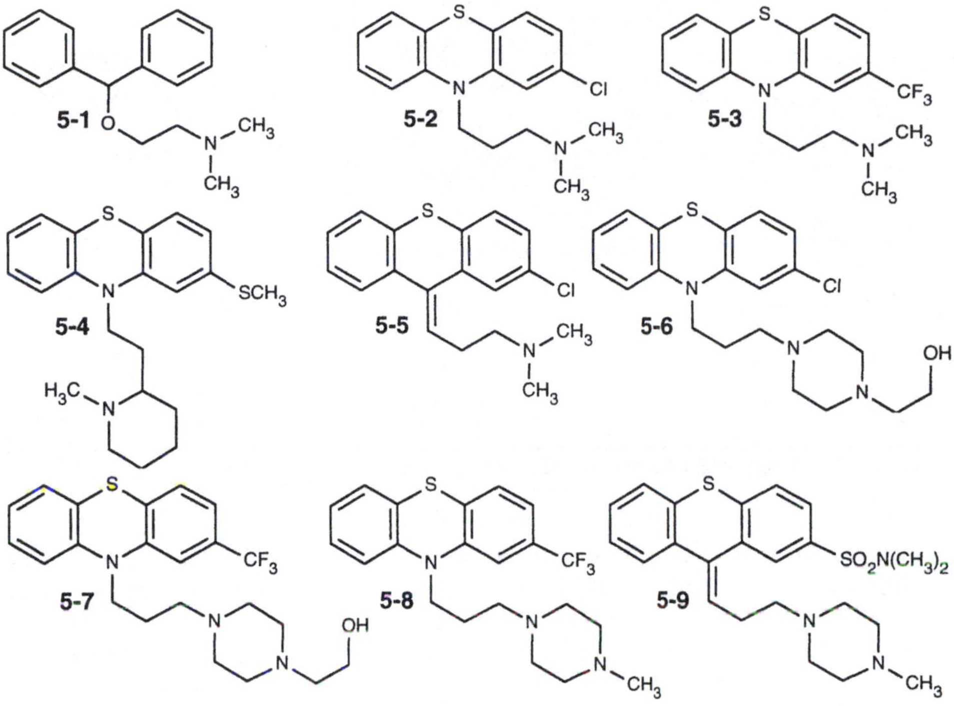

Figure 5.1 The antihistamine diphenhydramine and some antipsychotics based on its structure.

as potent as haloperidol (see next section) and if used (as the decanoate or enanthate salt) in an intramuscular depot form, maintain antipsychotic activity after one injection for as long as 2 to 3 weeks. thus minimizing problems with compliance.

Butyrophenones

A second class of major tranquilizers have in common a butyl phenyl ketone structure (called butyrophenone in an older terminology) within the molecule. *Haloperidol* (Haldol, **5-10,** Figure 5.2) is the most common of these. Haldol was developed by Dr. P. A. J. Janssen in the 1950s. (Compare the structure of this antipsychotic with the powerful opiate fentanyl and the antidiarrhetic loperamide, both discussed in Chapter 2, which were also developed by Janssen.) In the United States, this is the only butyrophenone antipsychotic prescribed; elsewhere, *trifluperidol* (Psychoperidol, **5-11**), *spiroperidol* (**5-12**), and *benperidol* (**5-13**) are also used. The butyrophenones are high-potency antipsychotics with activity similar to the piperazine phenothiazines such as trifluoperazine and fluphenazine. Haloperidol has the same pharmacological effects as the high-potency phenothiazines, even though its chemical structure is very different. Haloperidol is used to treat schizophrenia and active psychoses, paranoia, ballismus, and Tourette's syndrome, where it is the drug of choice.

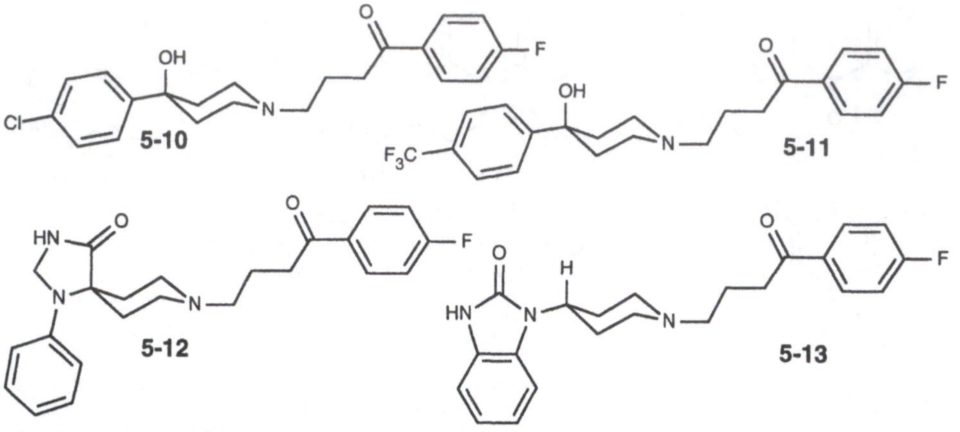

Figure 5.2 Some butyrophenone neuroleptics.

It is a powerful antiemetic. Like the piperazine phenothiazines, it is relatively nonsedating but likely to produce extrapyramidal reactions. Haloperidol is as potent as fluphenazine, the most potent of all the phenothiazines: 2 mg is the equivalent of 100 mg chlorpromazine. Like the piperazine drugs, haloperidol is less sedating than chlorpromazine; unfortunately, it also as likely to produce extrapyramidal reactions. A decanoate preparation is available for depot administration. Table 5.1 gives the relative potency and tendency to induce side effects for the common antipsychotics.

Table 5.1 Antipsychotic (Neuroleptic) Drugs

ANTIPSYCHOTIC	POTENCY	SEDATION	ANTICHOLINERGIC	HYPOTENSION	EPS
CHLORPROMAZINE (5-2)	LOW	HIGH	HIGH	HIGH	LOW
TRIFLUPROMAZINE (5-3)	LOW	HIGH	MEDIUM	HIGH	MEDIUM
THIORIDAZINE (5-4)	LOW	HIGH	HIGH	HIGH	LOW
CHLORPROTHIXENE (5-5)	MEDIUM	HIGH	HIGH	MEDIUM	MEDIUM
PERPHENAZINE (5-6)	MEDIUM	LOW	LOW	LOW	HIGH
FLUPHENAZINE (5-7)	HIGH	MEDIUM	LOW	LOW	HIGH
TRIFLUOPERAZINE (5-8)	HIGH	MEDIUM	LOW	LOW	HIGH
THIOTHIXENE (5-9)	HIGH	LOW	LOW	LOW	HIGH
HALOPERIDOL (5-10)	HIGH	LOW	LOW	LOW	HIGH

Atypical Antipsychotics

Clozapine. Clozapine (Clozaril, **5-14**, Figure 5.3) is a drug that has excited much controversy and even several class-action lawsuits. It is the first of the "atypical" antipsychotic drug—that is, it rarely causes Parkinson-like reactions and does not cause tardive dyskinesia. Indeed, clozapine is used to successfully treat the tremor of Parkinson's disease and the symptoms of tardive dyskinesia caused by other drugs. It also seems to be more effective than the classical antipsychotic drugs in alleviating the negative effects of schizophrenia. However, clozapine is usually prescribed only when severe schizophrenia fails to respond to at least two other standard antipsychotic medications; it ameliorates 30% of such cases. Clozapine causes two problems that the traditional drugs do not. The first, a tendency of the drug to evoke seizures in about 15% of those using it in a high regimen, is not so serious because it can be treated with antiepileptic drugs. A much more serious problem is that it can cause hematologic disturbances (such as agranulocytosis and leukopenia) that can be fatal. Consequently, the manufacturer, Sandoz, was requiring a weekly blood test costing about $9,000 per year so as to avoid liability lawsuits. In Europe, where doctors make their own monitoring arrangements, Sandoz sold the same drug for as little as $1,100 a year.

To be weighed against these drawbacks is the amazing success this drug has had in alleviating psychoses in patients for whom all other drugs have failed:

> At the Buffalo State hospital, 21 of 31 long-term patients treated were discharged; at Bronx State, 27 percent; at Elmira State, one-third. Dr. Sheikh Qadeer of Elmira was struck by how much clozapine reduced violent incidents. Of 785 state hospital patients treated with it, 170 were discharged.
>
> That should be good economic news. It costs $120,000 a year to keep a person at a state hospital versus $35,000 at a group home and $9,000 at a supervised single-room-occupancy [SRO] center. The savings should easily offset clozapine's cost—$5,000 per person a year.[16]

The above article goes on to say that unfortunately, in New York as elsewhere, nothing is easy: the lack of supervised group homes, combined with the need for weekly blood tests for those on clozapine, means that the overwhelming majority of schizophrenic patients in state hospitals—those who by definition have not responded to conventional drugs—have never been given Clozaril.

With increased awareness of these problems and more careful monitoring of patients,

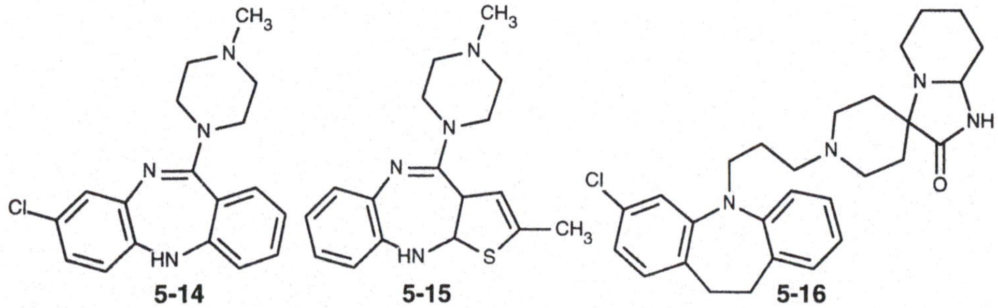

Figure 5.3 Clozapine, olanzepine, and clospipramine.

fatalities from agranulocytosis were avoided in the United States in recent years; and with increasing evidence that the drug is more effective against both positive and negative symptoms of schizophrenia[17] it is beginning to be used as a first-line agent. *Olanzapine* (Lanzac, **5-15**) and *mosapramine* (clospipramine, Cremin, **5-16**) are two newer antipsychotics of schizophrenia; the structures are analogous to that of clozapine.

Newer Antipsychotic Drugs. Merck has licensed *remoxipride* (Roxiam, **5-17**, Figure 5.4) from A. B. Astra of Sweden, which had marketed the drug earlier in Britain, Denmark, Germany, Ireland, and Sweden. Remoxipride seemed particularly promising because it cost only about a third as much as clozapine (Clozaril) and, unlike clozapine or haloperidol (Haldol), side effects seemed indistinguishable from placebo. Unfortunately, in 1993 the drug had to be withdrawn from clinical trials because of the occurrence of several cases of aplastic anemia.[18] A second new drug is *risperidone* (Risperdal, **5-18**), which was also developed by Dr. Paul Janssen, the creator of Haldol. Risperidone is said to have an approximately equal antagonist activity at both the dopamine D_2 and serotonin 5-HT_2 receptor subtypes. The anti-D_2 activity is similar to that of conventional antipsychotics such as haloperidol, but the antiserotonin activity is similar to that of the antidepressants. This is thought to explain the efficacy of risperidone in treating both the negative and positive symptoms of schizophrenia, since the negative symptoms are similar to those of clinical depression. It has been called a "quantitatively atypical" antipsychotic, in that extrapyramidal effects are not induced if the dosage kept below 6 mg/day.[19] Risperidone is also being explored for use in alcohol withdrawal and cocaine addiction.[20]

Mechanism of Action of the Antipsychotics

Schizophrenia and other psychoses represent some of the least understood diseases; moreover, the mode of action of the drugs that treat these diseases is also poorly understood. A major common feature of the phenothiazine and butyrophenone antipsychotic drugs is their ability to function as antagonists at the D_2 subset of receptors. The hypothesis that this common feature is also responsible for antipsychotic efficacy is strengthened by a high correlation between the clinical antipsychotic potency of a given drug and its in vitro binding to this receptor site. Although all the drugs bind to other NT receptor sites, only the D_2 subset exhibits this correlation.

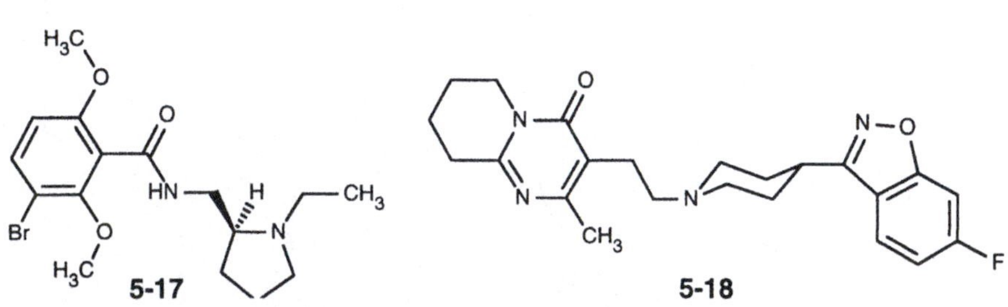

H_3C O O HN H CH_3 N Br O H_3C **5-17**

O N N CH_3 N N—O F **5-18**

Figure 5.4 Remoxipride and risperidone.

There are two pathways for the D_2 neurons from the midbrain. The first, which is responsible for extrapyramidal movement, leads from the substantia nigra to the basal ganglia. The blocking of these sites is presumed to be responsible for the undesirable EPS symptoms caused by the classical phenothiazine and butyrophenone antipsychotics. The second pathway leads from the ventral tegmental area to the nucleus accumbens and the frontal cortex, and it is thought that action by the antipsychotics at these D_2 sites is responsible for their beneficial effects. (This would imply that psychoses are the effect of excess dopamine in this area; a hypothesis supported by the fact that drugs like amphetamine, which release dopamine from the synapse, are psychotomimetic in overdose.) Because the atypical drug clozapine does *not* bind particularly strongly to the D_2 sites it does not cause EPS symptoms; but it has been a puzzle as to how it acts as an antipsychotic. Recently, a D_4 receptor subtype has been cloned that has a high affinity for clozapine, and presumably it is through these sites that the drug exerts its activity.[21] But other evidence suggests that it is the binding of clozapine to the serotonin 5-HT_{2A} receptors which accounts for its unique efficacy.[22]

To complicate the still evolving picture: a D_3 subtype that binds antipsychotic drugs much like the D_2 subtype has also been characterized.[23] And a D_5 receptor has also been cloned that binds the endogenous dopamine agonist ten times more strongly than the D_1 subtype.[24] Obviously, there is much yet to be learned both about schizophrenia and about the dopamine receptor class; and much more selective agents need to be developed.

Other Conditions Treated with Antipsychotic Drugs

In addition to the treatment of vertigo and nausea, there are several other conditions for which antipsychotic drugs such as haloperidol and *pimozide* (Orap, **5-19**, Figure 5.5) are used. *Singultus*, a condition of constant uncontrollable hiccuping, is one of these. *Gilles de la Tourette Syndrome* is still another. Haloperidol is often used to treat this syndrome, a rare but notorious disorder characterized by severe motor and vocal tics. The syndrome was often underdiagnosed in the past because of the erroneous belief that the cause was psychological and that such bizarre phenomena as *echolalia* (the nearly instantaneous and involuntary repetition of the words just spoken by another person) and *coprolalia* (the uncontrollable vocalization of obscene or scatological words in totally inappropriate contexts) were required for its diagnosis. In fact, these socially embarrassing symptoms

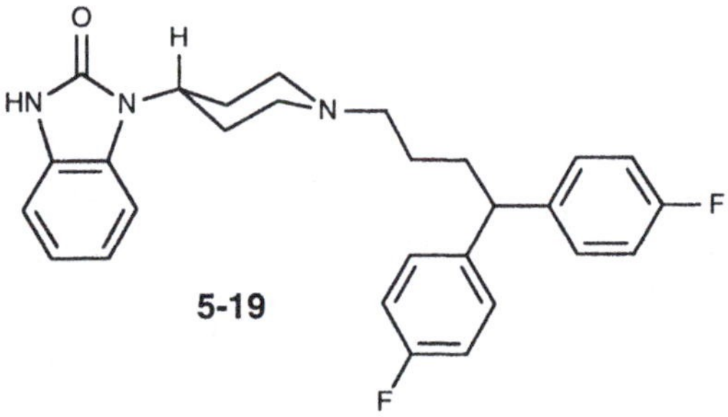

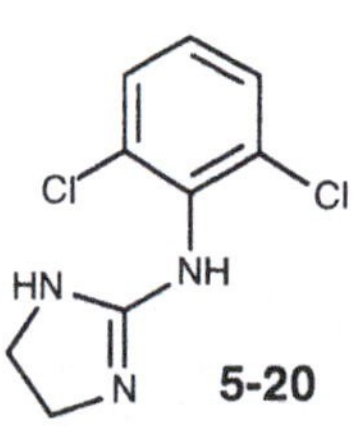

Figure 5.5 Pimozide and clonidine.

only occasionally accompany the syndrome.[25] Pimozide is slightly less successful[26] in treating Tourette's syndrome; and *clonidine* (Catapres, **5-20**), which is an α_2-adrenergic receptor agonist used primarily to lower blood pressure, sometimes helps where haloperidol has failed. (Other atypical uses of clonidine are mentioned in Chapters 2 and 3.) Another disorder helped by haloperidol or chlorpromazine is *ballismus*, which is a syndrome characterized by involuntary "flinging" movements of the arms or legs. Pimozide also alleviates the severe facial pain of *trigeminal neuralgia*.[27]

Syntheses of Some Representative Antipsychotics

Charpentier's original synthesis of chlorpromazine was carried out by heating 3-chlorodiphenylamine (**5-21**, Figure 5.6) with iodine and sulfur to form the phenothiazine **5-22**. The three-carbon side chain was appended by forming the anion with sodium amide and displacing the chloride from *N,N*-dimethyl-3-chloropropylamine, giving chlorpromazine, **5-23**.[28]

Haloperidol, **5-26**, is similarly formed from the nucleophilic displacement of the chlorine in **5-25** by the nitrogen of **5-24**.[29] (The piperidine **5-24** was itself first reported as the product of a series of somewhat unusual reactions.)[30]

Finally, the seven-atom diazepine ring of clozapine, **5-29**, is formed in an intramolec-

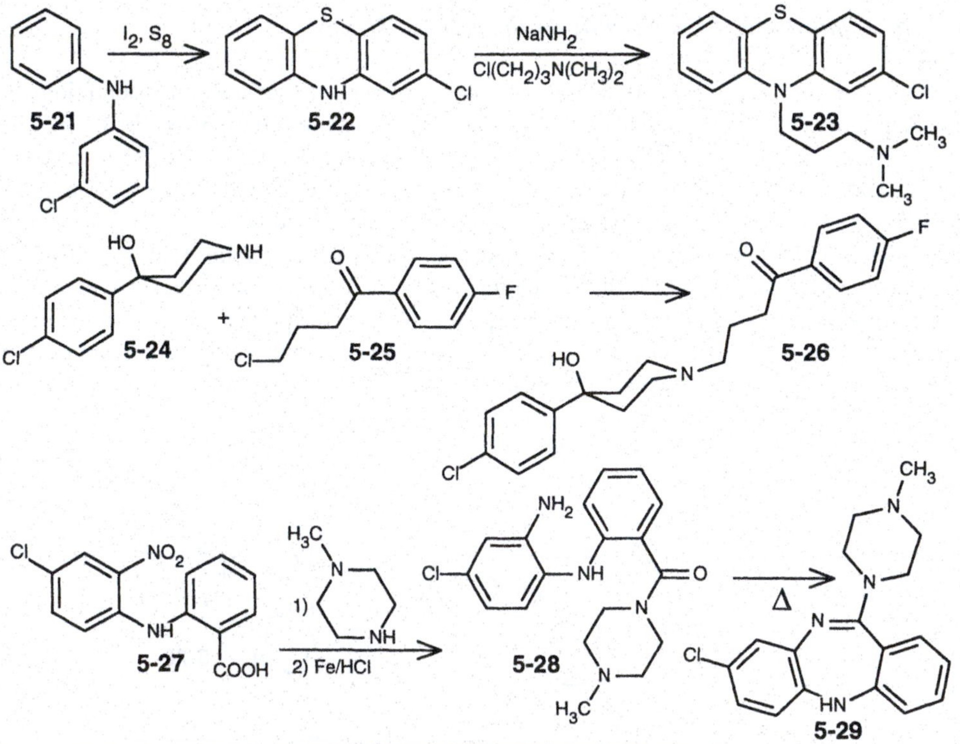

Figure 5.6 Syntheses of Chlorpromazine (5-23), haloperidol (5-26), and clozapine (5-29).

ular cyclization upon heating precursor **5-28**, itself synthesized in two steps from **5-27** (which in turn is formed from the Ullmann reaction of anthranilic acid and 2,5-dichloronitrobenzene).[31]

The Antidepressants

Affective Disorders

Clinical depression describes a *unipolar* affective disorder wherein the person is unaccountably (as opposed to uncomplicated grief or loss) subject to a dysphoric mood or loss of interest and pleasure in all usual activities; feelings of helplessness, worthlessness, and inappropriate guilt; indecisiveness, suicidal ideation, insomnia, and anorexia. This is to be distinguished from *bipolar* affective disorder, also called *manic depression,* which may afflict as much as 1% of the population,[32] wherein a depressed mood alternates with manic episodes in which the patient is unaccountably euphoric, exhibits racing speech and thought, is overactive and excessively sociable, may show extreme self-confidence with delusions of grandiose importance, and may engage in extreme behavior (e.g., gambling, investment, or buying sprees; sexual indiscretions; reckless driving; explosive outbursts of temper) that can lead to functional impairment manifested by loss of employment, indebtedness, divorce, or the like.

Depression, Mania, and Creativity. In her book *Touched with Fire: Manic-Depressive Illness and the Artistic Temperament*, Kay Jamison, a professor of psychiatry at Johns Hopkins University, argues that writers are 10 to 20 times more likely to suffer from depressive or manic-depressive disorders. Indeed, the roster of famous writers who are known to have had major bouts of depression, many to the point of suicide or attempted suicide, is impressive: Robert Lowell, Virginia Woolf, William Styron, Primo Levi, Anne Sexton, Sylvia Plath, Ernest Hemingway, Dorothy Parker, Yukio Mishima—the list goes on and on. According to Jamison, "The cognitive style of manic depression overlaps with the creative temperament. When we think of creative writers, we think of boldness, sensitivity, restlessness, discontent; this is the manic-depressive temperament."[33] In a mildly manic state there can be enhanced energy, original inspiration, and creativity; in a mildly depressed state, the person is self-critical even to the point of obsession, which can be helpful in revising and editing. It is, of course, claims such as these that fuel the debate about drugs like Prozac: if Hemingway had experienced the consolations of Prozac, would he have led a peaceful, productive life as a stockbroker in Oak Park, IL, instead of drinking his way through the Spanish Civil War and brooding for years in Paris at the *Café Deux Magots*?

There is also a common perception that writers are more subject than the average person to experience problems with drinking or other drug abuse. If Jamison's hypothesis is correct, it may predict this very correlation, for other studies have shown that there is a significantly increased prevalence of substance use disorders among patients with manic-depressant disease.[34]

Depression and CNS Serotonergic Neurons. Evidence has accumulated over the last two decades of atypical serotonin functioning in the CNS of depressed persons. It is possible that these features are the results of depression, but it is more likely that they are its cause. The correlations are chiefly these: (1) the major metabolite of serotonin (5-HT), 5-hydroxyindoleacetic acid (5-HIAA, Chapter 1) is found in lower concentration in the CNS of unmedicated depressed persons than in that of controls, (2) brain tissues of postmortem depressed patients or those who have committed suicide show abnormally low levels of both 5-HT and 5-HIAA, (3) patients who have responded to an antidepressant drug and are then subjected to a diet that is rigorously controlled to eliminate tryptophan (the precursor of 5-HT) relapse into a depression with similar features to their pre-medicated state,[35] and (4) there is a decreased number of 5-HT transporter binding sites in the postmortem brain tissue of suicide victims and in the platelets of drug-free depressed patients—a deficit not observed in mania, Alzheimer's disease, schizophrenia, or panic disorder.[36]

Drugs Used to Treat Affective Disorders

Lithium. *Discovery.* The surprising fact that lithium salts could dampen the harmonic cyclings of manic-depression was discovered in part by accident, as might be expected. In the late 1940s, John Cade, an Australian psychiatrist, noted that the symptoms of manic depression were sometimes mimicked by victims of thyroid disease. Hypothesizing that a hormonal imbalance was involved, he injected the concentrated urine of manic patients into guinea pigs and found that it was often three times as lethal as the urine of normal subjects. One possibility was that higher levels of uric acid from the manic patients somehow enhanced the toxicity of urea; in testing this hypothesis, Cade found that only the lithium salt of uric acid was sufficiently soluble to be useful. But he found that animals given lithium urate were protected from the convulsions caused by urea, and when they were administered lithium carbonate alone they became lethargic and sedated.

Cade eventually treated a patient who had been hospitalized 5 years for unremitting mania with large doses of lithium citrate: after 3 weeks the man was able to be discharged and soon returned to his former occupation. Nine other manic patients responded similarly.[37]

Psychopharmacology. Lithium has been considered the most specific drug for the treatment and prevention of bipolar mood disorder (however, valproic acid may be as effective). It is particularly effective in suppressing acute manic episodes; about 70% of the cases respond. Lithium must then be taken on a continual basis to prevent recurrence: persons who have experienced one episode of mania have a 90% chance of a second episode if they are not treated. The anticonvulsants *carbamazepine* (Tegretol) and *valproic acid* (see Chapter 3) help some patients for whom lithium is ineffective. Valproic acid recently[38] was shown to be as effective as lithium in controlling manic episodes in hospitalized patients. *Verapamil*, which is a calcium channel antagonist ordinarily used to control hypertension, also seems to be able to suppress acute mania in some patients.

Because carbamazepine and valproic acid are known to augment the action of γ-aminobutyric acid (GABA) inhibitory neurotransmission, it is not too surprising that they can inhibit mania. Similarly, blocking calcium ion channels suppresses neurotransmission generally, and could account for the activity of verapamil. Lithium probably works in a parallel fashion by blocking the phosphatase enzyme system necessary for generating the secondary messengers within the cell following the interaction of the neurotransmitter at the cell membrane. The cell becomes less sensitive generally to NT stimulation.[39] In normal individuals, lithium causes only mild sedation.

Side Effects. The most problematic side effect of lithium is a lowering of thyroid activity; in a very small proportion of patients, there can be kidney failure. Two more common side effects, weight gain and mental sluggishness, lead many patients to discontinue taking lithium; in about one-third of those administered lithium there is a demonstrable impairment of memory—something that many otherwise productive patients find intolerable. By contrast, fewer than 5% of patients taking valproate have any cognitive impairment.

Monoamine Oxidase Inhibitors. *Discovery.* Monoamine oxidase (MAO) inhibitors (MAOIs) were discovered by accident during clinical trials of the antitubercular drug *iproniazid* during the 1950s. Isolated from family, friends, and their ordinary occupations and confronted with a life-threatening disease, many of the patients in the tuberculosis sanatoria of the day were understandably depressed,[40] and it was soon noticed at several centers that the mood of those patients taking the antitubercular drug iproniazid was improved. It was hypothesized that iproniazid had an antidepressant effect because it blocks the enzyme MAO, which metabolizes excess epinephrine and serotonin (see Chapter 1) within the presynapse. Inhibiting the action of this enzyme would make more epinephrine and serotonin available.

Mode of Action. These early MAOIs are all "suicide substrate" irreversible inhibitors of MAO, a flavin-containing enzyme found in the mitochondria of neurons, intestinal mucosa, the liver, etc. Iproniazid contains a hydrazine ($-NHNH_2$) function that is oxidized by MAO to a reactive intermediate that binds irreversibly to the flavin prosthetic group of the enzyme, rendering it inactive. Maximal inhibition of MAO activity is usually achieved in a few days (although, as with all antidepressants, psychological effects are not seen for 2-3 weeks), and after the medication is discontinued, it requires 1-2 weeks for MAO activity to be restored to normal.

Side Effects. Iproniazid was replaced with better MAOIs, but all MAOIs still share common awkward side effects: orthostatic hypotension, sexual dysfunction (impotence, delayed ejaculation, loss of libido), and an unusual food/drug interaction: any foods or beverages containing large amounts of tyramine can cause a hypertensive crisis (headache, tachycardia, nausea and occasionally subarachnoid hemorrhage). Foods containing tyramine are aged cheeses, wines, beers, dried meats and fish, and fava beans. The tyramine (4-hydroxyphenethylamine, also called tyrosamine) in these foods does not ordinarily affect us because it is rapidly converted by MAO in the intestines to tyrosine, a common amino acid; absent this protective mechanism, tyramine enters the storage vesicles for norepinephrine (NE) and displaces the NE, causing a drastic rise in blood pressure.

Phenelzine, Isocarboxazid, and Tranylcypromine. Phenelzine (Nardil, **5-30**, Figure 5.7), isocarboxazid (Marplan, **5-31**), and tranylcypromine (Parnate, **5-32**), are all used for clinical

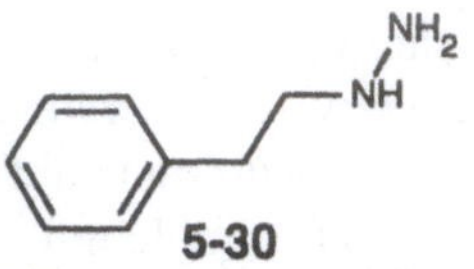

Figure 5.7 Three MAOI antidepressants.

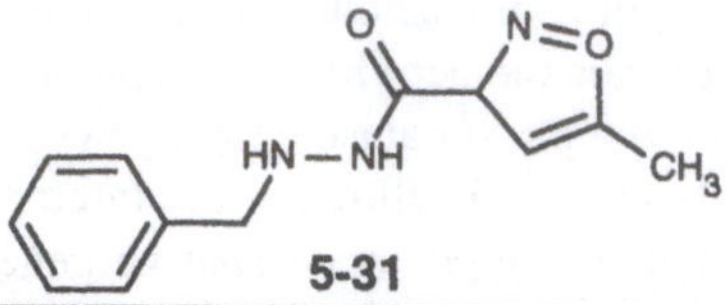

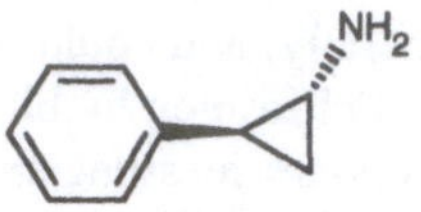

depression, panic disorder, and phobic disorders. Phenelzine, developed by Warner-Lambert, represents a simple substitution of hydrazine for the amine function of phenethylamine. Phenelzine also may improve eating behavior in some bulimics, but controlled clinical trials have not yet confirmed this. Isocarboxazid is an elaboration of the phenelzine structure, and is probably converted to it in vivo. Tranylcypromine can be seen either as a modification of phenethylamine or as an amphetamine with the side chain cyclized into a cyclopropane ring; in larger doses than those used to treat depression, it can cause psychomotor stimulation, and the development of dependence occasionally has been reported.

Reversible Inhibitors of Monoamine Oxidase A. More recent research shows that there are two types of MAO enzyme, MAOA and MAOB. Whereas MAOB is more active in the peripheral system and metabolizes dietary tyramine, it is the inhibition of MAOA enzyme in the CNS that is responsible for the antidepressant efficacy of MAOIs.[41] Older drugs such as phenelzine, isocarboxazid, and tranylcypromine irreversibly inhibit both enzyme types, but some newer drugs used in Europe are selective for MAOA. These reversible inhibitors of MAOA (RIMAs) are active antidepressants but do not induce a "cheese effect" hypertensive crisis if tyramine-containing foods are eaten.

Moclobemide. Moclobemide (Aurorix, **5-33**, Figure 5.8) was first marketed by Hoffmann-LaRoche in Sweden in 1990, and was subsequently approved for marketing in about 50 countries. It was the first of the RIMA class of drugs, and it shows good promise. In a randomized multicenter study comparing its efficacy to that of tranylcypromine, moclobemide was equally efficacious but had the clear advantage of a safer profile and a slightly better tolerability.[42] However, studies from the Danish University Antidepressant Group have indicated that moclobemide, citalopram, and paroxetine are all somewhat less effective than clomipramine.[43] In a comparative study of moclobemide and fluoxetine, tolerance of both drugs was quite good and efficacy statistically identical; whereas sedation, nausea, and vomiting were reported more frequently with fluoxetine, insomnia was more common with moclobemide.[44]

Brofaromine, **5-34**. In a randomized, double-blind Canadian study, this new RIMA

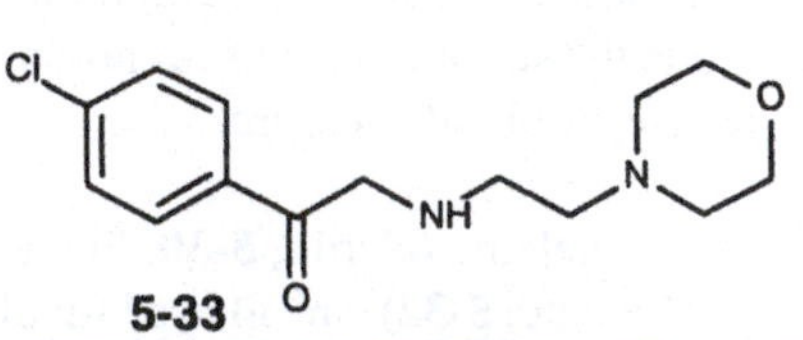

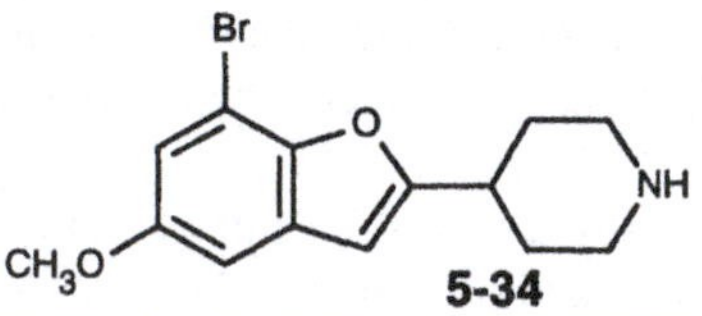

Figure 5.8 Two RIMA antidepressants.

drug was compared to clomipramine in patients with panic disorder with or without agoraphobia.[45] The drugs proved to be of equal efficacy, and the much more favorable safety and reduced side effects of the RIMA will most likely make it preferable to clomipramine. In a review of trials using brofaromine for patients with major depression who did not respond to current antidepressant medications, brofaromine was judged equally effective but better tolerated and safer to use.[46]

The Tricyclics. *History.* The term *tricyclic* is used to describe any system of three fused rings with (usually) the central ring seven-membered and the two side rings six-membered. As is the case with so many psychotropic medications, this class of drugs was discovered through a combination of luck and intelligent observation. During the mid-1950s, as the antipsychotics were being first introduced, Roland Kuhn at the Cantonal Psychiatric Clinic in Munsterlingen, Switzerland, noticed that chlorpromazine produced some effects he had observed earlier when testing an antihistamine from the J. R. Geigy Company of Basle. He wrote a long letter to the company and received a series of the compounds—which had been synthesized in the late 1940s by Häfliger and Schindler—for testing. When Kuhn gave the second of these compounds, G22355, to several depressed patients, it was obvious that it had a significant antidepressant effect. Seven months later, the drug, now named *imipramine*, was presented in Zurich at the Second International Congress of Psychiatry.[47] Other manufacturers soon produced similar tricyclic structures. The shape of these molecules is considerably altered in expanding the central ring to seven atoms, and the pharmacology of the system changes as well. These drugs have no effect on psychoses, but they cause dramatic mood improvement in the clinically depressed.

Side Effects. The tricyclics cause a variety of adverse reactions typical of their class, the most common of which are due to α-adrenergic-blocking activity: flushing, diaphoresis, constipation, and orthostatic hypotension. It is possible to have a fatal reaction to as little as 2 g of these drugs because of their anticholinergic and cardiac toxicity. Inasmuch as suicide attempts are not uncommon among clinically depressed individuals (and because all presently known antidepressants require about 2 weeks to become effective), the real possibility of these drugs being used as the successful means for suicide has to be taken into account; newer drugs such as the SSRIs represent an enormous step forward with regard to safety alone.

As a general rule, the tricyclics which are tertiary amines (i.e., those with no hydrogen atom on their amine nitrogen): such as *imipramine* (Tofranil, **5-35**, Figure 5.9), *clomipramine* (Anafranil, **5-37**), *amitriptyline* (Elavil, **5-38**), and *doxepin* (Adapin, Sinequan, **5-40**), have more extensive unwanted side effects than the secondary amines: *desipramine* (Norpramin, Pertofrane, **5-36**), *nortriptyline* (Pamelor, **5-39**), *maprotiline* (Ludiomil, **5-41**), and *amoxapine* (Asendin, **5-42**). But all of these drugs are cardiotoxic, are prone to induce weight gain, have sedative activity to a greater or lesser degree (though tolerance is soon acquired to this antihistaminic effect), have undesirable anticholinergic activity, and predispose to orthostatic hypotension. Table 5.2 summarizes the relative incidence of side effects with these drugs; the drugs are listed from top to bottom from roughly the most side-effect-prone to the least. Nonetheless, in practice a great degree of idiosyncratic interaction occurs, and one individual may respond better to a drug at the top of the list than to one at the bottom.[48]

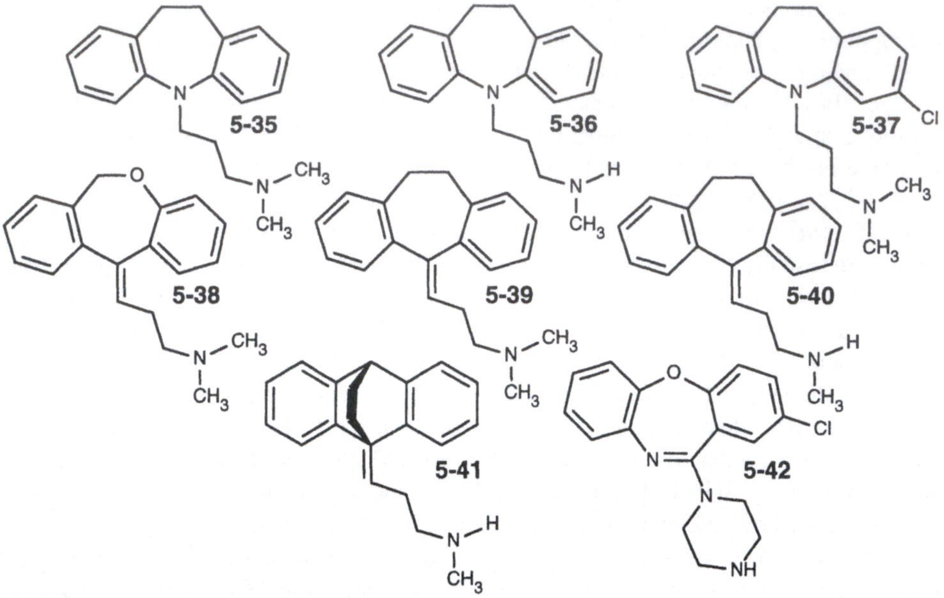

Figure 5.9 Some common tricycic antidepressants.

Table 5.2 Tricyclic Antidepressants

ANTIDEPRESSANT	WEIGHT GAIN	SEDATION	ANTICHOLINERGIC	HYPOTENSION	CARDIOTOXICITY
AMITRIPTYLINE	HIGH	HIGH	HIGH	HIGH	HIGH
CLOMIPRAMINE	MEDIUM	MEDIUM	HIGH	MEDIUM	HIGH
MAPROTILINE	MEDIUM	MEDIUM	MEDIUM	MEDIUM	MEDIUM
DOXEPIN	MEDIUM	HIGH	MEDIUM	HIGH	MEDIUM
IMIPRAMINE	MEDIUM	MEDIUM	MEDIUM	MEDIUM	HIGH
AMOXAPINE	MEDIUM	LOW	LOW	MEDIUM	MEDIUM
DESIPRAMINE	MEDIUM	LOW	LOW	LOW	MEDIUM
NORTRIPTYLINE	MEDIUM	LOW	LOW	LOW	MEDIUM

Other Uses of the Tricyclics. In addition to its use in depression, amitriptyline may control abnormal eating behavior in bulimic patients and is useful in the prophylaxis of migraine headache. Effective doses have a moderate to marked sedative action. Imipramine has also been used for panic disorder and phobic disorders. Clomipramine and fluoxetine are the only agents approved for treatment of obsessive-compulsive disorder (OCD). Controlled studies have shown that clomipramine has antiobsessional activity, and that this activity is not due to any further metabolite (e.g., the first metabolite of clomipramine, which is the desmethyl compound, where one of the two methyl groups on the nitrogen has been replaced with a hydrogen). The antidepressant activity and the antiobsessional activity are relatively independent, with the latter probably related to its ability to block reuptake of serotonin.

Atypical Antidepressants

Alprazolam. Alprazolam (Xanax, **5-43**, Figure 5.10) is a benzodiazepine that also is used as an anxiolytic and hypnotic. Compared with other antidepressants, alprazolam does not affect the uptake of NT; it has, like triazolam, a high affinity for the GABA receptor. Nonetheless, it lowers total NE turnover and causes the downregulation of β-adrenergic receptors—things traditional benzodiazepines such as diazepam do not do. It is effective in panic disorder and agoraphobia and seems to have antidepressant activity. However, there is concern about using a benzodiazepine for this indication because the required frequent dosing causes sedation and physical dependence.

Bupropion. The antidepressant action of bupropion (Wellbutrin, **5-44**) is comparable to that of the tricyclics and MAOIs. However, its structure is unrelated to any of the other antidepressants, it has a stimulant rather than a sedative activity, and it seems to be less likely than the tricyclics to precipitate mania when given to patients with bipolar disorder in their depressed phase. The structure of bupropion actually is quite close to that of amphetamine and the psychostimulants (see Chapter 4), and several studies have found it effective in treating attention-deficit hyperactivity disorder (ADHD).[49]

Just prior to its marketing in 1985 (it was then approved by the FDA for treatment of bulimia), there was an incidence of some seizures in patients using the drug. Marketing was delayed, but more careful analysis of the data seemed to indicate that bupropion at the recommended dosage was no more liable to induce seizures than many other medications.[50] Additionally, bupropion evidence indicates that it is effective in many individuals who do not respond to other antidepressants.

Selective Serotonin Reuptake Inhibitors. *Fluoxetine.* A newer class of antidepressants, the selective serotonin reuptake inhibitors (SSRIs), the first of which was fluoxetine (Prozac, **5-45**) has effected an almost revolutionary change in the treatment of depression.

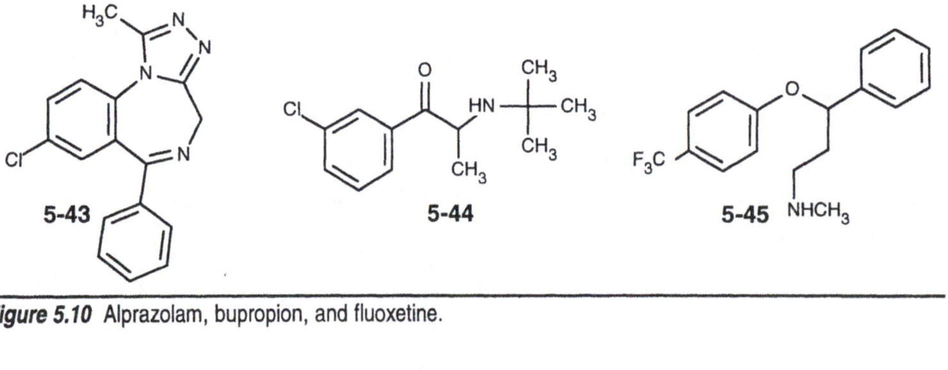

Figure 5.10 Alprazolam, bupropion, and fluoxetine.

The SSRIs inhibit serotonin reuptake, whereas the older antidepressants act primarily through the dopamine and norepinephrine pathways. These drugs do not seem to be any more effective than the classical antidepressants in the treatment of severe clinical depression, but their safety, low toxicity, and benign range of side effects has led them to be used increasingly for patients described not as clinically depressed but as "dysthymic." The change wrought on such people can be dramatic, almost a transformation of the personality. Dr. Peter Kramer writes about these people in his best-selling *Listening to Prozac* as "fairly healthy people who show dramatic good responses to Prozac, people who are not so much cured of illness as transformed"[51]:

> Prozac seemed to give social confidence to the habitually timid, to make the sensitive brash, to lend the introvert the social skills of a salesman. Prozac was transformative for patients in the way an inspirational minister or high-pressure group therapy can be—it made them want to talk about their experience. And what my patients generally said was that they had learned something about themselves from Prozac.[52]

The structure of fluoxetine is closer to that of a classical antihistamine than the tricyclic antidepressants. It rapidly became the most widely prescribed antidepressant drug in the United States because it has far fewer side effects (such as weight gain, sluggishness, dry mouth, and hypertension) than do the older tricyclics. But several lawsuits against the drug's maker, Eli Lilly, were filed in August 1990 after a report in the February 1990 American Journal of Psychiatry describing six patients who were free of suicidal tendencies until they were started on Prozac. The author of the article, Dr. Martin Teicher of Harvard Medical School, admits that all antidepressants can cause people to carry out their violent or suicidal fantasies, precisely by providing them for the first time with a sufficient energy level to enable them to act. Later studies have shown that the incidence of suicide among those taking Prozac is no greater than that among those using traditional antidepressants.

Such has been the success of Prozac (in 1994, it was the third most widely prescribed drug, with sales of $1.6 billion)[53] that it has prompted another criticism: that it has produced a "legal drug culture."[54] Indeed, Kramer raises this issue himself. Discussing a patient, Tess, who claimed she felt truly normal, truly herself for the first time when she started Prozac in her 40s, he questions "how does Prozac, in Tess's life, differ from amphetamine or cocaine or even alcohol? People take street drugs all the time in order to 'feel normal.' Certainly people use cocaine to enhance their energy and confidence."[55] But he concludes that the case with Prozac is different:

> Patients who undergo the sort of deep change Tess experienced generally say they never want to feel the old way again and would take quite substantial risks—in terms, for instance, of medication side effects—in order not to regress. This is not a question of addiction or hedonism, at least not in the ordinary sense of those words, but of having located a self that feels true, normal, and whole, and of understanding medication to be an occasionally necessary adjunct to the maintenance of that self.[56]

Not everyone has found fluoxetine helpful, and there has been a reaction in the popular culture to its extensive use. Two book titles perhaps summarize much of this criticism: *Talking Back to Prozac: What Doctors Won't Tell You About Today's Most Controversial Drug*,[57] and *Prozac Nation: Young and Depressed in America.*[58] Neither of these works is a scientific analysis, but rather a criticism of the perhaps unbalanced reliance in our culture on medication for what older generations would describe as a "spiritual" malaise. A third, more reflective such book is *Speaking of Sadness: Depression, Disconnection, and the Meanings of Illness,*[59] by David Karp, a sociologist at Boston College. The author laments the inadequacy of any one-dimensional approach to depression. A thoughtful review of the book in *The New York Times* by clinical psychologist Martha Manning[60] concludes with these words:

> No good is served by viewing the devastating experience of depression only in terms of neurotransmitters. Neither do we help people when we see them totally at the mercy of a sick society. *Speaking of Sadness* promises no answers. That's all right. I've developed a deep distrust of books about depression that make such promises. What *Speaking of Sadness* does provide is an open challenge to wrestle with the difficult questions.

Sertraline. Sertraline (Zoloft, Lustral, **5-46**, Figure 5.11), is an antidepressant drug available in Europe since 1991, and in the United States since 1992. It is a selective serotonin reuptake inhibitor that needs to be taken only once daily. Compared to fluoxetine, it has a shorter duration of action and fewer CNS activating side effects such as nervousness and anxiety. Sertraline is also being studied as an aid in treating obesity, obsessive-compulsive disorders, and to depression-prone individuals attempting to quit smoking.[61]

Jennifer Green Woodhull, a journalist and freelance writer, offers a first-hand account of how Zoloft has helped her.[62] Woodhull was a committed Buddhist for more than a decade when she began to encounter a depression so severe she found herself toying with the idea of suicide. But she feared that "giving in to the lure of a chemical fix meant giving up my faith in Buddhist practice as the ultimate remedy for suffering. . . . Popping pills would only postpone, even compound, the inescapable effects of my own actions." Then she realized that by joining the multitude of suffering souls on antidepressants, she would only grow in that compassion born of experience which Buddhists so prize; and she realized that the use of antidepressants was "no more profane than the use of digitalis for afflictions of the heart."[63]

> Before Zoloft, the self-absorption of extreme suffering had blotted out virtually all curiosity and compassion. I still suffer, of course; that's the human condition. But the pain of existence no longer absorbs my attention. . . . Recently a concerned Buddhist student questioned his teacher about the growing use of antidepressants. "In Tibet," the teacher answered, "we have a saying.

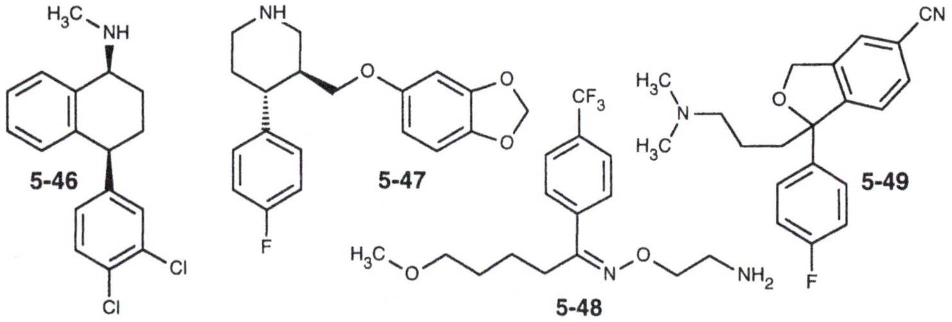

5-46 **5-47** **5-48** **5-49**

Figure 5.11 Some SSRI antidepressants.

When you are sick, pray for medication. But pray not to take the medication too long."[64]

Paroxetine. Studies comparing paroxetine (Seroxat, Paxil, **5-47**) to placebo,[65] to the tricyclic imipramine,[66] and to fluoxetine[67] have shown it to be an effective, well-tolerated, and safe antidepressant; the side effects from the tricyclic (including weight gain) were significantly greater than those from the SSRI; indeed, paroxetine seems to bring about some minor weight loss as one of its side effects. Like fluoxetine and sertraline, it occupies the serotonin transporter, inhibiting its reuptake;[68] and as with these SSRIs, the antidepressant effect involves the noradrenergic system as well.[69] Several extensive reviews judge paroxetine to be as effective as the traditional tricyclics and the other SSRIs, while its benign side effect profile favor its use with elderly patients.[70] A multicenter, long-term (up to 4 years) study in the United States showed that paroxetine was effective and that side effects were chiefly somnolence, nausea, headache and sweating, and that these tended to occur early in therapy and diminish with no new side effects as time went on.[71]

Fluvoxamine. Fluvoxamine (Avoxin, Faverin, Fevarin, Floxyfral, **5-48**) has been in use outside the United States since 1983, and has been tested extensively. A recent review concluded that the overall antidepressant efficacy of fluvoxamine is at least comparable to that of imipramine, and "similar to that of clomipramine, dothiepin, desipramine, amitriptyline, lofepramine, maprotiline, mianserin, and moclobemide."[72] Like fluoxetine, it has a much safer and more tolerable profile of side effects than the classical tricyclics, and there are some indications that it may have an earlier effect in alleviating suicidal ideation, anxiety, and sleep disturbances.

Citalopram. Citalopram (Cipramil, **5-49**) is an SSRI that has been in use in Europe since 1989; it has been shown to have efficacy and safety equal to that of fluvoxamine, fluoxetine, paroxetine, and sertraline.[73]

Trazodone. Trazodone (Desyrel, **5-50**, Figure 5.12) has no MAO inhibiting or amphetamine-like properties, and, like fluoxetine, is a SSRI. It and one of its metabolites, 3-chlorophenylpiperazine, are also both serotonin agonists at postsynaptic receptors. Controlled studies show that trazodone is as effective as amitriptyline and imipramine in patients with clinical depression. It is well tolerated, and while it can cause some drowsi-

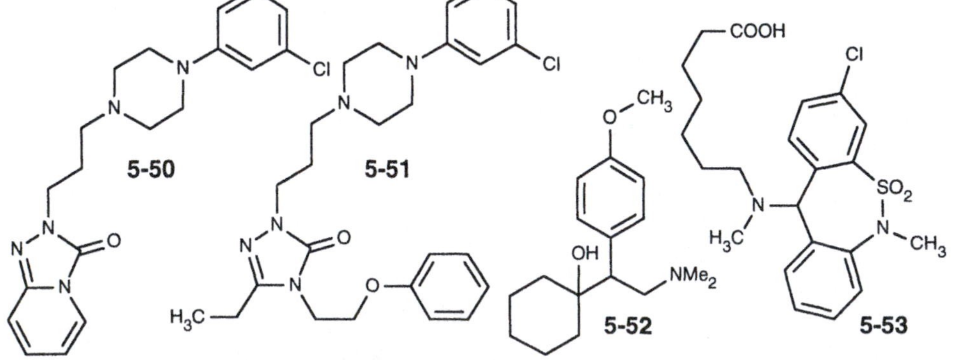

Figure 5.12 Trazodone, nefazodone, venlafaxine, tianeptine.

ness because of its antihistaminergic activity, it generally improves sleep quality and quantity.[74]

The classical tricyclics often cause a range of unwanted side effects: anticholinergic (dry mouth, constipation, blurred near vision, urinary hesitancy), anti-α_1-adrenergic (postural hypotension), and the inhibition of sexual functioning. The last problem is common with fluoxetine and the MAOIs as well—but not with trazodone, which has the quite opposite problem of occasionally inducing priapism. It has been considered for use as a treatment of impotence.[75]

Nefazodone. Nefazodone (Serzone, **5-51**) is quite similar in structure to trazodone, and its major metabolite is also 3-chlorophenylpiperazine; it is perhaps not an SSRI in the ordinary dose range, but rather exerts its antidepressant effect (comparable to imipramine)[76] through blockade of serotonin 2A receptors.[77] It improved the sleep pattern of responding patients by decreasing wake and movement time with no suppression, and even some enhancement, of REM sleep time.[78] In direct contrast to the effects of imipramine, nefazodone (which has minimal effect on α-adrenergic and cholinergic receptors) enhances learning, memory, and motor performance and does not potentiate the sedative-hypnotic effects of alcohol.[79]

Like trazodone, and unlike the tricyclics or the older SSRIs, nefazodone seems to enhance sexual interest and achievement. Dr. Alan Feiger, of the Feiger Psychmed Center in Wheat Ridge, CO, and his colleagues at four other centers treated 80 men and women with sertraline for 6 weeks. The drug successfully relieved their depression without altering sexual desire. But when their medication was changed to nefazodone, Feiger said there was a "robust improvement in sexual interest in both sexes [sic], which has continued now for 36 weeks."[80]

Unexpectedly, in animal tests nefazodone exhibits a significant analgesic potency that is not reversed or blocked by naloxone. Even more interestingly, it potentiates the analgesia of morphine, and inhibits the development of tolerance to morphine's analgesic effects, without increasing the opioid's toxicity or its undesirable gastrointestinal effects. An antidepressant drug with these additional properties would obviously be of great benefit in the treatment of the pain of terminal cancer or AIDS.[81] At the same time, nefazodone showed no tendency to provoke self-administration in monkeys (nor to substitute for cocaine dependence), and would appear to have no abuse liability.[82]

Venlafaxine. Although the structure of venlafaxine (Effexor, **5-52**) resembles fluoxetine,[83] it is thought to act not only as a serotonin reuptake inhibitor but also as a norepinephrine reuptake inhibitor and to have little or no affinity for muscarinic, histaminergic, or adrenergic postsynaptic receptors. Venlafaxine inhibits reuptake of dopamine to a moderate degree, but less than sertraline.[84] In a double-blind, placebo-controlled trial lasting over a year, the efficacy of venlafaxine was compared with trazodone in the treatment of major depression.[85] Both drugs were significantly superior to placebo, but venlafaxine improved alertness and cognitive status, while trazodone improved sleep. Patients on venlafaxine were less likely to discontinue treatment because of side effects. There is some evidence that some patients begin responding to venlafaxine within the first 2 weeks of treatment, showing it may be a lead compound for the desired goal of a short-acting antidepressant.[86]

Tianeptine: A Non-SSRI Atypical Antidepressant. Tianeptine (Stablon, **5-53**) represents a variation on the basic tricyclic structure with some considerable differences (position of the halogen, length of the alkyl chain, and the carboxylic acid function); indeed, it is not only an antidepressant, but a psychostimulant with antiulcer and antiemetic properties. In an intriguing study, it was shown that the memory deficits induced in mice by chronic, yearlong ethanol consumption could be substantially reversed by tianeptine and, to a lesser degree, by paroxetine.[87]

Efficacy of SSRIs versus Tricyclics. The extreme enthusiasm in the popular press about the benefits of Prozac has provoked a corrective process. Dr. Roger Greenberg, a psychiatrist at the SUNY Health Center at Syracuse, conducted a large meta-analysis of 13 studies in a comparison of the efficacy of fluoxetine with the tricyclic antidepressants. "Despite all the talk of this being a wonder drug, it doesn't seem to produce any better effects than other antidepressants. We're concerned that what's presented to the public is an exaggeration."[88] The perception that SSRIs reduced suicidal ideation more rapidly than traditional tricyclics has been refuted by a double-blind study comparing paroxetine and amitriptyline that found no difference in effect.[89] Overall, the contest is a dead heat, with perhaps a slight increase in benefit on the part of the tricyclics. However, efficacy alone is not the whole story: there are considerably fewer and less severe side effects with the SSRIs, and this has led to their being widely used among patient populations with a severity of symptoms that would not warrant the use of the more toxic tricyclics. Indeed, Emory University's Melvin Konner—a Harvard-trained physician and anthropologist and himself a beneficiary of Zoloft—puts it bluntly enough: "Doctors are giving antidepressants to patients who are not mentally ill. And the patients, who like the effects, ask: Why not?"[90] Conversely, the tricyclics have mostly been used with more severely depressed patients. "The most severely depressed patients are more likely to respond to medications of all kinds," says Dr. Matthew Rudorfer of the National Institute of Mental Health.[91]

A British study that compared two older tricyclics (amitriptyline and dothiepin) with three newer drugs (lofepramine, fluoxetine, and paroxetine) attempted to take into account such variables as drug price (the newer drugs are considerably more expensive) and the diminishment in quality of life from side effects, ranging from safety in overdose to degree of impairment of psychological functioning, concluded that the "prescription of older tricyclic compounds should be the exception rather than the rule in the pharmacotherapy of depression."[92] A recent and extensive metastudy of treatment outcomes has shown that the triggering of mania among depressed patients is much more likely to occur with the tricyclics (11%) than with SSRIs (4%), where the frequency is no greater than with placebo (4%).[93]

Mechanism of Action of the Antidepressants

In animal studies, imipramine and its tertiary-amine congeners block the reuptake of both NE and serotonin; clomipramine is particularly active as a serotonin reuptake inhibitor, and its unique activity in the treatment of obsessive-compulsive disorder may be related to this. The secondary-amine tricyclics block reuptake of NE, with no effect on serotonin; amoxapine additionally blocks the reuptake of dopamine. These blockings result in initial higher concentrations of monoamines in the synaptic cleft. However, over a period of time the cell adjusts to this change by downregulating the number of receptor sites at the postsynaptic membrane. One explanation for the observation that antidepressants require a minimum of about 2 weeks on average to exert an effect is that this downregulation of receptor sites is involved in their efficacy. More nuanced studies suggest that desensitization of 5-HT_A presynaptic autoreceptors of two kinds are involved in producing the antidepressant effect: those on the cell body of the presynapse (somatodendritic), which respond to serotonin in a feedback mechanism to inhibit cell firing; and those on the presynaptic terminal, which operate in a second feedback mechanism to diminish release of serotonin from the synaptic vesicles. Desensitization of both of these classes of receptor, which takes about 2 weeks to have its effect, leads to an overall increase of serotonin in the synaptic cleft for each impulse reaching the terminal. Both the tricyclics and the SSRIs (below) produce this effect on these autoreceptors, and it explains why response to both classes of drugs is never immediate.[94] But it does not explain the antidepressant efficacy of bupropion. At this writing, despite a large body of experimental data on the antidepressants, it must be said that "a coherent accounting of their mechanisms of action remains elusive."[95]

Other Conditions Treated with Antidepressants

Obsessive-Compulsive Disorder (OCD). It is estimated that as many as 2% of the population is afflicted with OCD, which usually starts in early adulthood or adolescence[96]. There are two aspects to the phenomenon: (1) an internal obsession with obtrusive ideas, images, or impulses that continually recur against the will of the victim; (2) compulsive behavior, such as washing one's hands or taking a shower many times daily, counting the number of steps when walking, checking every locked door exactly seven times, and so on. Persons afflicted with this disorder are usually perfectly aware of the irrationality of their actions, and indeed this constitutes no small part of the torment. They are often ashamed and secretive about their problem and never admit it to even their spouse or closest friends.

OCD was previously thought to be caused by psychological maladaptation; the Freudian school in particular stressed the likely effects of punitive potty training. But it may well be that OCD originates in specific parts of the brain and is traceable to genetic neurological deficits. This argument has been strengthened by the improvement many OCD sufferers experience when given certain drugs. Three have proved particularly helpful in OCD: the clomipramine,[97] fluoxetine, sertraline,[98] and fluvoxamine. In contrast to ordinary depressive illness, OCD shows a selective response to SSRIs.[99] But conventional psychotherapy is still of value: a study of OCD patients employing brain imaging using positron-emission tomography showed that there were changes in the orbital frontal cortex on the part of those who benefited from SSRI drugs—but a quite similar change was shown on the part of those who had benefited from 10 weeks of cognitive-behavioral therapy.[100]

Some famous individuals have most likely suffered from OCD, among them Samuel Johnson, the foremost literary critic of 18th-century England. While there are no traces of OCD in Johnson's written work, several of his contemporaries (including his biographer James Boswell) describe the whirling gyrations he would execute whenever he passed through a doorway, how he invariably touched every post and pole as he walked down the street—returning if he missed one the first time—while simultaneously avoiding any cracks in the pavement.

> On entering Sir Joshua's [Sir Joshua Reynolds] house with poor Mrs. Williams, a blind lady who lived with him, he would quit her hand, or else whirl her about on the steps as he whirled and twisted about to perform his gesticulations; and as soon as he had finish'd, he would give a sudden spring and make such an extensive stride over the threshold, as if he were trying for a wager how far he could stride, Mrs. Williams standing groping about outside the door unless the servant or mistress of the house more commonly took hold of her hand to conduct her in, leaving Dr. Johnson to perform at the Parlor Door much the same exercise over again.

So writes Miss Frances Reynolds, a close friend of Dr. Johnson who otherwise deeply admired him. That Johnson, otherwise excessively polite to women, would subject a blind woman he deeply respected and loved to this bizarre behavior, shows how beyond the control of his (extraordinarily powerful) will these rituals were. There is also ample evidence from his private diaries that he was tormented with religious scruples.[101]

The 19th-century composer Anton Bruckner (1824-1896) was probably a victim of OCD: during his nervous breakdown in 1867 and oftentimes afterward he "felt an inner compulsion to count everything: the windows of houses, the leaves on trees, the stones on the street, the logs in woodpiles, the stars in the sky. . . ."[102] His obsession with numbers led him to peculiar ritualized revisions of the scores of several of his works; he would meticulously readjust the number of bars, lengthening or shortening the melodies, until they achieved a "regular" pattern: at times this required him to insert several bars of rests (no instrument playing) *before* the beginning of a movement!

Compulsive Buying, Kleptomania, Pathological Gambling. Compulsive buying (technically known as *oniomania*[103]), kleptomania,[104] pathological gambling,[105] and even some forms of binge-eating disorder are considered by some authorities as variants of OCD and within a spectrum of mood disorders.[106] Often patients suffering from these unusual conditions are helped by one of the antidepressant drugs.[107]

Trichotillomania. This condition may be a variant of OCD.[108] Persons suffering from trichotillomania compulsively pluck out their hair, usually of their scalp or eyebrows, one hair at a time, until they are completely bald (sometimes on only one side of their head). It seems to equally afflict girls and boys when it starts as it usually does, in early childhood. But the late-onset variety seems more prevalent among women. There may be an inherited component—but another possibility is that men with the syndrome escape notice because of the frequency of male pattern baldness. There are an estimated 2 to 4 million persons in the United States who are trichotillomaniacs. Group therapy, hypnosis, and either Prozac or Anafranil have helped many people.[109]

Seasonal Affective Disorder. It is thought that seasonal rhythms in mood and behavior occur normally in the general population. But more recently, attention has been given to the syndrome of recurrent major winter depression, which has been termed seasonal affective disorder (SAD). One study concludes that there is a genetic contribution to this syndrome accounting for at least 29% of the variance in seasonality in men and women.[110] The syndrome has been successfully treated by intense light therapy, but these results have been questioned since it appears impossible that a "placebo" alternative can be provided.[111] Nonetheless, on the assumption that the tryptophan-serotonin system is operative in the depression and its remission, a placebo alternative of sorts can be employed: in a placebo-controlled, double-blind, crossover study, patients in remission from SAD from light therapy underwent rapid dietary depletion of tryptophan (the biologic precursor of serotonin). The tryptophan depletion reversed the antidepressant effect of the light therapy, making it at least very plausible that the light therapy works by altering serotonin levels.[112] Although melatonin is affected by light exposure, and synthesized endogenously from serotonin, the authors of this study feel that melatonin levels are not the operative factor in either SAD or its light-induced remission.[113]

A study that used moclobemide, one of the RIMA antidepressants, to treat SAD in Norway had an ambiguous outcome.[114] If the serotonin system is critical in SAD, it would seem that SSRIs would help. And, indeed, a recent placebo-controlled study using fluoxetine seemed to provide relief, but the sample was too small to provide conclusive results.[115]

Bulimia. In 1994, the FDA issued a letter supporting the use of fluoxetine for treating bulimia; although several antidepressants have been used to treat eating disorders, this is the first to receive specific authorization.[116]

Syntheses of Some Representative Antidepressants

The synthesis of the prototypical tricyclic antidepressant imipramine is outlined in Figure 5.13. The base-catalyzed self-condensation of 2-nitrobenzylchloride provides an inexpensive source of **5-54**. Sequential reduction of the nitro groups in this intermediate and then

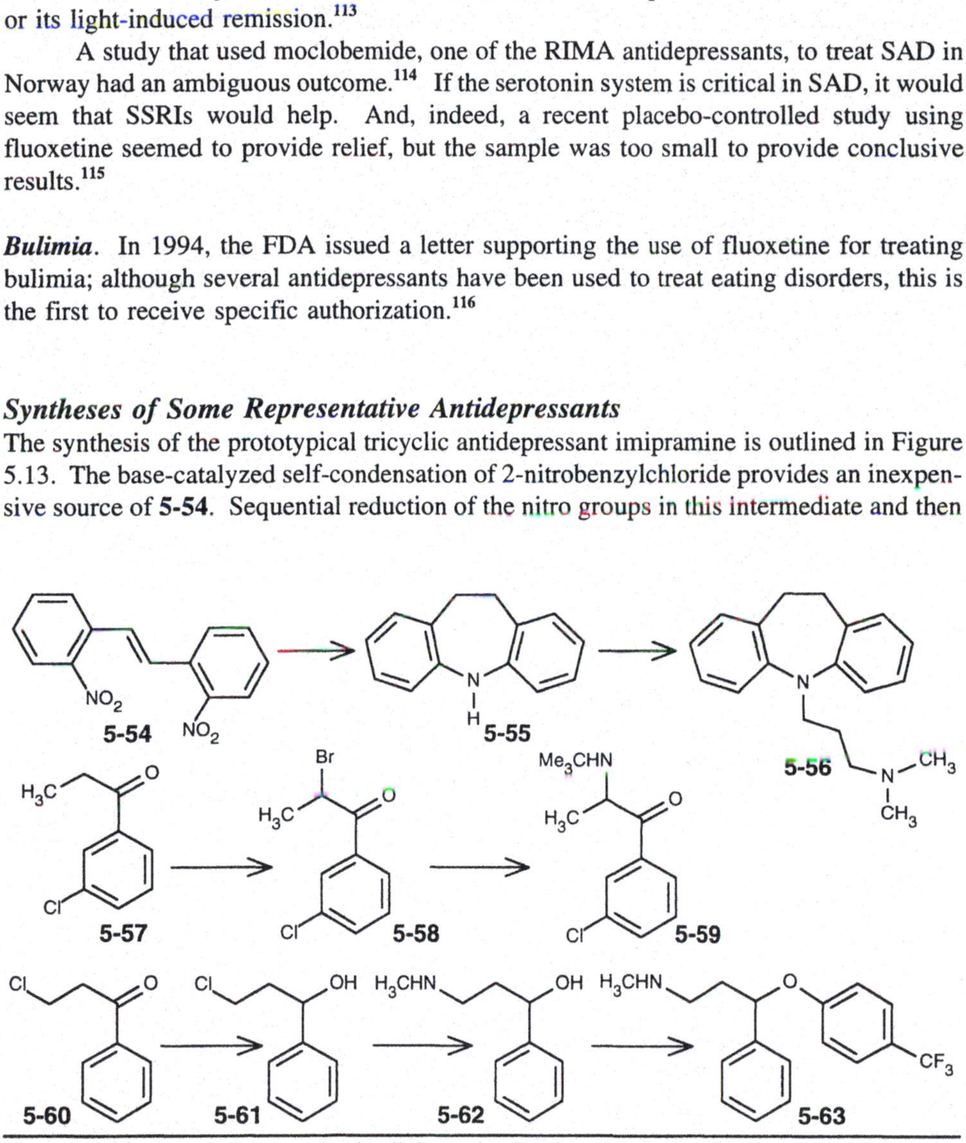

Figure 5.13 Syntheses of imipramine (5-56), bupropion (5-59), and fluoxetine (5-63).

the double bond allows a simple thermal process to form the tricyclic **5-55**. This is easily converted to imipramine, **5-56**,[117] by the same procedure used to synthesize chlorpromazine (see Figure 5.6). Intermediate **5-55** can also be used to produce the antiseizure medication carbamazepine (Chapter 3).[118]

Despite the astonishingly subtle yet powerful effects of the newer antidepressants, many of them have surprisingly simple structures and are readily synthesized from common starting materials. Bupropion, **5-59,** can be made in a few simple steps: alpha-bromination of 3-chloropropiophenone, **5-57,** yields **5-58**; and nucleophilic displacement of the bromine by *tert*-butylamine gives the antidepressant.[119] Only a little more complex is the synthesis of fluoxetine: 3-chloropropiophenone, **5-60**, is reduced to the alcohol, **5-61**, and halide displaced with methylamine. Upon treatment with sodium hydride, **5-62** forms an anion sufficiently nucleophilic to displace the chloride from 4-chlorobenzotrifluoride, giving fluoxetine, **5-63.**[120]

In the next two chapters, we turn to the consideration of the psychedelics and the cannabinoids, drugs that are among the most baffling and mysterious in the altered state of consciousness they produce—at once subtle and overwhelming, often rendering a subject somehow simultaneously lucid and lunatic—and among the most controversial as to the legitimacy of their use. Indeed, they are controversial not only among those concerned with social and political policy, but even among that staid group of generally conformist individuals, the scientists.

References and Notes

The following abbreviations are used for frequent sources in these References/Notes:

ARMC: *Annual Reports in Medicinal Chemistry*; Academic Press, New York.

DE95: *Drug Evaluations: Annual 1995*; American Medical Association Division of Drugs and Toxicology, Department of Drugs; Bennett, D. R., Ed.; AMA: Chicago, 1995.

G&G8: *Goodman and Gilman's The Pharmacological Basis of Therapeutics*, 8th ed., Gilman, A. G.; Rall, T. W.; Nies, A. S.; Taylor, P., Eds.; Pergamon: New York, 1990.

G&G9: *Goodman and Gilman's The Pharmacological Basis of Therapeutics*, 9th ed., Hardman, J. G.; Limbird, L. E.; et al., Eds.; McGraw-Hill: New York, 1996.

HPDT: Arana, G. W.; Hyman, S. E., *Handbook of Psychiatric Drug Therapy*, 2nd ed., Little, Brown: Boston, 1991.

L&ID: *Licit and Illicit Drugs: The Consumers Union Report on Narcotics, Stimulants, Depressants, Inhalants, Hallucinogens, and Marijuana—Including Caffeine, Nicotine and Alcohol*, Brecher, E. W. and the Editors of Consumer Reports, Eds., Little, Brown: Boston, 1972.

LTP: Kramer, Peter D., *Listening to Prozac: A Psychiatrist Explores Antidepressant Drugs and the Remaking of the Self*, Viking: New York, 1993.

MI11: *The Merck Index*, 11th ed.; Budavari, S.; O'Neil, M. J.; et al., Eds.; Merck: Rahway, NJ, 1989. References are given by monograph number.

MI12: *The Merck Index*, 12th ed.; Budavari, S.; O'Neil, M. J.; et al., Eds.; Merck: Rahway, NJ, 1996. References are given by monograph number.

MM16: *The Merck Manual of Diagnosis and Therapy*, 16th ed., Berkow, R.; Fletcher, A. J., Eds., Merck: Rahway, NJ, 1992.

OCDS: *Organic Chemistry of Drug Synthesis*, Lednicer, D.; Mitscher, L. A.; Georg, G. I., Wiley: New York, 1977, 1980, 1984, 1990, 1995. The **bold** number following *OCDS* indicates the volume number (1-5); this is followed by the page number.

PHP: Kaplan, H. I.; Sadock, B. J., *Pocket Handbook of Psychiatric Drug Therapy*, Williams & Wilkins: Baltimore, 1993.

SA: *Substance Abuse: A Comprehensive Textbook,* 2nd ed.; Lowinson, J. H.; Ruiz, P.; Millman, R. B.; Langrod, J. G., Eds.; Williams & Wilkins: Baltimore, 1992.

1. Italics added. From Boswell, J., *The Life of Samuel Johnson, L.L.D.*, [Aet. 67], Modern Library: New York, p. 618.
2. Hopkins, G. M., "To R. B.," *The Poems of Gerard Manley Hopkins*, 4th ed., Gardner, W. H.; MacKenzie, N. H., Eds., Oxford: New York, 1970, p. 108.
3. *The New York Times*, 7 September 1994.
4. Kety, S. S.; Wender, P. H.; et al., "Mental illness in the biological and adoptive relatives of schizophrenic adoptees. Replication of the Copenhagen study in the rest of Denmark," *Arch. Gen. Psychiatry.*, **1994**, *51*, 442-455.
5. Cannon, T. D.; Zorilla, L. E., et al, "Neuropsychological functioning in siblings discordant for schizophrenia and healthy volunteers," *Arch. Gen. Psychiatry*, **1994**, *51,* 651-661.
6. Nasar, S., "The lost years of a Nobel laureate," *The New York Times*, 13 November 1994, Section 3, pp. 1, 8.
7. *HPD*, 41.
8. *Tardive Dyskinesia: A Task Force Report of the American Psychiatric Association,* American Psychiatric Association: Washington, DC, 1992.
9. Fenton, W. S.; Wyatt, R. J.; McGlashan, T. H., "Risk factors for spontaneous dyskinesia in schizophrenia," *Arch. Gen. Psychiatry*, **1994**, *51*, 643-650.
10. *HPDT*, pp. 29-30.
11. *PHP*, pp. 32-34.
12. *PHP*, pp. 96-97.
13. Report on studies by Dr. Preben Bo Mortensen of the Danish Institute of Basic Psychiatric Research, *The New York Times*, 31 August 1994, p. C9.
14. Sneader, W., *Drug Discovery: The Evolution of Modern Medicines*, Wiley: New York, 1985, pp. 177-178.
15. Winter, J. C., "Antipsychotic drugs (neuroleptics)." In: *Textbook of Pharmacology*, Smith, C. M., Reynard, A. M., Eds., p. 298-308.
16. Michael Winerip, *The New York Times,* 9 August 1992.
17. Kane, J.; Honigfeld, G., "Clozapine for the treatment-resistant schizophrenic," *Arch. Gen. Psychiatry*, **1988**, *45*, 789-796.
18. *ARMC* **1994**, *29*, p. 48.
19. *G&G9*, p. 406.
20. *ARMC* **1994**, *29*, p. 344.
21. Van Tol, H. H. M.; et al., "Cloning of the gene for a human dopamine D_4 receptor with high affinity for the antipsychotic clozapine," *Nature*, **1991**, *350*, 610-614.

22. Arranz, M.; Collier, D., "Association between clozapine response and allelic variation in 5-HT_{2A} receptor gene," *The Lancet*, **1995**, *346*, 281-282.

23. Sokoloff, P.; et al., "Molecular cloning and characterization of a novel dopamine receptor (D_3) as a target for neuroleptics," *Nature*, **1990**, *347*, 146-151.

24. Sunahara, R. K.; et al., Cloning of the gene for a human dopamine D_5 receptor with higher affinity for dopamine than D_1, *Nature*, **1991**, *350*, 614-619.

25. Shapiro, A. K.; Shapiro, E.; et al., *Gilles de la Tourette Syndrome*, Raven Press: New York, 1987.

26. Shapiro, E.; Shapiro, A. K.; et al., "Controlled study of haloperidol, pimozide, and placebo for the treatment of Gilles de la Tourette's syndrome," *Arch. Gen. Psychiatry*, **1989**, *46*, 722-730.

27. Lechin, F.; et al., "Pimozide therapy for trigeminal neuralgia," *Arch. Neurol. (Chicago)*, **1989**, *46*, 960-963.

28. Charpentier, P.; Gailliot; et al., *Compt. Rend.*, **1952**, *235*, 59.

29. Janssen, P. A. J.; VanDeWesteringhe, C.; et al., *J. Med. Chem.*, **1959**, *1*, 281.

30. See: (a) *OCDS* **I**, 306. (b) Schmiddle, C. S.; Mansfield, R. C., *J. Am. Chem. Soc.*, **1956**, *78*, 1702.

31. (a) *OCDS* **II**, 425. (b) Hunziker, F.; Kunzle, F.; et al., *Helv. Chim. Acta*, **1967**, *50*, 1588.

32. Weissman, M. M.; Leaf, P. J., "Affective disorders in five United States communities," *Psychol. Med.*, **1988**, *18*, 141-153.

33. Interview by Grimes, W., "Exploring the links between depression, writers and suicide," *The New York Times*, 14 November 1994, pp. C11, C13. Cited material p. C11.

34. Brady, K. T.; Sonne, S. C., "The relationship between substance abuse and bipolar disorder," *J. Clin. Psychiatry*, **1995**, *56*, Suppl. 3, 19-24.

35. Delgado, P. L.; Charney, D. S.; et al., "Serotonin function and the mechanism of antidepressant action: reversal of antidepressant-induced remission by rapid depletion of plasma tryptophan," *Arch. Gen. Psychiatry*, **1990**, *47*, 411-418.

36. Owens, M. J.; Nemeroff, C. B., "Role of serotonin in the pathophysiology of depression: focus on the serotonin transporter," *Clin. Chem.*, **1994**, *40*, 288-295.

37. *Drug Discovery: The Evolution of Modern Medicines*, Sneader, W.; Wiley: New York, 1985, pp. 185-186.

38. Bowden, C. L.; Brugger, A. M.; Calbrese, J. R., "Efficacy of Divalproex vs lithium and placebo in the treatment of mania," *JAMA*, **1994**, *271*, 918-924. Divalproex is a patented formulation of Abbott Laboratories consisting of a 1:1 molar mixture of sodium valproate and valproic acid. Abbott provided partial funding for the study, and several of the many contributors to the study own stock in Abbott. It is unlikely that Divalproex has effects significantly different from generic valproic acid or generic sodium valproate.

39. Avissar, S.; Schreiber, G., "Muscarinic receptor subclassification and G-proteins: Significance for lithium action in affective disorders and for the treatment of the extrapyramidal side effects of neuroleptics," *Biol. Psychiatry*, **1989**, *26*, 113.

40. Thomas Mann's *The Magic Mountain* provides an unforgettable impression of the psychological atmosphere pervading these institutions.

41. It is perhaps the reversibility of the newer MAO agents rather than their selectivity that leads to their improved profile: the predominant enzyme active in metabolizing tyramine the human intestine is MAO-A: Anderson, M. C.; Hasan, F.; et al., "Monoamine oxidase inhibitors and the cheese effect," *Neurochem. Res.*, **1993**, *18*, 1145-1149.

42. Heinze, G.; Rossel, L.; et al., "Double-blind comparison of moclobemide and tranylcypromine in depression," *Pharmacopsychiatry*, **1993**, *26*, 240-245.

43. (a) Vestergaard, P.; Gram, L. F.; et al., "Therapeutic potentials of recently introduced antidepressants," *Psychopharmacol. Ser.*, **1993**, *10*, 190-198. (b) Danish University Antidepressant Group, "Moclobemide: a reversible MAO-A inhibitor showing weaker antidepressant effect than clomipramine in a controlled multicenter study," *J. Affect. Disord.*, **1993**, *28*, 105-116.

44. Williams, R.; Edwards, R. A.; et al., "A double-blind comparison of moclobemide and fluoxetine in the treatment of depressive disorders," *Int. Clin. Psychopharmacol.*, **1993**, *7*, 155-158.

45. Bakish, D.; Saxena, B. M.; et al., "Reversible monoamine oxidase-A inhibitors in panic disorder," *Clin. Neuropharmacol.*, **1993**, *16*, Suppl. 2, S77-S82.

46. Nolen, W. A.; Hoencamp, E.; et al., "Reversible monoamine oxidase-A inhibitors in resistant major depression," *Clin. Neuropharmacol.*, **1993**, *16*, Suppl. 2, S69-S76.

47. (a) Sneader, W., *Drug Discovery: The Evolution of Modern Medicines*, Wiley: New York, 1985, pp. 187-188. (b) *G&G9*, p. 432.

48. *DE93*, pp. 280-281.

49. (a) Simeon, J. G.; et al., "Bupropion effects in attention deficit and conduct disorders," *Can. J. Psychiatry*, **1986**, *31*, 581-585. (b) Casat, C. D.; et al., "A double-blind trial of bupropion in children with attention deficit disorder," *Psychopharmacol. Bull.*, **1987**, *23*, 120-122.

50. Davidson, J., "Seizures and bupropion: A review, " *J. Clin. Psychiatry*, **1989**, *50*, 256-261.

51. *LTP*, p. xix.

52. *LTP*, p. xv.

53. *SCRIP*, 7 July 1995, p. 23.

54. *The New York Times*, 13 December 1993.

55. *LTP*, p. 16.

56. *LTP*, p. 20.

57. Breggin, P. R.; Breggin, G. R., *Talking Back to Prozac: What Doctors Won't Tell You About Today's Most Controversial Drug*, St. Martin's Press: New York, 1994.

58. Wurtzel, E., *Prozac Nation: Young and Depressed in America*, Houghton Mifflin: Boston, 1994.

59. Karp, D. A., *Speaking of Sadness: Depression, Disconnection, and the Meanings of Illness*, Oxford University Press: New York, 1996.

60. Manning, M., "Invisible wounds: A scholar investigates depression, as experienced by the depressed," *The New York Times Book Review*, 21 January 1996, 29.

61. *The New York Times*, 31 August 1994, C9.

62. In *Living with Prozac [And Other Selective Serotonin-Reuptake Inhibitors]: Personal Accounts of Life on Antidepressants*, Elfenbein, D., Ed., Harper: San Francisco, 1995, pp. 202-205.

63. *Ibid.*, p. 205.

64. *Ibid.*, pp. 204-205.

65. Claghorn, J. L.; Klev, A.; et al., "Paroxetine versus placebo: a double-blind comparison in depressed patients," *J. Clin. Psychiatry*, **1992**, *53*, 434-438.

66. Ohrberg, S.; Christiansen, P. E.; et al., "Paroxetine and imipramine in the treatment of depressive patients in psychiatric practice," *Acta Psychiatr. Scand.*, **1992**, *86*, 437-444.

67. Tignol, J., "A double-blind, randomized, fluoxetine-controlled, multicenter study of paroxetine in the treatment of depression," *J. Clin. Psychopharmacol.*, **1993**, *13* (6 Suppl. 2), 18S-22S.

68. Scheffel, U.; Kim, S.; Cline, E. J.; Kuhar, M. J.,"Occupancy of the serotonin transporter by fluoxetine, paroxetine, and sertraline: in vivo studies with [^{125}I]RTI-55," *Synapse (N.Y.)*, **1994**, *16*, 263-268.

69. Lundmark, J.; Waalinden, J.; et al., "The effect of paroxetine on cerebrospinal fluid concentrations of neurotransmitter metabolites in depressed patients," *Eur. Neuropsychopharmacol.*, **1994**, *4*, 1-6.

70. (a) Nemeroff, C. B., "The clinical pharmacology and use of paroxetine, a new selective serotonin reuptake inhibitor," *Pharmacotherapy*, **1994**, *14*, 127-138. (b) Nemeroff, C. B., "Paroxetine: an overview of the efficacy and safety of a new selective serotonin reuptake inhibitor in the treatment of depression," *J. Clin. Psychopharmacol.*, **1993**, *13* (6 Suppl. 2), 10S-17S.

71. Duboff, E. A., "Long-term treatment of major depressive disorder with paroxetine," *J. Clin. Psychopharmacol.*, **1993**, *13*, 6 Suppl. 2, 28S-33S.

72. Wilde, M. I.; Plosker, G. L.; Benfield, P., "Fluvoxamine. An updated review of its pharmacology, and therapeutic use in depressive illness," *Drugs*, **1993**, *46*, 895-924. Quoted material p. 896.

73. (a) Kasper, S.; Hoeflich, G.; et al., "Safety and antidepressant efficacy of selective serotonin reuptake inhibitors," *Hum. Psychopharmacol.*, **1994**, *9*, 1-12. (b) Hyttel, J., "Pharmacological characterization of selective serotonin reuptake inhibitors," *Int. Clin. Psychopharmacol.*, **1994**, *9* Suppl. 1, 19-26.

74. Mouret, J.; Lemoine, P.; et al., "Effects of trazodone on the sleep of depressed subjects: a polygraphic study," *Psychopharmacology*, **1988**, *95*, S37.

75. (a) Thompson, J. W. Jr.; Ware, M. R.; et al., "Psychotropic medication and priapism: a comprehensive review," *J. Clin. Psychiatry*, **1990**, *51*, 430. There have been some clinical observations that psychostimulants such as dextroamphetamine or pemoline (see Chapter 4) can reverse serotonin-active antidepressant-induced erectile and ejaculatory dysfunction: (b) Gitlin, M. J., "Treatment of sexual side effects with dopaminergic agents," [letter] *J. Clin. Psychiatry*, **1995**, *56*, 124.

76. Fontaine, R., "Novel serotonergic mechanisms and clinical experience with nefazodone," *Clin. Neuropharmacol.*, **1993**, *16*, Suppl. 3, S45-S50.

77. Hemrick-Luecke, S. K.; Snoddy, H. D.; et al., "Evaluation of nefazodone as a serotonin uptake inhibitor and a serotonin antagonist in vivo," *Life Sci.*, **1994**, *55*, 479-483.

78. Armitage, R.; Rush, A. J.; et al., "The effects of nefazodone on sleep architecture in depression," *Neuropsychopharmacology*, **1994**, *10*, 123-127. According to a Bristol-Myers patent (*Chem Abstracts*, **117**: 10273) subjects given 400 mg nefazodone experienced 28% REM sleep time as compared to 22% for placebo, 16% for trazodone, and 17% for buspirone.

79. Frewer, L. J.; Lader, M., "The effects of nefazodone, imipramine and placebo, alone and combined with alcohol, in normal subjects," *Int. Clin. Psychopharmacol.*, **1993**, *8*, 13-20.

80. Presumably, Dr. Feiger means "on the part of" both sexes. From: Brody, J. E., "Personal health: When depression lifts but sex suffers," *The New York Times*, 15 May 1996, C8.

81. Pick, C. G.; Paul, D.; et al., "Potentiation of opioid analgesia by the antidepressant nefazodone," *Eur. J. Pharmacol.*, **1992**, *211*, 375-381.

82. Gold, L. H.; Balster, R. L., "Evaluation of nefazodone self-administration in rhesus monkeys," *Drug Alcohol Depend.*, **1991**, *28*, 241-247.

83. For the synthesis of venlafaxine, see: Yardley, J. P.; Husbands, G. E. M.; et al., *J. Med. Chem.*, **1990**, *33*, 2899-2905.

84. Bolden-Watson, C.; Richelson, E., "Blockade by newly developed antidepressants of biogenic amine uptake into rat brain synaptosomes," *Life Sci.*, **1993**, *52*, 1023-1029.

85. Cunningham, L. A.; Borison, R. L.; et al., "A comparison of venlafaxine, trazodone, and placebo in major depression," *J. Clin. Psychopharmacol.*, **1994**, *14*, 99-106.

86. Khan, A.; Fabre, L. F.; Rudolph, "Venlafaxine in depressed outpatients," *Pyschopharmacol Bull.*, **1991**, *27*, 141-144.

87. Beracochea, D; Deslandes, A.; et al., "Comparison between the effects of tianeptine and paroxetine on memory deficits induced by long-term ethanol consumption in mice," *Encephale*, **1994**, *20*, 13-16.

88. Cited in Goleman, D., "New view of Prozac: It's good but it's not a wonder drug," *The New York Times*, 19 October 1994, p. C11. The study can be found in *The Journal of Nervous and Mental Disease*.

89. Moller, H. J.; Steinmeyer, E. M., "Are serotonergic reuptake inhibitors more potent in reducing suicidality? An empirical study on paroxetine," *Eur. Neuropsychopharmacol.*, **1994**, *4*, 55-59.

90. Konner, M., "Out of the darkness," *The New York Times Magazine*, 2 October 1994, pp. 70-73.

91. Cited in Goleman, D., "New view of Prozac: It's good but it's not a wonder drug," *The New York Times*, 19 October 1994, p. C11.

92. Sherwood, Neil; Hindmarch, Ian, "A comparison of five commonly prescribed antidepressants with particular reference to their behavioral toxicity," *Hum. Psychopharmacol.*, **1993**, *8*, 417-422.

93. Peet, M., "Induction of mania with selective serotonin reuptake inhibitors and tricyclic antidepressants," *Br. J. Psychiatry*, **1994**, *164*, 549-550.

94. Blier, P.; de Montigny, C., "Current advances and trends in the treatment of depression," *Trends Pharmacol. Sci.*, **1994**, *15*, 220-226. For arguments suggesting postsynaptic 5-HT_A receptors are involved, see: Lucki, I.; Singh, A.; Kreiss, D. S., "Antidepressant-like behavioral effects of serotonin receptor agonists," *Neurosci. Biobehav. Rev.*, **1994**, *18*, 85-95.

95. For a discussion with references: (a) *G&G9*, pp. 437-439. For an overview of ongoing developments in the design of new antidepressant drugs: (b) Broekkamp, C. L. E.; Leysen, D.; et al., "Prospects for improved antidepressants," *J. Med. Chem.*, **1995**, *38*, 4615-4633. A recent review of the mode of action of fluoxetine versus other SSRIs and the tricyclics concludes: "Ironically, fluoxetine might be distinctive in that it is the least selective SSRI and has marked effects on catecholamine function in the brain." (c) Stanford, S. C., "Pro-

zac: panacea or puzzle," *TIPS*, **1996**, *17*, 150-154.

96. For a fascinating description of numerous case studies and the dramatic improvement Anafranil brings to some victims of OCD: Rapoport, J. L., *The Boy Who Couldn't Stop Washing,* Penguin: New York, 1991.

97. Greist, J. H.; Jefferson, J. W.; et al., "Efficacy and tolerability of serotonin transport inhibitors in obsessive-compulsive disorder: A meta-analysis," *Arch. Gen. Psychiatry*, **1995**, *52*, 53-60.

98. Greist, J., "Double-blind parallel comparison of three dosages of sertraline and placebo in outpatients with obsessive-compulsive disorder," *Arch. Gen. Psychiatry*, **1995**, *52*, 289-295.

99. Zohar, J.; Insel, T. R., "Obsessive-compulsive disorder: Psychobiological approaches to diagnosis, treatment, and pathophysiology," *Biol. Psychiatry,* **1987**, *22*, 667-687.

100. Goleman, D., "Psychotherapy found to produce changes in brain function similar to drugs," *The New York Times*, 15 February 1996.

101. Bate, W. J., *Samuel Johnson*, Harcourt Brace Jovanovich: New York, 1977, p. 382.

102. According to Dr. Wolfgang Dömling, writing in the brochure accompanying the Deutsche Grammophone collected recording of Bruckner's *Geistliche Chorwerke*, **1987**, # 423 127-2, p. 15.

103. From the Greek ωνιος, meaning "for sale," but with an unintended pun on the English word "own." (The word seems to have been coined by E. Kraepelin writing in German: *Psychiatrie*, 8th ed., Verlag von Johann Ambrosius Barth: Leipzig, 1915, pp. 408-409).

104. McElroy, S. L.; Keck, P. E.; et al., "Kleptomania, compulsive buying, and binge-eating disorder," *J. Clin. Psychiatry,* **1995**, *56*, Suppl. 4, 14-26.

105. (a) Hollander, E.; Wong, C. M., "Body dysmorphic disorder, pathological gambling, and sexual compulsions," *J. Clin. Psychiatry,* **1995**, *56*, Suppl. 4, 7-12. (b) Hollander, E.; Frenkel, M.; et al., "Treatment of pathological gambling with clomipramine," *Am. J. Psychiatry,* **1992**, *149*, 710-711.

106. A fascinating series of articles on these topics (from the symposium "Obsessive-Compulsive Spectrum Disorders" held at the 147th meeting of the American Psychiatric Association in Philadelphia, 21 May 1994) can be found in *J. Clin. Psychiatry*, **1995**, *56*, Suppl. 4.

107. Lejoyeux, M.; Hourtané, M; et al., "Compulsive buying and depression," *J. Clin. Psychiatry,* **1995**, *56*, 38.

108. Stein, D. J.; Simeon, D.; et al., "Trichotillomania and obsessive-compulsive disorder," *J. Clin. Psychiatry,* **1995**, *56*, Suppl. 4, 28-34.

109. According to Jane E. Brody, writing for "Personal health" in *The New York Times*, 4 January 1995, there is a clearinghouse for information on compulsive hair-pulling called the Trichotillomania Learning Center, 1215 Mission St., Suite 2, Santa Cruz, CA 95060; phone: 408-457-1004.

110. Madden, P. A. F.; Heath, A. C.; et al., "Seasonal changes in mood and behavior: The role of genetic factors," *Arch. Gen. Psychiatry,* **1996**,*53*, 47-55.

111. Eastman, C.I., "What the placebo literature can tell us about light therapy for SAD," *Psychopharmacol. Bull.,* **1990**, *26*, 495-504. Brown, W. A., "Is light treatment a placebo?" *Ibid.* 527-530.

112. Lam, R. W.; Zis, A. P.; et al., "Effects of rapid tryptophan depletion in patients with seasonal affective disorder in remission after light therapy," *Arch. Gen. Psychiatry,* **1996**, *53*, 41-44.

113. Blehar, M. C.; Lewy, A. J., "Seasonal mood disorders: Consensus and controversy," *Psychopharmacol. Bull.*, **1990**, *26*, 465-494.

114. Lingjærde, O.; Reichborn-Kjennerud, T.; et al., "Treatment of winter depression in Norway: II. A comparison of the selective monoamine oxidase A inhibitor moclobemide and placebo," *Acta Psychiatr. Scand.*, **1993**, *88*, 372-380.

115. Lam, R. W.; Gorman, C. P.; et al., "Multicenter, placebo-controlled study of fluoxetine in seasonal affective disorder," *Am. J. Psychiatry,* **1995**, *12*, 1765-1770.

116. *The New York Times*, 13 November 1994.

117. (a) *OCDS*, **I**, 401. (b) Schindler, W.; Hafliger, *Helv. Chim. Acta*, **1954**, *37*, 472.

118. *OCDS*, **I**, 403f.

119. *OCDS*, **II**, 124.

120. The original patent of 2 February 1982 uses a more cumbersome approach: (a) *OCDS*, **III**, 32f. The synthesis outlined here is based on later, enantioselective syntheses: (b) Gao, Y.; Sharpless, K. B., *J. Org. Chem.*, **1988**, *53*, 4081-4084. (c) Corey, E. J.; Reichard, G. A., *Tetrahedron Lett.*, **1989**, *30*, 5207-5210. (d) Mitchell, D.; Koenig, T., *Synth. Commun.*, **1995**, *25*, 1231-1238.

6
Psychedelics: LSD to XTC

O world invisible, we view thee,
O world intangible, we touch thee,
O world unknowable, we know thee,
Inapprehensible, we clutch thee!

Not where the wheeling systems darken,
And our benumbed conceiving soars!—
The drift of pinions, would we hearken,
Beats at our own clay-shuttered doors.

The angels keep their ancient places;—
Turn but a stone, and start a wing!
'Tis ye, 'tis your estrangèd faces,
That miss the many-splendoured thing.

—Francis Thompson[1]

Introduction

The drugs discussed in this chapter are variously called *hallucinogens, psychotomimetics,* or *psychedelics*. There are problems with each of these terms, and to some extent the controversy over the use of this class of drugs is reflected in the nomenclature. "Psychedelic" was coined by Humphry Osmond,[2] the Canadian psychiatrist who introduced Aldous Huxley to mescaline and who made early and extensive use of these drugs in psychotherapy. Osmond's word is the Greek for a substance which would "make the mind visible," hence "mind opening." (Huxley himself was later to object that the proper coinage for this would be psych*o*delic, a term sometimes seen in older literature.) At perhaps the other extreme is the 1970 Controlled Substances Act (CSA), in which all these drugs are relegated to the Schedule I category as lacking accepted medical use or safety and having a high potential for abuse. According to the CSA they are "hallucinogens"; but as we shall see, none of the drugs in this chapter (nor Δ^9-THC, the active principle in marijuana, which the CSA likewise categorizes as a hallucinogen) causes true hallucinations. However, they do modify perception, at times to the point of a pseudohallucination; but "pseudohallucinogen" would obviously be a strange term, defining a class of drugs by what it did not do. The term "psychotomimetic" has largely fallen out of use, since it is now recognized that none of these drugs induces a state of consciousness that usefully mimics a psychosis, although this was at first thought to be the case with LSD, and was one of the indications suggested by Sandoz when it provided the drug on an experimental basis.

Semantically, it is felt that people who call these drugs psychedelics tend to see them as having positive uses, while those calling them hallucinogens tend to see them as useless or dangerous. But the terms are often used more or less interchangeably and without prejudice, and they are used in this way here.[3]

We will consider any drug a psychedelic (psychotomimetic, hallucinogen) *to the extent that it has effects on the human psyche that resemble those of LSD.* Using this somewhat arbitrary but simple definition, the only drugs that are genuine psychedelics as ordinarily used are LSD, psilocybin, and mescaline—and a dozen or so less common drugs whose chemical structures (lysergic acids, tryptamines, and phenethyl amines) are very closely based on those of this classic trinity.[4] Part of the evidence suggesting that these three drugs have a similar biochemical mode of action is the fact that there is cross tolerance among them in all sequences of administration.[5]

Biochemistry is only part of the picture; there is also set and setting. And other substances that do not act at the same receptor sites as LSD occasionally seem to produce similar states of mind (just as we have seen in Chapter 5 that antipsychotic and antidepressant activity can be produced by agents acting at different receptor classes): concentrated forms of cannabis like hashish or hashish oil when taken orally; phencyclidine and ketamine; inhaled solvents or anesthetic gases. These substances will be considered in the next chapter.

The Varieties of Psychedelic Experience

The quality that uniquely characterizes LSD and the other psychedelic drugs is not the hallucinations and paranoid ideation they can occasionally induce (many other classes of drugs, including steroids and antimalarials, can do this in high doses), but "their capacity reliably to induce states of altered perception, thought, and feeling that are not experienced otherwise except in dreams or at times of religious exaltation."[6]

This too should probably be qualified: when someone dreams, they are unaware they are dreaming, but persons undergoing psychedelic experiences are normally quite aware that what they are perceiving and feeling is not actually "real," but is, at the least, reality dramatically altered or intensified. As the Angelic Doctor wrote many years ago, "those who make syllogisms in their sleep always realize on awakening that they have made an error";[7] however, psychedelic syllogisms are usually recognizably Aristotelian. There is often a sense that the true self is only a passive observer or a spectator ego rather than a real participant. A dreamer, on awakening, is usually not aware of the contrast between the lengthy detailed sequence of events in the dream and its short duration in clock time (as can be shown by the length of REM sleep); a person using a psychedelic drug is almost always struck by the seemingly endless flood of feelings and ideas that take place in an astonishingly few moments of clock time.

Like beauty, religious exaltation lies in the I of the exalted; but psychedelic drugs often dissolve or merge the boundaries of the ego and the environment, leading to a sense of oneness with the universe that numerous users—including serious individuals like Hofmann and Huxley—have found to be suggestive of religious mysticism.[8] But the truism of the medieval philosophers holds here as well: whatever is perceived is perceived according to the capacity of the recipient.[9] Or, as a more modern mind has put it: "A psychedelic drug might be compared to television. It can be—with thoughtful care in the selection of channels—the means by which extraordinary insights can be achieved. But to many people, psychedelic drugs are simply another form of entertainment: nothing profound is looked for, thus—usually—nothing profound is experienced."[10]

Like the world of dreams, the world of psychedelic experiences is astonishingly varied. We will cite some of the descriptions under each of the major drug types, but the number of descriptions in the literature is immense, and each is as different as the person undergoing the experience.[11] Some of the more common features are the following:

- A heightened sensitivity to sensory perception. Vision is usually the most affected. Aldous Huxley, after taking mescaline: "The books . . . like the flowers . . . glowed, when I looked at them, with brighter colors, a profounder significance. Red books, like rubies; emerald books; books bound in white jade . . . whose color was so intense, so intrinsically meaningful, that they seemed to be on the point of leaving the shelves to thrust themselves more insistently on my attention."[12] The objects in perception may change shape and size, or undulate and dissolve. True hallucinations are uncommon, but often when the eyes are closed a panorama of vividly iridescent geometric figures may swim before the eyes—a phenomenon many people perceive in a less vivid manner before falling asleep (the hypnagogic state),[13] in the delirium of a high fever, or in the aura preceding an epileptic seizure. Fantastic and elaborate dreams may occur with the eyes closed but with the observant ego experienced as fully awake. All the other senses can likewise be similarly affected: background sounds become present in the foreground, and music can be perceived with preternatural clarity. The phenomenon of *synesthesia* is common: perceptions proper to one sense are perceived through another: most commonly, sounds are "seen."
- A sense that these perceptual heightenings have some profound import. Still looking at a bouquet of flowers, Huxley continues: "I was seeing what Adam had seen on the morning of his creation—the miracle, moment by moment, of naked existence. . . . The Beatific Vision, *Sat Chit Ananda*, Being-Awareness-Bliss—for the first time I understood, not on the verbal level, not by inchoate hints or at a distance, but precisely and completely what those prodigious syllables referred to."[14]
- The sense of profound meaning attaches both to the perceived and the perceiver. In 1986, Dr. Weir Mitchell provided one of the first descriptions of the use of

mescaline in the form of an extract from the peyote cactus. He writes, "I had a certain sense of the things about me as having a more positive existence than usual. It is not easy to define what I mean, and at the time I searched my vocabulary for phrase or word which should fitly state my feeling. It was vain. At this time, also, I had a decisive impression that I was more competent in mind than in my every-day moods . . . a mere consciousness of power, with meanwhile absolute control of every faculty."[15]

- Changes in body feeling ranging from the impression that an arm or leg is dissolving to an out-of-body experience of viewing one's own body from above and outside. There can be a sense of loss of the ego: either a death-like experience of leaving the body, or a sense that one's ego is merging with the entire cosmos, or an identification of the ego with some inanimate object. These impressions are not so alien to ordinary experience as it may at first seem; perceptive observers have noted that they often accompany hypnagogic state. Proust's immense tome *A la recherche du temps perdu* opens with the famous passage: "For a long time I used to go to bed early. Sometimes . . . my eyes would close so quickly that I had not even time to say to myself: 'I'm falling asleep.' And half an hour later the thought that it was time to go to sleep would awaken me; . . . I had gone on thinking, while I was asleep, about what I had just been reading, but these thoughts had taken a rather peculiar turn; *it seemed to me that I myself was the immediate subject of my book: a church, a quartet, the rivalry between François I and Charles V.*"[16]
- More perceptive—or more metaphysically inclined—observers such as Alan Watts, who was both a scholar of Zen Buddhism and an Episcopal priest, find the "occasional and incidental bizarre alterations of sense perception"[17] of marginal significance compared to the "cosmic" or "mystical" dimensions of the experience. Watts describes how "in the course of two experiments [with LSD] I was amazed and somewhat embarrassed to find myself going through states of consciousness which corresponded precisely with every description of major mystical experiences I had ever read."[18] He points to four dominant characteristics of the experience: (1) a slowing down of time, a concentration in the present, (2) an awareness of polarity, (3) an awareness of relativity, and (4) an awareness of eternal energy.[19]

It should also be said that a rare individual may either experience little change in perception or affect upon taking LSD, or be quite resistant. Richard Alpert, later to be known as Ram Dass, is said to have given 900 μg of LSD to his guru in the Himalayan mountains and the guru reportedly did not bat an eye.[20] It is possible that the guru was more or less constantly in a mode of consciousness quite similar to that induced by LSD. And most groups working in more scientific settings have found that the dosage of LSD generally needs to be adjusted upward if the person to be treated is of an obsessive-compulsive personality type.[21]

LSD and the Lysergic Acid Amides

The Discovery of LSD

During unusually wet harvest seasons a fungus, *Claviceps purpurea*, can develop on rye kernels: in Europe before the 20th century, epidemics of ergotism caused death by convulsions and gangrene among the peasant population when bread made from ergot-contaminated rye was eaten. Because the victim's gangrenous extremities seemed to be charred black, the disease was known as *ignis sacer*, or St. Anthony's (holy) fire. The toxic principles responsible for these outbreaks are the dozen or so ergot alkaloids that occur in the fungus, most of which are peptide derivatives of lysergic acid.

From time immemorial, ergot was used by midwives to promote labor (an oxytocic), and for a brief period in the early 1800s, called *pulvis ad partum* ("powder for parturition"), it was used for this purpose by professional physicians. However, the uncertainty of dosage from this crude product often led to uterine spasms and the death of the fetus (skeptical physicians called it *pulvis ad mortem*); use as an oxytocic was discontinued, but ergot continued to be used to arrest postpartum bleeding.

The separation and identification of the many alkaloids contained in ergot proved to be a slow and difficult process. A key step was made by Arthur Stoll of the Swiss pharmaceutical company Sandoz, when he isolated ergonovine (ergometrine), **6-1**,[22] which proved to be a relatively simple amide of lysergic acid and propanolamine (Figure 6.1). More typical of the ergot alkaloids was ergotamine, **6-2**, which was only isolated and characterized later (it is still used today for treatment of migraine). For use in obstetrics, ergonovine seemed an obvious improvement over crude ergot. Since lysergic acid could be obtained by hydrolyzing the amide linkages in the more abundant ergot alkaloids such as ergotamine and its many isomers, Albert Hofmann, a young chemist beginning his career at Sandoz under Stoll, began to work on the production of ergonovine and other amides from lysergic acid. In 1938 he produced the 25th substance in this series, lysergic acid diethylamide, **6-3**, which was given the lab code LSD-25 (for the German *L*yserg*s*äure-*d*iäthylamid).[23]

Hofmann had synthesized LSD-25 in analogy to the diethylamide of nicotinic acid, *coramide* (Nikethamide, **6-4**), hoping that, like coramide, it would be a circulatory and respiratory stimulant. But when the new compound was tested on animals, it showed only about 70% of the activity of ergonovine in inducing uterine contractions and no activity

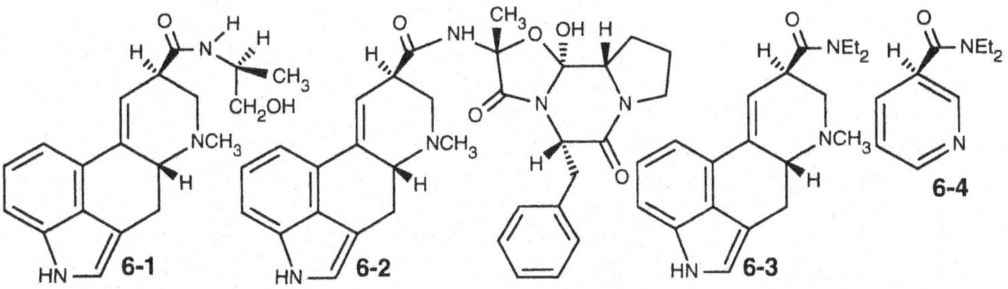

Figure 6.1 Ergonovine, ergotamine, LSD-25, coramide.

at all as a respiratory stimulant (although the test animals, mice in this case, appeared "restless" under what we can now see in retrospect to have been enormous hallucinogenic doses). As is the case in most pharmaceutical houses, the material was discarded and no further testing performed.

Five years passed, and Hofmann scored several notable successes: a semisynthetic process that made ergonovine readily available, and two new drugs, methylergonovine (with the same uses as ergonovine) and dihydroergotamine (a cerebral vasoconstrictor for migraine); all three drugs are still in use today. However, he was haunted by "a peculiar presentment"[24] with regard to his discarded creation of 5 years past: "the feeling that this substance could possess properties other than those established in the first investigations."[25]

Contrary to the protocol at Sandoz for compounds found wanting in pharmacological activity, Hofmann set out to repeat the synthesis of LSD-25. On a Friday afternoon, April 16, 1943, he was making a last recrystallization of the tartrate salt when he suddenly felt strangely dizzy and restless. He left work and went home, where, he afterwards wrote, "I sank into a not unpleasant intoxicated-like condition . . . in a dreamlike state, with eyes closed . . . I perceived an uninterrupted stream of fantastic pictures, extraordinary shapes with intense, kaleidoscopic play of colors. After some two hours this condition faded away."[26]

He reasoned that the LSD-25 he had been working with had somehow been absorbed through his skin and was responsible for these strange effects. On the following Monday he measured out the smallest amount he could conceive of as having any effect, 0.25 mg, dissolved it in water, and drank the tasteless solution. With his laboratory assistant ready to provide help, he waited, prepared to increase the dose until he noticed an effect. In about a half hour, the dizziness and visual distortions of the previous Friday began to return. He asked his assistant to accompany him, and they began to bicycle home. (Due to the war, there was strict rationing of gasoline.) Despite Hofmann's perception that everything was distorted as though seen in a curved mirror, and that he was frozen in time and space, his assistant later assured him that he cycled along perfectly normally and at a quite rapid pace; and—as a further indication that he retained some degree of rational judgment—when he arrived at his home he immediately requested milk from a neighbor to be used as a nonspecific antidote for poisoning.

> My surroundings had now transformed themselves in more terrifying ways. Everything in the room spun . . . and assumed grotesque, threatening forms. Even worse than these demonic transformations of the outer world, were the alterations that I perceived in myself, in my inner being. Every exertion of my will, every attempt to put an end to the disintegration of the outer world and the dissolution of my ego, seemed to be wasted effort. . . . I was seized by the dreadful fear of going insane. . . . Was I dying? Was this the transition? At times I believed myself to be outside my body, and then perceived clearly, as an outside observer, the complete tragedy of my situation. I had not even taken leave of my family (my wife, with our three children had traveled that day to visit her parents, in Lucerne). Would they ever understand that I had not experimented thoughtlessly, irresponsibly, but rather with the utmost caution, and that such a result was in no way foreseeable . . . ?
>
> By the time the doctor arrived, the climax of my despondent condition had already passed. . . . Pulse, blood pressure, breathing were all normal. . . . I became more confident that the danger of insanity was conclusively past.

> Now, little by little I could begin to enjoy the unprecedented colors and plays of shapes that persisted behind my closed eyes. Kaleidoscopic, fantastic images surged in on me, alternating, variegated, opening and then closing themselves in circles and spirals, exploding in colored fountains, rearranging and hybridizing themselves in constant flux. It was particularly remarkable how every acoustic perception, such as the sound of a door handle or a passing automobile, became transformed into optical perceptions. Every sound generated a vividly changing image, with its own consistent form and color.[27]

To Hofmann's surprise and gratification, he awoke the next morning with no hangover but with the world seeming as clear and fresh as the first day of creation. He was surprised to realize that he could remember all of what occurred during his LSD inebriation in every detail, and that part of the horror he experienced was precisely due to his being at the time able to compare it to a clear notion of what normal reality was like.

When he reported the results of his experiment to Stoll he was greeted with incredulous astonishment that so tiny a dose could have such an effect: but three members of the Sandoz pharmacology department each ingested a third of Hofmann's dose and confirmed his experience. It seemed that Sandoz had stumbled on one of the most potent drugs ever discovered—but it was a drug in search of a use. After some preliminary studies by psychiatrist Werner Stoll (the son of Arthur Stoll) at the University of Zurich, the drug was released by Sandoz for use as an experimental drug under the trade name Delysid (generic name lysergide). It was made available in tablets (0.025 mg) and in an injectable solution with two indications: (1) in psychotherapy as an aid to eliciting repressed material, and (2) as an agent for the induction of a model psychosis: "by taking Delysid himself, the psychiatrist is able to gain an insight into the world of ideas and sensations of mental patients."[28] In the United States, the first research using Delysid took place in 1949 at the Massachusetts Mental Health Center in Boston.

Therapeutic Uses of LSD

We do not realize that psychiatric research using LSD and other psychedelics was carried out for many years by numerous respectable psychiatrists who were in no sense cultural rebels or hippie freaks. In the words of Lester Grinspoon, associate professor of psychiatry at Harvard Medical School: "This was not a quickly rejected and forgotten fad. Between 1950 and the mid-1960s there were more than a thousand clinical papers discussing 40,000 patients, several dozen books, and six international conferences on psychedelic drug therapy."[29] The extent of this research can be outlined only briefly here.

Psychotomimesis. From about 1950 to 1960, LSD was used exclusively in research settings in an attempt to find some answers to questions still unsettled today: are the causes of schizophrenia to be found in nurture (maladaptive learning or malignant circumstances) or nature (neurologic or metabolic abnormalities) or a combination of both? If nature is to blame, could it be because the schizophrenic brain is exposed to perhaps minute quantities of a substance with effects like LSD? Results were mixed: some groups reported that there was no difference in the response to LSD between normal and schizophrenic patients; but the studies were not double-blind, and later work has shown an enormous potential for placebo effects in these circumstances. Others reported that schizophrenics seemed resistant to the effects of LSD—could it be that an endogenous toxin already present in the schizophrenic brain blocked access to LSD?

Psychotherapy. The emphasis turned to using LSD as an aid to psychotherapy; nonpsychotic but unhappy individuals might find freedom through LSD from their inhibitions and repressions and make more rapid progress in psychoanalysis. Two types of LSD psychotherapy emerged: psychedelic therapy, extensively explored in the United States and Canada,[30] which used LSD in a carefully planned set and setting to produce a single overwhelming healing experience; and psycholytic therapy, widely employed in Europe during the middle 1960s,[31] which used moderate or increasing doses of psychedelic drugs combined with traditional psychoanalytic techniques to uncover the unconscious roots of neuroses.[32] Unfortunately, the "infusion of irrationality from the psychedelic crusaders and their enemies" and the "revolutionary proclamations and religious fervor of the nonmedical advocates of LSD began to evoke hostile incredulity."[33] The consequence was that "psychedelic drug therapy did not die a natural death from loss of interest; it was killed by the law. Even though many of the researchers who devoted a large part of their careers to psychedelic drugs have retired or died, and many more now ignore them entirely, there are still others who would like to use the drugs if they could."[34]

Experimental Mysticism. In the address to the 1957 conference of the New York Academy of Sciences in which he coined the word psychedelic, Humphry Osmond opined that "those who have not themselves taken the particular substance with which they wish to work, preferably several times, would be wise not to use these agents in therapy."[35] Not all followed this practice, but many of the psychiatrists and psychologists who did came to feel that the experience was much more than a brief immersion in the pathology of insanity. Osmond: "for myself, my experiences with these substances have been the most strange, most awesome, and among the most beautiful things in a varied and fortunate life. These are not escapes but enlargements, burgeonings of reality. Insofar as I can judge . . . the brain, although its functioning is impaired, acts more subtly and complexly than when it is normal. . . . The psychedelics help us to explore and fathom our own nature. . . . We may also be something more, 'a part of the main,' a striving sliver of a creative process, a manifestation of Brahma in Atman, an aspect of an infinite God imminent and transcendent within and without us."[36] Impressions such as these, as well as the growing realization that psychedelic ethnobotanicals such as the psilocybe mushrooms had played a central role in many ancient native religions, soon led to the use of psychedelics to explore states of chemically induced religious exaltation.

Alcoholism. The literature on the use of LSD to treat alcoholism is extensive, but the conclusions are still ambiguous, despite the initial hopes of many authorities, including Bill Wilson, co-founder of AA, who declared that his LSD trip resembled the original mystical experience that had first moved him to sobriety.[37] Indeed, the transforming religious experiences of many LSD conversions sound similar to narrations heard in AA meetings. Here is the story of a 47-year-old man who had been an alcoholic and thief most of his life:

> I finally began to realize that this session was centered around the fact that I had to make a choice, a choice as to whether I was the greater power or whether there was a God which I had to recognize and accept. . . . I realized that I had attempted to bargain with God all my life. I can see now why I have struggled in vain all my life, refusing to accept anything but myself. Suddenly out of nowhere came the decision, I would make the choice, I would accept and hope to be accepted by Him. I could write for years and not be able to describe that exquisite moment of accepting and being accepted. It was without a doubt the most beautiful moment of my life and as I write this I am still amazed at the exquisite felling of release, peace of mind, and complete realization which took place at that moment.[38]

A 40-year-old unskilled laborer, diagnosed as severely anxious and depressed, was brought to Kurland's group at Spring Grove Hospital from jail after an episode of drinking uncontrollably for 10 days. Of his LSD session he said:

> I prayed to the Lord. Everything looked better all around me. The rose was beautiful. My children's faces cleared up. I changed my mind from alcohol toward Christ and the rose came back into my life. I pray that this rose will remain in my heart and my family forever. As I sat up and looked in the mirror, I could feel myself growing stronger. I feel now that my family and I are closer than ever before, and I hope that our faith will grow forever and ever.[39]

A week later the man's score on a questionnaire testing neurotic traits had dropped from the 88th to the 10th percentile. Kurland notes how difficult it would have been to help this illiterate, culturally deprived man with traditional psychotherapy.

Other early studies were also dramatic. At Hollywood Hospital in British Columbia more than 500 alcoholics and other patients were treated with psychedelic drugs, usually one high dose of LSD or mescaline after 2 to 4 weeks of preparation. About 50% of the alcoholics, many who had failed in AA, recovered and were sober a year later.[40]

It is true that later, controlled studies[41] and careful reviews of previous studies[42] have shown that LSD therapy is neither more nor less effective than many other accepted treatment modalities. It seems to be the case that many alcoholics will improve after any treatment, particularly treatment given at a "bottomed out" period of their lives, which of course imprisonment or hospitalization represents. But there is no proven treatment for alcoholism; all that can be said of LSD therapy is what can be said of other therapies that are often but not always effective: when it works, it works dramatically, and it has often worked. The issue is complex, and there is some indication that psychedelic therapy might be reinforcing were the psychedelic sessions repeated at regular intervals. This is something no group has tried, all attempting rather a one-shot cure. Again, the analogy of traditional religious conversion experiences seems helpful: the pentecostal groups, who endorse a moment of "born-again" ecstasy, still rely on regular churchgoing and periodic "revival" experiences.

Use as a Euthanatogenic-Viaticum and in Counseling the Terminally Ill. It was Aldous Huxley who made the first association of the use of psychedelic drugs with the *Tibetan Book of the Dead*, which is meant to be a spiritual guide to the art of dying. Passages from the book were read aloud during his first encounter with mescaline, described in *The Doors of Perception*, and because of this the book later became one of the bibles of

Timothy Leary's Millbrook and the Haight-Ashbury scene. In Huxley's last book, *Island*, he describes the use of a psychedelic drug called *moksha* (the Sanskrit for freedom) in a Utopian setting where science and technology are used in a humane way under the influence of a Buddhist religious tradition.[43] Unlike the *soma* of his earlier anti-Utopian *Brave New World*, the *moksha*-medicine does not enslave but enlightens. And it is used to help the dying die with freedom, vision, and hope.

On November 22, 1963, Huxley's long battle with throat cancer came to an end, and in his dying hours, faithful to his own convictions, he asked his wife, Laura, for the *moksha*-medicine. Here is the story as Laura tells it:

> When he realized that the labor of his body leaving this life might lessen his awareness, Aldous prescribed his own medicine or—expressed in another way—his own sacrament.
>
> "The last rites should make one more conscious rather than less conscious," he had often said, "more human rather than less human." In a letter to Dr. Osmond [he wrote] ". . . My own experience with Maria [Aldous's first wife, who had died of cancer] convinced me that the living can do a great deal to make the passage easier for the dying, to raise the most purely physiological act of human existence to the level of consciousness and perhaps even of spirituality."
>
> Then, I don't know exactly what time it was, he asked me for his tablet and wrote, "Try LSD 100 mm intramuscular [mm = milli-milli =μ: 100 micrograms]. . . ." I read it aloud and he confirmed it. . . . I went quickly to fetch the LSD, which was in the medicine chest in the room across the hall. There is a TV set in that room [and] . . . when I entered the room, Ginny, the doctor, the nurse, and the rest of the household were all looking at television. The thought shot through my mind: "This is madness, these people looking at television when Aldous is dying." A second later, while I was opening the box containing the LSD vial, I heard that President Kennedy had been assassinated. . . .
>
> I said, "I am going to give him a shot of LSD—he asked for it."
>
> The doctor had a moment of agitation—you know very well the uneasiness in the medical mind about this drug. But no "authority," not even an army of authorities, could have stopped me then. I went into Aldous's room with the vial of LSD and prepared a syringe. The doctor asked me if I wanted him to give the shot—maybe because he saw that my hands were trembling. . . . I said "No, I must do this." I quieted myself, and when I gave him the shot my hands were firm.
>
> Then, somehow, a great relief came to us both. . . . Then we were quiet. . . . Suddenly he had accepted the fact of death; now, he had taken this *moksha*-medicine in which he believed. Once again he was doing what he had written in *Island*, and I had the feeling that he was interested and relieved and quiet. . . .
>
> After half an hour, . . . the expression on his face was beginning to look as it did when he had taken the *moksha*-medicine, when this immense expression of complete bliss and love would come over him. . . . I let another half hour pass, and then I decided to give him another 100 mm. I told him I was going to do it, and he acquiesced. I gave him another shot, and then I began to talk to him. He was very quiet now; he was very quiet and his legs were getting colder; higher and higher I could see purple areas of cyanosis. Then I began to talk to him, saying, . . . "Light and free you let go, darling; forward and up. You are going forward and up; you are going toward the light. Willingly and consciously you are going, . . . you are going toward a greater love—you are going forward and up. It is so easy—it is so beautiful. . . . You are going toward Maria's love with my love. You are going toward a greater love than you have ever known. You are going toward the best, the greatest love, and it is easy, it is so easy, and you are doing it so beautifully."
>
> Once I asked him, "Do you hear me?" He squeezed my hand; he was hearing me.

> Later on I asked the same question, but the hand didn't move any more. I repeated these or similar words for the last three or four hours. . . . The breathing became slower and slower, and there was absolutely not the slightest indication of contraction, of struggle. It was just that the breathing became slower—and slower—and slower; the ceasing of life was not a drama at all, but like a piece of music just finishing so gently in a sempre più piano, dolcemente . . . and at five-twenty the breathing stopped. . . .
>
> If the way Aldous died were known, it might awaken people to the awareness that not only this, but many other facts described in *Island* are possible here and now. . . .
>
> Now, is his way of dying to remain for us, and only for us, a relief and consolation, or should others also benefit from it? Aren't we all nobly born and entitled to nobly dying?[44]

It was not this event that initiated the use of LSD to alleviate the psychological and physical pain of terminal illness, however. The most extensive and carefully planned use of LSD for this purpose was undertaken some years later by the group directed by Albert Kurland at the Maryland Psychiatric Center at Spring Grove Hospital in Catonsville, which had been using LSD to treat alcoholics. They were well into this work, and had seen many of their patients undergo transforming experiences under the influence of the drug, when one of the social workers in the group was diagnosed with metastatic cancer. She asked to undergo an LSD session herself as a way of dealing with her disease.[45] After lengthy consultation and extensive preliminary counseling, her request was granted. By this time she had become withdrawn from her family and uncommunicative, depressed by the unrelenting progress of the disease. But the change after her religious psychedelic experience was dramatic; she became outgoing and cheerful, embarking on a vacation trip to Europe shortly before her death.

The group began to plan the use of LSD in the counseling of other terminal cancer patients. By 1967, increased funding from the State of Maryland had allowed expansion of the group. It was joined by William Richards, who, as a Yale divinity student studying theology in Göttingen in 1963, had volunteered for a psilocybin experiment conducted by psychiatrist Hanscarl Leuner and was astonished by the totally unexpected mystical experience that resulted. And by Walter Pahnke, who had just completed his psychiatric residency wearing a tiara of three Harvard degrees—a bachelor of divinity, an M.D., and a Ph.D. in the history and philosophy of religion. (We will encounter Pahnke's Miracle of Marsh Chapel experiment in the next section, discussing the use of psilocybin.) Pahnke and Richards were already close friends, having met in Göttingen, where Pahnke had gone on a Sheldon fellowship to observe Leuner's work.[46] And by Stanislav Grof, who had been working with LSD since 1954 in Prague and had joined the group for a year's sabbatical, soon to be extended indefinitely when the Soviets invaded Czechoslovakia.

As a result of the group's work with dying patients, Pahnke was invited in 1968 to give the Harvard Divinity School's distinguished Ingersoll Lecture on Immortality, where he described about a third of the cancer patients as achieving a "psychedelic-mystical experience" that enabled them to bear the pain of their cancer with lower doses of narcotics or even with no narcotics at all: the experience both distanced them from their pain, and removed their fear of death.[47]

Here is a report of an LSD therapy session of a 58-year-old Jewish married woman who had suffered from cancer of the breast for 12 years. Despite numerous surgical procedures, including hysterectomy, oophorectomy, and adrenalectomy, the disease had metastasized to her spine, causing partial paralysis of the lower half of her body. Her anxiety and depression had been considerably relieved by two earlier LSD sessions.[48]

> The session began smoothly but the patient became frightened when she saw a huge wall of flames. After support and encouragement by the therapist, the patient was able to go through the middle of the flames, and at this point experienced positive ego transcendence. She felt that she had left her body, was in another world, and was in the presence of God which seemed symbolized by a huge diamond-shaped iridescent Presence. She did not see Him as a Person but knew He was there. The feeling was one of awe and reverence, and she was filled with a sense of peace and freedom. Because she was free from her body, she felt no pain at all. She was quiet during most of the day and emerged from the session with a deep feeling of peace and joy. when her family had arrived, she radiated a psychedelic afterglow of peace and beauty which all remarked upon.[49]

LSD Moves to the Street

During the early 1960s, Timothy Leary, who had been researching the effects of psilocybin at Harvard, acquired a colleague, Richard Alpert, and moved definitively from psilocybin to LSD. After a "shattering ontological confrontation" with the drug, Leary said he was never able again "to take myself, my mind, and the social world around me seriously. . . . From the date of this session it was inevitable that we would leave Harvard, that we would leave American society . . . tenderly, gently disregarding the parochial social insanities."[50] What had started out as informal academic group sessions exploring LSD began to be perceived as chaotic drug parties, and the Harvard administration was aghast. As Leary mischievously observed, "psychedelic drugs cause panic and temporary insanity in people who have not taken them." January 1963 found Leary, through the Harvard University Department of Social Relations, applying to Sandoz for 100 g of LSD and 25 kg of psilocybin—enough to provide 1 million doses of LSD and 2.5 million doses of psilocybin! But Sandoz declined to send the materials upon learning that Leary was being dismissed from Harvard. No matter: they sent 43 cases of LSD to Al Hubbard, the millionaire maverick uranium mogul, who willingly dispensed it to any religious seeker; and lysergic acid was still legally available and being bought in kilogram lots for transformation to the diethylamide by the able chemists of California.

Parturient montes, et nascetur . . . ?[51] LSD was intersecting with several powerful social forces in American culture: the leaden proprieties of the conformist '50s, Leary's "parochial social insanities," were being cast aside to the magical music of the Beatles; the presumptive authorities of church and state were being revealed as the corrupt proponents of an immoral war; and the ugly gangrene of American racism was being exposed by the civil rights movement. Another chemical breakthrough, oral contraceptives, had suddenly provided adolescents with hitherto unheard-of possibilities for carefree sexual expression. A revolution was slouching towards California to be born, and Leary would be there to midwife it. Indeed, the humble *Claviceps* fungus was to provide him with a *pulvis ad partum* unlike any ever seen before.

But the story of Leary's charismatic call to "turn on, tune in, drop out"; of his conversion to Hinduism and his founding of the League for Spiritual Discovery (LSD); of his conviction for drug smuggling, escape from San Luis Obispo prison, and political asylum in Switzerland; of amateur chemist Augustus Owsley Stanley III and his girlfriend Melissa (a chem grad student at Berkeley), and how his enormous profits from the manufacture of LSD went in part to support the Grateful Dead rock group; of poet Alan Ginsberg, novelist Ken Kesey, the Pranksters and the Kool-Aid acid test; of Sergeant Pepper's lonely heart and Lucy's skyward diamonds; of Haight-Asbury and the First Human Be-In—all this fascinating social history of the '60s, with its lurid sensationalizing by the media of the day, cannot be told here, and the interested reader must be referred elsewhere.[52]

The backlash was inevitable. By August 1965, Sandoz, the only legal manufacturer, had stopped all distribution of LSD and psilocybin except to the National Institute of Mental Health; in 1966, California and New York made its possession a criminal offense. All too late: the demon was out of the flask and into the streets. Unfortunately, while uncontrolled street use was to increase, finally declining only in the mid '70s, legitimate research was essentially banned:

> It is difficult to understand how the work of Masters and Houston, of Myron Stolaroff and Oscar Janiger, of the dozens of other legitimate researchers, could have been ignored by the therapeutic community, but it was. . . . A curious, almost Kafkaesque situation arose whereby those who knew most about psychedelics were relegated to the sidelines of the debate, while those who knew the least were elevated to the status of "expert." . . . None of the hundreds of questions raised by psychedelics, many of them fundamental to the way the mind processes information, have been answered. Rather, the powers that be have performed a holding action comparable to the one the Papal Curia tried with Galileo, when they confined him to a house in Arcetri and forbade him the right to continue his research.[53]

In 1965, all researchers were required to return any LSD and psilocybin to Sandoz and to reapply to the FDA for an Investigational New Drug (IND) permit. Few were granted, and in 1974 the last grant for use of LSD on human subjects was terminated. For almost two decades, Albert Kurland at Taylor Manor Hospital in Ellicott City, Maryland, has been the only researcher in the United States with an active IND for LSD.[54]

However, in recent years there has been a cautious reevaluation by the medical community of the role that psychedelic drugs might play in psychotherapy:

> . . . The prospect of new insights into the molecular mechanisms of hallucinogens is excellent. Not unlike epilepsies, hallucinogens sit at the crossroads of the mind-brain interaction. There are two advantages to their use experimentally. In humans a relatively clear sensorium lends insight into psychological processes. In animal models there is accessibility to neurons, membranes, messengers, and genes. The challenge will be to integrate each approach into a single experimental paradigm. . . . Such explorations may illuminate the mechanisms of psychoses, affective disorders, and hallucinoses. Techniques of verifiable effects of drugs on psychotherapy may be applicable in selected human populations. Such new strategies may uncover at least some of the keys of mental illness.[55]

Psychopharmacology

Dosage. LSD is one of the most powerful psychoactive drugs known: as little as 10 μg, or 0.010 mg, can produce some symptoms. The lowest dose that can produce psychedelic effects is about 0.050 mg; effects are proportional to dosage up to a limit of perhaps 0.500 mg. Higher doses do not intensify the psychological effect, although they may prolong it. In one careful study, it was found that the intensity of the psychological effects of LSD on a given individual was linearly proportional to the dosage for doses between 1 μg/kg through 16 μg/kg (i.e., for doses of 0.070 mg through 1.120 mg on the usual assumption of a 70-kg individual). But *between* individuals, the effects of a given dose could vary widely, particularly for those effects that were judged to be a function of personal predisposition (such as paranoid ideation), but also for those effects such as hallucinatory distortions of colors, patterns, and objects.[56]

There is one record of a death possibly due to LSD overdose, and the amount ingested seems to have been an extraordinary 320 mg taken intravenously.[57] Viewed as a traditional drug, LSD's very high therapeutic index (ratio of toxic dose to effective dose) makes it one of the safest drugs known.

In the '60s, LSD was distributed absorbed on sugar cubes or mixed into a sugar paste; but because of its astonishing potency, it can be applied via a few drops of an aqueous solution to sheets of paper—sometimes blank, often imprinted with designer trademark cartoon figures—and perforated for easy separation into quarter-inch-square single-dose "tabs." This is the form in which it is currently found as a street drug, and forensic reports indicate there is usually about 20-50 μg of LSD per tab.

Physiological Effects. These vary considerably but are not very significant. There is always dilation of the pupils; usually some increase in blood pressure and body temperature; and occasionally some trembling, nausea, or loss of appetite. These are typical effects of mild sympathomimetic stimulation.

Tolerance to LSD develops rapidly; after several sequential days of usage, the dosage needed to produce the same effect is markedly increased. Cross-tolerance develops to mescaline and psilocybin, which induce a similar reciprocal cross-tolerance to LSD; and this is one indication that their mechanism of action is similar. There is no cross-tolerance to any other of the common drugs such as amphetamines, phencyclidine, or marijuana. The acquired tolerance is lost just as rapidly as it is gained, after 3 to 4 days of abstention. In practice, the great majority of those who use LSD do so only every few weeks or months, so the acquisition of tolerance is not likely to be noticed.

Duration of Action. The sympathomimetic symptoms are usually detectable within a few minutes of an oral dose. Typical psychedelic effects begin at about the second hour and reach a peak during the fourth or fifth hours. Although the half-life in the body is only 3 hours, effects are felt up to 8 hours; the entire syndrome, including pupillary dilation, clears in 12 hours.[58]

"Bad Trips." The possibility of a negative or painful experience, which seems best described as an acute anxiety or panic reaction, following ingestion of a psychedelic drug, seems always to be present, although the majority of "trips" are experienced as positive. Because they are usually self-revelatory, even bad trips, according to one study, are considered by 50% of those who have had them as being beneficial.[59] There is conflicting, mostly anecdotal, evidence correlating the incidence of panic attacks with dosage, set, or setting; common sense suggests that lower doses, a positive outlook (set), and a familiar environment (setting) favor a more benign outcome. But the very suggestibility and claimed in-depth confrontation with Self and Other intrinsic to the action of the drug probably makes the exclusion of all painful encounters an impossibility; the same can be said of psychoanalysis and religious mysticism: in Christian tradition, the Dark Night of the Soul is an essential step in the progress of the mystic, and in St. Ignatius Loyola's program of spiritual guidance, periods of "desolation" are expected to alternate with periods of "consolation."

Suicide, Accidents, Murder. As Grinspoon emphasizes, the extent of these incidents was greatly exaggerated by a sensationalist press in the late '60s: "If deaths caused by alcohol intoxication had been given attention proportional to that devoted to deaths allegedly caused by LSD, there would have been room for no other news."[60] LSD users are much more likely to be absorbed with their own thoughts or lost in the contemplation of a bouquet of flowers than wielding an axe. They are enthralled by the mantra of St. Augustine: "ne in foras ire, sed in teipsum redire; quia in interiore homine habitat veritas (Do not go out into the marketplace, but rather enter into yourself: for in the inner self dwells the truth)."

The most famous case of suicide induced by an LSD trip was that of Frank Olson, who was a biological warfare expert working for the CIA. In an obviously unethical "experiment" by the Agency, he was given LSD in a cocktail without his knowledge in 1953. He had a psychotic reaction and jumped out of a 10-story window 2 weeks later; the story behind the suicide was suppressed for 20 years.[61]

Psychotic Breakdowns. Occasionally, users of LSD have a bad trip that changes into an acute psychotic episode. At a 1992 rock concert in Northern California, 14 patients were treated at the medical area for LSD intoxication. Twelve were discharged after "talking down" their panic; but two required medication. Here is a description of one of them:

> An 18-year-old Japanese male was brought to the medical area in four-point leather restraints, screaming hysterically in Japanese. He was accompanied by a bilingual male friend who indicated that the patient was seeing "bad demons." The friend stated that the patient had ingested two "hits" of LSD approximately two hours earlier. . . . This was the patient's first experience with LSD. Initial attempts to establish rapport and provide talk-down were unsuccessful. . . The patient was medicated with 2 mg lorazepam and 2 mg haloperidol intramuscularly. Thirty minutes later the leather restraints were removed. . . . Over the next hour the patient improved and was able to converse in Japanese with his friend. . . . Two and one-half hours after admission the patient was discharged to the care of his friend. He was alert, oriented, spoke English, and apologized sheepishly to the medical staff for "causing a scene." The patient reported no further medical problems the next day at a follow-up visit.[62]

The other incident requiring medication had a similarly benign outcome. However, numerous street users of LSD and the other psychedelics have entered a lengthy or permanent psychotic state after ingesting the drug. Much like the controversy surrounding incidents of suicide among patients prescribed Prozac, these psychotic breakdowns have been attributed directly to the effect of LSD or other psychedelics, but the preponderant weight of evidence now seems to support the contention that chronic psychotic breakdowns have occurred only in those in whom the drug has precipitated an already existing underlying pathology. A 1983 study comparing hospitalized patients whose schizophrenia began after taking LSD with those whose schizophrenia began in other circumstances concludes (in the rather dense language favored by the profession):

> In summary, the pattern of results argues against regarding LSD psychosis as a nosologic entity distinct from acute schizophrenia or attributable fundamentally and specifically to toxic drug effects. The emergent clinical features may be understood as an interaction of the acute psychotic decompensation precipitated by the LSD, or whatever crisis situation had motivated its use, with the chronic pathologic premorbid traits of these patients. Preexisting sociopathic characteristics, together with a genealogic predisposition for substance abuse, may have contributed to the involvement with LSD, perhaps as a route of escape from other pressures or, . . . as an attempt at self-medication. Considering the high rates of parental psychosis in these patients, it is quite possible that the pharmacologic stressor induced a schizophreniform reaction that in most respects is indistinguishable from non-drug-precipitated schizophrenic conditions.[63]

Although the very possibility that an LSD trip could trigger a latent psychosis is what makes its classification as a Schedule I drug seem justified to many people, this potential should be placed in perspective. LSD has been used by thousands of individuals; and the frequency of chronic psychotic episodes, even in totally uncontrolled circumstances, seems no greater than in the population at large. And it must be recalled that even traditional and edifying religious practices, such as the observance of monastic silence, will catalyze a psychotic breakdown in labile persons.[64]

Flashbacks. The reports on these phenomena are confusingly contradictory.[65] Part of the problem is the definition of what a flashback is: does it include simply a vivid recollection of some aspect of an intense psychedelic experience (i.e., is it just the association of a present nondrug event with a similar past drug-influenced event) or is it restricted to an actual reexperiencing of the psychedelic state? Some sort of flashback occurrence in the broadest sense probably happens with about 25% of the user population, but real recurrences of the psychedelic state including time distortion and vivid visual distortions under drug-free conditions seem to occur rarely if ever.

There may be idiosyncratic differences in susceptibility to these phenomena. Alexander Shulgin, who probably has taken more diverse psychedelic and potentially psychedelic substances than anyone else on the planet, feels that flashbacks are only vivid memories of a past traumatic event, whether the event was drug-influenced or not. We all have such deja vu-like incidents: coming home after years away and entering the room in which one's mother died will predictably excite memories so vivid as to be almost a reliving of the original experience. Shulgin never had a flashback, but he describes three occurrences of a spontaneous altered state, one lasting about 5 hours, that he has undergone.[66] He refers to them as "fugues."

A careful study of the visual aspects of flashbacks (that nonetheless is limited because most of the subjects were psychiatric outpatients and the exact nature and dosage of the street drugs they had taken cannot be known) found that the most common phenomena were geometric pseudohallucinations (geometric figures coming and going before closed or open eyes); objects in the peripheral field appearing to move or flow; trailing images like the frames of a movie film following moving objects; and flashes of color. Use of marijuana was one of the two most common events precipitating flashbacks; the other was entering a dark room after coming from bright daylight. There was no association between incidents of flashback and anxiety and no indication that the frequency of the incidents decreased in time. Low doses of antipsychotic drugs intensified the occurrence of flashbacks, and benzodiazepines reduced their occurrence. This is the same pattern as with epileptic seizures, and the author speculates there is a form of kindling taking place, a locking of visual circuitry into an "on" position following perception of a visual stimulus. Additionally, the subjects appear to fall into three groups—one very sensitive to LSD and experiencing flashbacks after a few uses of the drug, a second that requires 40 to 50 lifetime exposures, and a third that seems quite resistant. "The most intriguing hypothesis this observation generates is that . . . there exists a genetic basis for variable sensitivities to LSD."[67]

Other Hallucinogenic Lysergic Acid Amides

Ololiuqui. This morning-glory plant was one of the most important sacred hallucinogenic plants in use among Mexican Indians at the time of the Conquest. It is also called *coaxihuitl*, or snake plant, and has been variously classified botanically as *Ipomoea siaefolia* and *Turbina corymbosa*. It was only in 1939 that Schultes obtained identifiable specimens of the plant and it was assigned the name *Rivea corymbosa*. In 1955, Osmond described a state of intoxication that followed ingesting the seeds of this plant, and careful analysis of the seeds by Hofmann showed that they contained lysergic acid amide and several of its derivatives, including ergonovine. As is often the case, nature here was imitating art, since Hofmann previously synthesized most of these compounds.[68] There was at first great skepticism about his report, since the morning-glory *Rivea* and the fungus *Claviceps* were botanically so distant it seemed implausible that they would both produce as unique and complex a molecule as lysergic acid. It was suggested that the numerous lysergic acid samples in Hofmann's lab had contaminated his seed extracts, or that the ergot fungus itself had somehow infected the seeds. But other workers eventually confirmed Hofmann's findings.

Later a second morning-glory species was found, the seeds of which contain about twice the lysergic alkaloid content of *Rivea*. This second plant, *I. tricolor* or *I. violacea*, is used by the Zapotecs of Oaxaca, who call it *badoh negro* because its seeds are black, while those of ololiuqui are brown. And it was found that the seeds of the Hawaiian baby woodrose, *Argyreia nervosa*, that has been used in India, contained lysergic acid amide and several of its derivatives—not, however, including ergonovine.

Lysergamide, Ergonovine, Witch-hunts. Further investigation by Hofmann showed that lysergic acid amide (lysergamide) was hallucinogenic, although it had only about one-tenth the potency of LSD; ergonovine in doses much larger than was used for obstetrical purposes turned out also to be hallucinogenic. Apparently the medical profession was always so cautious in using ergonovine that its hallucinogenic properties never were noticed. Since ergonovine occurs in ergot, it is possible that it caused the psychotic symptoms exhibited by some victims of ergotism in the Middle Ages. Mary Kilbourne Matossian has argued that weather patterns likely to encourage ergot infestation of rye and the incidence of witch persecutions in medieval Europe (and 17th century Salem) can be linked; possibly many so-called witches were suffering from ergot psychoses.[69]

The Eleusian Mysteries. The ancient mysteries of Eleusis, said to be established by the goddess of agriculture, Demeter, were highly praised by Plato and Cicero.[70] During the initiation, before the final revelation in the holiest chamber, all drank from a potion containing an extract of barley and mint, the κυκεών. Barley and some of the wild grasses common to the Mediterranean region are frequently infected by ergot; and ergonovine, unlike the other ergot alkaloids, can be easily extracted into water. It is possible that this extract contained ergonovine, and Wasson, Ruck, and Hofmann propose that this is the most likely candidate for the secret of the Eleusian mysteries.[71]

Structure-Activity Relationships. In the first two decades after its discovery, many modifications of the LSD structure were made in an effort to determine a structure-activity relationship for its hallucinogenic action,[72] but none of these variations had an activity greater than about one-third that of LSD. However, more recently Nichols'[73] group at Purdue University synthesized several compounds that, when tested by a group of rats trained to discriminate the effects of intraperitoneal injections of saline from LSD (a method that seems to be reliable[74]), show comparable or greater activity. Previous modifications of the LSD structure involved changing the alkyl groups on the amide nitrogen; for the first time, the Purdue workers changed the alkyl group on the 6-N amine nitrogen. They found that the 6-propyl modification, **6-5**, Figure 6.2, was equipotent to LSD, while the 6-ethyl compound, **6-6**, was about twice, and the 6-allyl, **6-7**, about three times as potent.

Hofmann was the first to observe that even minor modifications of the alkyl groups on the amide nitrogen of LSD cause significant loss of activity, and that only the *d* enantiomer was active. In a recent study by Nichols' group,[75] it was shown that activity is also significantly affected by the chirality of the *N*-alkyl function. They prepared the (*R*)-2- and (*S*)-2-aminobutylamides of lysergic acid (Figure 6.3, **6-9, 6-10**) and compared their activity with LSD, **6-8**, and with the pyrrolidyl amide of LSD, **6-11**, a compound that had been studied earlier[76] and shown then to have only about one-tenth the hallucinogenic activity of LSD in human subjects. The activity was evaluated both by rat behav-

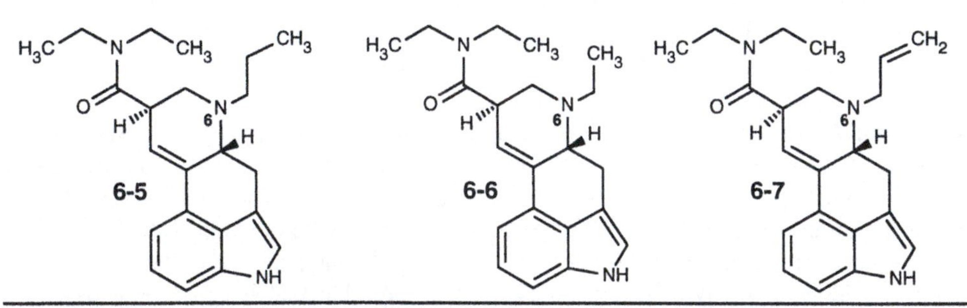

Figure 6.2 Three modifications of the LSD structure with equal or greater activity.

ior and by displacement of a radioactive ligand from rat 5-HT_2 receptors. The relative potency of the pyrrolidyl amide in these tests corresponded to its activity as a hallucinogen in humans, arguing for the validity of this approach. While the (*R*)-2-butyl substituent produced an amide of approximately equal potency to LSD, the (*S*)-2-butyl group was significantly less potent. Molecular mechanics showed that the conformations of these two diastereomers were quite different, the more active (*R*) isomer more closely resembling that of LSD. The Purdue workers conclude that the alkyl group of the amide function interacts with a hydrophobic region on the receptor, and this interaction affects binding either directly or indirectly by inducing conformational changes elsewhere in the molecule.

Mechanism of Action

Although a great deal has been learned, and continues to be learned, about the mechanism of action of LSD and the other hallucinogenic drugs, there is much that is yet unclear, as is indeed the case with all psychotropic drugs. A brief recapitulation of some of what has been discovered to date follows:

The structure of LSD, like psilocybin and dimethyltryptamine, incorporates much of the structure of the NT 5-hydroxytryptamine, or serotonin, and early workers were excited by the discovery that LSD blocked the contractile effect of serotonin on isolated smooth muscle: LSD was a serotonin antagonist. However, 2-bromo-LSD, which lacks any hallucinogenic action in humans, turned out to have the same effect.[77] Later, Freedman found that LSD increased serotonin concentrations in the rat brain and decreased the level of serotonin metabolites, while 2-bromo-LSD did not have this effect: LSD decreased serotonin turnover (synthesis and release) in the brain.[78] Aghajanian specified this property further: LSD bound to presynaptic autoreceptors, thereby inhibiting the firing of serotonin neurons.[79] But later work showed that the phenethylamine hallucinogens such as mescaline do *not* bind to presynaptic serotonergic receptors; and that although tolerance is rapidly established to the hallucinogenic effects of LSD (with cross tolerance to mescaline), there is *no* tolerance established to the action of LSD on the presynaptic receptors, so binding at these sites could not account for its hallucinogenic activity.[80] (It ought to be noted in general that hypothetical mechanisms claiming to account for psychotropic

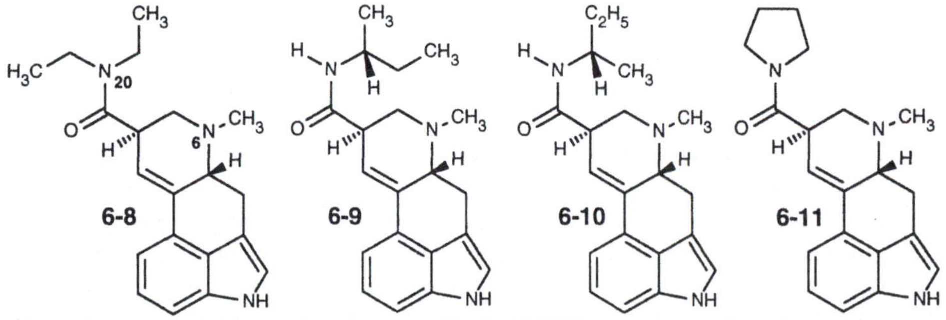

Figure 6.3 LSD and its (*R*)-butyl, (*S*)-butyl, and pyrrolidyl congeners.

activity share with mechanisms for organic reactions a common epistemological weakness: they are "right" only because they have not yet been proved wrong; nonetheless, they can stimulate much fruitful research.)

Since this work, numerous subtypes of the serotonin NT system have been discovered and continue to be discovered. Both the tryptamine and phenethylamine classes of hallucinogen have been found to bind as agonists or partial agonists to one of these, the 5-HT_2 subtype ("HT" because serotonin is 5-*h*ydroxy*t*ryptamine), which is abundant in the cerebral cortex. And there is a high correlation ($r = 0.924$) for relative potency of both tryptamine and phenethylamine classes of psychedelics and their binding affinity to these receptors, both in rat[81] and in human[82] cortical homogenates. Further: 5-HT_2 receptors are downregulated by chronic administration of LSD (hence accounting for acquired tolerance), but not of 2-bromo-LSD;[83] monoamine oxidase inhibitors (MAOIs) downregulate serotonin receptors and inhibit the behavioral effects of LSD,[84] but reserpine, which depletes stores of serotonin and upregulates postsynaptic receptors, potentiates LSD's behavioral effects. Note that if LSD's behavioral effects came from its acting on the presynapse to trigger dumping of stored serotonin, pretreatment with reserpine would have the opposite effect.[85] Finally, selective 5-HT_2 receptor antagonists inhibit LSD behavioral effects.[86]

But LSD also interacts with 5-HT_{1c} receptors[87] as well as many other 5-HT receptors, including some which have been recently cloned and whose functions have not yet been determined.[88] And while all the classic hallucinogens act at the 5-HT_2 receptors, all of them act elsewhere as well: for example, LSD acts at the 5-HT_{1a}, the 5-HT_{1c}, dopamine, and α-adrenergic receptors.[89] In an *in vitro* study of the binding of four hallucinogens with nine neurotransmitter binding sites in the human cortex, Pierce and Peroutka found that while all the agents bound to all the receptors, there was a marked difference between the relative affinities of the indolealkylamine and the phenalkylamine for three of the binding sites.[90]

Syntheses of Lysergic Acid

Kornfeld-Woodward. The first successful synthesis of lysergic acid was achieved in 1956 by the collaborative efforts of a group headed by Edmund Kornfeld at Eli Lilly and Robert B. Woodward at Harvard. The synthesis began with indole 3-propionic acid, **6-12** (Figure 6.4), which was reduced to the indoline propionic acid in order to avoid the possibility of an isomerization of the **C** ring in intermediates like **6-14** to an aromatic naphthalenoid system, with irreversible loss of the 2,3-indole double bond. This expedient has been employed in all subsequent syntheses except that of Oppolzer. Restoring the indole system proved somewhat problematic: when the *N*(1)-acetyl ethyl ester of **6-18** was treated with palladium-charcoal, ring **C** aromatized forming the naphthalenoid compound; when the sodium salt of the acid was treated with Raney nickel, disproportionation occurred with the double bond of the **D** ring migrating to the **B** ring. Finally, the indole system was restored, and the first total synthesis of lysergic acid achieved, by using Raney nickel and sodium arsenate.[91] The entire sequence took 15 steps, and resulted in racemic material with two chiral centers: because only one of the four isomers is physiologically

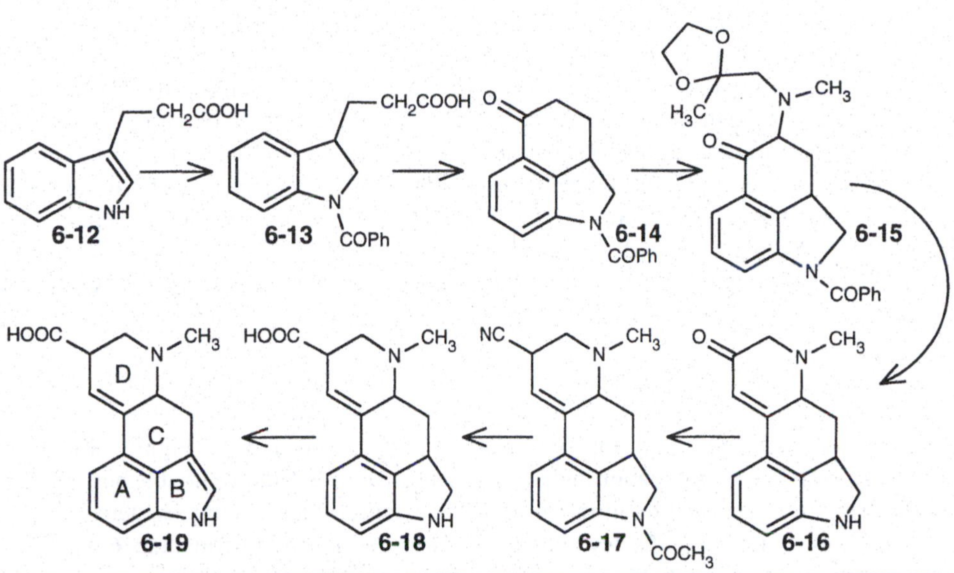

Figure 6.4 The Kornfeld-Woodward synthesis of lysergic acid.

active, this synthesis was in no way able to compete with fermentation of ergot alkaloids for the commercial production of lysergic acid. However, it represents a milestone in the progress of organic synthetic methodology.

Oppolzer's Intramolecular Condensation. In 1981, Oppolzer reported an elegant synthesis of lysergic acid, in which the **C** and **D** rings are formed in a single intramolecular process. The essential steps are given in Figure 6.5. The phosphonium bromide **6-20**

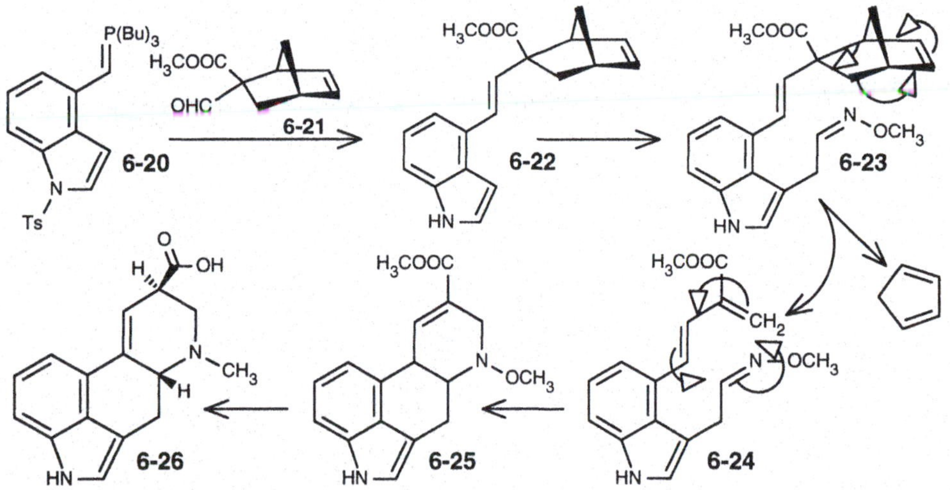

Figure 6.5 Oppolzer's synthesis of lysergic acid.

could be made in near quantitative yield, and bicycloheptene **6-21** in about 60% yield, from readily available starting materials. A Wittig reaction between these provided **6-22**, after removal of the tosyl protecting group, in 62% yield. Condensing the Mannich salt of **6-22** with nitromethane and reduction with $TiCl_3$ gave **6-23** in an overall yield of about 35%. The thermolysis of this crucial intermediate required the incremental addition of a 1% solution over a period of 5 hours into 200° trichlorobenzene. Under these conditions, a retro-Diels Alder reaction extruded cyclopentadiene and exposed the labile diene system of **6-24** to the imine dienophile; **6-25** was formed as mixture of diastereomers in 67% yield. This could be converted to (±)-lysergic acid in 33% yield. A remarkable feature of the synthesis is that the indole nucleus is preserved throughout.[92]

Rebek's Synthesis from Tryptophan. In 1984, Rebek published an efficient synthesis of lysergic acid from its biosynthetic precursor, tryptophan, which is available as the naturally occurring L-amino acid [the (*S*)-(−) isomer]. Tryptophan, **6-27** (Figure 6.6), is first protected both by dibenzoylation and hydrogenation of the 2,3-indole double bond to form **6-28**. This is dehydrated to an intermediate azlactone, which undergoes stereoselective intramolecular Friedel-Crafts acylation to give tricyclic ketone intermediate **6-29**. (If *L*-tryptophan is employed, **6-28** is a diastereomeric pair; if the racemic amino acid is used, the enantiomers of this pair are also present. In either case, upon formation of the azlactone, rapid racemization occurs at the α-carbon, and during cyclization the γ-center determines the stereochemistry of this carbon so that the resulting **6-29** is always an enantiomeric pair with the hydrogens *cis*). A Reformatsky reaction with ethyl α-(bromomethyl)acrylate produces spiro methylene lactone **6-30**, after which addition of hydrogen bromide, methylation of the amine, and removal of the protective benzyl groups results in spontaneous ring closure to form **6-32** in 44% overall yield from **6-30**. This could be converted to the methyl ester of dihydroisolysergic acid in 95% yield; after rebenzoylation (accompanied by some epimerization), oxidation of the indolane ring with MnO_2 gave a mixture of the methyl esters of lysergic and isolysergic acid, **6-33**, in 45% yield.[93]

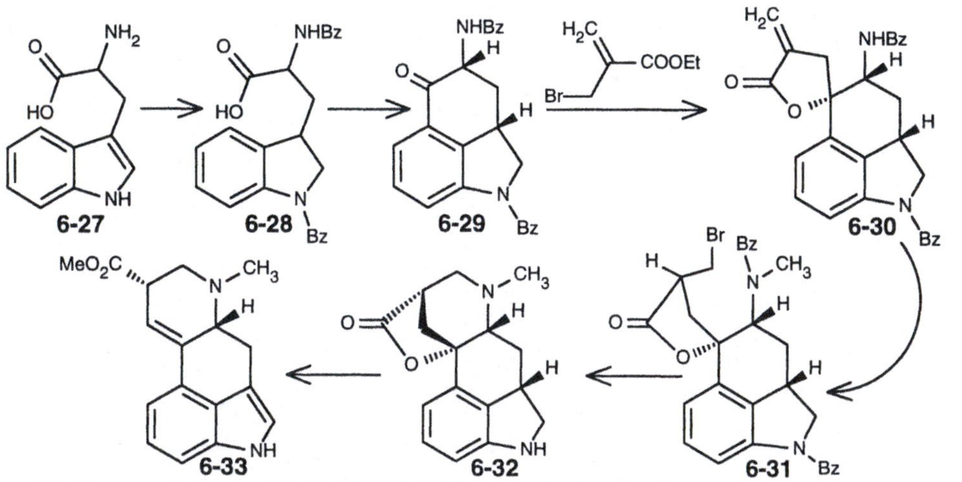

Figure 6.6 Rebek's synthesis of lysergic acid.

Other Methods of Preparation. There have been other total syntheses of lysergic acids, some displaying equally elegant chemistry.[94] Of particular interest is Ramage's synthesis, reported in 1976, which was based on an idea explored with Woodward just before his death, and involves an insight into the mechanism whereby the naturally occurring (+) forms of both lysergic and isolysergic acids (**6-34**, **6-36**, Figure 6.7) could be converted into racemic lysergic acid by aqueous barium hydroxide at high temperatures. This involves loss of stereochemical integrity not only at position 8 (which is not surprising, since it is α to the carbonyl) but at position 5 as well. Woodward proposed that the racemization proceeded through acyclic intermediate **6-35**. Following up on this suggestion, Ramage synthesized the 1,2-dihydro analog of the methyl ester of **6-35** and found that it did indeed cyclize spontaneously to the racemic methyl ester of dihydrolysergic acid, from which lysergic acid could be produced in the usual way.[95]

Meanwhile, the biologists had found a better way: in 1964, they observed that a strain of *Claviceps paspali* under saprophytic cultivation in fermenters would produce an abundant quantity of paspalic acid, **6-37**.[96] (*C. purpurea* will only grow on grain in a field.) And soon it was discovered that paspalic acid easily isomerized under basic conditions to lysergic acid; conjugation with the indole ring being under these circumstances more thermodynamically favored than with the carbonyl.[97] To the present day, this probably remains the most efficient and economical way of obtaining lysergic acid.

Illegal Production. As for the illegal routes of production, total synthesis is out of the question. One source of lysergic acid is the diversion from pharmaceutical houses or drug stores of ergotamine tartrate, **6-2**, which is used intravenously in legitimate medicine for the arrest of acute migraine attacks. This can be easily hydrolyzed to lysergic acid. Ergonovine and methylergonovine (both used to induce uterine contractions; methylergonovine is the lysergic acid amide of butanolamine; ergonovine, **6-1**, of propanolamine) could be used in the same way. It is remotely possible that two other prescription drugs, methysergide (which is the 1(*N*)-methyl derivative of methylergonovine) or dihydroergotamine (the 9,10-dihydro derivative of ergotamine) could be employed. In the former case, the 1-*N*-methyl group would have to be removed to make LSD (however, 1-*N*-methyl-LSD is a known hallucinogen with a potency of about one-tenth that of LSD); in the latter case, the double bond would have to be reintroduced.

In one major incident of clandestine LSD production, the "Operation Julie" case in the United Kingdom in which 120 persons were arrested in 1977, ergotamine tartrate was

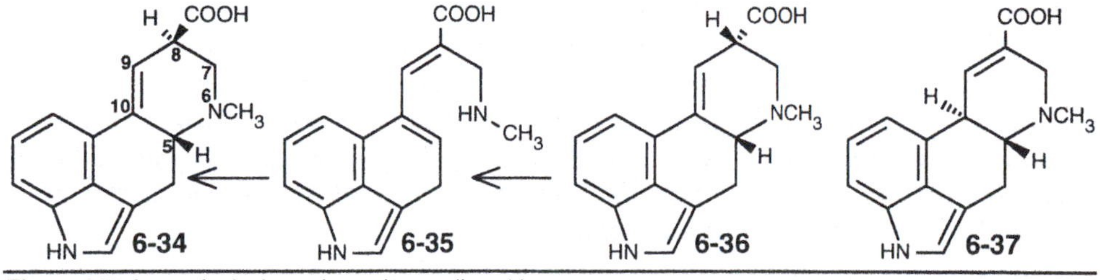

Figure 6.7 Lysergic, isolysergic, and paspalic acids.

the confiscated starting material, and recovered containers indicated that *N,N*-carbonyldiimidazole was used to form the diethyl amide. It was estimated that almost two-thirds of the world was supplied by this group, which operated small but efficient laboratories in London and Wales.[98]

It is likely that the LSD so widely available in the United States (at least on college campuses and at rock concerts) comes from a few clandestine laboratories in San Francisco and, perhaps, on the East Coast, each with a few tanks of fermenting *C. paspali*—cultivation of which is only one or two orders of magnitude more sophisticated than fermenting yoghurt.[99]

Psilocybin and the Tryptamines

The Mushroom Flesh of the Gods

When the Spanish naturalists came to Mexico and Central America after its conquest by Cortez, they described three "intoxicating" plants used and worshiped by the "witch doctors" of the native tribes: peyote and ololiuqui, which we have discussed earlier, and a third, a mushroom that the native people described as "teonanácatl," Aztec for "flesh of the gods." Nonetheless, in subsequent centuries the stories about this mushroom were largely dismissed as superstitious imaginings, despite the continued discovery, in the highlands of Guatemala, of pre-Columbian stone artifacts a foot high or more dating as far back as 800-1000 BC, and portraying a human or animal face or figure merged into the stalk of a giant mushroom, whose cap protectively shielded the effigy like an umbrella.

In the 1930s, however, several American scientists reported that indigenous peoples in the remote Sierra Mazateca in Oaxaca, Mexico, were still employing mushrooms for divination and healing. This intrigued two amateur mycologists, R. Gordon Wasson, a vice president of J. P. Morgan Bank, and his wife, Valentina, a Russian-born pediatrician, who had been studying the interaction of mushrooms and culture for some years.[100] They visited the area twice, in the summers of 1953 and 1955, and by patient efforts won the trust of the native people; on their second visit they became the first nonindigenous persons ever to partake of the teonanácatl, an experience that profoundly impressed them: "As your body lies there, your soul is free, with no sense of time, alert as never before, living an eternity in a night, seeing infinity in a grain of sand. . . . The divine mushroom introduces ecstasy to us. Your very soul is seized and shaken until it tingles, until you fear that you will never recover your equilibrium. After all, who will choose to feel undiluted awe, or to float through that door yonder into the Divine Presence?"[101]

On still a third expedition in the following year, they were accompanied by Roger Heim, a mycologist from the Muséum National d'Histoire Naturelle, Paris. Heim found that there were at least a dozen hallucinogenic mushrooms used in Oaxaca; *Psilocybe mexicana* was probably the most common, although *Stropharia cubensis* was said to be even more potent. The former was able to be cultivated in Paris, and soon samples were sent to Hofmann in Basel to be analyzed.

Teonanácatl in Native Religion

A 1993 cover story in *Time* magazine describes a solemn ceremony of the ancient Maya as reconstructed by recent archeology:

> The crowd at the base of the enormous bloodred pyramid has been standing for hours in the dripping heat of the Guatemalan jungle. No one moves; every eye stays fixed on the building's summit, where the king, his head adorned with feathers, his scepter a two-headed crocodile, is about to emerge from a sacred chamber with instructions from his long-dead ancestors. . . . Lifted into the next world by hallucinogenic drugs, the king will take an obsidian blade or the spine of a stingray, pierce his own penis, and then draw a rope through the wound, letting the blood drip onto bits of bark paper. Then he will take the bark and set it afire, and out of the rising smoke a vision of a serpent will appear to him.
>
> When the king finally emerges, on the verge of collapse, he reaches under his loincloth, displays a bloodstained hand and announces the ancestors' message—the same message he has received so many times in the past: "Prepare to go to war." The crowd erupts in wild cheers.[102]

The "hallucinogenic drugs" that lifted this king into the next world, fortifying him to endure the pains of royalty, were almost certainly one of the varieties of psychedelic mushrooms identified by Heim. While cultivation and use of these mushrooms survives to this day among the inhabitants of southern Mexico, some of them direct descendants of the Maya, the ceremony is much simpler and relatively painless. Communicants lie on the floor of a one-room hut, while the priestess or *curandera* chants and sings. According to Wasson, "The singing is good, but under the influence of the mushroom you think it is infinitely tender and sweet. . . . In the darkness and stillness, that voice hovers through the hut, coming now from beyond your feet, now at your very ear, now distant, now actually underneath you, with strange ventriloquistic effect."[103] Midway through the service, the priestess asks each participant to express their needs—healing, guidance, divination—and she answers with advice and prophecy.

The Alkaloids of Psilocybe mexicana Heim: Isolation and Synthesis

From the mushrooms brought back from Mexico by Heim and cultivated in Paris, Hofmann and his co-workers in the Sandoz laboratories in Basel attempted to establish an animal model for activity, but no clear-cut response on the part of mice or dogs could be observed. Doubt began to arise as to whether the dried material from Paris had lost its activity. Finally, Hofmann sampled 32 dried specimens of *Psilocybe mexicana*—about 2.4 g, a medium dose by Indian standards:

> Thirty minutes after taking the mushrooms, the exterior world began to undergo a strange transformation. Everything assumed a Mexican character. . . . Whether my eyes were closed or open I saw only Mexican motifs and colors. When the doctor supervising the experiment bent over me to check my blood pressure, he was transformed into an Aztec priest and I would not have been astonished if he had drawn an obsidian knife. . . . At the peak of the intoxication, about 1.5 hours after ingestion of the mushrooms, the rush of interior pictures, mostly

abstract motifs rapidly changing in shape and colour, reached such an alarming degree that I feared that I would be torn into this whirlpool of form and colour and would dissolve.[104]

The group thenceforth used each other and a group of student volunteers to taste small quantities of the different chromatographic fractions until they had isolated the two active alkaloidal ingredients, which turned out to be 4-hydroxy-*N,N*-dimethyltryptamine and its phosphate ester, which they named *psilocin* (**6-41**, Figure 6.8) and *psilocybin*, **6-43**. The former is unstable and present in only trace amounts; the latter is present in dried mushrooms at a concentration of about 0.2-0.4%.[105]

The structures were quickly confirmed by synthesis.[106] The protected 4-hydroxyindole **6-39** was fortuitously already available in the Sandoz laboratory from previous work, having been made a few years earlier in four steps from **6-38**.[107] Reaction with oxalyl chloride and dimethylamine gives **6-40**, which was reduced with LAH and deprotected with H_2/Pd to give psilocin, **6-41**. Psilocybin, **6-43**, could be made in two steps from psilocin. Later work confirmed that psilocin was the active species in the CNS; it is formed by the metabolic dephosphorylation of psilocybin in vitro. The similarity in structure between psilocin/psilocybin and LSD was immediately noted: common to all three structures is a *N,N*-dimethyltryptamine moiety. Because the brain NT serotonin is 5-hydroxytryptamine, it is perhaps not surprising that all these structures interact at the serotonin receptors. It was not long before a family of synthetic psychedelics was developed based on this structure.

Uses of Psilocybin in Counseling and Religion

Behavior Change. During a visit to Mexico in 1960, a bored Harvard research psychologist named Timothy Leary bought some psilocybe mushrooms from a shaman and entered into a state of ecstasy that he said was the most profound experience of his life. He returned to Harvard to initiate a program to treat prisoners at the Massachusetts Correctional Institution with the active ingredient in the mushrooms, psilocybin,[108] which had just been synthesized by Hofmann and was now marketed experimentally by Sandoz.

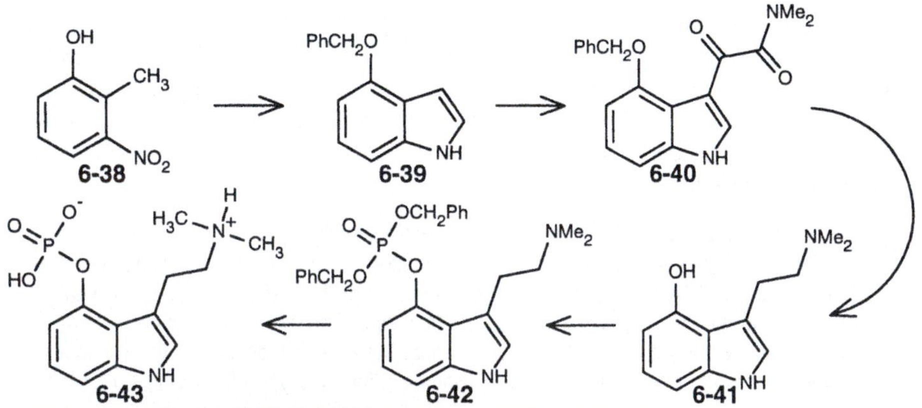

Figure 6.8 Hofmann's synthesis of psilocin and psilocybin.

Psychology of Religious Experience: The "Miracle of Marsh Chapel." In April 1962, Harvard-trained psychiatrist Walter N. Pahnke carried out the central research project for his Ph.D. in the history and philosophy of religion from Harvard Divinity: a Good Friday service in Boston University's Marsh Chapel, with psilocybin as the "sacrament." The participants, 19 divinity students and Andover Newton Professor Walter Houston Clark, were divided into groups of four; each group was to be guided by two of Leary's followers joined by MIT Professor Huston Smith, all of whom were experienced in the effects of psilocybin on themselves and others. (Pahnke himself had never taken a psychedelic drug and remained throughout a nonparticipating observer.)[109] The experiment was double-blind, with half of each group and one of each group's two leaders receiving an active placebo of nicotinic acid to produce a perceptible peripheral vasodilation. But within half an hour, it was fairly evident who had gotten the psychedelic and who the nicotinic acid.[110] The response was measured during the service, shortly after, and at a 6-month follow-up. Using a typology of mystical experience including such categories as unity, transcendence of time and space, deeply felt joy and peace, and sacredness, those receiving psilocybin had a significantly more profound religious experience than the controls.

> The experience of the experimental subjects was certainly more like mystical experience than that of the controls, who had the same expectation and suggestion from the preparation and setting. . . .
>
> After an admittedly short follow-up period of only six months, life-enhancing and -enriching effects similar to some of those claimed by mystics were shown by the higher scores of the experimental subjects when compared to the controls. In addition, after four hours of follow-up interviews with each subject, the experimenter was left with the impression that the experience had made a profound impact (especially in terms of religious feeling and thinking) on the lives of eight out of ten of the subjects who had been given psilocybin. Although the psilocybin experience was unique and different from the "ordinary" reality of their everyday lives, these subjects felt that this experience had motivated them to appreciate more deeply the meaning of their lives, to gain more depth and authenticity in ordinary living, and to rethink their philosophies of life and values. The data did not suggest that any "ultimate" reality encountered had made "ordinary" reality no longer important or meaningful. . . .
>
> The results of our experiment would indicate that psilocybin (and LSD and mescaline, by analogy) are important tools for the study of the mystical state of consciousness. . . .
>
> Possibilities for further research with these drugs in the psychology of religion . . . . would be the establishment of a research center where carefully controlled drug experiments could be done by a trained research staff which would consist of psychiatrists, clinical psychologists, and professional religious personnel.[111]

Pahnke anticipated that—perhaps from a Puritan bias in American religiosity—many would be scandalized to think that there was a biochemical basis to religious experience, and would judge a drug-induced mystical experience unauthentic because "unearned."

> Although a drug experience might seem unearned when compared with the rigorous discipline that many mystics describe as necessary, our evidence has suggested that careful preparation and expectation play an important part, not only in the type of experience attained but in later

> fruits for life. Positive mystical experience with psychedelic drugs is by no means automatic. It would seem that the "drug effect" is a delicate combination of psychological set and setting in which the drug itself is the trigger or facilitating agent—i.e., in which the drug is a *necessary* but not *sufficient* condition. . . . "Gratuitous grace" is an appropriate theological term, because the psychedelic mystical experience can lead to a profound sense of inspiration, reverential awe, and humility, perhaps partially as a result of the realization that the experience *is* a gift and not particularly earned or deserved.[112]

Psychopharmacology of Psilocybin

It is difficult to quantify something as intrinsically variable as a psychedelic experience; however, the effects of psilocybin seem to begin at about 5 mg. The usual dose is 10 to 20 mg, which corresponds to about 5 to 10 g of dried *Stropharia cubensis*. On this basis it is approximately 30 times more potent than mescaline and about 0.01 as potent as LSD. Psilocin is 1.4 times as potent as psilocybin, but this is simply the ratio of their molecular masses. The psychic effects of psilocybin seem to be the same as those of LSD, although some users believe that psilocybin is more gentle and euphoric than LSD, inviting rather than compelling self-analysis. Psilocybin is noticeably shorter in its effects than LSD; the total experience lasts only 4 to 6 hours, whereas LSD lasts nearly twice as long.

Other Psychedelic Tryptamines

Psilocin and Psilocybin Analogues. The synthetic psychedelics formed by substituting two ethyl groups for the methyl groups of psilocin or psilocybin are known as *CZ-74* (4-hydroxy-*N,N*-diethyltryptamine) and *CEY-19* (the phosphate ester of CZ-74). Each is as potent as the corresponding natural compound, but the ethyl homologues are somewhat shorter acting. Both were employed in clinical studies in Europe during the 1960s. Here is the response of a young theology student to his fifth encounter with CZ-74:

> It is meaningful to say that I, John Robertson, ceased to exist, becoming immersed in the ground of Being, in Brahman, in God, in "nothingness," in Ultimate Reality, or in some similar religious symbol for oneness. . . . The feelings I experienced could best be described as cosmic tenderness, infinite love, penetrating peace, eternal blessing and unconditional acceptance on one hand, and on the other, as unspeakable awe, overflowing joy, primeval humility, inexpressible gratitude and boundless devotion. Yet all of these words are hopelessly inadequate and . . . it is misleading even to use the words "I experienced," as during the peak of the experience (which must have lasted at least an hour) there was no duality between myself and what I experienced. Rather I *was* these feelings, or ceased to be in them and felt no loss at the cessation.[113]

The N,N-Dialkyltryptamines. In 1954 *dimethyltryptamine* (DMT, **6-44**, Figure 6.9) was found with several other indole alkaloids in a variety of plants used in hallucinogenic snuffs or drinks by the indigenous peoples of the Amazon forest and the Caribbean: *Prestonia amazonica, Virola calophylla, Mimosa hostilis, Piptadenia peregrina, Banisteriopsis rusbyana*, and others.[114] Because the *N,N*-dimethyltryptamine structure is embedded in

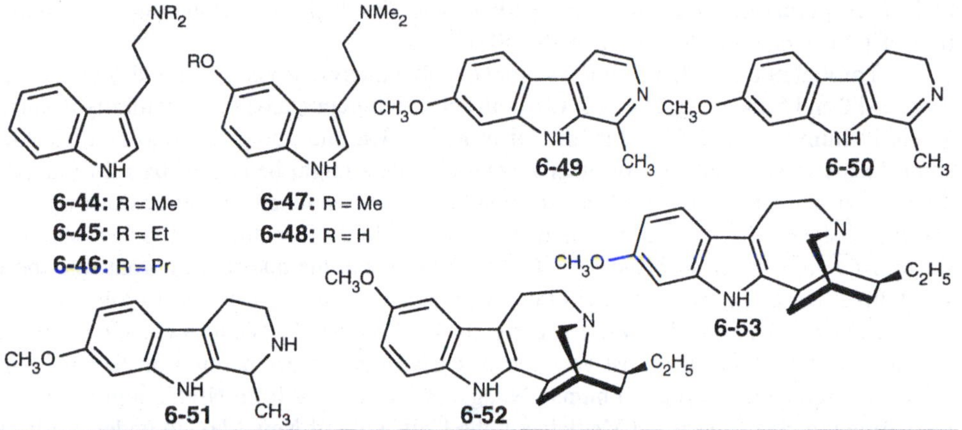

6-44: R = Me
6-45: R = Et
6-46: R = Pr
6-47: R = Me
6-48: R = H

Figure 6.9 Some tryptamine psychedelics; the harmala alkaloids.

both LSD and psilocybin, it was natural to investigate this relatively simple structure for hallucinogenic activity. Szára was the first to report testing DMT in humans, administering it by IM injection.[115] It proved to be inactive if taken orally because of its rapid degradation by MAO in the digestive tract (it is nonetheless active if taken as a snuff or as it occurs in the *Virola* plants—and in certain concoctions of tribal shamans using *Banisteria* plant materials—that contain harmaline and other MAO inhibitors). In IM doses of 30 mg or more the effects proved to be similar to LSD, and even more intense, but with an extremely prompt onset and proportionately brief duration: typical hallucinogenic activity begins in seconds, reaches a peak in 15 minutes, and is entirely resolved within less than an hour. Szára determined that DMT was metabolized to 6-hydroxy-DMT and excreted in this form in the urine; he hypothesized that some or all of the biological activity of the drug was due to this form—a plausible hypothesis in view of the known activity of psilocin, which is 4-hydroxy-DMT. But human subjects receiving 6-hydroxy-DMT, DMT, and placebo in a single-blind crossover study responded only to DMT.[116]

DMT experienced a brief popularity among the psychedelic devotees of the '60s as a smoked or insufflated[117] drug, being called "businessman's LSD" because of its capacity to provide a speedy, efficient, round-trip passage to the Beyond accessible during one's lunch break. But soon thereafter, DMT acquired the mythical reputation for producing more bad trips than LSD. Leary's group at Harvard, convinced that "psychedelic drugs had no specific effect on consciousness, except to expand it," and that "expectation, preparation, spiritual climate, and the emotional contract with the drug-giver, accounted for specific differences in reaction,"[118] administered 50 to 60 mg IM injections of the drug on one another and proved to their own satisfaction that DMT was like unto LSD in all things, duration of effect alone excepted.

Szára also prepared the diethyl (DET, **6-45**) and dipropyl (DPT, **6-46**) analogues of DMT.[119] These were found to have a milder and longer action than DMT—about 2 hours for the diethyl and 4 hours for the dipropyl derivative.[120] Richards and co-workers at the Maryland Psychiatric Research Center found they were able to use DPT in place of LSD in psychotherapy with cancer patients and obtain similar results (although reduc tion of pain seemed more lasting with LSD).[121]

Throughout the 1970s, there was considerable interest aroused by the discovery that both DMT and 5-methoxy-DMT (**6-47**), both active hallucinogens, occur naturally at small levels in human spinal fluid, blood, and urine.[122] The intriguing hypothesis arose that some forms of schizophrenia or other mental disorders might be caused by idiosyncratic production of or metabolism of either or both of these endogenous hallucinogens. But careful analysis of their concentration in the CNS fluid of controls and schizophrenics showed no significant differences.[123] On the other hand, the question of what role these substances actually play in normal brain chemistry remains unanswered to this day.

Strassman's DMT Studies. Recently, the FDA initiated a cautious change of course and for the first time in almost two decades gave researchers permission to study the effects of psychedelic drugs on human beings. With funding from NIDA, a group at the Department of Psychiatry and Medicine at the University of New Mexico under Dr. Rick Strassman administered DMT intravenously in doses from 0.05 to 0.4 mg/kg (about 3.5 to 28 mg) to 11 experienced hallucinogen users. IV rather than IM injection was chosen after a subject in preliminary testing who was familiar with smoked DMT complained that the "rush" from IM injection of 1.0 mg/kg was not equal to a "full dose" of smoked street drug; however, he expressed complete satisfaction with 0.4 mg/kg IV. Indeed, at the highest dose, the subjects were all "overwhelmed at the intensity and speed of onset" of DMT so administered; most lost awareness of their bodies and environment for a minute or two; all had vivid visual imagery that often overlaid even their eyes-open perceptions. Some of the imagery was quite striking: "a fantastic bird," "a little round creature with one big eye and one small eye," "beautiful, colorful pink cobwebs; an elongation of light." Some of the ideation was bizarre ("prankish and ornery elves," "alien beings"); some quasireligious ("the personal self and consciousness are just slowed down and less refined versions of 'pure consciousness'"). The subjects were monitored throughout for fundamental biological parameters such as blood pressure, heart rate, and body temperature, all of which rose in a dose-dependent manner, as well as for blood concentrations of β-endorphin, corticotropin, cortisol, and prolactin, which were also elevated.[124] The extreme intensity of the experience was matched by its brevity: it peaked at about 100 seconds after injection and was entirely over within 30 minutes. The same group later performed similar experiments with oral psilocybin, and developed a new Hallucinogen Rating Scale that might allow a more quantifiable comparison of the subjective effects of the many diverse drugs in the psychedelic family.[125]

Street Usage. There seems to be a resurgence of interest in DMT among the psychonaut population, inspired in part by the books of Terence McKenna.[126] What reports are available from users have in common a fairly bizarre claim to be communicating with disincarnate entities, usually referred to as "elves":

> Impression of basic colors, unmuted blue, yellow, and red, shimmering into being, depth imperceptible yet defined within the space, endlessly recurring back from/into the corner when, slowly, from around the edges they peer toward me, watching, eyes bright and watching in small faces, then small hands to pull themselves, slowly, from behind and into view; they are small white-blond imp-kids, very old in bright, mostly red, togs and caps; candy-store, shiny, teasing and inquisitive, very solemn and somewhat pleased (ah, here you are!) watching me as I meet their eyes bright and dark without any words (look!) . . .[127]

Many first-hand accounts describe a sense of terror at being close to death or dying, and one is solemnly advised that "yielding to the temptation to believe that one has died is not helpful when navigating psychedelic states since the resultant anxiety will usually distract one from a scientific observation of what is going on."[128] Another experienced users comments: "Almost everyone I've known who has used . . . DMT repeatedly, eventually encounters deeper fear than they've ever felt before."[129] Here are the effects of a "full-strength dose" on one user:

> The stuff hit me instantaneously. MILLIONS of brilliantly colored little "skull clowns" swarmed me in a most visionary way while emitting crickling, tinging sounds which looked like violet sparks coming out of their mouths. These tiny skull clowns were laughing most musically as I died in the light. Melt down—feels like drowning and being electrocuted at the same time. . . . The glowing, ember-like afterimage instantly swirls and shatters into blue and red sizzling domes that pinwheel ecstatically into a Creative, God-Thing with a trillion jeweled eyes that dissolves into an atomic ocean. This is the multi-eyed God that is my Creator, Master, Destroyer. I am nothing compared to this Thing which has no ego boundaries whatsoever. . . .[130]

The World Wide Web, source of much that is strange in today's world, tells us through one Mescalito Ted that DMT "is easy to obtain from licit sources,"[131] specifically several easily grown plants such as *Phalaris arundinacea* and *P. aquatica*. These are grasses, not very different in appearance from those in middle-America's front yards, that contain significant quantities of DMT that can be extracted and smoked.[132]

5-Methoxy-N,N-dimethyltryptamine. The 5-methoxy derivative of DMT (5-Methoxy-DMT, *O*-Methylbufotenine, **6-47**), is said to by some to resemble DMT or psilocybin in its effects. But anecdotally (there seem to be no controlled studies), there are significant differences. Thus, according to LordNose!,[133] DMT provokes an "initial overwhelming rush of colorful kaleidoscopic imagery with substantial ego loss. Auditory tones. Spectacular visions. Calm and centered psychedelic afterglow." But the same source says of 5-methoxy-DMT: "Immediate ego loss. No colorful imagery. Feeling of being overwhelmed by emotional impressions. Psychedelic afterglow slight or absent." Peter Stafford's *Psychedelics Encyclopedia* concurs: "Many people don't like it. . . . [One user] compares its effects to having an elephant sit on one's head."[134] 5-Methoxy-DMT occurs in many plant[135] (and some toad) species, often—as in *P. peregrina*—accompanied by DMT and bufotenine.

A Tale of Two Toads. Although the 5-hydroxytryptamine derivative, *bufotenine*, **6-48**, has been claimed to be a hallucinogen,[136] it probably does not even cross the blood-brain barrier, although it may produce a toxic delirium due to its violent cardiovascular effects in the peripheral system.[137] It was originally found in 1934 by Heinrich Wieland[138] in the venom of the parotid glands of toads (hence the name, from the Latin *bufo*, toad), and nearly 20 years later by his son Theodor Wieland in toadstools,[139] but it was only studied as a substance with possible psychotropic activity when it was isolated in 1954 from the seeds of *Piptadenia peregrina,* a flowering tree of the Amazon valley.[140] Columbus's expeditions observed these seeds used by the shamans of Haiti in a snuff called *cohoba.*

Not since *The Wind in the Willows* and Toad of Toad Hall have these gentle, warty amphibians been the recipients of such attention. It all may have started in the '60s, when the DEA outlawed bufotenine under the mistaken apprehension that it was hallucinogenic. This prompted a brief foray into licking toads, but the results were so unpleasant as to be self-limiting. Much later, in 1988, Darryl S. Inaba, director of drug programs at the Haight-Ashbury Free Medical Clinic in San Francisco, in a passing effort to rouse his audience from their dogmatic slumbers, mentioned this quaint '60s phenomenon of toad-licking during a lecture. Soon there were reports of rampant cane-toad licking in California.[141] The cane toad, *bufo marinus*, is a large amphibian that could be purchased in California pet stores. It excretes a venom containing bufotenine as well as several highly toxic cardiac glucosides; the venom has been known to cause convulsions and death simply from skin contact.

Meanwhile, on an independent track, serious ethnopharmacologists had been studying the venom of an alternate species of toad, the Colorado River toad, *Bufo alvarius*. This fellow looks a lot like his cousin, and his venom if ingested orally or even absorbed through the skin is just as lethal. However, his venom is chemically distinct in that it contains a rich supply of *N,N*-dimethyl-5-methoxytryptamine, **6-47**. If the venom of *B. alvarius* is dried and smoked, the toxic ingredients are destroyed and the smoker is subject to a powerful hallucinogenic experience. In 1984, a pamphlet was published by Venom Press under the authorship of one Al Most (?) entitled *Bufo alvarius: The Psychedelic Toad of the Sonoran Desert*. Andrew Weil of the University of Arizona and Wade Davis of the New York Botanical Garden suggest that a snuff or smoking material composed of this toad's venom may lie behind the numerous iconographic representations of toads found in many Mayan and Olmec sites. Following Al Most's prescriptions, they performed a series of self-experiments smoking the dried venom of *B. alvarius* and experienced an intense, short-lived intoxication "marked by auditory and visual hallucinations," commencing within 15 seconds of a single deep inhalation. Bizarre as all this may seem, it is actually of some significance for Mesoamerican archeology and anthropology, since this is probably the unique instance of the use of an animal (as opposed to a plant) toxin in aboriginal ritual for its psychoactive effects.

Harmine and Harmaline. Thousands of South American Indians in Columbia, Ecuador, and Brazil—and some middle-class Chileans—use a hallucinogen elaborated from jungle plants of the *Banisteriopsis* family. The drink prepared from the bark of these plants is variously called *caapi, ayahuasca* (a Quechua term meaning "vine of the soul"), *yagé, hoasca, natema, vegetal, daime*, and *pinde*. The chief alkaloid responsible for its effects is harmine, **6-49**; but harmaline (dihydroharmine), **6-50**, tetrahydroharmine, **6-51**, and DMT (which is orally active because of the MAO-A inhibitory action of harmine) are also present. There is a great deal of myth with regard to the effects of this drink: that it endows the user with telepathic powers (the drink was once called *telepathine*), that it induces specific archetypical visions of jaguars, eagles, and so forth.[142] Here are some first-hand accounts from William S. Burroughs, writing to Allen Ginsberg from South America in 1953:

> [April 12] I have taken it three times. (1st time came near dying.) . . . . A large dose of Yage is sheer horror. I was completely delirious for four hours and vomiting at 10-minute intervals. As to telepathy, I don't know. All I received were waves of nausea. The old bastard

who prepared this potion specializes in poisoning gringos who turn up and want Yage. . . . I later took smaller doses. Very similar to weed including aphrodisiac results. . . .

[June 18] Hold the presses! Everything I wrote about Yage subject to revision in the light of subsequent experience. It is *not* like weed, nor anything else I have ever experienced. . . . I experienced first a feeling of serene wisdom so that I was quite content to sit there indefinitely. What followed was indescribable. It was like possession by a *blue spirit.* (I could *paint* it if I could paint.) Blue purple. And definitely South Pacific, like Easter Island or Maori designs, a blue substance throughout my body, and an archaic grinning face. At the same time a tremendous sexual charge, but *heterosex.* This was not in any way unpleasant, but shortly I felt my jaws clamping tight, and convulsive tremors in arms and legs, and thought it prudent to take phenobarbital and Codeine.[143]

The barbiturate, according to Burroughs, immediately interrupted all the effects of the yage, and he later advises that one should never take the drug without ample supplies of a barbiturate at hand to treat the convulsions that are, he says, likely to occur.[144] At the same time he describes a rapid tolerance that develops to these physical effects of the drug. All in all, it is hard to tell where the rich imagination of Burroughs stops and the drug effects begin, but extreme suggestibility is in any case characteristic enough of a psychedelic state.

According to one scientific study, the yage tea as it is concocted in Brazil by the syncretic religious movement União do Vegetal (UDV), the psychedelic activity is entirely due to the DMT from the leaves of *Psychotria viridis* which are brewed with *B. caapi.* The harmala alkaloids are "essentially devoid of psychedelic activity at doses obtained from the tea (1 to 3 mg/kg),"[145] but the harmine at this dosage provides reversible inhibition of MAO-A, rendering the DMT orally active and greatly prolonging and intensifying its effects.

Ibogaine. In the west African sub-Sahara, several tribes (Fang, MPongwe) utilize the root bark of *Tabernanthe iboga*, called *eboka*, during healing rituals or all-night ceremonies in which the spirits of the ancestors are encountered and in which communicants experience the passage over to the afterlife.[146] The alkaloids in eboka constitute a family of similar structures including ibogaine, **6-52**, and tabernanthine (**6-53**). Of these, only ibogaine seems to be psychoactive. The Mexican psychiatrist Claudio Naranjo has used this drug extensively in psychotherapy.[147] It is hallucinogenic in oral doses of about 300 mg and has a duration of action of about 10 hours.

There has been considerable interest aroused in the use of ibogaine (a Schedule I drug) to cure drug addiction and alcoholism. At this time most reports are anecdotal; but they appear to be so genuine that they have attracted the attention of workers at NIDA's Addiction Research Center in Baltimore, and the University of Miami has received FDA approval to conduct limited human trials. The central figure behind the claim that ibogaine cures addiction is Howard Lotsof, who holds several domestic and foreign patents for this use of ibogaine (as well as tabernanthine and ibogamine) under the trade name Endabuse.[148] His discovery occurred in 1962, when he was a 19-year-old college dropout and heroin addict who had experimented with LSD and mescaline. Then he tried ibogaine:

"The first stage was like watching a slide show of your life coming on at a thousand frames a second," Mr. Lotsof recalled. "No euphoria, all intellectual and distant." He said the next

> phase was like "eight to ten years of psychoanalysis" compressed into a few hours. When it was over, he was exhausted. But there was a final jolt: Although he had gone almost two days without heroin, he didn't crave it.
>
> "Ibogaine immediately interrupted my chemical dependency," he said. Curious to see how others would react, he gave the drug to 20 friends, seven of them addicts. Five of the addicts stopped doing drugs, he said. . . .
>
> In 1987, Mr. Lotsof and an old friend set up an underground railroad that has ushered 30 American addicts to Holland, a country whose relaxed attitudes toward drugs meant ibogaine treatments could go on quietly. Treatment, in apartments and hotel rooms, was supervised by psychiatrists in some cases, by lay workers in others.[149]

At the Johns Hopkins University School of Medicine, Mark Molliver and Elizabeth O'Hearn have found that ibogaine in large doses destroys part of the rat brain that may be implicated in repetitive tasks or addiction. Stanley Glick at the State University of New York, Albany College of Medicine has claimed that addicted rats stopped pushing a lever to deliver morphine after being given ibogaine. "In several cases, it almost eliminated and usually decreased their drug intake. . . . It did so for at least a day after administration of ibogaine, and in some rats it did so for several days or weeks. I can't say it totally halted it."[150] It is possible that a metabolite of ibogaine, 12-hydroxyibogamine, is responsible for some of ibogaine's effects;[151] there is also evidence that ibogaine and this metabolite are active at NMDA receptors.[152]

Mescaline and the Phenethylamines

The Peyote and San Pedro Cacti

Peyote. In 1896, Heffter for the first time isolated the active principle, mescaline, from the peyote (Aztec *peyotl*) cactus.[153] (The cactus is now botanically named *Lophophora williamsii*, but earlier references call it *Anhalonium Lewinii.*) However, the use of this cactus by the indigenous peoples of what is now Mexico and the southwestern United States precedes recorded history. The cactus grows in a quite limited ecological niche within Texas along the Rio Grande from approximately Laredo to Rio Grande City and southward about 100 miles into Mexico, mostly within the Chihuahuan Desert. Within this niche is a narrow region in Texas where the cactus is grown commercially and legally to supply members of the Native American Church (NAC) with sacramental peyote. Until recently, it has not been successfully cultivated outside these areas.[154] Peyote is a pale green plant about 2 inches across which grows in clusters close to the ground from a long taproot. It is harvested by cutting off the top, which is called the button; if this is done carefully, the roots will then produce a cluster of new buttons.

Longstanding confusion exists from the incorrect use of the word *mescal* to describe the peyote cactus. In Mexican Spanish the word is used only to refer to the *agave tequilana* plant, called the century plant in English, which can be fermented and distilled to produce tequila, the liquor used in margaritas. The century plant grows throughout Central America. Still a third plant, an evergreen bush, *Sophora secundiflora*, grows in almost exactly the same small region as the peyote cactus, and produces woody pods in which there are bright red hard beans called *frijolillo* by the Mexicans. The frijolillos are usually used as a decoration, and a necklace of the beans is often worn during peyote meetings. The beans are very poisonous: half of a bean causes 48 hours of delirium and stupor, and an entire bean is lethal. Although there has been speculation about an ancient ritual use of this bean among the Amerindians, it is never deliberately consumed by them now and it is questionable if it ever was. Perhaps because all three of these plants were known to produce some sort of intoxication, English-speaking settlers managed to call them all mescal; the peyote buttons were called mescal buttons; and the most active alkaloid in peyote, 3,4,5-trimethoxyphenethylamine, received the name mescaline. But neither in Spanish nor in any of the indigenous Amerindian languages is the peyote cactus ever called mescal.[155]

The San Pedro Cactus. The *curanderos* of the Peruvian peasantry use an aqueous extract of the San Pedro cactus, *Trichocereus pachanoi*, which is known to contain mescaline,[156] in the practice of their shamanic healing and spiritual guidance. Use of this cactus is thought to date back at least 3,000 years, judging from its portrayal with jaguars and spirit beings on ceremonial textiles and ritual burial ceramics from the Chavín period.[157] The cactus grows in many areas of Ecuador, Peru, and Bolivia; but its use, though widespread, has not been studied as thoroughly as peyote.

Peyote in Native Religion

At the time of the Spanish conquest of Mexico, peyote was known, traded, and used far beyond its natural habitat. In 1620, the Holy Inquisition felt obliged to issue an edict against its use:

> We, the Inquisitors against heretical perversity and apostasy in the City of Mexico, states and provinces of New Spain, New Galicia, Guatemala, Nicaragua, Yucatan, . . by virtue of apostolic authority . . . .
>
> Inasmuch as the use of the herb or root called peyote has been introduced into these Provinces for the purpose of detecting thefts, of divining other happenings, and of foretelling future events, it is an act of superstition condemned as opposed to the purity and integrity of our Holy Catholic Faith. . . .
>
> Said abuse has increased in strength. . . . We order that henceforth no person of whatever rank or social condition can or may make use of the said herb, peyote, nor any other kind under any name or appearance for the same or similar purposes, nor shall he make the Indians or any other person take them, with the further warning that disobedience to these decrees shall cause us . . . to take action against such disobedient and recalcitrant persons as we would against those suspected of heresy to our Holy Catholic Faith.[158]

The splendidly vague inclusion of "any other kind under any name or appearance for the same or similar purposes" makes this the first designer-drug law to be promulgated in the Western Hemisphere; and the threat of an *auto-da-fé* would satisfy the punitive spirit of the most zealous drug abolitionist. In pursuit of a drug-free empire, the Holy Office carried out nearly a hundred hearings over the next 265 years, the most distant from the natural growth area of peyote being Santa Fe, in what is now New Mexico, and Manila in the Philippines. Most of these trials concerned cases of peyote use by individuals for visions, charms, or healing; but some communal ceremonies were observed, and these involved all-night dancing and singing with the participants arranged in a circle. Some of the surviving peyote rituals retain this feature.

The Huichol. Over the centuries since the time of the conquistadors, the imposition of Catholicism and Spanish culture on the aboriginal tribes of Mexico led to the cessation of any use of peyote for ceremonial use except by a few tribes, chief amongst them the Huichol and the Tarahumara, living in the western mountains of the high Sierra Madre Occidental. These tribes probably were originally located in the southeastern Chihuahua area where peyote abounds, but withdrew into the mountains as Spanish colonization advanced. For more than a century, members of the Huichol people have made yearly religious pilgrimages of over 250 miles to harvest *híkuri* (peyote) in the sacred fields of *Wirikúta* near San Luis Potosi: here grows the híkuri, and here live the *kakauyaríxi*, the spirits of the Huichol ancestors.[159] The layers of symbolism are complex: peyote is the Elder Brother, the Deer, Maize (reflecting a transition from hunting-gathering to cultivation). A description follows of the "hunting" (harvesting) of the peyote after the land of Wirikúta is reached:

> Lupe . . . almost at once discovered a thicket of cactus and mesquite so rich in peyote that in a couple of hours she had filled her tall collecting basket. Occasionally she would stop to admire and speak quietly to an especially beautiful *híkuri* and to touch it to her forehead, face, throat, and heart before adding it to the others. We also saw people exchanging gifts of peyote. This seemed to us a very beautiful aspect of the pilgrimage. No ceremony in which peyote was eaten communally went by without this kind of ritual exchange, in which each participant is expected to share his peyote with every companion. . . . Sometimes an older participant would place his gift directly into the mouth of a younger one, urging him to "chew well, younger brother, chew well, so that you will see your life."
>
> The night was passed in singing and dancing around the ceremonial fire, chewing peyote in astounding quantities, and listening to the ancient stories. Considering the lack of food, the long days on the road, the bitterly cold nights with little sleep (by now, Ramón had not closed his eyes to sleep for six days and nights!), and above all the high emotional pitch of the sacred drama, . . . one might have expected them to . . . lapse into a dream state induced by the considerable quantities of *híkuri* they had already consumed. . . . But most . . . were wide awake, in varying states of exaltation, supremely happy and possessed of seemingly boundless energy. If the dancing and singing stopped it was only because Ramón laid down his fiddle to commune quietly with the ceremonial fire or to chant the stories of the first pilgrimage. Neither he nor Lupe was ever without a piece of peyote in the mouth. Yet they were never out of control—indeed none of them was—and neither they nor any of the others, little Francisco [a ten-year-old] included, showed the slightest adverse effect, then or later. . . .
>
> The Huichol regard their peyote experiences as private and do not, as a rule, discuss them with anyone, except in the most general terms ("there were many beautiful colors," "I saw maize in brilliant colors, much maize," or simply, "I saw my life").[160]

There is little in all of this that, to the external observer, would seem to involve any drug usage or drug effects. This is as true of today's NAC services. Jesuit priest Paul Steinmetz, who worked among the Oglala Lakota on the Pine Ridge Reservation in South Dakota for 20 years, writes of the many NAC peyote services he attended: "I have been unable to observe any significant change in outward behavior other than an added fervor in praying and singing. . . . This is not to say that Peyote does not facilitate visions but rather that it is only one influence in a total religious setting."[161]

Peyotism in the Native American Church. *Origins.* A large proportion of the Indian population of the United States and Canada participates in peyote ceremonies (although they may also be members of traditional Christian denominations as well); they are loosely represented in this by the NAC. Several factors combined to bring about this intertribal pan-Indian movement in the last half of the 19th century. The U.S. government as early as 1825 began deporting all Indian tribes east of the Mississippi (most of which were village and farming communities) to a newly established "Indian Territory," which included much of present Oklahoma and Kansas. Following the end of the Civil War, the remaining nomadic Western Plains Indians were subjugated in a series of conflicts resulting from the rapid expansion of the white population westward. The government attempted to force these tribes to join the others within a single Indian Territory, and made considerable efforts to teach English on these reservations; Christian missionary efforts were likewise encouraged. As a result, several dozen different linguistic and cultural groups found themselves with a crudely learned but common language and religion, and a shared experience of enforced displacement and oppression. Peyotism probably spread from the Lipan Apache Indians to the other tribes (among other evidence, the music sung at the peyote meetings is Lipan Apache in origin); when the first published reference to its use occurs in 1886, it is said to be widespread on the Oklahoma reservations. By then, the Indians were considered "pacified" in the territories, and they were allowed to use a railroad link through Laredo to Mexico to transport peyote to the reservations.

The Ghost Dance. With a common language and several decades of interaction, a pan-Indian religious movement could arise for the first time. In 1889, the Ghost Dance spread rapidly through the reservations. The Ghost Dance originated in the syncretic teachings of Wovoka, a Northern Paiute whose teachings included an ancient Northwestern Plateau Indian belief in a periodic world renewal that could be brought about by group dancing, combined with Christian teachings of the second coming of Jesus. "Faithful dancing, clean living, peaceful adjustment with whites, hard work, and following God's chosen leaders would hasten the resurrection of dead relatives"[162] and world renewal; the world renewal included a return to the happy days gone by when the Indians roamed the plains in freedom—and included an unspecified but presumably nonviolent disappearance of the intruding whites by God's action.

The Ghost Dance was spread by frequent revival meetings that attracted large crowds of fervent worshippers on the reservations; in South Dakota the army overreacted on the Sioux reservation, and the massacre at Wounded Knee resulted. In Oklahoma the army pursued a different policy of noninterference; the movement ran its course and was over in a few years. But it helped to spread the peyote religion, which gradually replaced it as the only pan-Indian religion, retaining however many of the Christian elements in the Ghost Dance.

The Peyote Ceremony. In a series of writings in the 1890s, James Mooney, a reporter for the Richmond, IN, newspaper *Palladium* and a student of Indian customs, was hired by the Smithsonian Institution as an ethnologist. He spent most of the next decade traveling throughout the reservations observing in particular the peyote religious rituals.[163]

The anthropologist Omer Steward has made an extensive study of peyote ceremonies as practiced today. In his view the rituals have changed little in the intervening century since Mooney's observations except for the inclusion of women as direct participants—although men are usually the leaders, called "roadmen." Furthermore, "the peyote ritual in its two variations—Cross Fire and Half Moon—is everywhere the same throughout the United States and Canada. . . . Many of the differences between the two sects are related to smoking. Besides not using tobacco in the peyote meeting, the Cross Fire ritual is distinctive by greater reference to the [Christian] Bible, by displaying a Bible in the peyote meeting, and by ending the all-night ceremony with a sermon based on a text taken from the Bible."[164] Here is a description of a typical service:

> It generally begins about nine o'clock on a Saturday night. . . . The peyotists sit on the floor in a circle, the roadman facing east, the chief drummer to his right and the cedarman to his left. The ceremony begins when the roadman takes from a special case his ceremonial paraphernalia consisting of a staff, usually carved or ornamented with beads and dyed horsehair, . . . a decorated rattle; a special fan, probably of eagle feathers; and an eagle bone or reed whistle. He also produces a large peyote button to be the Chief Peyote, which he places on the moon altar. He also brings forth a sack of peyote buttons and usually a container of peyote tea to be consumed during the night. . . .
>
> The sacks of peyote and tea are then passed clockwise around the circle, and each partakes either of tea or four buttons. The . . . first four peyote buttons eaten, the roadman, holding the staff and fan in one hand and vigorously shaking the rattle with the other, accompanied by the quick beat of the chief drummer, sings the ceremonial Opening Song, "Na he he he na yo witsi nai yo," syllables which go back to the Lipan Apache, to the songs of the first U.S. peyotists. He sings the hymn four times. He sings three more songs four times, and then the staff, fan, and rattle are passed to the next participant and the drum to that person's next neighbor. The person receiving the staff, fan, and rattle then sings four songs four times, and passes it to the next participant. . . . The singing continues in this way only interrupted a few times when the peyote is circulated again [or when]. . . . individuals . . . request a prayer cigarette from the roadman, and the singing will stop while all listen to the prayer. These prayers are sometimes confessionals, sometimes testimonials, sometimes supplications for help and guidance. This is also the time for a special curing ceremony, if such has been planned.
>
> As the first rays of sunrise appear, the roadman sings another special song. . . . This is followed by the entrance of the peyote woman, usually the wife of the host, who brings in water which is again blessed and circulated. She also brings the ceremonial breakfast of corn, meat, and fruit, which is circulated, with each person taking a bit of each food. Following the ceremonial breakfast, the roadman may give a little talk. If this has been a Cross Fire service, the roadman may read a text from the Bible and interpret its meaning to the congregation. In full daylight the roadman sings the last of the ceremonially determined songs, the Quitting Song. . . . It is then that the meeting is over and all go outside to "welcome the sun."[165]

The Legal Status of Sacramental Peyote Use. From the beginning of peyote use on the reservations, there was opposition. Dozens of arrests and confiscations were made under an 1897 federal law prohibiting the provision of intoxicating liquors to Indians on the reservations. In a trial of Harry Black Bear of Pine Ridge Reservation in 1916, chemist E. B. Putt of the North Dakota Agricultural College testified that he had eaten peyote under a doctor's observation and found it harmful and dangerous. The jury convicted Black Bear, but the judge eventually granted a motion of dismissal, ruling that the intent of the law was to control alcoholic beverages only. From 1916 to 1918, several attempts were made in the U.S. Congress to pass a bill outlawing peyote; testifying in favor of its passage were several missionaries and the Utah state chemist, Herman Harms, whose five-page report concluded that peyote "produces a demoralizing, harmful, and depraved condition." The Omaha Indians, aided by interpreter and fellow native American Francis La Flesche, an ethnologist of the Smithsonian, testified in person in Washington, arguing that the practice of peyotism helped them resist using liquor. La Flesche had never used peyote but said he felt only gratitude for its introduction. He described the first years after the restriction of the Omahas to their reservations as "a continuous drunken orgy," with nearly complete disintegration of family structures. The Indian Office, he said, did nothing about the situation.

> But suddenly there came a lull in all this drunkenness and lawlessness. I had a sister who was a physician, . . . and she wrote me regularly about the conditions of the Omaha people. . . . "A strange thing has happened among the Omahas. They have quit drinking, and they have taken to a new religion, and members of that new religion say that they will not drink; and the extraordinary part of the thing is that these people pray, and they pray intelligently, they pray to God, they pray to Jesus. . ."
>
> She regarded that as something very strange, because the Indian, although they had missionaries for many years, could not understand the white man's religion. . . . But the teaching of this new religion was something they could understand. . . . This peyote, they said, helped them not only to stop drinking, but it also helped them to think intelligently of God and of their relations to Him. At meetings of this new religion is taught the avoidance of stealing, lying, drunkenness, adultery, assaults, the making of false and evil reports against neighbors. People are taught to be kind and loving to one another and particularly to the little ones.[166]

James Mooney also strongly argued against the bill, which was passed by the House but failed in the Senate. He urged the Indians to incorporate their religion as the Native American Church; in a countermove, the Bureau of Indian Affairs succeeded in banning Mooney from the reservations.

By 1930, almost every western state had passed a law prohibiting peyote; however, since state laws had no jurisdiction on reservation lands, the laws had little effect on American Indian religious practices. With the New Deal, Harold Ickes was appointed Roosevelt's secretary of the interior, and a more enlightened Indian Bureau now declared itself strongly opposed to any interference with Indian ritual or religion.[167] But in 1965, the U.S. Drug Abuse Control Act included peyote in its list of prohibited narcotics, probably because hippies had been invading the Texas peyote gardens. Nonetheless, the courts tended generally to find members of the NAC exempt from federal or state laws against peyote, subjecting such laws to the strict scrutiny of a government's need to show a compelling state interest to warrant interference with First Amendment religious liberties. In 1978, the American Indian Religious Freedom Act (42 USC 1996, P.L. 95-341) gave legal protection to the practice of peyotism by American Indians. However, although this act protected bona fide members of the NAC from *federal* prosecution for the sacramental use of peyote, it did not necessarily protect them from *state* prosecution.[168]

In 1990, the U.S. Supreme Court made its final ruling in *Employment Division v. Smith*, a case with a long history of going up and down the appellate ladder for several years between the U.S. and Oregon Supreme Courts:

> Simply stated, the case was brought by American Indian substance abuse counselors who were dismissed for ingesting peyote at a worship service of the Native American Church. The counselors were denied unemployment compensation and related benefits and they sued the Oregon Employment Division of the Department of Human Services.... The Oregon Supreme Court held that, while the state statute that specifies controlled substances includes peyote and makes no exception for sacramental use, the prohibition against peyote's sacramental use by adult members of the Native American Church would violate the First Amendment. The most recent U.S. Supreme Court decision now reverses this ruling and holds that regulating the sacramental use of peyote by Church members does not offend the First Amendment's protection of religious freedom....
>
> The Supreme Court has served notice that religious practices that run counter to the criminal law may not enjoy the First Amendment protections of religious liberty.
>
> Second, this decision counters the substantial trend of state courts, which have allowed sacramental peyote use. In reaching the *Smith* decision, the Supreme Court did not require Oregon to prove that its prohibition of peyote was a "compelling state interest" that should overshadow the Church's religious liberty....
>
> The thrust of the ruling was a "states' rights" determination that allows states to criminalize activity they deem too hazardous, even if that activity is sacramental to a few of its citizens.[169]

This decision drew a reaction from the U.S. Congress. Senators Kennedy and Hatch cosponsored a bill, the Religious Freedom Restoration Act of 1993 [S.578], which permitted federal or state governments to "burden a person's exercise of religion . . . only if it demonstrates that application of the burden to the person—(1) is in furtherance of a compelling governmental interest; and (2) is the least restrictive means of furthering that compelling governmental interest" (Section 3). In introducing the measure, Kennedy said that the *Smith* decision had created fears that "dry communities could ban the use of wine in communion services, Government meat inspectors could require changes in the preparation of kosher food, and school boards could force children to attend sex education classes" contrary to the dictates of their religion.[170] The bill overturned the Supreme Court's decision by using the authority granted Congress by the 14th Amendment, a tactic the court has yielded to in the past. The bill was passed by both houses of Congress and was signed into law by President Clinton in late 1993.[171]

The members of the NAC are now afforded full legal protection in every state, but are ironically faced with a more practical issue, how to obtain supplies of peyote. As previously mentioned, the ecological niche for this little cactus in the United States is limited to a small region in Texas along the border with Mexico east of Laredo and south to McAllen. This area was once unfenced ranch land, but recently was converted to pasture in a process known as root-plowing, where all vegetation is periodically scraped up. One might suppose the problem could be overcome by importation from Mexico, where peyote grows "like apples, three or four times the size it grows in Texas."[172] But peyote remains a Schedule I substance, and importation is forbidden by federal law.

The Alkaloids of Lophophora Williamsii

Heffter was the first to succeed in isolating any of the alkaloids contained in peyote. In 1896, he determined the molecular formula for the most abundant of them, mescaline (**6-**

54, Figure 6.10), as well as for *pellotine*, **6-58**.[173] Proof for the structure of mescaline and its synthesis was first achieved by Ernst Späth in 1919.[174] In 1935, Späth and Becke showed that in addition to pellotine and mescaline, peyote contained *anhalamine*, **6-55**; *anhalonidine*, **6-56**; *anhalonine*, **6-59**; and *lophophorine*, **6-60**.[175] In more recent times, *peyonine*, **6-57**, a related carboxylic acid, has also been found in peyote.[176] The pharmacological or psychic effects of these substances is largely unknown, but some of them may account for the nausea and vomiting experienced by many novice users of peyote (a problem only rarely experienced by those using pure mescaline in a clinical setting). Additionally, many users share William Burroughs' conviction (yet to be confirmed in controlled studies) that "peyote intoxication causes a peculiar vegetable consciousness or identification with the plant. Everything looks like a peyote plant. It is easy to understand why the Indians believe there is a resident spirit in the peyote cactus."[177]

Uses of Mescaline

Psychotomimesis. Mescaline has been used in psychiatry for most of the purposes to which LSD and psilocybin have been put, but generally less frequently because of its inconveniently long duration of action. In a recent study, one of the first to employ psychedelics on human subjects in many years, 500 mg of mescaline administered orally to twelve adult normal males was said to produce an acute psychotic state 3.5 to 4 hours after consumption as measured by the Brief Psychiatric Rating Scale and the Paranoid Depression Scale; the results of single-photon emission functional brain imaging were more ambiguous. The authors speculate that despite the known differences (visual pseudohallucinations rather than true auditory hallucinations, and residual presence of an insightful ego) the similarities between mescaline-induced schizophreniform states and endogenous schizophrenia may quite possibly point to similar pathogenetic pathways.[178]

Mysticomimesis. As the active ingredient of one of the oldest hagiobotanicals and as the first psychedelic to be synthetically available in pure form, mescaline has enjoyed considerable use by the cognoscenti before and even after the discovery of LSD. Here is Aldous Huxley, writing to Humphry Osmond in October of 1955:

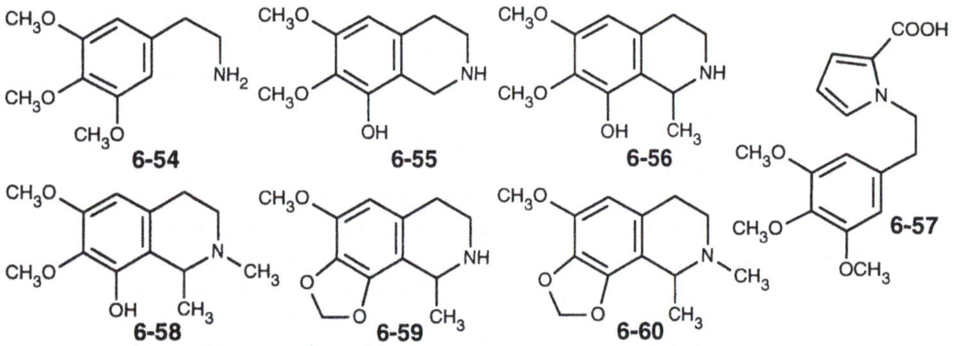

Figure 6.10 Alkaloids found in peyote *(L. Williamsii)*.

> I had another most extraordinary experience with mescaline the other day . . . . I took half the contents of a 400 mg capsule at ten and the other half about forty minutes later, and the effects began to be strong about an hour and a half after the first dose. There was little vision with the eyes closed, as was the case during my experiment under your auspices . . . . Instead there was something of incomparably greater importance; for what came through the opened door was the realization—not the knowledge, for this wasn't verbal or abstract—but the direct, total awareness, from the inside, so to say, of Love as the primary and fundamental cosmic fact. The words, of course, have a kind of indecency and must necessarily ring false, seem like twaddle. But the fact remains. (It was the same fact, evidently, as that which the Indians discover in their peyote ceremonies.) I was this fact; or perhaps it would be more accurate to say that this fact occupied the place where I had been. . . .
>
> Among the by-products of this state of being in the given fact of love was a kind of intuitive understanding of other people, a "discernment of spirits" in the language of Christian spirituality. . . . I now understood such previously incomprehensible events as St. Francis's kissing of the leper. Explanations in terms of masochistic perversion etc. are merely ridiculous. This sort of thing is merely the overflow of a cosmic fact too large, so to speak, for the receptacle. . . Another thing I remember saying and feeling was that I didn't think I should mind dying; for dying must be like this passage from the known (constituted by life-long habits of subject-object existence) to the unknown cosmic fact. . . .
>
> What emerges as a general conclusion is the confirmation of the fact that mescaline does genuinely open the door, and that everything including the Unknown in its purest, most comprehensive form can come through. After the theophany it is up to the momentarily enlightened individual to "cooperate with grace"—not so much by will as by awareness.[179]

And here is the distinguished Huston Smith, first humanities professor at MIT and author of *The Religions of Man*, describing his experience of mescaline in Timothy Leary's living room on New Year's Day, 1961:

> I was experiencing the metaphysical theory known as emanationism, in which, beginning with the clear, unbroken and infinite light of God or the void, the light then breaks into forms and decreases in intensity as it diffuses through descending degrees of reality. My friends in the study were functioning in an intelligible wave band, but one which was far more restricted, cramped and wooden than the bands I was now privileged to experience. Bergson's notion of the brain as a reducing valve seemed accurate. . . . The emanation theory and elaborately delineated layers of Indian cosmology and psychology had hitherto been concepts and inferences. Now they were objects of direct, immediate perception. I saw that theories such as these were required by the experience I was having. . . . But beyond accounting for the origin of these philosophies, my experience supported their truth. As in Plato's myth of the cave, what I was now seeing struck me with the force of the sun in comparison with which normal experience was flickering shadows on the wall. How could these layers upon layers, these worlds within worlds, these paradoxes in which I could be both myself *and* my world . . . be put into words? I realized how utterly impossible it would be for me to describe them on the morrow or even right then to Tim or Eleanor. . . .
>
> It should not be taken from what I have written that the experience was pleasurable. The accurate words are significance and terror—or awe, in Rudolf Otto's understanding of a peculiar blend of fear and fascination. The experience was positive in that it unfolded range upon range of reality I hadn't known existed. Whence, then, the terror? In part, from my sense of the utter freedom of the psyche and its dominion over the body. . . . I had the sense that the body could function only if my spirit chose to return to it, infuse it and animate it. Should it choose to

return? There seemed to be no clear reason to do so. Moreover, *could* it return if it chose? Can man see God and live, or is the vision too much for the body to stand—like plugging a toaster into a trunk power line. . . ?

Later . . . I said to Tim, "I trust you know what you are playing around with here. . . . It looks to me like you are taking an awful chance in these experiments. . . . There is such a thing as people being frightened to death. I feel like I'm in an operating room, having barely squeaked through an ordeal in which for two hours my life hung in the balance."[180]

Poetomimesis. Havelock Ellis was a late Victorian physician best known for his path-breaking (and initially banned) *Studies in the Psychology of Sex.* He was also (like Freud) interested in the phenomena of dreaming, and this led him in the 1890s to the study of peyote. In 1896 he published his observations of the effects of peyote on himself and on his friend William Butler Yeats[181] who provided the following description of his experience under its influence:

I have never seen a succession of absolutely pictorial visions with such precision and such unaccountability. . . . For instance, I saw the most delightful dragons, puffing out their breath straight in front of them like rigid lines of steam, and balancing white balls at the end of their breath! When I tried to fix my mind on real things, I could generally call them up, but always with some inexplicable change. Thus, I called up a particular monument in Westminster Abbey, but in front of it, to the left, knelt a figure in Florentine costume, like some one out of a picture of Botticelli; and I *could not* see the tomb without also seeing this figure. . . . I played the piano with closed eyes, and got waves and lines of pure color, almost always without form, though I saw one or two appearances which might have been shields or breastplates—pure gold, studded with small jewels in intricate patterns. All the time I had no unpleasant feelings whatever, except a very slight headache, which came and went. I slept soundly and without dreams.[182]

Some Noted Failures. William James, whose interest in the psychic and religious effects of drugs extended to numerous instances of self-experimentation (see Chapter 7) was either the victim of a "bad trip" or perhaps cacti that had deteriorated in transit. We will unfortunately never know. Perhaps it was the setting—William's son, Henry, describes the circumstances as "gratuitously heroic," not to say outright dangerous:

William James and his wife and the youngest child were alone in the Chocorua cottage a few days, picnicking by themselves without any servant. They had no horse; at that season of the year hours often went by without any one passing the house; there was no telephone, no neighbor within a mile, no good doctor within eighteen miles. It was quite characteristic of James that he should think such conditions ideal for testing an unknown drug on himself. There would be no interruptions. He had no fear. He was impatient to satisfy his curiosity about the promised hallucinations of color. But the effects of one dose were, for a while, much more alarming than his letter [to James' brother Henry, below] would give one to understand.[183]

Perhaps he suffered what we would describe as a panic attack. Henry *fils* suggests that James was much more "alarmed" than his nonchalant commentary to Henry *frère* (whom he addresses as "Dear Heinrich") conveys:

I had two days spoiled by a psychological Experiment with *mescal*, an intoxicant used by some of our S. Western indians in there [sic] religious ceremonies, a sort of cactus bud, of which the U.S. government had distributed a supply to certain medical men including Weir Mitchell who sent me some to try. He had himself been "in fairyland." It gives the most glorious visions of

> colour—every object thought of appears in a jeweled splendor unknown to the natural world. It disturbs the stomach somewhat but that, according to W.M., was a cheap price, etc. I took one bud 3 days ago, vomited and spattered for 24 hours and had no other symptom whatever except that and the Katzenjammer the following day. I will take the visions on trust.[184]

Another prominent soul upon whom the powers of Brother *híkuri* proved inefficacious—or at least unrewarding—was Carl Djerassi, who tested some mescaline sulfate on himself at a pizza party attended by his graduate students. He found that "the flowers did not look any different, with eyes closed or open, nor were the colors any more intense than in my previous, drugless life. . . . I felt only the nausea that crested in my stomach every time one of my pepperoni-smelling students asked, 'What's it like?'"[185] He concludes somewhat ambivalently:

> Even as I, the lifelong teetotaler, behaved like a drunk, another wave of the mescaline experience descended upon me. The "real" me seemed to be sitting in one corner of the room observing, coolly, the spectacle of the other me acting without inhibition. Aldous Huxley quoted this line in William Blake's *The Marriage of Heaven and Hell:* "If the doors of perception were cleansed every thing would appear to man as it is, infinite." I reached a different conclusion: it takes more than mescaline, or indeed chemistry, to change a man's persona.[186]

Psychopharmacology of Mescaline

Mescaline is a relatively weak psychedelic as measured by the minimum dosage, about 200 mg of the sulfate, needed to produce a psychic response. Nonetheless, it has been used as a standard for comparison of psychedelic activity, an effective acute dose of 350 mg of mescaline sulfate being considered a "mescaline unit," abbreviated ME. A dose of 200 mg corresponds to about three to five peyote buttons. Mescaline probably has a relatively greater human toxicity than LSD or psilocybin[187] as estimated by extrapolation from animal studies; but no human fatalities from overdose have been reported. (Burroughs claims knowledge of peyote fatalities, but this is most likely myth.)[188] The initial effects of orally ingested mescaline are felt, as with LSD, in about 20 to 40 minutes; they reach a peak between the third and fifth hour, and last somewhat longer than those of LSD, tapering off within 10 to 11 hours.

Syntheses of Mescaline

The first synthesis of mescaline was that of Späth in 1919. Gallic acid, **6-61** (Figure 6.11), was converted to trimethoxybenzoic acid, **6-62**, with base and dimethylsulfate; this was converted to its acyl chloride, **6-63**, which underwent a Rosenmund reduction to the aldehyde **6-64**. This was condensed with nitromethane and the resulting nitrostyrene, **6-65**, successively reduced to the oxime, **6-66**, with zinc dust and acetic acid, and to mescaline, **6-67**, with sodium amalgam.[189] The synthesis is essentially that used today except for the substitution of LAH for the two final reduction steps.[190]

Other routes to mescaline include the conversion of trimethoxybenzaldehyde in several steps to the trimethoxyphenylpropionamide, which is converted to mescaline by the Hofmann rearrangement;[191] LAH reduction of the trimethoxyphenylacetonitrile,[192]

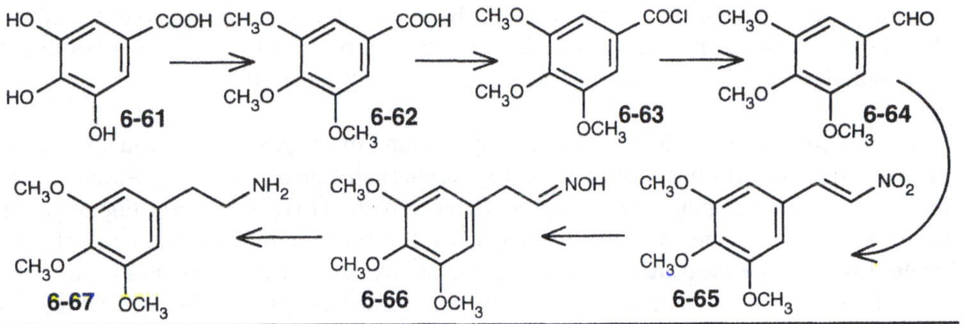

Figure 6.11 Späth's synthesis of mescaline.

which can also be synthesized from 2,6-dimethoxyphenol via a Mannich addition;[193] and treatment of **6-202** with diazomethane followed by LAH reduction of the diazoketone.[194]

Structure-Activity Relationships of Mescaline

The structure of mescaline, with the ready possibility of creating positional isomers, varying the number of methoxy groups, altering the aliphatic chain, and so on, invites the investigation of its structure-activity relationships; and extensive work in this area has been carried out, chiefly by Shulgin.[195] Numerous methoxylated phenethylamines and methoxylated phenylisopropylamines (amphetamines) have been synthesized and tested on human volunteers. In contrast to the case with LSD, many of these synthetic variations have proved more active than mescaline itself, and some of them, as we shall see below, produce psychic effects that are qualitatively different from the classical psychedelics, arguably creating new categories of psychoactivity. The major results of this research have been extensively reviewed by Shulgin, and only a few of the more significant structures can be described here.

The six isomers of mescaline that result from simple positional rearrangement of the three methoxy groups are either much weaker in psychedelic activity than mescaline itself or essentially inactive. However, extending the middle 4-methoxy group of mescaline to ethoxy (*escaline*) or to propoxy (*proscaline*) results in an approximately sixfold enhancement of psychedelic activity; and when the 4-methoxy group is replaced by methylthio, CH_3S (*TM, 4-TM*), there is a tenfold increased in potency.

2,5-Dimethoxy-4-X-Phenethylamines

Alexander Shulgin is a sometime research scientist of the faculty of the University of California, San Francisco, and lecturer in toxicology at the University of California, Berkeley. In 1992 he and his wife, Ann, published a book, *PIHKAL: A Chemical Love Story*, in which he describes the psychotropic action of more than a hundred variations of the mescaline structure as tested over a period of many years by the Shulgins and a group of their friends (PIHKAL stands for Phenethylamines I Have Known And Loved).[196] Most of the compounds discussed in this chapter can be found in Shulgin's book; the book includes many others that had not been previously reported.

2C-D. One of the richest lodes in this Shulgin mining of the psychedelic underground is the group of structures obtained by attaching various substituents in the 4-position of 2,5-dimethoxyphenethylamine. The simplest of these is 2C-D (LE-25, **6-68**, Figure 6.12), which Shulgin describes as "pharmacological tofu" because of its ambiguous effects. While it is by some measures about 20 times as active as mescaline, and has been found useful by therapists in Germany, it has been experienced by some as nearly inert and by still others as useful only in relatively large doses or in catalyzing the effects of other psychedelics.[197] However, its ethyl analogue, *2C-E,* **6-69**, is one of what Shulgin calls "the magical half-dozen," and seems able to enhance self-insight at the 15-mg level.[198]

2C-B. When the 4-position is occupied by a halogen, a significant enhancement of psychedelic potency results; in the case of 2C-B (NEXUS, **6-70**), there are some aphrodisiac overtones.[199] Shulgin describes the considerable underground use of this drug:

> Here is a simple phenethylamine . . . that is, even today [March, 1994], right up front. This is an active, and relatively potent phenethylamine (16-24 mg dosage) that is extremely widespread in the States today. About a year ago it was introduced into the street scene as NEXUS, and was promoted as "brominated cathinine." The suggestion was that this was a product, somehow, from the *Catha edulis* plant of Africa, that was "natural and legal." There is a cathine (norpseudoephedrine) and there is a cathinone (the corresponding primary amine ketone), and these are both from the Khat plant. But there is no such thing as cathinine, and therefore there is no such thing as brominated cathinine. I had looked at samples from three sources, South Africa, Los Angeles, and Florida, and each of them was 2C-B. This quickly became quite available on the street, and the DEA made it a Schedule I drug. So another rather safe psychedelic has become a felony to possess and to use.[200]

2,5-Dimethoxy-4-X-amphetamines

DOB, DOI. As was presaged by the high activity of 2,4,5-trimethoxyamphetamine and the 4-substituted 2,5-dimethoxyphenethylamines, the correspondingly substituted amphetamines proved to have the highest activity; indeed, DOB, **6-71**, and DOI, **6-72**, have an activity of 400 ME, making them by far the most potent yet discovered in the mescaline class. Their psychological effects in humans seem to be much like LSD, and rats trained to discriminate LSD find them a satisfactory substitute. The *R* chiral form of DOI is about twice as active as the *S* enantiomer. The fact that DOI contains an iodine atom as an intrinsic feature of its activity has allowed in vivo autoradiographic studies in rats using [^{125}I]*R*-DOI. The radioactive tagged drug was found to bind to a subset of the receptors that are labeled by radioactive [^{125}I]LSD, and to show a distribution in the rat brain very close to that shown by [^{3}H]LSD and [^{3}H]ketanserin, both of which bind to $5HT_2$ receptors.[201]

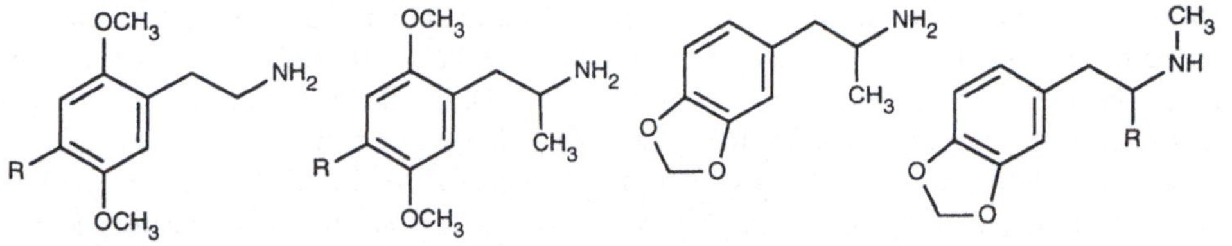

6-68: R = Me
6-69: R = Et
6-70: R = Br

6-71: R = Br
6-72: R = I
6-73: R = Me

6-74

6-75: R = Me
6-76: R = Et

Figure 6.12 Dimethoxyphenethylamines, dimethoxyamphetamines, methylenedioxyamphetamines.

DOM. This compound, **6-73**, goes by the peculiar moniker *STP*, the origin of which has always been something of a puzzle. It has been suggested that it was named after STP, a brand of motor fuel additive used by a hypothetical drug-taking motorcycle gang, or as an acronym for "Serenity, Tranquility, Peace," or Super Terrific Psychedelic, or Stop The Police, or (riposte by the police) too Stupid To Puke.[202] It unexpectedly appeared in abundance on the streets of San Francisco in 1967, often distributed free and in wallop doses: tablets averaged first 20 mg; later editions were lowered to 10 mg. In small doses under 5 mg, DOM produces mild euphoria, enhanced self-awareness, and talkativeness with few perceptual distortions or hallucinogenic effects.[203] One of Shulgin's experimenters reports enthusiastically from 4 hours into a 4-mg experience:

> I saw the clouds towards the west. THE CLOUDS!!! No visual experience has ever been like this. The meaning of color has just changed completely, there are pulsations, and pastels are extremely pastel. And now the oranges are coming into play. It is a beautiful experience. Of all past joys, LSD, mescaline, cannabis, peyote, this ranks number one. Normally I have no color effects with mescaline. A dynamic experience.[204]

But at higher doses of 10 to 30 mg, it produces an effect said to be like amphetamine combined with LSD but of much longer duration than either—16 to 24 hours or even longer. It has a slower onset and longer duration of action than LSD, and its effects tend to fade recur in wavelike fashion. Such phenomena could confuse the unwary user, which is probably why DOM use was associated with a greater incidence of bad trips and flashbacks.[205]

Methylenedioxyamphetamines

MDA. MDA (methylenedioxyamphetamine, "Love Drug," "Hug Drug," "Mellow Drug of America," **6-74**) has been used as a street drug since 1967. At moderate doses, (100 to 150 mg) MDA appears not to be psychedelic; rather, it enhances feelings of esthetic delight, empathy, insight, and self-awareness while eliminating anxiety and emotional inhibition. This often results in a great childlike yearning for affection, closeness, and touching, climaxing in a pleasant merging of egos from the perceived dissolving of the boundaries of the self. It appeared on the street scene in the mid-'60s, where it was called the "Love Drug," and, in a patriotic interpretation of the chemical acronym, the

"Mellow Drug of America."[206] Several studies[207] have shown that the amphetaminelike properties of MDA lie with the *S*-(+) enantiomer, and the psychedelic properties with the *R* (-). In high doses it has been said to resemble LSD, but there is usually some trismus (jaw-clenching) typical of the amphetamines. The effects last 8 to 10 hours; peak activity is at 1 to 2 hours. Although Kurland's group,[208] Naranjo,[209] and Richards[210] have used the desirable effects of moderate-dosage MDA to lower the barriers to self-disclosure and help patients engage in constructive psychotherapy, all legal use was abruptly curtailed when it was placed in the 1970 Controlled Substances Act.

MDMA. The addition of an *N*-methyl group to MDA produces MDMA (methylenedioxymethamphetamine, MDM, "ecstasy," XTC, "Adam," **6-75**). It is argued that with MDMA a new class of drugs is produced that have been variously called *empathogens* or *entactogens* because of their ability to enhance empathy or psychological communication. MDMA is the best known of these drugs. Although it was prepared and patented by Merck of Darmstadt[211] as long ago as 1912, it never was commercially produced. Its reintroduction and extensive use in psychotherapy is largely due to the persistent exploration of the structure-psychoactivity relationships of the phenethylamines related to mescaline carried out over many years by Shulgin. But Shulgin might not have tried MDMA himself had he not met a young chemistry graduate student named Merrie Kleinman, who told him that she had synthesized it and tried it with a few of her close friends with profound results. Shulgin found that several of his friends and acquaintances found themselves able to resolve even quite painful psychological trauma or blockages with surprising grace through the insights they obtained when using MDMA—an almost "snake-oil" effect, as he describes it. "It was not a psychedelic in the visual or interpretive sense, but the lightness and warmth of the psychedelic was present and quite remarkable."[212] By the mid-1970s, it had made its way into the therapeutic community, where it was used quite extensively: one popular California psychologist called "Adam" is said to have treated over 4,000 patients with MDMA, and it was widely used by other psychotherapists on both the east and west coasts.[213]

Part of the argument that MDMA represents a new class of drugs is the fact that the activity it shares with MDA is not lost on *N*-methylation, whereas all other known psychedelic drugs of the mescaline class lose their activity when transformed to their *N*-methyl congener. Additional evidence that drugs like MDMA are not psychedelic is given by *N*-methyl-1-(1,3-benzodioxol-5-yl)-2-butanamine (MBDB, **6-76**) in which the carbon chain of MDMA has been extended from propyl to butyl. This alteration abolishes psychedelic activity in all the phenylisopropyl amines that have been studied to date, but the entactogenic effect remains in MBDB.[214] A recent animal study also supports the claim that a unique facilitation of learning is effected by MDMA: in rabbits both conditioned and unconditioned responding was enhanced by the drug.[215]

What is the "entactogenic" effect? Most accounts are anecdotal, since the drug was used without the usual FDA procedures (most likely because the psychiatrists and psychologists involved feared MDMA would be banned as was LSD). However, there are hundreds of these reports, with a few clinical studies, and they demonstrate considerable internal consistency. "They include words such as *ecstasy, empathy, openness, acceptance, forgiveness,* and *emotional bonding.*"[216]

Some first-hand accounts convey it best. Kathy Tamm, a San Francisco marriage and family counselor, was abducted, beaten, and tortured by unknown assailants one night after leaving a meditation class. She suffered for almost a year from intense posttraumatic stress syndrome with repeated nightmares and fear of leaving the house. She described her condition at this point as "suicidal."

> As a last resort, Tamm and her psychiatrist (Dr. Joseph Downing) decided to treat her with MDMA, an experimental drug that some psychiatrists had found effective with traumatized patients. "I've taken it several times, and each time I felt a little less fearful," Tamm said. "The drug helped me regain some measure of serenity and peace of mind and enabled me to begin living a normal life again. For the first time, I was able to face the experience, go back and piece together what had happened. By facing it, instead of always burying it, I was able to sort of slowly discharge a lot of horror."[217]

The drug has been used extensively in marital counseling; estranged couples seem to discover from direct experience a truth more tediously excogitated by Fichte: "at the heart even of hatred there lies a longing for love, and enmity only springs from the soil of a failed friendship."[218] Tentative lovers became sure of their eternal devotion on MDMA and rushed into marriage, some soon to be disabused by the grim realities of life and time. T-shirts appeared in California with the colophon: *Don't get married for six weeks after X-T-C!* But Timothy Leary was game, as always: he and wife-to-be Barbara took MDMA, dined at a French restaurant with a bottle of Maison Pierre Grolau, "chatted away like newborn Buddhas just down from Heaven," married 3 days later, and were still happily espoused a decade later.[219]

There appear to be other intriguing (if somewhat less ennobling) contexts in which this effect has been employed to advantage:

> Respondents described how the reduced inhibitions and other effects of MDMA assisted them in their sexually oriented employment. Interviews of gay male prostitutes . . . revealed that many found MDMA helpful in creating a beneficial atmosphere with their clients. Similarly, a respondent described a topless dancer who found that MDMA was really helpful in her work: "It's a crude emotional situation so you can be more loving, you can accept the more gross behavior and make more tips."[220]

But MDMA, like the exaltation of Schiller's *Freude*,[221] is able to empower both worm and mystic alike:

> One of the experiences that people hesitate to talk about with MDMA, but which is actually the basic and most important experience that MDMA can give anybody, usually comes spontaneously. . . . It goes under many names, some of them very far-out spiritual sounding sorts of nonsense, but those are the only names you can use. It's an experience of the core self or [what is] sometimes called the God-space or the peaceful center, or somebody once described it as experiencing themselves being held in the hands of God, this feeling that something in the universe is totally accepting of them as a whole, you know, bumps and warts and all, no matter what they have done or not or what they are or are not. It is a deeply spiritual experience, and it is perhaps most valuable for the person who is in a severe depression because it is the core of the experience of self-loving and self-acceptance.[222]

The drug continued to be used in therapeutically oriented groups until the early 1980s, when the recreational possibilities of the drug began to be commercially exploited. What had been a low-level production from chemists in the Boston area was superseded when one of their Southwest distributors started his own operation in Texas, and renamed the drug "Ecstasy":

> The Texas entrepreneurs used blatant promotional tactics that shocked therapeutically oriented users and more established dealer networks. They swiftly built a pyramid referral system rivaling that of Amway. Ecstasy was sold openly at bars in Austin and Dallas. Purchases could be charged to Visa cards and proprietors paid taxes on their sales. The Texas Group circulated posters announcing "Ecstasy parties" at bars and discos with MDMA billed as a "fun drug" that was "good to dance to."[223]

"Raves" and "House Parties." It *was* good to dance to. In the United States, there were Grateful Dead concerts, the Acid House, the Rave. All became scenes where MDMA was widely used. Especially Raves (which may have originated on the island of Ibiza, already known by the summer of '86 as "XTC island," and moved from there to England, then to the United States)[224] were known from the beginning as Ecstasy parties. In the Rave, young people engage in marathon nonstop dancing to ear-splitting and mind-damaging music, usually under the influence of MDMA; always under the influence of excessive youth. In England, where initial attempts to outlaw the events drove them underground into overcrowded and poorly ventilated quarters, there have been over a dozen deaths from what appears to be heat stroke; some Ravers have arrived at hospital comatose with temperatures of 110 °F.[225]

In contrast, the Dutch from the beginning pursued their policy of "harm reduction,"[226] as they have tried to do with most drugs, and have succeeded in avoiding many of these problems.[227] In favor of the general safety of the drug, it must be said that overall usage in the United States, even under Rave conditions (which are almost precisely those calculated to induce a panic reaction, and diametrically opposite of those recommended by therapists) has resulted in almost no acute problems: in 1986, Newmeyer conducted a review showing that adverse reactions to MDMA averaged fewer than 20 per year from all emergency rooms nationwide, despite literally millions of individuals using the drug.[228]

In July 1984, the DEA responded to the lurid publicity about Ecstasy by recommending that MDMA be placed in Schedule I of the Controlled Substances Act. There was a prompt challenge from a group of psychiatrists and therapists, surprising the DEA, which had no idea it was being used in a therapeutic context. A reprise of the prohibition of LSD ensued. Despite the testimony of numerous psychiatrists (Downing, Greer, Lynch, Strassman, Wolfson), and despite the ruling of the DEA administrative law judge recommending that MDMA be placed in Schedule III so that use and research could continue, it was put in Schedule I on November 13, 1986. There were appeals, but in 1988 the drug was permanently placed in the most prohibitive schedule and has remained there since; nonetheless, in 1990 the Texas School Survey, in a random sampling of over 30,000 students, found that nearly 10% of high school seniors had tried MDMA, with nearly 3% having used it in the past month. There was a significantly greater usage among white students, with nearly 15% of white students having tried it.[229] And the dance goes on: quasi-underground raves continue to take place throughout the United States, doubtless in far more dangerous circumstances than the quasi-supervised environment provided in Holland.

Neurological Damage from MDMA Use. Part of the argument used by the DEA to put MDMA in Schedule I were initial reports of neurological damage in animals from both MDA and MDMA. Since then, extensive work by Ricaurte seems to validate the claim that some fairly permanent change to human serotoninergic neurons is caused by MDMA; still unresolved is how serious these changes are. Here is a summary of the more important results:

- All night polysomnograms of 23 MDMA users were compared to those of matched controls. The MDMA users averaged 37 minutes less stage 2 sleep than controls.[230]
- In animals, preadministration of serotonin reuptake inhibitors like the antidepressant fluoxetine (Prozac) prevents MDMA from causing neural damage. Reports from a study of four people who took fluoxetine before using MDMA seem to indicate that their response to MDMA was not affected. Hence the psychoactive effects of MDMA may be separable from its neurotoxicity.[231]
- There seem to be clear species-specific differences in the effects of MDMA. While serotonin neurodamage in monkeys from MDMA appears to be permanent, in rats function seems slowly to return to normal.[232] In an earlier study of humans, there seemed to be no correlation between either the number of MDMA exposures or the time since the last exposure and the extent of injury to serotonin neural function as measured by the amount of 5-hydroxyindoleacetic acid in the spinal fluid.[233]
- A study of nine people who had used MDMA extensively indicated no patent memory deficit on clinical examination nor any subjective or objective indication of mood disorder; but subtle impairments were noted in subjects in at least one test in the Wechsler Memory Scale.[234]
- Male rats given very high doses of MDMA (40 mg/kg) every 12 hours for 4 consecutive days exhibited disruption in sexual behavior with lengthening of ejaculation latency. But after a week, sexual function was normal despite severe depletion in brain 5-HT.[235]

The neurotoxicity does not seem to be actual neural cell death, but diminishment of serotonin presynaptic functioning, and that this alteration does not seem to be manifested in any irreversible long-term behavioral deficits in either animals or humans (and tens of thousands of people have used the drug). Proponents of the use of the drug in psychiatric counseling argue that one must weigh benefits to risks, and that similar neurotoxicity has been found[236] with regard to a prescription drug, dexfenfluramine, which the FDA allows to be prescribed on a daily basis for weight loss. Indeed, this fact was raised by the presiding DEA administrative law judge in recommending that the DEA assign MDMA to Schedule III. And the extent or significance of neural damage from dexfenfluramine is a matter of considerable controversy: "There is zero neurotoxicity," says Richard J. Wurtman, distinguished professor of neuroscience at MIT. "There's a change in the neurons. That's interesting, but it has nothing to do with toxicity."[237]

There remains considerable interest in the potential of entactogenic drugs in psychotherapy.[238] As Nichols and Oberlender point out:

> There are, presently, a number of psychopathologies that are extremely difficult to treat, for which a drug that would facilitate communication between patient and therapist might find great utility. Two examples are borderline personality disorder, and antisocial personality disorder. There is no reason to believe that psychotherapeutic agents cannot be developed to facilitate treatment of these, or other psychiatric disorders, but no one has apparently attempted to do this.[239]

And some limited progress in this direction has been made. Nichols has synthesized a number of MDMA analogues that do not cause neurotoxicity (in animals) but substitute with high potency in tests with rats trained to discriminate MDMA.[240] And the FDA recently gave Dr. Charles Grob, a psychiatrist at the University of California at Irvine Medical Center, approval for a limit study of the effects of MDMA on volunteers who used the drug before it was banned. He is hoping to obtain approval to use the drug in facilitative psychotherapy with terminal cancer patients. Still more progressive steps were taken in Switzerland, which in 1988 became the first country to permit psychiatrists—at first only five in number—to use MDMA (as well as LSD and psilocybin) at their discretion in psychotherapy.[241]

Spinal Tap? There are persistent rumors among young people, usually in the colleges and universities, that MDMA "drains your spinal fluid," or "causes Alzheimer's." Where these myths originated is difficult to say. It may be that the identification of MDMA as a "designer drug," and hence its association with the MPTP, a contaminant of an earlier designer opiate that caused Parkinson's (see Chapter 2) led to the myth. Or perhaps the rumors developed from reports of Ricaurte's group, which was at one time involved in testing the cerebrospinal fluid of people who had used MDMA to determine neurological damage.

To many people, the hallucinogens will probably seem the most peculiar and frightening class of drugs to be considered in this book. True, they have little or none of the addictive properties ascribable to drugs like heroin and cocaine—but then the sort of compulsive behavior that constitutes addiction, however frightening it might be, is something that most people feel they understand at least to some degree as an extrapolation of patterns we all find in normal life. To those who went to college in the '60s and were familiar with psychedelic drugs in their adolescence—a familiarity the present college-age generation is revisiting—these drugs will probably not seem so strange. A few glasses of wine, they may say, will go almost as far in distorting perception as a few psilocybin mushrooms or a tab of LSD—and quite a bit farther in distorting judgment.

To others, the most intriguing feature of the hallucinogens is the similarity of the state they induce to some aspects of religious experience. Timothy Leary memorably overstated the case in claiming that religion without psychedelics is like astronomy without telescopes; and Albert Hofmann's conclusion that LSD is a "sacred drug,"[242] rings odd in our modern ears, especially coming from a distinguished scientist. Nonetheless, the psychedelics seem to offer an opportunity for an objective study of one aspect of the psychology of religious experience. This at any rate is what one highly respected scholar in the field of comparative religions, the late Walter Houston Clark, opined nearly a quarter century ago:

> The issues that the psychedelics pose seem to most people to be in the realm of therapy, health, and the law. They may be more importantly religious. . . . LSD is a tool through which religious experience may, so to speak, be brought into the laboratory that it may more practically become a matter for study. It is important that religious institutions face the issues raised so that any decisions they may have to make will derive from sound knowledge rather than prejudice, ignorance, and fear . . . . If such decisions are to be sound, they must be based on thorough information, freedom from hysteria, and above all, open-mindedness to what may reliably be learned both of the great promise and the dangers of these fascinating substances.[243]

And we have seen that, while the hallucinogens are relatively new and even frightening to our culture, they constitute the active principle in dozens, perhaps hundreds of botanical potions used in native religious rituals for millennia. There is only one other plant that to any degree approaches this extent of use in religious ritual as an "ethnobotanical entheogen," and that is marijuana. The last chapter of this book is concerned with marijuana and a potentially useful class of drugs based on its active principle, the cannabinoids.

But we will begin the next chapter with a brief look at the dissociative anesthetics, which perhaps occupy in their psychic effects a middle position between the hallucinogens and the cannabinoids.

References and Notes

The following abbreviations are used for frequent sources in these References/Notes:

BMC4: *Burger's Medicinal Chemistry*, 4th ed., Wolff, M. E., Ed., Wiley: New York, 1981. The **bold** number following *BMC4* indicates the volume number (1-3); this is followed by the page number.

DE95: *Drug Evaluations: Annual 1995*; American Medical Association Division of Drugs and Toxicology, Department of Drugs; Bennett, D. R., Ed.; AMA: Chicago, 1995.

G&G8: *Goodman and Gilman's The Pharmacological Basis of Therapeutics*, 8th ed., Gilman, A. G.; Rall, T. W.; Nies, A. S.; Taylor, P., Eds.; Pergamon: New York, 1990.

G&G9: *Goodman and Gilman's The Pharmacological Basis of Therapeutics*, 9th ed., Hardman, J. G.; Limbird, L. E., et al., Eds.; McGraw-Hill: New York, 1996.

L&ID: *Licit and Illicit Drugs: The Consumers Union Report on Narcotics, Stimulants, Depressants, Inhalants, Hallucinogens, and Marijuana—Including Caffeine, Nicotine and Alcohol*, Brecher, E. W. and the Editors of Consumer Reports, Eds., Little, Brown: Boston, 1972.

MI11: *The Merck Index*, 11th ed.; Budavari, S.; O'Neil, M. J., et al., Eds.; Merck: Rahway, NJ, 1989. References are given by monograph number.

MI12: *The Merck Index*, 12th ed.; Budavari, S.; O'Neil, M. J., et al., Eds.; Merck: Rahway, NJ, 1996. References are given by monograph number.

PER: *Ecstasy: The Clinical, Pharmacological and Neurotoxicological Effects of the Drug MDMA*, Peroutka, S. J., Ed., Kluwer Academic: Boston, 1990.

PIHK: Shulgin, A.; Shulgin, A., *PIHKAL: A Chemical Love Story*, Transform Press: Berkeley, CA, 1992.

PsyDR: Grinspoon, L.; Bakalar, J. B., *Psychedelic Drugs Reconsidered*, Basic Books: New York, 1979.[244]

PURX: Beck, J.; Rosenbaum, M., *Pursuit of Ecstasy: The MDMA Experience* [*SUNY Series in New Social Studies on Alcohol and Drugs*, Levine, H. G.; Reinarman, C., Eds.], SUNY Press: Albany, NY, 1994.

SA: *Substance Abuse: A Comprehensive Textbook,* 2nd ed.; Lowinson, J. H.; Ruiz, P.; Millman, R. B.; Langrod, J. G., Eds.; Williams & Wilkins: Baltimore, 1992.

SH: Stevens, J., *Storming Heaven: LSD and the American Dream*, Atlantic Monthly Press: New York, 1987.

1. Thompson, F., "In no strange land." In *The New Oxford Book of English Verse: 1250-1950*, Gardner, H., Ed., Oxford University Press: New York, 1972.
2. Osmond, H., "A review of the clinical effects of psychotomimetic agents," *Annals N. Y. Academy Sci.* **1957**, *66*, 417-435.
3. Gordon Wasson has coined the term *entheogen* (engendering an [awareness of] the divine within) "for those plant substances that inspired Early Man with awe and reverence for their effect on him. By 'Early Man' we mean mankind in prehistory or proto-history, before he could read and write, whether long ago or since then or even living today in remote regions of the earth. 'Entheogen' . . . has the advantage that it does not carry the odor of 'hallucinogen,' 'psychedelic,' 'drug,' etc. of the youth of the '60s." Wasson, G., "The last meal of the Buddha," in Wasson, G.; Kramrisch, S.; Ott, J.; Ruck, C. A. P., *Persephone's Quest: Entheogens and the Origins of Religion*, Yale University Press: New Haven, 1986, p. 124. [Originally published in the *Journal of the American Oriental Society*, **1982**, *102*, (4). See also (a) Wasson, G., "Persephone's Quest," *ibid*., pp. 30-31. (b) *J. Psychedelic Drugs*, **1979**, *11*, pp. 145-146.
4. In a double-blind test, subjects were unable to distinguish mescaline from LSD: Hollister, L. E.; Sjoberg, B. M., "Clinical syndromes and biochemical alterations following mescaline, lysergic acid diethylamide, psilocybin, and a combination of the three psychotomimetic drugs," *Comprehensive Psychiatry*, **1964**, *5*, 170-178.
5. Rech, R. H.; Tilson, H. A.; Marquis, W. J., "Adaptive changes in behavior after repeated administration of various psychoactive drugs." In: *Neurological Mechanisms of Adaptation and Behavior*, Mandell, A., Ed., Advances in Biochemical Psychopharmacology, Vol. 13, pp. 263-286.
6. Jaffe, J. H., "Drug addiction and drug abuse." In: *G&G8*, 553.
7. "Illi, qui dormiendo syllogizant, cum excitantur, semper recognoscunt se in aliquo defecisse," Aquinas, T., *Summa Theologiae*, Ia. 84. 8 ad 2.
8. Freedman, D. X., " The use and abuse of LSD," *Arch. Gen. Psychiatry*, **1968**, *18*, 300-347.
9. Quidquid recipitur per modem recipientis recipitur.
10. Shulgin, A. T., "Confessions of a psychedelic alchemist," *Whole Earth Review*, **1991**, *72*, 25.
11. The interested reader is directed to (a) *Psychedelics: The Uses and Implications of Hallucinogenic Drugs*, Aaronson, B.; Osmond, H., Eds., Anchor Doubleday: Garden City, NY, 1970, pp. 21-66; (b) *PsyDR*, pp. 89-156; (c) Masters, R. E. L.; Houston, J., *The Varieties*

of Psychedelic Experience, Holt, Rinehart & Winston: New York, 1966.

12. Huxley, A., *The Doors of Perception*, Harper & Row: New York, 1990 [original publication 1954], p. 19. There is a striking resemblance to the passage in Hermann Hesse's *Siddhartha* wherein the hero has his first *satori*, or awakening:

    Er blickte um sich, als sähe er zum ersten Male die Welt. Schön war die Welt, bunt war die Welt, seltsam und rätselhaft war die Welt! Hier war Blau, hier war Gelb, hier war Grün, Himmel floß und Fluß, Wald starrte und Gebirg, alles schön, alles rätselvoll und magisch, und inmitten er, Siddhartha, der Erwachende, auf dem Wege zu sich selbst. . . . Blau war Blau, Fluß war Fluß, . . . so war es doch eben des Göttlichen Art und Sinn, hier Gelb, hier Blau, dort Himmel, dort Wald und hier Siddhartha zu sein. Sinn und Wesen waren nicht irgendwo hinter den Dingen, sie waren in ihnen, in allem.

    (He looked around himself as though he were seeing the world for the first time. The world was splendid with color and mystery! There was azure, gold, and green; the heavens flowed with the rivers, while the forest was rooted in the mountains! Everything was beautiful, mysterious, and magical, and in the midst of it all there was Siddhartha, the Enlightened, the Awakened, finally on the path which led to himself. . . . Blue was blue, the river was the river . . . it was the very nature of the divine to be here yellow, here blue, there sky, there forest, and here Siddhartha. The meaning and essence of everything was not to be found somewhere behind or outside of things, but as their very life, living within them.) Hesse, H., *Siddhartha: Eine indische Dichtung*, Suhrkamp Verlag: Frankfurt am Main, 1950, p. 40 [original edition 1922]. The translation is my own.

13. The unusual and fascinating "spiral" experience described by "Alice" (presumably Ann Shulgin) in Chapter 16 of *PIHK* seems to have usually occurred under hypnagogic circumstances—but not always. That even two other individuals should have experienced nearly identical phenomena invites some sort of Jungian archetypical analysis.

14. Huxley, *ibid.*, 17-18.

15. Mitchell, S. W., "Remarks on the effects of Anhalonium Lewinii (the mescal button)," *British Med. Journal* **1896**, *2*, 1625-1629.

16. I have added the emphasis to the translation by C. K. S. Moncrieff and T. Kilmartin, Vintage: 1982, vol I, p. 3. "Il me semblait que j'étais moi-même ce dont parlait l'ouvrage: une église, un quatour, la rivalité de François Ier et de Charles-Quint. Cette croyance survivait pendant quelques secondes à mon réveil; elle ne choquait pas ma raison. . . ." In the ordinary experience of being half-awake, it does not affront our reason to be a church, a quartet, a rivalry. One is reminded of Aristotle's definition of the agent intellect: *potentia omnia facere et fieri*'—a potentiality to create and become all things.

17. Watts, Alan, "Psychedelics and religious experience," in *Does It Matter? Essays on Man's Relation to Materiality,* Pantheon Books (Random House): New York, 1968, pp. 78-95 (quoted material p. 81). This excellent essay (original publication: *California Law Review,* **1968**, *56*, No. 1) illuminates the differences between Western (Christian, Judaic, Muslim) and Eastern (Hindu, Buddhist) perspectives on mystical experience, a phenomenon they share in common.

18. *Ibid.*, p. 81.

19. *Ibid.*, pp. 82-87.

20. *SH*, pp. 349-350.

21. Grof describes an obsessional who was "bored and a little hungry" after 1500 μg LSD injected IM; it took 38 high-dose sessions before he responded. Grof, S., *Realms of the Human Unconscious: Observations from LSD Research*, Viking: New York, 1975, pp. 30-31.

22. Ergonovine is consistently referred to in the British literature as ergometrine. The nomenclature reflects a controversy among the four groups that more or less simultaneously reported the isolation of the alkaloid. For a more detailed history, see Sneader, W., *Drug Discovery: The Evolution of Modern Medicines*, Wiley: New York, 1985, pp. 105-110.

23. Hofmann, A., *LSD, My Problem Child*, J. P. Tarcher (St. Martin's): Los Angeles, 1983, p. 12.

24. "Eine merkwürdige Ahnung, dieser Stoff könnte noch andere als nur die bei der ersten Untersuchung festgestellten Wirkungsqualitäten besitzen," reads the original German (Hofmann, A., *LSD—Mein Sorgenkind*, Klett-Cotta (Ullstein Taschenbuch): Frankfurt am Main, 1982 (original edition 1979), pp. 27-28. In a May 1993 interview, he says "I decided to prepare a new batch of the lysergic acid diethylamide for more extensive testing because from the very beginning I thought this substance was something special. It was just a feeling I had, when I was working and preparing substances." Gorman, P., "Albert Hofmann, the father of LSD," *High Times*, Psychedelics: Best of *High Times* #17 (1995), p. 59. [Original publication was in the May 1993 issue.]

25. Hofmann, *My Problem Child*, 14.

26. Hofmann, *ibid.*, 15.

27. Hofmann, *ibid.*, 17-19.

28. Hofmann, *ibid.*, 47.

29. Grinspoon, L., *PsyDR*, p. 192.

30. (a) Hoffer, A.; Osmond, H. *The Hallucinogens*, Academic Press: New York, 1967. (b) Unger, Sanford M., "Mescaline, LSD, psilocybin, and personality change: a review," *Psychiatry: Journal for the Study of Interpersonal Processes*, **1963**, *26*, 111-125.

31. In England: Sandison, R. A.; Spencer, A. M.; Whitlaw, J. D. A., "The therapeutic value of lysergic acid diethylamide in mental illness," *Journal of Mental Science*, **1954**, *100*, 491-507. In Germany: Leuner, Hanscarl, *Die Experimentelle Psychose*, Springer Verlag: Berlin, 1962.

32. For more extensive references see Grinspoon, *PsyDR*, pp. 192-237.

33. *Ibid.*, p. 232.

34. *Ibid.*, pp. 232-233.

35. Osmond, H., "A review of the clinical effects of psychotomimetic agents," *Annals N. Y. Academy Sci.* **1957**, *66*, 417-435. Cited material on p. 423.

36. Osmond, H., *ibid.*, pp. 428, 430.

37. "He had an experience [that] was totally spiritual, [like] his initial spiritual experience." So writes Nell Wing, Wilson's secretary from 1950 until his death. Cited in *'Pass It On': The Story of Bill Wilson and How the A.A. Message Reached the World,* Alcoholics Anonymous World Services, Inc.: New York, 1984, p. 370. Wilson became aware of LSD through his friendship with Gerald Heard and Aldous Huxley (who called Wilson "the greatest social architect of the century"). Heard introduced Wilson to the efforts of psychiatrists Humphry Osmond and Abram Hoffer, who were attempting to cure alcoholics at a mental hospital in

Saskatoon, Saskatchewan, using LSD and mescaline. Osmond relates that Wilson was at first "extremely unthrilled" by the very idea of giving alcoholics drugs (*ibid.*, 369). But he was won over eventually and took the drug for the first of several times on 29 August 1956.

Wilson describes his "original mystical experience" of December 1934 as an electric transformation from the depths of despair: "Suddenly, my room blazed with an indescribably white light. I was seized with an ecstasy beyond description. Every joy I had known was pale by comparison. The light, the ecstasy—I was conscious of nothing else" (*ibid.*, p. 121).

Wilson felt that LSD experiences "helped him eliminate many barriers erected by the self, or ego, that stand in the way of one's direct experience of the cosmos and of God" (*ibid.*, p. 371). Many of his closest friends, including his longtime confidant, Fr. Edward Dowling, S.J., shared LSD experiences with Wilson and concurred in his evaluation of its spiritual potential. But most members of the growing A.A. fellowship, on hearing of his drug use, "were violently opposed to his experimenting with a mind-altering substance" (*ibid.*, p. 372). Wilson's sense of responsibility for the group led him eventually to withdraw from the experiments.

38. Jensen, S. E.; Ramsay, R., "Treatment of chronic alcoholism with lysergic acid diethylamide," *Canadian Psychiatric Association Journal*, **1963**, *8*, 184-185.

39. Kurland, Albert A., "The therapeutic potential of LSD in medicine." In: *LSD, Man, and Society*, DeBold, R.; Leaf, R., Eds., Wesleyan University Press: Middletown, CT, 1967, pp. 20-35.

40. Maclean, J. R; Macdonald, D. C.; Ogden, F.; Wilby, E., "LSD-25 and mescaline as therapeutic adjuvants." In: Abramson, H., Ed., *The Use of LSD in Psychotherapy and Alcoholism*, Bobbs-Merrill: New York, **1967**, pp. 407-426. In the same work, see also Ditman, K. S.; Bailey, J. J., "Evaluating LSD as a psychotherapeutic agent," pp. 74-80; and Hoffer, A., "A program for the treatment of alcoholism: LSD, malvaria, and nicotinic acid," pp. 353-402. Dipropyltryptamine (DPT) has also been employed in the treatment of alcoholics: Grof, S.; Soskin, R. A.; Richards, W. A.; Kurland, A. A., "DPT as an adjunct in psychotherapy of alcoholics," *International Pharmacopsychiatry*, **1973**, *8*, pp. 104-115.

41. Ludwig, A. M.; Levine, J.; Stark, L. H. *LSD and Alcoholism: A Clinical Study of Treatment Efficacy*, Charles C. Thomas: Springfield, IL, 1970.

42. Abuzzahab, F. S.; Anderson, B. J., "A review of LSD treatment in alcoholism," *International Pharmacopsychiatry*, **1971**, *6*, 223-235. McCabe, O. L.; Hanlon, T. E., "The use of LSD-type drugs in psychotherapy: Progress and Promise." In: *Changing Human Behavior: Current Therapies and Future Directions*, McCabe, O. L., Ed., Grune and Stratton: New York, 1977, pp. 221-253.

43. According to Huxley's widow, Laura, every use of psychedelic drugs described in *Island* reflects his own first-hand experience: Huxley, L., *This Timeless Moment: A Personal View of Aldous Huxley*, Farrar, Straus & Giroux: New York, 1968, p. 146.

44. *Ibid.*, pp. 295-308.

45. Albert Kurland, personal communication.

46. I am grateful to Bill Richards for sharing with me in the summer of 1993 some of his memories from those distant days. Dr. Richards continues to live in Baltimore, where he is a clinical psychologist in private practice.

47. Published as "The psychedelic mystical experience in the human encounter with death," *Harvard Theological Review*, **1969**, *62*, 1-21. The publication includes two responding critiques, one by a psychiatrist and one by a theologian. See also Pahnke, W. N., "Implications of LSD and experimental mysticism," *Journal of Religion and Health*, **1966**, *5*, 175-208.

    Pahnke drowned at the age of 40 in 1971 while scuba diving in Maine. At the time he was the Director of Clinical Sciences at the Maryland Psychiatric Research Center and an assistant professor of clinical psychiatry at The Johns Hopkins School of Medicine. A personal note: the secretary of the chemistry department here at Loyola College in Baltimore told me that she remembers Pahnke as a "very caring doctor," who from 1967 to 1970 treated her mother, then dying of breast cancer with multiple bone metastases, with LSD. The transformation, she says, was remarkable: for several subsequent months her mother was peaceful and nearly pain-free, whereas before she had been "like a caged animal" from pain. Over the course of the disease, as the pain returned, Pahnke held several further LSD sessions with similar results. (At one time hypnotism was attempted, but her mother did not respond.) In 1970 she died of metastases to the brain.

48. Pahnke, W. N.; Kurland, A. A., "The experimental use of psychedelic (LSD) psychotherapy," *JAMA*, **1970**, *212*, 1856-1863.

49. *Ibid.*, p. 1861.

50. Leary, T. *High Priest*, New York: College Notes & Texts [New American Library], 1968, 2nd printing, pp. 256-257.

51. "Parturient montes, nascetur ridiculus mus (the mountains will be in labor, and give birth to a ridiculous mouse)." Horace, *Ars Poetica*, 139.

52. A fascinating and thoroughly researched history of the era can be found in Stevens, *SH*. A shorter history, relying much on Stevens, is given by Ulrich, R. F., Patten, B. M., "The Rise, Decline, and Fall of LSD," *Perspectives in Biology and Medicine*, **1991**, *34*, 561-578. Also: Wolfe, T., *The Electric Kool-Aid Acid Test*, Farrar, Straus, and Giroux: New York, 1968; Draper, R., *Rolling Stone Magazine: The Uncensored History*, Doubleday: New York, 1990; Krassner, P., *Confessions of a Raving, Unconfined Nut: Misadventures in the Counter-Culture,* Simon & Schuster: New York. Recollections from more academic types can be found in *Psychedelic Reflections*, Grinspoon, L.; Bakalar, J., Eds., Human Sciences Press: New York, 1983.

53. *SH*, pp. 370-371.

54. Research is conducted in Baltimore under the auspices of the Orenda Institute.

55. (a) Abraham, H. D.; Aldridge, A. M.; et al., "The psychopharmacology of hallucinogens," *Neuropsychopharmacology,* **1996**, *14,* 285-298; cited material p. 294. See also the following articles from one of the few active researchers in this field: (b) Strassman, R. J., "Hallucinogenic drugs in psychiatric research and treatment," *J. Nerv. Ment. Dis.,* **1995**, *183*, 127-138. (c) Strassman, R. J., "Human hallucinogenic drug research in the United States: A present-day case history and review of the process," *J. Psychoactive Drugs,* **1991**, *23,* 29-38.

56. Klee, G. D.; Bertino, J.; et al., "The influence of varying dosage on the effects of lysergic acid diethylamide (LSD-25) in humans," *Journal of Nervous and Mental Disorders*, **1961**, *132*, 404-409.

57. Griggs, E. A.; Ward, M., "LSD toxicity: A suspected cause of death," *Journal of the Kentucky Medical Association*, **1977**, *75*, 172-173.

58. Jaffe, J. H., "Drug addiction and drug abuse." In: *G&G8*, 555-556.

59. McGlothlin, W. H.; Arnold, D. O., "LSD revisited: A ten-year follow-up of medical LSD use," *Arch. Gen. Psych.*, **1971**, *24*, 35-49.

60. *PsDR*, p. 171.

61. Marks, J., *The Search for the Manchurian Candidate*, Times Books: New York, 1971, pp. 73-86.

62. Miller, P. L.; Gay, G. R.; et al., "Treatment of acute, adverse psychedelic reactions: 'I've tripped and I can't get down,'" *J. Psychoactive Drugs*, **1992**, *24*, 277-279.

63. (a) Vardy, M. M.; Kay, S. R., "LSD psychosis or LSD-induced schizophrenia?" *Arch. Gen. Psychiatry*, **1983**, *40*, 877-883. See also the very thorough review of all documented cases of adverse reactions to psychedelic drugs: Strassman, R. J., "Adverse reactions to psychedelic drugs: A review of the literature," *J. Nerv. Ment. Dis.*, **1984**, *172*, 577-595.

64. The author here speaks from his own observation. A traditional part of Jesuit novice training is the 30-day Spiritual Exercises of St. Ignatius, wherein complete silence is observed for four weeklong periods. Despite psychological screening of candidates, almost every year in the '60s when the author was a novice, one or more of the 40 or so souls making the Exercises would flip out and be bundled quietly off to a psychiatric hospital. Most recovered after a month or so, but they were no longer considered suitable prospects for religious life. That such incidents are less frequent in the '90s is probably because the average age of such candidates is no longer the late teens but the late twenties: what used to be called *dementia praecox* is characteristic of the younger group. Significantly, those like Albert Hofmann who advocate the use of LSD for spiritual and personal enlightenment caution that it should not be used by people too young to have established ego boundaries.

65. That is, there are reports that flashbacks are more likely with heavy users, and reports denying this; that the incidence of flashbacks diminishes with increasing time since the last drug experience and other reports that the incidence is unchanged even years after drug usage; that anxiety triggers the phenomenon and that it does not. A summary of these results with references can be found in Abraham, H. D., "Visual phenomenology of the LSD flashback," *Arch. Gen. Psychiatry*, **1983**, *40*, 884-889.

66. *PIHK*, pp. 358-364.

67. Abraham, H. D., "Visual Phenomenology of the LSD Flashback," *Arch. Gen. Psychiatry*, **1983**, *40*, 884-889. Quoted material on p. 889.

68. Hofmann, A., "Teonanácatl and Ololiuqui, two ancient magic drugs of Mexico," *Bulletin on Narcotics*, **1971**, *23*, 3-14.

69. Matossian, M. K., *Poisons of the Past: Molds, Epidemics, and History*, Yale University Press: New Haven, CT, 1989. In the Middle Ages, well-educated and wealthy classes did not eat rye but wheat bread, with one exception: monastic communities observant of the vow of poverty. Since this is just where such proofs of holiness as mystical illumination were to be found, one is tempted to indulge the irreverent speculation that some saints were no more than upper-class (unburned) witches/wizards.

70. "Not only have we been given reason thereby to live in joy, but still more, reason to die with hope," writes Cicero. The full passage, with the translated section italicized, reads: "tum nihil melius illis mysteriis, quibus ex agresti immanique vita exculti ad humanitatem et mitigati sumus, initiaque ut appellantur, ita re vera *principia vitae cognovimus; neque solum cum laetitia vivendi rationem accepimus, sed etiam cum spe meliore moriendi.*" *De legibus* II, XIV, 36. *Cicero: de re publica, de legibus*, Clinton W. Keyes, Ed., Harvard University Press: Cambridge, 1928. See also Kerényi, C., *Eleusis: Archetypal Image of*

Mother and Daughter, trans Ralph Manheim, Princeton University Press: New Haven, 1967.

71. Wasson, R. G.; Ruck, C. A. P.; Hofmann, A., *The Road to Eleusis: Unveiling the Secret of the Mysteries*, Harcourt Brace Jovanovich: New York, 1978.

72. For summaries of the activity of these compounds and references to their investigation, see (a) Shulgin, A. T., "Hallucinogens." In: *BMC4*, **3**, 1109-1137 (tables on pp. 1126-1127). (b) Fanchamps, A., "Some compounds with hallucinogenic activity." In: *Ergot Alkaloids and Related Compounds*, Berde, B.; Schild, H., Eds.; Springer: Berlin, 1978, pp. 567-583.

73. Hoffman, A. J.; Nichols, D. E., "Synthesis and LSD-like discriminative stimulus properties in a series of *N*(6)-alkyl norlysergic acid *N,N*-diethylamide derivatives," *J. Med. Chem.*, **1985**, *28*, 1252-1255.

74. Cunningham, K. A.; Appel, J. B., "Neuropharmacological reassessment of the discriminative stimulus properties of *d*-lysergic acid diethylamide (LSD)," *Psychopharmacology*, **1987**, *91*, 67-93.

75. (a) Oberlender, R.; Pfaff, R. C.; Johnson, M. P.; Huang, X.; Nichols, D. E., "Stereoselective LSD-like activity in *d*-lysergic acid amides of (*R*)- and (*S*)-2-aminobutane," *J. Med. Chem.*, **1992**, *35*, 203-211. See also (b) Monte, A. P.; Marona-Lewicka, D.; Kanthasamy, A.; Sanders-Bush, E.; Nichols, D. E., "Stereoselective LSD-like activity in a series of *d*-lysergic acid amides of (*R*)- and (*S*)-2-aminoalkanes," *J. Med. Chem.*, **1995**, *38*, 958-966.

76. Murphree, H. B.; DeMarr, E. W. J.; Williams, H. L.; Bryan, L. L., "Effects of lysergic acid derivatives on man; antagonism between *d*-lysergic acid diethylamide and its 2-bromo congener," *J. Pharmacol. Exp. Ther.*, **1958**, *122*, 55A-56A.

77. More recent work has shown that LSD is a partial agonist, and 2-bromo-LSD a full antagonist at the 5-HT_2 receptors, but both induce an unsurmountable antagonism to serotonin: Burris, K. D.; Sanders-Bush, E., "Unsurmountable antagonism of brain 5-hydroxytryptamine$_2$ receptors by (+)-lysergic acid diethylamide and bromo-lysergic acid diethylamide," *Molecular Pharmacology*, **1992**, *42*, 826-830.

78. Freedman, D. X., "Effects of LSD-25 on brain serotonin," *J. Pharmacology and Experimental Therapeutics*, **1961**, *134*, 160-166.

79. Aghajanian, G. K.; Sprouse, J. S.; Rasmussen, K., "Physiology of the midbrain serotonin system." In: *Psychopharmacology: The Third Generation of Progress*, Meltzer, H. Y., Ed., Raven Press: New York, 1987, pp. 141-149.

80. Trulson, M. E.; Heym, J.; Jacobs, B. L., "Dissociations between the effects of hallucinogenic drugs on behavior and raphe unit activity in freely moving cats," *Brain Res.*, **1981**, *215*, 275-293.

81. (a) Glennon, R. A.; Titeler, M.; McKenney, J. D., "Evidence for 5-HT_2 involvement in the mechanism of action of hallucinogenic agents," *Life Sci.*, **1984**, *35*, 2505-2511. (b) Titeler, M.; Lyon, R. A.; Glennon, R. A., "Radioligand binding evidence implicates the brain 5-HT_2 receptor as a site of action for LSD and phenylisopropylamine hallucinogens.," *Psychopharmacology*, **1988**, *94*, 213-216.

82. Sadzot, B.; Baraban, J. M.; Glennon, R. A.; Lyon, R. A.; Leonhardt, S.; Jan, C. R.; Titeler, M, "Hallucinogenic drug interactions at human brain 5-HT2 receptors: implications for treating LSD-induced hallucinogenesis," *Psychopharmacology*, **1989**, *98*, 495-499.

83. Buckholtz, N. S.; Zhou, D.; Freedman, D. X.; Potter, W. Z., "Lysergic acid diethylamide (LSD) administration selectively downregulates serotonin receptors in rat brain," *Neuropsychopharmacology*, **1990**, *3*, 137-148.

84. (a) Grof, S.; Dytrych, Z., "Blocking of LSD reaction by premedication with Niamid," *Activitas Nervosa Superior*, **1965**, *7*, 306.

85. Ungerleider, M. T.; Pechnick, R. N., "Hallucinogens." In: *SA*, 281-289.

86. (a) Meert, T. F.; de Haes, P.; Janssen, P. A. J., "Risperidone (R 64 7660), a potent and complete LSD antagonist in drug discrimination by rats," *Psychopharmacology*, **1989**, *97*, 206-212. (b) Glennon, R. A.; Teitler, M.; Sanders-Bush, E., *Hallucinogens and Serotonergic Mechanisms* [NIDA Research Monograph], **1992**, *119*, 131-135.

87. Burris, K. D.; Breeding, M.; Sanders-Bush, E., "(+)-Lysergic acid diethylamide, but not its nonhallucinogenic congeners, is a potent serotonin 5-HT_{1c} receptor agonist," *Journal of Pharmacology and Experimental Therapeutics*, **1991**, *258*, 891-896. Perhaps supporting this view is the finding that DMT has opposing actions on 5-HT receptor subtypes, being an agonist at the 5-HT_{1a} and an antagonist at the 5-HT_2: Deliganis, A. V.; Pierce, P. A.; Peroutka, S. J., *Biochemical Pharmacology*, **1991**, *41*, 1739-1744.

88. Sanders-Bush, E.; Mayer, S. E., "5-Hydroxytryptamine (serotonin) receptor agonists and antagonists," *G&G9*, pp. 249-263 (LSD, pp. 258-259).

89. Cunningham, K. A.; Appel, J. B., "Neuropharmacological reassessment of the discriminative stimulus properties of *d*-lysergic acid diethylamide (LSD)," *Psychopharmacology*, **1987** *91*, 67-73.

90. Pierce, P. A.; Peroutka, S. J., "Hallucinogenic drug interactions with neurotransmitter receptor binding sites in human cortex," *Psychopharmacology*, **1989**, *97*, 118-122.

91. Kornfeld, E. C.; Fornefeld, E. J.; Kline, G. B.; Mann, M. J.; Morrison, D. E.; Jones, R. G.; Woodward, R. B., "The total synthesis of lysergic acid," *J. Am. Chem. Soc.*, **1956**, *78*, 3087-3114. A review of the early syntheses of lysergic acid and other ergot alkaloids: Horwell, D. C., *Tetrahedron*, **1980**, *36*, 3123-3149.

92. Oppolzer, W.; Francotte, E.; Bättig, K., *Helv. Chim. Acta*, **1981**, *64*, 478-481. Oppolzer, W.; Grayson, J. I; Wegmann, H.; Urrea, M., *Tetrahedron*, **1983**, *39*, 3695.

93. Rebek, J.; Tai, D. F., *Tetrahedron Letters*, **1983**, *24*, 859-860. Rebek, J.; Tai, D. F.; Shue, Y.-K., *J. Am. Chem. Soc.*, **1984**, *6*, 1813-1819.

94. In chronological order: (a) Julia, M.; Le Goffic, F.; Igolen, J.; Baillarge, M., *Tetrahedron Letters*, **1969**, 1569-1571. (b) Ramage, R.; Armstrong, V. W.; Coulton, S., *Tetrahedron*, **1981**, *37*, 157-164. (c) Ninomyia, I.; Hashimoto, C.; Kiguchi, T.; Naito, T., *J. Chem. Soc. Perkin Trans. I*, **1985**, 941-948. (d) Barton, D. J. R.; Fekih, A.; Lusinchi, X., *Bull. Soc. Chim. France*, **1988**, 681-687.

95. Ramage, R.; Armstrong, V. W.; Coulton, S., *Tetrahedron*, **1981**, *37*, 157-164.

96. Kobel, H.; Schreier, E.; Rutschmann, J., *Helv. Chim. Acta*, **1964**, *47* 1052.

97. Troxler, F., *Helv. Chim. Acta*, **1968**, *51*, 1372.

98. Scaplehorn, A. W., "Illicit synthesis of phencyclidine (PCP), lysergic acid diethylamide (LSD), methaqualone and their analogues." In: *Clandestinely Produced Drugs, Analogues and Precursors: Problems and Solutions*, Klein, M.; Sapienza, F.; McClain, Jr. H., Khan, I., Eds., U.S. DEA: Washington, DC, 1989, pp. 91-94. Two-thirds of the world's supply sounds implausibly high—there seems to have been no noticeable shrinkage of supply. It should be said that, despite the unimpeachable source of *Clandestinely . . .* , it is at times astonishingly inaccurate. For example, on p. 216 it is said that "DET and DMT are shown to have a slower onset and longer duration of effects than LSD." DMT is actually the shortest-acting hallucinogen known, with the quickest onset (4 seconds if smoked or taken IV)

and DET is not far behind.

99. After all, it can be done by biologists. Instructions on how to cultivate *Claviceps* are given in Smith, M. V., *Psychedelic Chemistry*, Loompanics Unlimited: Mason, MI 48854, 1981. A more reliable source of information, chemist Alexander Shulgin, tells me that specimens of *C. paspali* are readily available from the U.S. Department of Agriculture. And, of course, as with yoghurt, once you have a sample of the fungus it reproduces indefinitely. You can endow other clandestine daughter labs. This and the fact that LSD is so astonishingly potent (1 g = 10,000-20,000 doses) probably accounts for the fact that there has been no discovery of any clandestine LSD production in the United States by any police agency in almost two decades.

100. Wasson, S. H.; Wasson, V. P., *Mushrooms, Russia, and History*, Pantheon Books, New York, 1957, Vol. II, pp. 215-322. Also: Wasson, S. H., "The divine mushroom of immortality." In: *Flesh of the Gods: The Ritual Use of Hallucinogens*, Furst, P. T., Ed., Praeger Publishers: New York, 1972, pp. 185-200.

101. Wasson, S. H., "The divine mushroom of immortality," *loc. cit.*, pp. 198-199.

102. Lemonick, M. D., "Lost secrets of the Maya: what new discoveries tell us about their world—and ours," *Time*, 9 August 1993, pp. 44-46.

103. Wasson, S. H., "The divine mushroom of immortality," *loc. cit.*, p. 197.

104. Hofmann, A., "Teonanácatl and Ololiuqui, two ancient magic drugs of Mexico," *Bulletin on Narcotics*, **1971**, *23*, p. 4.

105. Hofmann, A.; Heim, R.; Brack, A.; Kobel, H., "Psilocybin, ein psychotroper Wirkstoff aus dem mexikanischen Rauschpilz *Psilocybe mexicana* Heim," *Experientia*, **1958**, *14*, 107-109.

106. Hofmann, A.; Heim, R.; Brack, A.; Kobel, H.; Frey, A.; Ott, H.; Petrzilka, Th.; Troxler, F., "Psilocybin und Psilocin, zwei psychotrope Wirkstoffe aus mexikanischen Rauschpilzen," *Helv. Chim. Acta*, **1959**, *42*, 1557-1572.

107. Stoll, A.; Troxler, F.; Peyer, J.; Hofmann, A., *Helvetica Chimica Acta*, **1955**, *38*, 1452.

108. Leary, T.; Metzner, R.; et al., "A new behavior change program using psilocybin," *Psychotherapy*, **1965**, *2*, 61-72.

109. Pahnke had indeed preserved thus far his "psychedelic virginity," to use Leary's pixyish phrase (*High Priest*, College Notes & Texts: New York, 1968, p. 292). But this was not because Pahnke "didn't have a rebellious bone in his body. . . he was an establishment man, a good boy, right down the line . . . And stubborn about his virtue. Your classic, old-fashioned Protestant type" (*ibid.*, p. 307), but because he quite rightly wanted to be free from bias. Leary is sometimes casual with the facts and their interpretation. He describes Pahnke's "first conversation with God" as follows:

> Walter Pahnke got his thesis uneasily approved, and his degree was awarded. Walter went to Germany on a fellowship and arranged to have his first conversation with God in a mental hospital in the Rhineland. He had a clinical examination room converted into a shrine and got a Yale theologian to be his guide, and played sacred music on his record player, and to the shocked amazement of the German psychiatrists (who . . . [were] using LSD to produce dirty psychoanalytic experiences), Walter made the eternal voyage and laughed in gratitude and wept in reverence. And only then, a year later, did he realize the wondrous miracle he had wrought in Marsh Chapel (*ibid.*, p. 315).

Among several peculiar errors in this account is placing Hanscarl Leuner's Göttingen clinic in the "Rhineland." Göttingen is in Lower Saxony, on the Leine River, about 15 miles from what was then the East German border. Additionally, I am told on the best authority (the only other person who was with Pahnke in the room) that it was an ordinary hospital room that had not been "converted into a shrine," and that the German psychiatrists were not particularly "shocked" since they had by now seen this sort of reaction quite frequently.

110. The best account of the protocol is found in: Pahnke, W. N., "Summary of the thesis: drugs and mysticism: An analysis of the relationship between psychedelic drug experience and the mystical state of consciousness," Harvard Graduate School of Arts and Sciences (unpublished typescript kindly provided me by William Richards). See also Pahnke, W. N.; Richards, W. A., "Implications of LSD and experimental mysticism," *Journal of Religion and Health,* **1966**, *5*, 175-208. The event is also described by Timothy Leary in: Leary, T., *High Priest*, pp. 290, 304-318.

111. Pahnke, Walter N., "Drugs and mysticism," *International Journal of Parapsychology*, **1966**, *8*. Reprinted in *Psychedelics: The Uses and Implications of Hallucinogenic Drugs*, Aaronson, B., Osmond, H., Eds., Anchor Doubleday: Garden City, NY, 1970, pp. 145-165. Cited material from pp. 158-160.

112. Pahnke, *ibid.*, pp. 161-162.

113. Robertson, J., "Uncontainable Joy." In: *The Ecstatic Adventure: Reports of Chemical Explorations of the Inner World by Philosophers, Theologians, Scientists, Architects, Writers, Artists, Businessmen, Students, Housewives, Musicians, Mothers, Children, Patients, Convicts, Addicts, Secretaries,* Metzner, R., Ed., Macmillan: New York, 1968, pp. 84-91. Quoted material p. 88.

114. (a) Stromberg, *J. Am. Chem. Soc.*, **1954**, *76*, 1701. (b) Fish, M.; Johnson, N.; Horning, E., "Piptadenia alkaloids: indole bases of *P. peregrina* (L.) Benth. and related species," *J. Am. Chem. Soc.*, **1955**, *77*, 5892-5895. (c) Pachter, I. J.; Zacharias, D. E.; Ribeiro, O., "Indole alkaloids of *Acer saccharinum* (the Silver Maple), *Dictyoloma incanescens, Piptadenia colubrina,* and *Mimosa hostilis*,"*J. Org. Chem.*, **1959**, *24*, 1285-1297. (d) Hochstein, F. A.; Paradies, A. M., "Alkaloids of *Banisteria caapi* and *Prestonia amazonicum*," *J. Am. Chem. Soc.*, **1957**, *79*, 5735-5736.

115. Szára, S., "Dimethyltryptamine: Its metabolism in man; the relation of its psychotic effect to the serotonin metabolism," *Experientia*, **1956**, *12*, 441-442. The first synthesis of DMT had been achieved some 20 years earlier: Hoshino, Shimodaira, *Annalen*, **1935**, *78*.

116. Rosenberg, D. E.; Isbell, H.; Miner, E. J., "Comparison of a placebo, *N*[sic]-dimethyltryptamine, and 6-hydroxy-*N*[sic]-dimethyltryptamine in man," *Psychopharmacologia,* **1963**, *4*, 39-42. On the other hand, Szara administered to himself 10 mg of the 6-hydroxy derivative of diethyltryptamine (whether orally or IM is unclear) and experienced "typical psychotomimetic disturbances" which lasted several hours: Szara, S.; Hearst, E., "The 6-hydroxylation of tryptamine derivatives: A way of producing psychoactive metabolites," *Ann. N. Y. Acad. Sci.*, **1962**, *96*, 134-141.

117. Smoking DMT can leave the smoker with a "strong indole ('burnt plastic') aftertaste." Thus spake *Xochi Speaks*, a colorful poster describing the proper approach to be taken to the dozen most popular psychedelics. Further particulars concerning this poster, which has been tastefully and accurately designed after consultation with several eminent scientists and provided with a properly reverential obeisance to the Aztec deity Xochipilli, the "Prince of Flowers," by the noted San Francisco artist of nom de brush "LordNose!®," can be obtained in: Fraser, L., "Xochipilli: a context for ecstasy," *The Whole Earth Review*, **1991**, *75*, 38-43, or by writing LordNose! PO Box 170473X, SF, CA 94117-0473. "Snorting" DMT requires

some heroics, too: its aroma is said to lie somewhere between that of senescent armpits and fresh mothballs.

118. Leary, *High Priest*, p. 266. Their testing of DMT is narrated in "Trip 13," pp. 264-279.

119. Szara, S., "DMT (*N,N*-dimethyltryptamine) and homologues: Clinical and pharmacological considerations." In: *Psychotomimetic Drugs*, Efron, D. H., Ed., Raven Press: New York, 1970, pp. 275-286.

120. (a) Szára , S., "The comparison of the psychotic effects of tryptamine derivatives with effects of mescaline and LSD-25 in self-experiments." In: *Psychotropic Drugs*, Garattini, S.; Ghetti, V., Eds., Elsevier: Amsterdam, 1957, pp. 460-466. (b) Szara, S.; Rockland, L. H.; et al., "Psychological effects and metabolism of *N,N*-diethyltryptamine in man," *Arch. Gen. Psychiatry*, **1966**, *15*, 320-329. (c) Böszörményi, Z.; Dér, P.; Nagy, T., "Observations on the psychotogenic effect of *N,N*-diethyltryptamine, a new tryptamine derivative," *J. Ment. Sci.*, **1959**, *105*, 171-181. (d) Faillace, L. A.; Vourlekis, A.; Szara, S., "Clinical evaluation of some hallucinogenic tryptamine derivatives," *J. Nerv. Ment. Dis.*, **1967**, *145*, 306-313. DET is said by most authorities to be orally inactive: Grinspoon, *PsyDR*, p. 20; Shulgin, *BMC4*, **3**, 1122; Stafford, P., *Psychedelics Encyclopedia*, 3rd ed., Ronin: Berkeley, CA, p. 324. But Szara (*Psychotomimetic Drugs*, Efron, D. H., Ed., Raven Press: New York, 1970, p. 276) indicates it is orally active, and Shulgin (personal communication) has confirmed that his forthcoming *TIHKAL* contains stories of fairly intense effects from DET in oral doses of about 100 mg.

121. (a) Richards, W. A.; Rhead, J. C.; DiLeo, F. B.; Yensen, R.; Kurland, A. A., "The peak experience variable in DPT-assisted psychotherapy with cancer patients," *J. Psychedelic Drugs*, **1977**, *9*, 1-10. (b) Richards, W. A.; Rhead, J. C.; Grof, S.; Goodman, L. E.; DiLeo, F.; Rush, L., "DPT as an adjunct in brief psychotherapy with cancer patients," *Omega*, **1979**, *10*, 9-26. (c) Richards, W. A., "Mystical and archetypal experiences of terminal patients in DPT-assisted psychotherapy," *J. Religion and Health*, **1978**, *17*, 117-126.

122. Gillin, J. C.; Kaplan, J.; Stillman, R.; Wyatt, R. J., "The psychedelic model of schizophrenia: the case of *N,N*-dimethyltryptamine,"*Am. J. Psychiatry*, **1976**, *133*, 203-208.

123. Corbett, L.; Christian, S. T.; et al., "Hallucinogenic *N*-methylated indolealkylamines in the cerebrospinal fluid of psychiatric and control populations," *Brit. J. Psychiat.*, **1978**, *132*, 139-144.

124. (a) Strassman, R. J.; Qualls, C. R., "Dose-response study of *N,N*-dimethyltryptamine in humans: I. Neuroendocrine, autonomic, and cardiovascular effects," *Arch. Gen. Psychiatry*, **1994**, *51*, 85-97. (b) Strassman, R. J.; Qualls, C. R.; et al., "Dose-response study of *N,N*-dimethyltryptamine in humans: Subjective effects and preliminary results of a new rating scale," *Arch. Gen. Psychiatry*, **1994**, *51*, 98-108.

125. Strassman, R., personal communication. The psilocybin studies, which include data on the highest dosage to be administered in a clinical setting, are still unpublished as of this writing (June 1996).

126. (a) McKenna, T., *Archaic Revival: Collected Essays and Conversations*, Citadel: New York, 1991. (b) McKenna, T., *The Archaic Revival: Speculations on Psychedelic Mushrooms, the Amazon, Virtual Reality, UFOs, Evolution, Shamanism, the Rebirth of the Goddess, and the End of History*, Harper: San Francisco, 1992. (c) McKenna, T., *Food of the Gods*, Bantam: New York, 1993.

127. Subject O, cited in Meyer, P., "Apparent communication with disincarnate entities induced by dimethyltryptamine (DMT)," in *Psychedelics: A Collection of the Most Exciting New Material on Psychedelic Drugs,* Lyttle, T., Ed., Barricade Books: New York, 1994, p. 173.

128. *Ibid.*, p. 172.

129. Turner, D. M., *The Essential Psychedelic Guide,* Panther Press: San Francisco, 1994, p. 54.

130. *Ibid.*, pp. 55-56.

131. Ted, M., "Smokable dimethyltryptamine from organic sources," http://hyperreal.com/drugs-/psychedelics/tryptamines/smokable.dmt.

132. Indeed, the grasses are a well-known problem among farmers who raise ruminants; the toxins in these grasses are blamed for the occurrence of such diseases as perennial ryegrass staggers. As is the case with toad venom, it is likely that the process of smoking the extract destroys the more dangerous toxins in these grasses but volatilizes the relatively harmless DMT. Two review articles: (a) Bouke, C. A., "The clinical differentiation of nervous and muscular locomotor disorders of sheep in Australia," *Aust. Vet. J.*, **1995**, *72*, 228-234. (b) Cheeke, P. R., "Endogenous toxins and mycotoxins in forage grasses and their effects on livestock," *J. Anim. Sci.,* **1995**, *73*, 909-918.

133. Xochi speaking (see note 117).

134. Of course, there is no accounting for tastes, and some people are fond of elephants. Stafford, P., *Psychedelics Encyclopedia*, 3rd ed., Ronin Publishing: Berkeley, CA, 1992, p. 322.

135. It occurs in several *Virola* species, in *Piptadenia peregrina, Psychotria viridis* and others. Indians in Colombia, Brazil, and the West Indies make psychoactive snuffs from the bark and seeds of these plants. See *PsyDR*, p. 18, and references cited therein.

136. All reports of the psychedelic activity of bufotenine seem to originate with some experiments carried out on volunteers in the Ohio State Penitentiary in the 1950s: Fabing, H. S.; Hawkins, J. R., "Intravenous bufotenine, injection in the human being," *Science*, **1956**, *123*, 886. The authors had an odd idea of what constituted an "LSD-like" experience. The subjects' skin turned the color of an eggplant, they vomited and saw red-purple spots passing before their eyes. For complete references to the evidence for and against the hallucinogenic activity of bufotenine, see Weil, A.; Davis, W., "*Bufo alvarius:* a potent hallucinogen of animal origin," *J. Ethnopharmacology*, **1994**, *41*, 1-8.

137. (a) Fischer, R., *Nature,* **1968**, *220*, 411. (b) Turner, W. J.; Merlis, S., "Effect of some indole alkylamines on man," *Arch. Neur. Psychiatry,* **1959**, *81*, 121-29.

138. Wieland, H.; Konz, W.; Mittasch, H., "Die Konstitution von Bufotenin und Bufotenidin. Über Kröten-Giftstoffe. VII," *Ann*, **1934**, *513*, 1-25. The exact species of toad is not given, save that it is described as the "einheimische" variety. This "domestic" or "household" variety seems to have been abundant enough: in the detailed description of the research given in those more chatty days, it is said that 27,000 of them were collected in the region of Freiburg im Breisgau during a ten-day period in April. The venom was expressed by squeezing the glands behind their eyes with pincers, and then the gentle creatures were once again set free. Wieland (1877-1957), who had received the Nobel Prize in Chemistry in 1927 for his work on bile acids and was in 1934 the director of chemical research at the University of Munich, shows a commendable sensitivity to what we would now call "animal rights," since he assures us that "die Tiere erleiden durch den Entzug der Giftstoffe keinen Schaden," (the beasts suffer no injury from the removal of their venom). To make sure, the region was checked after a few days for dead toads, and none were found. Wieland further points out that they had caught and released 36,000 toads the spring of 1933, and the fact

they were able to bag 1,000 more in the spring of 1934 showed that the population could not have been harmed. (The reader is begging us to make a cynical remark relative to the quality of mercy of future politics in the region, but the author is above indulging this.)

Bergtod am Matterhorn. There is yet further evidence of the humanity of Bavarian Wissenschaft. The dangers of toad venom and psychedelic drugs are nothing to the dangers of the Alpine recreations prized in the region, and the third author of this work, Heinz Mittasch, had experienced a fatal misadventure while indulging these passions. Wieland dedicates the work to this "exceptionally gifted and likeable (ungewöhnlich begabten und sympathischen) student," who had "am 11. August 1932 am Matterhorn den Bergtod erlitten." I could not find this amazing word, *Bergtod*, in any German dictionary, but its meaning is obvious enough: he fell off a mountain.

139. The pun does not exist in German, where toads are *Kröte* and the mushrooms *Knollenblätterpilzen*, in this case *Amanita mappa.* Wieland, T.; Motzel, W.; Merz, H., "Über das Vorkommen von Bufotenin im gelben Knollenblätterpilz," *Ann*, **1953**, *581*, 10-16. Theodor had gotten his Ph.D. under his father at Munich and had published a paper with him about the isolation of bufotenine from toads (*Ann*, **1937**, *528*, 239); his familiarity with the substance made the identification of bufotenine in toadstools much easier, but he by no means expected to find a toad venom in fungi.

140. Stromberg, V., "The isolation of bufotenine from *Piptadenia peregrina*," *J. Am. Chem. Soc.*, **1954**, *76*, 1707-1710. The alkaloid does not seem to exist in either leaves or branches but only in the seeds. Since, as we have seen, the resulting snuff contains DMT and 5-Methoxy-DMT as well, it would conceivably produce quite a trip.

141. Horgan, J. "Bufo abuse: a toxic toad gets licked, boiled, teed up and tanned," *Scientific American*, **1990**, *262 (8)*, August, pp. 26-27. There is supposedly a rock band called *Mojo Nixon and the Toad Lickers* and a 1989 documentary film, "Cane Toads," describing toad-venom smokers in Australia, which has become a cult-video favorite, according to Bill Richards, a staff reporter for the *Wall Street Journal* ("Toad-smoking gains on toad-licking among drug users: toxic, hallucinogenic venom, squeezed, dried and puffed, has others turned off," p. A1, A8). From the most reliable study of the matter: Weil, A.; Davis, W., "*Bufo alvarius:* a potent hallucinogen of animal origin," *J. Ethnopharmacology*, **1994**, *41*, 1-8.

142. See the amusing conversation between Richard Schultes (the Harvard botanist who classified *Banisteriopsis caapi*) and Claudio Naranjo (who was convinced that archetypical images were provoked by the drug) as narrated by Shulgin (*PIHK*, pp. 68-69). Schultes had taken the decoction more than a dozen times, but to Naranjo's great dismay had never seen anything but "wiggly lines." A note on a web bulletin board (http://hyperreal.com/drugs/-psychedelics/tryptamines/practical.yage) claims of yage that "there seems to be a propensity for this chemical to produce visions of cats and serpents . . . and a smokey, bluish haze that seems to drift about. McKennah [sic, Terence McKenna?] has given both the beta carbolines and the raw plant to Eskimos who have never seen either large cats OR serpents in the Arctic [but maybe on TV?] . . . and they saw them. He was proposing a sort of Periodic Table of Hallucinogens at one time." The same source advises that yage, doubtless due to its MAOI effect, "is EXCELLENT for increasing the intensity of *P. cubensis* intoxication. As little as 25 mg or so can VASTLY increase within minutes, the 'levels' one is traversing. I usually just put a pinch between my cheek and gum."

143. *The Letters of William S. Burroughs: 1945-1959*, O. Harris, Ed., Viking: New York, 1993, pp. 155, 171. There are further descriptions throughout this interesting volume of letters, many of which appear later with elaborations in: Burroughs, W. S.; Ginsberg, A., *The Yage*

Letters, City Lights Books: San Francisco, CA, 1963. In the latter are some interesting sketches by Burroughs of what he saw under the influence of yage ("The Great Being," and "The Vomiter") as well as a poem by Ginsberg.

144. Nonetheless, he concludes that "Yage is it. It is the drug really does what the others are supposed to do. This is the most complete negation possible of respectability." *The Letters of William S. Burroughs*, p. 180.

145. Callaway, J. C.; Airaksinen, M. M.; et al., "Platelet serotonin uptake sites increased in drinkers of *ayahuasca*," *Psychopharmacology*, **1994**, *116*, 358-387. It was found that healthy members of the religious group who had been drinking the tea at least biweekly for 10 years or more had an increased number of serotonin uptake sites relative to controls, but the authors conclude that this "is not indicative of an undesirable neurological or psychiatric state. On the contrary, the religious use of *ayahuasca* in Brazil is now protected by law because of its apparent benefits to the individual, their families and community."

146. Fernandez, J., "Tabernanthe iboga: narcotic ecstasis and the work of the ancestors." In: *Flesh of the Gods: The Ritual Use of Hallucinogens*, Furst, P. T., Ed., Praeger: New York, 1972, pp. 237-260.

147. Naranjo, C., *The Healing Journey*, New York: Ballantine, 1975.

148. *Chemical Abstracts* **116**:100980b; **116**:17031x; **102**:160426w.

149. Bor, J., "Hallucinogen touted as stopping addicts' craving," *The Baltimore Sun*, 9 August 1993, p. 1A, 6A.

150. *Ibid.*

151. Mash, D. C.; Staley, J. K.; et al., "Identification of a primary metabolite of ibogaine that targets serotonin transporters and elevates serotonin," *Life Sci.*, **1995**, *57*, 45-50.

152. Mash, D. C.; Staley, J. K.; et al., "Properties of ibogaine and its principal metabolite (12-hydroxyibogamine) at the MK-801 binding site of the NMDA receptor complex," *Neurosi. Lett.*, **1995**, *192*, 53-56.

153. Heffter, A. *Berichten*, **1896**, *29*, 216-227.

154. But the ingenuity of gardeners is unlimited. A recent article (King, W., "Perpetual Peyote," *High Times*, March 1996, pp. 54-55, 66) shows photographs of dozens of large peyote cacti in bloom, and concludes: "I've filled my journal with more notes than I could ever print here—entries dealing with the medicinal, spiritual and universal concepts I've been blessed with while working with this sacred plant. Many truths have been revealed to me. And one truth is that anyone can grow peyote." The trick seems to be grafting the peyote onto larger, hardier, more rapidly growing cacti species.

155. This is the view with regard to these terms of Omer C. Stewart, emeritus professor of anthropology at the University of Oklahoma: Stewart, O. C., *Peyote Religion: A History*, University of Oklahoma Press: Norman, OK, 1987; undoubtedly, it is correct for the Indian tribes of Canada, the United States, and northern Mexico. The shaman of Carlos Castaneda—Don Juan, the Sonoran Indian who (supposedly) was born in southwest Mexico and lived most of his life in central Mexico—refers to the "spirit" of peyote as "Mescalito," and occasionally refers to peyote itself in the same way. See *The Teachings of Don Juan: A Yaqui Way of Knowledge*, Ballantine: New York, 1968, pp. 34-35 and p. 89 in particular. Other writings: *A Separate Reality: Further Conversations with Don Juan*, Simon & Shuster: New York, 1971. Castaneda claimed to have apprenticed himself to a shaman of Yaqui Indian origin while doing field work for his doctoral degree in anthropology. His apprenticeship supposedly lasted several years and included acquaintance with several

hallucinogenic plants including psilocybin and datura.

Castaneda's descriptions of psychedelic experience (which often sound much more like the effects of excessive tequila) should be taken, like tequila itself, with generous pinches of salt: they are probably part of a wholly fictional, albeit gloriously creative, spoof. See *Seeing Castaneda: Reactions to the "Don Juan" Writings of Carlos Castaneda,* D. C. Noel, Ed., Perigee (Putnam): New York, 1976; and *The Don Juan Papers: Further Castaneda Controversies*, R. de Mille, Ed., Ross-Erickson: Santa Barbara, 1980.

156. Poisson found 1.2 g mescaline/kilo of cactus. Poisson, J., "The presence of mescaline in a Peruvian cactus," *Annales Pharmaceutiques Françaises*, **1960**, *18*, 764-765. Poison, J., "Note sur le 'natem,' boisson toxique peruvienne et ses alcaloides," *Annales Pharmaceutiques Français*, **1965**, *23*, 241-244. Turner, W. J.; Heyman, J. J., "The presence of mescaline in *Opuntia cylindrica* (*Trichocereus pachanoi*)," *J. Org. Chem*, **1961**, *25*, 2250.

157. Sharon, D., "The San Pedro Cactus in Peruvian Folk Healing." In: *Flesh of the Gods: The Ritual Use of Hallucinogens*, Furst, P. T., Ed., Praeger: New York, 1972, pp. 114-135.

158. Ramo de Inquisición, tomo 289, Archivo General de la Nación. Quoted in Stewart, O. C., *Peyote Religion*, University of Oklahoma Press: Norman, OK, 1987, pp. 20-21.

159. The actual name of the people for themselves is *Wixárika*. Many of the group know Spanish, but they speak among themselves only in the Huichol language, which is in the Uto-Aztecan family along with that of the Hopi of Arizona and the 16th-century Aztecs. The anthropologist Peter Furst accompanied the Huichol on two pilgrimages to collect the sacred cactus, and on the second he was allowed to film much of the journey. "To Find Our Life: Peyote Among the Huichol Indians of Mexico." In: *Flesh of the Gods: The Ritual Use of Hallucinogens*, Furst, P. T., Ed., Praeger: New York, 1972, pp. 136-184. For a sensitive portrayal of the religious dimensions of the Huichol peyote ritual: Ruland, V., *Eight Sacred Horizons: The Religious Imagination East and West*, Macmillan: New York, 1985, pp. 1-24.

160. Furst, P. T., "To find our life: Peyote among the Huichol Indians of Mexico." In: *Flesh of the Gods: The Ritual Use of Hallucinogens*, Furst, P. T., Ed., Praeger: New York, 1972, pp. 180-181.

161. Steinmetz, P. B., *Pipe, Bible, and Peyote Among the Oglala Lakota: A Study in Religious Identity,* University of Tennessee Press: Knoxville, TN, 1990, p. 99.

162. Stewart, O. C., *Peyote Religion: A History*, University of Oklahoma Press: Norman, OK, 1987, pp. 54-67. Quoted material is from p. 66.

163. For a summary of his description of the ceremony as practiced in the 1890s and for references, see Stewart, O. C., *Peyote Religion: A History*, University of Oklahoma Press: Norman, OK, 1987, pp. 36-39.

164. *Ibid.,* p. 339.

165. *Ibid.,* pp. 328-330.

166. Stewart, O. C., *Peyote Religion . . .* , Chapter 8, "Efforts to pass a federal law," pp. 213-238. Cited material from pp. 220-221.

167. There was strong opposition to peyote from some unexpected quarters. Mabel Dodge Luhan, a rich and sophisticated New Yorker, had moved to Taos in 1920, married a leading Taos Indian peyotist, Tony Luhan, and lived near the pueblo in a harmoniously designed adobe house. Soon a colony of artists and intellectuals sprang up, attracting such luminaries as D. H. Lawrence. But she had been revolted by peyote since its use in a pseudo-peyote ceremony that took place in her New York apartment in 1914: it had a revolting taste, and under its influence some of her guests seemed to be losing their minds. Mabel made Tony

quit peyote when he married her, and she tried to get all the other Taos indians to abandon it too.

168. Nor does it protect non-American Indians. The possibility that a non-Indian member of the NAC or a non-Indian member of another church [the Peyote Way, e.g.] could claim the right to use peyote sacramentally, appealing to the 14th Amendment for equal protection, has been ruled out. The courts have held that the exemption granted American Indians is not racial but political and hence cannot be extended to non-Indians. Bullis, R. K., "Swallowing the scroll: legal implications of the recent supreme court peyote cases," *Journal of Psychoactive Drugs*, **1990**, *22*, 325-332. In a practice oddly reminiscent of the racial laws of the Third Reich, persons wishing to practice this religion in the United States must prove that they have a minimum of 25% Native American racial identity.

169. Bullis, R. K., "Swallowing the scroll: legal implications of the recent supreme court peyote cases," *Journal of Psychoactive Drugs*, **1990**, *22*, 325-332. Quoted material on pp. 325, 327, 331, 332.

170. *The New York Times*, 10 May 1993.

171. A more detailed bill, the Native American Free Exercise of Religion Act of 1993 [S.1021], introduced by Senators Inouye and Baucus, did not pass. It explicitly stated that "the traditional ceremonial use by Indians of the peyote cactus is integral to a way of life that plays a significant role in combating the scourge of alcohol and drug abuse among some Indian people," and would have declared that the use by Indians of peyote for ceremonial purposes in Native American religions "is lawful and shall not be prohibited by the Federal Government or any State" (Section 202).

172. "For Indian church, a critical shortage," *The New York Times*, Monday 10 March 1995, p. A10. Quoted is Emerson Jackson, a president of the Native American Church of North America from 1978 to 1991.

173. Heffter, A., "Ueber Cacteenalkaloïde," *Berichten*, **1896**, *29*, 216-227.

174. Späth, E., "Über die Anhalonium-Alkaloide," *Monatshefte für Chemie*, **1919**, *40*, 129-154.

175. Späth, E., Becke, F., "Über die Trennung der Anhaloniumbasen," *Monatshefte für Chemie*, **1935**, *66*, 327-336.

176. Kapadia, Shah, *Lloydia*, **1967**, *30*, 287.

177. Burroughs, W. S., "Letter from a master addict to dangerous drugs." Appendix to Burroughs, W. S., *Naked Lunch*, Grove Weidenfeld: New York, 1959, pp. 229. [Originally printed in *The British Journal of Addiction*, *53* (2).] See also (a) Burroughs, W. S., *Junky*, Penguin Books: New York, 1977, p. 145-147. (First published as *Junkie*, Ace Books: New York, 1953.) (b) *The Letters of William S. Burroughs: 1945-1959*, O. Harris, ed., Viking: New York, 1993, p. 130.

178. Hermle, L.; Fünfgeld, M.; Oepen, G.; Botsch, H.; Borchardt, D.; Gouzoulis, E.; Fehrenbach, R. A.; Spitzer, M., "Mescaline-induced psychopathological, neuropsychological, and neurometabolic effects in normal subjects: experimental psychosis as a tool for psychiatric research," *Biological Psychiatry*, **1992**, *32*, 976-991. Citation on p. 977. It is also possible, of course, that these volunteers were as convinced that they would become "psychotic" when they consumed mescaline as William Burroughs was convinced that everything would look like peyote plants if he consumed peyote. Psychedelics make one very suggestible. In 1956, when mescaline was considered a "psychotomimetic," doctoral psychology students taking quite modest dosages (200 mg of the sulfate) sometimes behaved in a thoroughly psychotic way, just as they were expected to. See Sinnett, E. R., "Experience and reflec-

tions," in *Psychedelics: The Uses and Implications of Hallucinogenic Drugs*, Aaronson, B.; Osmond, H., Eds., Anchor Doubleday: Garden City, NY, 1970, pp. 29-35.

179. Huxley, L., *This Timeless Moment: A Personal View of Aldous Huxley*, Farrar, Straus & Giroux: New York, 1968, pp. 139-141.

180. Smith, Huston, "Empirical metaphysics." In: *The Ecstatic Adventure: Reports of Chemical Explorations of the Inner World . . .* , Metzner, R., Ed., Macmillan: New York, 1968, pp. 72-74. Huston Smith's experience was that of a Plotinian emanationist monism; for a description of the experience of a "Death of God" undifferentiated monism on the part of an amateur theologian taking mescaline, see Braden, William, *The Private Sea: LSD and the Search for God*, Quadrangle Books, Chicago, 1967.

181. Yeats (1865-1939) was probably the greatest lyric poet that Ireland has produced; in 1923 he would receive the Nobel Prize in literature. At the time of his peyote experiments with Ellis, he was involved with several occultist groups including the theosophists and the Rosicrucians. Yeats had previously experimented with hashish and opium.

182. Ellis, H., "Mescal: A new artificial paradise." In: *The Hashish Club: An Anthology of Drug Literature,* Vol. I., Haining, P., Ed., pp. 176-188. Cited material pp. 195-186. Ellis' work was originally published in 1898 in *The Contemporary Review.*

183. James, H., *The Letters of William James Edited by His Son Henry James*, Atlantic Monthly Press: Boston, 1920, Vol. II, p. 35.

184. James, W., *The Correspondence of William James, Vol. 2: William and Henry, 1885-1896*, Krupskelis, I. K.; Berkeley, E. M., Eds., University Press of Virginia: Charlottesville, 1993, p. 403. Letter of June 11, 1896. For Weir Mitchell's vivid descriptions of the phantasmata he experienced under mescaline, see Mitchell, S. W., "Remarks on the effects of Anhalonium Lewinii (the mescal button)," *British Med. Journal* **1896**, 2, 1625-1629.

185. Djerassi, C., *Steroids Made It Possible* [*Profiles, Pathways, and Dreams: Autobiographies of Eminent Chemists*, Seeman, J. I., Ed.], American Chemical Society: Washington, DC, 1990, p. 15-16.

186. *Ibid.*, p. 17. Djerassi gives a more detailed description of this event in the *Michigan Quarterly Review* (**1989**, *28(1)*, 123-129), where he states that they each took one-tenth of the acute toxic dose for mice, adjusted for their body weight. The *Merck Index* gives an acute toxic dose of the free base i.p. for rats, and of the hydrochloride salt, presumably p.o., for mice, rats, and guinea pigs; Djerassi says they used the sulfate. In any case, using even the smallest of these numbers, which range from 132 to 370 mg/kg, and using a 70 kg human body mass, the dosages must have been enormous: 0.10 x 132 x 70 = 924 mg, a very hefty dose of mescaline, whether sulfate or hydrochloride. And much of the "personality alteration" Djerassi describes in this earlier article (e.g., "I announced to my wife that with mescaline I seemed to be shedding one protective layer of my personality after another . . . soon, I declared, I would arrive at my ultimate truth") seem to this reader anything but trivial, if somewhat tinged with ambivalence and denial.

187. That is, as a ratio of toxicity in mice (LD_{50} = 240 mg/kg) to effective dose in humans (200 mg); the LD_{50} of psilocybin in mice is only slightly higher, 285 mg/kg, but it has a much smaller effective human dose of 10 mg.

188. "I know of a man who substituted peyote during late withdrawal, claimed to lose all desire for morphine, ultimately died of peyote poisoning." Burroughs, W. S., "Letter from a master addict to dangerous drugs." Appendix to Burroughs, W. S., *Naked Lunch*, Grove Weidenfeld: New York, 1959, p. 224. [Originally printed in *The British Journal of Addiction*, *53*

(2).] "By the way, Peyote poisoning presents symptoms similar to polio." Letter to Allen Ginsberg, March 20, 1952. *The Letters of William S. Burroughs: 1945-1959*, O. Harris, ed., Viking: New York, 1993, p. 105. Both these remarks sound like the usual street myths that accompany the use of many illegal drugs.

189. Späth, E., "Über die Anhalonium-Alkaloide," *Monatshefte für Chemie*, **1919**, *40*, 129-154.

190. Benington, F.; Morin, R. D., *J. Am. Chem. Soc.*, **1951**, *73*, 1353. There has been little reported illicit synthesis of mescaline, but the Wiesbaden prosecutor's office has reported confiscation of mescaline contaminated by its α-hydroxy derivative, (a known hallucinogen of weak potency), evidently from a botched condensation of nitromethane with trimethoxybenzaldehyde: Bernhauer, D.; Fuchs, E-F.; Gloger, M.; Vordermaier, G., *Archiv für Kriminologie*, **1983**, *171*, 151-160.

191. Slotta, K. H.; Heller, H., *Berichten*, **1930**, *63*, 3029-3044.

192. Tsao, M. U., *J. Am. Chem. Soc.*, **1951**, *73*, 5495-5496.

193. (a) Short, J. H.; Dunnigan, D. A.; Ours, C. W., *Tetrahedron*, **1973**, *29*, 1931-1939. (b) Aboul-Enein, M. N.; Eid, A. I., *Acta Pharm. Suecica*, **1979**, *16*, 267-270.

194. Banholzer, K.; Campbell, T. W.; Schmid, H., *Helvetica Chimica Acta*, **1952**, *35*, 1577-1581.

195. Shulgin, A. T., "Hallucinogens." In: *BMC4*, **3**, 1109-1137; in particular, pp. 1117-1122, and the tables on pp. 1118, 1120, and 1121. For numerous vivid first-hand accounts of the psychic activity of many of these compounds, see Alexander and Ann Shulgin's *PIHK*. As of 1996, the Shulgins have nearly completed a companion volume entitled *TIHKAL* (Tryptamines I Have Known and Loved).

196. The volume is several things at the same time, which at first seem to be an unlikely combination, but then eventually somewhat surprisingly hang together: the first 500 or so pages of the work, called Book I, include a number of first-hand accounts of the use of a dozen or so psychedelics with enough detail for the reader to get a good feeling for the set and setting. Woven with this as an integral part is the story of how Ann and Sasha (under the rather thin pseudonymous cloak of Alice and Shura) met, loved, and eventually married. (The plot line of this triangular love story involving a strange German lady with an oddly constructed psyche becomes intriguing enough after a while to absorb the reader's interest in itself.) In the second 500 or so pages, called Book II, structures and syntheses of 179 compounds are given in "recipe" form followed in most cases by brief reports on self-experimentation by various anonymous volunteers. With drugs, sex, and Beethoven (Shura plays the viola), there should be something here for everyone.

197. *PIHK*, pp. 511-515.

198. *PIHK*, pp. 515-518. In Book I, Chapter 15, "Tennessee," and Chapter 40, "Mortality" exploit the effects of 2C-E.

199. *PIHK*, p. 506: "If there is anything ever found to be an effective aphrodisiac, it will probably be patterned after 2C-B in structure." See Book I, Chapter 24, "2C-B."

200. Personal communication. Letter of March 1994. Other particularly interesting drugs in this class are those in which the 4 position (R of **6-68, 69, 70**) is substituted by various alkyl thio groups such as ethylthio (2C-T-2), isopropylthio (2C-T-4), and propylthio (2C-T-7). Shulgin describes these as being "marvelous tools of psychotherapy" with "broad acceptance in the therapeutic community." A psychotherapist who has used these drugs extensively is Myron Stolaroff, who describes them as eliciting "empathic qualities, which lead to free communication and feelings of well-being." Stolaraff, M. J., *J. Psychoactive Drugs,* **1990**, *22*, 379 [letter].

201. (a) McKenna, D. J.; Mathis, C. A.; Shulgin, A. T.; Sargent, T.; Saavedra, J. M., "Autoradiographic localization of binding sites for [^{125}I]DOI, a new psychotomimetic radioligand, in the rat brain," *Eur. J. Pharmacology,* **1987**, *137*, 289-290. (b) McKenna, D.J.; Nazarali, A. J.; Hoffman, A. J.; Nichols, D. E.; Mathis, C. A.; Saavedra, J. M., "Common receptors for hallucinogens in rat brain: a comparative autoradiographic study using [^{125}I]LSD and [^{125}I]-DOI, a new psychotomimetic radioligand," *Brain Research*, **1989**, *476*, 45-56. (c) Pazos, A.; Cortes, A.; Palacios, J. M., "Quantitative autoradiographic mapping of serotonin receptors in the rat brain. II: Serotonin receptors," *Brain Research*, **1985**, *346*, 231-249.

202. Shulgin, *PIHK*, p. 641. As Shulgin explains in Chapter 9, pp. 53-56, of *PIHK*, he had originally developed DOM (*des*oxy*m*ethyl TMA-2) as one of his first variations back in 1963; its appearance as a street drug 4 years later at the height of the psychedelic era took him completely by surprise—and dismay. STP appeared as a street drug while Shulgin was in medical school and just a few months after he had given a talk on DOM at Johns Hopkins. It is remotely possible that the STP acronym was an act of homage from someone attending this lecture who started manufacturing the drug, christening it with the patronymic Shulgin's Terrific Psychedelic. Shulgin had nothing to do with its production or distribution, and he laments that his association with the drug became for him something of a "hair shirt."

203. Jaffe, *G&G8*, p. 556.

204. *PIHK*, p. 639. For a more extended report on DOM in context, see Chapter 25, "Dragons," pp. 223-229.

205. Snyder, S. H.; Faillace, L. A.; Weingartner, H., "DOM (STP), a new hallucinogenic drug, and DOET: Effects in normal subjects," *American J. of Psychiatry*, **1968**, *125*, 357-364. In 1967, many Haight-Ashbury folks were used to LSD, and took a second 20 mg tablet of DOM when no effects were felt half an hour after taking the first; thus they were stoking up an enormous overdose and were liable to panic.

206. Weil, A., "The love drug," *J. Psychedelic Drugs*, **1976**, *8*, 335-337. See also *PIHK*, pp. 714-719.

207. (a) Glennon, R. A.; Young, R. Y., "MDA: an agent that produces stimulus effects similar to those of 3,4-DMA, LSD and cocaine," *Eur. J. Pharm.*, **1984**, *99*, 189-196. (b) Nichols, D. E.; Hoffman, A. J.; et al., "Derivatives of 1-(1,3-benzodioxol-5-yl)-2-butanamine. Representatives of a novel therapeutic class," *J. Med. Chem.*, **1986**, *29*, 2009-2015.

208. Yensen, R.; DiLeo, F.; Rhead, J. C.; Richards, W. A.; Soskin, R. A.; Turek, B.; Kurland, A. A., "MDA-assisted psychotherapy with neurotic outpatients: A pilot study," *J. Nerv. Ment. Dis.*, **1976**, 233-245.

209. Naranjo, C., *The Healing Journey*, Ballantine: New York, 1975.

210. Richards, R. N., "Experience with MDA," *Canadian Medical Association Journal*, **1972**, *106*, 256-259.

211. (a) Shulgin, A. T., "The background and chemistry of MDMA," *J. Psychoactive Drugs*, **1986**, *18*, 291-304. (b) Shulgin, A. T., "History of MDMA," in *PER*, pp. 1-20.

212. *PIHK*, p. 69.

213. (a) *PURX*, pp. 14-15. (b) *PIHK*, 73-74. (c) Greer, G. R.; Tolbert, R., "The therapeutic use of MDMA," in *PER*, pp. 21-35.

214. (a) Nichols, D. E., "Differences between the mechanism of action of MDMA, MBDB, and the classic hallucinogens. Identification of a new therapeutic class: entactogens," *J. Psychoactive Drugs*, **1986**, *18*, 305-312. (b) Nichols, D. E.; Oberlender, R., "Structure-Activity relationships of MDMA and related compounds: A new class of psychoactive agents?" in *PER*, pp. 105-131.

215. Romano, A. G.; Harvey, J. A., "MDMA enhances associative and nonassociative learning in the rabbit," *Pharmacol. Biochem. Behav.* **1994**, *47*, 289-293.

216. Riedlinger, T. J.; Riedlinger, J. E., "Psychedelic and entactogenic drugs in the treatment of depression," *J. Psychoactive Drugs,* **1994**, *26*, 41-55. Cited material p. 52.

217. Cited in Eisner, B., *Ecstasy: The MDMA Story*, Ronin: Berkeley, CA, 1989, p. 59.

218. I have freely translated. "Selbst ihrem Hasse liegt der Durst nach Liebe zum Grunde, und es entsteht keine Feindschaft, außer aus versagter Freundschaft."—Fichte, *Bestimmung der Menschen*, 189.

219. Eisner, B., *Ecstasy: The MDMA Story*, Ronin: Berkeley, CA, 1989, pp. 48-50. But upon Leary's death on 31 May 1996, it was reported "His family life was not so blessed. After his first wife's suicide and two divorces, he separated from his fourth wife, Barbara, a few years ago. His daughter, Susan, committed suicide in 1990." Mansnerus, L., "Timothy Leary, pied piper of psychedelic 60s, dies at 75," *The New York Times*, 1 June 1996, 1, 12.

220. *PURX*, pp. 76-77.

221. "Wohllust ward dem Wurm gegeben, und der Cherub steht vor Gott"—joy is dispensed even to worms, and the Cherubim stand before the face of God.

222. *PURX*, p. 39.

223. *PURX*, p. 19.

224. *PURX*, p. 50.

225. Randall, T., "Ecstasy-fueled 'rave' parties become dances of death for English youths," *JAMA*, **1992**, *268*, 1505-1506; "'Rave' scene, ecstasy use, leap Atlantic," *ibid.,* 1506.

226. The Dutch Ministry of Justice views illicit drug use as "a limited and manageable social problem rather than an alien threat forced upon an otherwise innocent society." (Leuw, E., "Dutch penal law and policy: notes on criminological research from the research and documentation centre," Ministry of Justice: The Hague, Netherlands, 4 November 1991.) Cited in *PURX*, p. 136. In an editorial in the October 1994 *Jellinek Quarterly* (the news bulletin of the Jellinek Institute and the Amsterdam Institute for Addiction Research), Jaap Jamin writes: "There is no point to conducting anti-Ecstasy campaigns. Forbidden fruits are the most delectable. . . . The history of subcultures has taught us that in general, repression is counterproductive. It polarizes and unites the members of the subculture. The consumer goes underground, and is difficult for prevention workers to reach. Marginalizing the consumer is apt to promote problematic use. . . . Let us have more faith in our youngsters and be willing to learn from them."

227. Although XTC is not legally sold in the Netherlands, the Dutch have developed a network of auxiliary services to minimize the likelihood of the sort of tragedy that occurred in Britain. The police will not prosecute anyone for possession of small amounts of MDMA, and this allows the house parties (as the raves are called in the Netherlands) to take place in the open, where the organizers of the rock groups are subject to the usual regulations for large public gatherings: ample facilities for ventilation, crowd control, sanitation, etc. A private group provides free analyses of samples of XTC submitted to them by prospective ravers,

and any of the underground-manufactured trade logos (usually cartoon animals imprinted on the tablets) that proves not to be MDMA are quickly blackballed by word of mouth among users. At the house party itself, Rotterdam city police and health officials are present, and any large-scale merchandising of XTC or any other drugs are curtailed. Jaap de Vlieger, coordinator for drug policy for the city of Rotterdam, described to me the circumstances of a large house party that took place in June 1994. About 15,000 young people attended the party, which took place on a large recently created polder. He showed me 50 bags of XTC confiscated from dealers at the party; each bag containing a few hundred tablets had its own idiographic imprimatur. (As a synthetic chemist, I found it interesting to note the considerable variation in the gross appearance of the samples, from near-chocolate brown to sparkling white.) Also confiscated were about 200 joints, 1,000 grams of marijuana, and 72 doses of cocaine. There was no heroin use noted. Three youths were hospitalized briefly and released after a few hours. Almost 200 participants were briefly treated at the mobile clinic at the rave scene for the usual range of psychological or physical injuries to be expected at a rock concert that lasted from Friday afternoon to Sunday morning. Alcoholic drinks were available but not much was used: pamphlets widely distributed by the Odysee-Hadon foundation advised users to "know how to party" with XTC, drinking at least three glasses of water or fruit juice, but not alcoholic drinks, per hour, etc. There was very little intoxication, according to Officer de Vlieger, who was present at the scene, and no violence. He remarked that this was in startling contrast to a typical rock concert where alcohol is the chief drug employed, where there would have been a very considerable amount of violence. While these arrangements still leave much to be desired, they represent a pragmatic compromise with the contradictory realities that have to be dealt with.

228. Newmeyer, J. A., "X at the crossroads," *J. Psychoactive Drugs,* **1993**, *25*, 341-342.

229. Fredlund, E.; Kavinsky, M.; et al., *The 1990 Texas School Survey of Substance Abuse*, Texas Commission on Alcohol and Drug Abuse: Austin, 1990. Referenced in *PURX*, p. 211.

230. Allen, R. P.; McCann, U. D.; Ricaurte, G. A., "(±)-3,4-methylenedioxymethamphetamine (MDMA, 'ecstasy') on human sleep," *Sleep*, **1993**, *16*, 560-564.

231. McCann, U. D.; Ricaurte, G. A., "Reinforcing subjective effects of (±)-3,4-methylenedioxymethamphetamine ('ecstasy') may be separable from its neurotoxic actions: clinical evidence," *J. Clin. Psychopharmacol.* **1993**, *13*, 214-217.

232. Scanzello, C. R.; Hatzidimitriou, G., et al., "Serotonergic recovery after (±)-3,4-(methylenedioxy)methamphetamine injury: observations in rats," *J. Pharmacol. Exp. Ther.*, **1993**, *264*, 1484-1491.

233. Ricaurte, G. A.; Finnegan, K. T.; et al., "Aminergic metabolites in cerebrospinal fluid of humans previously exposed to MDMA: preliminary observations," *The Neuropharmacology of Serotonin*, Whitaker-Azmitia, P. M.; Peroutka, S. J., Eds. [*Annals of the New York Academy of Sciences*, **1990**, *600*], 699-710.

234. Krystal, J. H.; Price, L. H., et al., "Chronic 3,4-methylenedioxymethamphetamine (MDMA) use: effects on mood and neuropsychological function?" *Am. J. Drug Alcohol Abuse,* **1992**, *18*, 331-341.

235. Dornan, W. A.; Katz, J. L.; Ricaurte, G. A., "The effects of repeated administration of MDMA on the expression of sexual behavior in the male rat," *Pharmacol. Biochem. Behav.*, **1991**, *39*, 813-816.

236. McCann, U.; Hatzidimitriou, G., et al., "Dexfenfluramine and serotonin neurotoxicity: further preclinical evidence that clinical caution is indicated," *J. Pharmacol. Exp. Ther.*, **1994**, *269*, 792-798. It is of interest that MDMA generalizes to a fenfluramine cue in discrimination

studies: Schechter, M. D., "Discriminative profile of MDMA," *Pharmacol. Biochem. Behav.*, **1986**, *24*, 1533-1536.

237. Cited in Voelker, R., "Obesity drug renews toxicity debate," *JAMA*, **1994**, *272*, 1087-1088. For further reviews of MDMA neurotoxicity, see (a) Gibb, J. W.; Stone, D.; et al., "Neurochemical effects of MDMA," *PER*, pp. 133-150. (b) Schmidt, C. J.; Taylor, V. L., "Neurochemical effects of methylenedioxymethamphetamine in the rat: Acute versus long-term changes," *PER*, pp. 151-169. (c) Battaglia, G.; Zaczek, R.; et al., "MDMA effects in brain: Pharmacologic profile and evidence of neurotoxicity from neurochemical and autoradiographic studies," *PER*, pp. 171-199. (d) Whitaker-Azmitia, P. M.; Azmitia, E. C., "A tissue culture model of MDMA toxicity," *PER*, pp. 201-211.

238. For a survey of 20 psychiatrists who had taken MDMA and two commentaries with contrasting views: (a) Liester, M. B.; Grob, C. S.; et al., "Phenomenology and sequelae of 3,4-methylenedioxymethamphetamine use," *J. Nerv. Ment. Dis.*, **1992**, *180*, 345-352. (b) Kosten, T. R.; Price, L. H., "Commentary: Phenomenology and sequelae of 3,4-methylenedioxymethamphetamine use," *ibid.*, pp. 353-354. (c) Grob, C. S.; Bravo, G. L.; et al., "Commentary: The MDMA-neurotoxicity controversy: Implications for clinical research with novel psychoactive drugs," *ibid.*, pp. 355-356.

239. Nichols, D. E.; Oberlender, R., *The Neuropharmacology of Serotonin*, Whitaker-Azmitia, P. M.; Peroutka, S. J., Eds. [*Annals of the New York Academy of Sciences*, **1990**, *600*], 613-625. Cited material p. 613-614.

240. (a) Nichols, D. E.; Brewster, W. K.; Johnson, M. P.; Oberlender, R.; Riggs, R. M., *J. Med. Chem.*, **1990**, *33*, 703-710. (b) Johnson, M. P.; Frescas, S. P.; Oberlender, R.; Nichols, D. E., *J. Med. Chem.*, **1991**, *34*, 1662-1668.

241. Zanger, R., "Dr. Peter Baumann: LSD therapy in Switzerland," *The Albert Hofmann Foundation Newsletter*, **1989**, *1*, 3-11. Referenced in *PURX*, pp. 151-152.

242. "The transformation of the objective world view into a deepened and thereby religious reality consciousness can be accomplished gradually, by continuing practice of meditation. It can also come about, however, as a sudden enlightenment; a visionary experience. It is then particularly profound, blessed, and meaningful. . . . *I see the true importance of LSD in the possibility of providing material aid to meditation aimed at the mystical experience of a deeper, comprehensive reality. Such a use accords entirely with the essence and working character of LSD as a sacred drug.*" Hofmann, A., *LSD, My Problem Child*, Ott, J., Trans., J. P. Tarcher (St. Martin's): Los Angeles, 1983, pp. 208-209. The translation seems infelicitous (philosophical and theological texts are notoriously difficult to translate into English), and in the German original the book ends only after a further paragraph omitted in the English. Here are the last two paragraphs of the German edition:

In der Möglichkeit, die auf mystisches Erleben einer zugleich höheren und tieferen Wirklichkeit ausgerichtete Meditation von der stofflichen Seite her zu unterstützen, sehe ich die eigentliche Bedeutung von LSD. Eine solche Anwendung entspricht ganz dem Wesen und Wirkungscharakter von LSD als sakraler Droge.

Die Unterstützung der Meditation durch LSD beruht auf den gleichen Wirkungen, die auch seiner Verwendung als medikamentöses Hilfsmittel in der Psychoanalyse und Psychotherapie zugrundeliegen, auf seiner Fähigkeit, die Ich-Du-Schranke, die Trennung von der Außenwelt vorübergehend zu lockern oder gar aufzuheben. Das begünstigt die Lösung aus einem ichhaftfixierten Problemkreis und das Finden einer bergenden Wirklichkeit. Hofmann, A., *LSD—Mein Sorgenkind*, Klett-Cotta (Ullstein Taschenbuch): Frankfurt am Main, 1982 (original edition 1979), p. 230.

243. Clark, W. H., "The psychedelics and religion." In: *Psychedelics: The Uses and Implications of Hallucinogenic Drugs,* Aaronson, B., Osmond, H., Eds., Anchor Doubleday: Garden City, NY, 1970, pp. 182-195. Cited material from pp. 192-195.

244. "The most balanced and authoritative work on the psychiatric aspects of these drugs to date. Its low profile in current psychiatric education is lamentable." (Strassman, R., *Am. J. Psychiatry*, **1986**, *143*, 249.)

7

Dissociatives and Cannabinoids: PCP, THC, ETCs

Hashish has nothing of that ignoble heavy drunkenness about it which the races of the North obtain from wine and alcohol: it offers an intellectual intoxication.

—Theophile Gautier[1]

There is a struggle in nature against this divine substance,—in nature, which is not made for joy, and clings to pain. Nature, subdued, must yield in the combat; reality must succeed to the dream; and then the dream reigns supreme. . . . You desire to live no longer, but to dream thus for ever. When you return to this mundane sphere from your visionary world, you seem . . . to quit paradise for earth, heaven for hell! Taste the hashish, guest of mine,—taste the hashish!

—Alexandre Dumas[2]

Heaven bless hashish, if its dreams end like this!

—Louisa May Alcott[3]

Introduction

The drugs we will consider in this chapter are difficult to categorize; each is to a large extent *sui generis*. That is not to deny their importance: marijuana has for many decades held the place of honor as the world's most widely used, most widely available, and safest illegal drug. PCP, although it is rarely heard of nowadays, seemed quite dangerous when it was popular; and ketamine, the drug that has replaced it in both legitimate and clandestine use, is intriguingly similar in many respects to the psychedelic drugs of the previous chapter, although it is known to act at an entirely different set of receptors. Finally, under the somewhat whimsical category of "ETCs" are grouped such substances as nitrous oxide, amyl nitrite, ether, chloroform, nutmeg, and absinthe, which fall into no obvious category. Most of these are of minor or mainly historical interest.

PCP and the Dissociative Anesthetics

Phencyclidine

History. Phencyclidine (Sernyl, Sernylan, "HOG," "PCP," "angel dust," "peace pill," "krystal," "ozone," "earth," **7-4**, Figure 7.1) was originally marketed by Parke-Davis as Sernyl for use as an anesthetic, but it was soon withdrawn due to the peculiar psychic symptoms of delirium and hallucinations that frequently, although not always, accompanied its use. Later it was reintroduced as Sernylan for use as a veterinary anesthetic; since then it has been largely supplanted by ketamine. However, it remains a Schedule III substance, and could in principle be used in legitimate medicine. It was not until the early 1970s that PCP began to appear as a street drug in capsule form for oral ingestion; since 1980, it is almost exclusively smoked, usually sprinkled on tobacco, marijuana, or dried parsley.[4] From the beginning users seem to have found the drug problematic, for it soon began to be smoked, either by sprinkling on tobacco, marijuana, or even parsley, in the belief that undesirable effects were minimized when it was taken by this route. By 1974 it was widely used; for a period it was deceptively marketed as "THC." It is easily and cheaply synthesized in underground labs and several novel "designer drug" variations of the PCP structure with essentially identical properties were legally sold at one time. In 1986, the Controlled Substances Act was amended to include the Controlled Substances Analogue Enforcement Act. The Attorney General can unilaterally declare that a new compound is in Schedule I or II if it has a chemical structure "substantially similar" to an existing drug of that schedule.[5]

In the early 1980s, yearly PCP sales in the Baltimore area alone were thought to be around $70 million. But since then, usage has steadily declined, and by the 1990s it was almost never reported as a street drug. However, tragic incidents still involve the drug: in March 1994, Lisa Bongiorno, a 28-year-old resident of Queens, was charged with two counts of murder when she lost control of her car and killed two leading officials of the Greek Orthodox Church. She admitted she was smoking PCP after stopping to buy two $10 bags in East Harlem while driving home from her job as a legal secretary in Manhattan, and that she had been using the drug for the previous 10 years.[6]

The relative decrease in usage of PCP probably represents in part the market phenomenon of customers finding that the product was less attractive in retrospect than in prospect. It appears that "PCP abuse may decrease when other more desired drugs like crack cocaine are widely available and inexpensive. This may account for the decreases

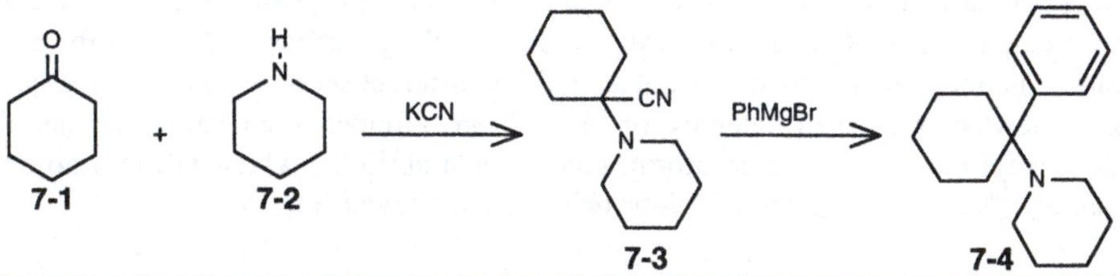

7-1 + 7-2 —KCN→ 7-3 —PhMgBr→ 7-4

Figure 7.1 Synthesis of PCP.

in PCP abuse in recent years in both Los Angeles and New York, since both cities showed an increase in the abuse of crack cocaine during the same period."[7]

Psychopharmacology. In small quantities (5 to 10 mg), phencyclidine causes a floating euphoria, stupor, an "eyes-open coma" and zombielike responses. Nystagmus (jerky eye movements) are common. There have been frequent claims that users become violent and exhibit superhuman strength; but "as with past drug horror stories, this is not often well documented."[8]

The earliest studies of the effects of PCP on humans, which date to 1958, reported that its action was unlike that of any other known anesthetic. Although it produced complete analgesia, the patients were still awake, with eyes open.[9] Soon after, it was noticed that transient psychotic states often developed in the postoperative period. Further trials in humans produced psychiatric abnormalities in about one of every six patients. Symptoms included hallucinations, echolalia, noisy delirium lasting as long as 12 hours, violent behavior (especially in younger males), and "maniacal excitement."[10] Because of this, the drug was never marketed commercially for human use, and clinical trials were discontinued in 1962. A careful study of its effects on nine normal (medical students or psychiatric residents) and nine schizophrenic patients, which was conducted in 1959, before the emergence of PCP as a street drug, reported the following symptoms:[11]

- *Body-Image Changes.* These were seen in all subjects. A normal subject reported "I was standing, but didn't know where my feet were." Another: "My arm feels like a 20-mile pole with a pin at the end." The experience of floating and flying was repeatedly described. The physical sensations were accompanied by a psychic sense of unreality, ego loss, and depersonalization.
- *Estrangement.* A sense of aloneness or isolation, of loneliness and detachment from everything in the environment was seen in all subjects. It was not necessarily accompanied by fear or anxiety; often it made the subjects feel passive and relaxed.
- *Disorganization of Thought.* This was seen in all subjects. There were neologisms, word salad, echolalia. Asked to interpret the proverb "A drowning man will clutch at a straw," one subject responded without the drug: "A person who is desperate will grab at anything regardless of its value to him." In the drug state, the same person responded: "I think it is—a drowning man . . . will clutch at a straw. It means a drowning man will clutch."
- *Negativism and Hostility.* Two thirds of the subjects experienced some form of this, ranging from catatonic withdrawal to responding "no" when asked if they could hear. A medical student in recovery said he had felt hostility to the examiner but was unable to express it.
- *Drowsiness and Apathy.* This symptom was relatively late to develop, but it occurred in all subjects.
- *Inebriation and Euphoria.* 78% of the normal subjects reported a sense of inebriation like that from alcohol.

The authors conclude that the PCP state is distinct from that induced by LSD and mescaline and resembles most closely an intense form of sensory deprivation. Grinspoon similarly concludes that, while some of the effects of PCP resemble those of the psychedelics, "it should probably be described as a tranquilizer, analgesic, or euphoriant rather than a psychedelic drug."[12]

The drug is very soluble in lipids, and tissue levels are much higher than those in plasma or blood; the drug and its metabolites can persist for long periods in the brain and other fatty tissue after a single dose, and even accumulate after several widely spaced doses. The half-life of PCP in humans is variously reported as ranging from 11 to 89 hours: the variability probably is in part a function of body fat. PCP, like cocaine and the opioids, but unlike any of the psychedelics, is self-administered by experimental animals.[13]

Because the drug is weakly basic, its excretion can be hastened by acidifying the urine to pH 5.5 or less and encouraging diuresis. Cranberry juice, which contains benzoic acid, and ascorbic acid can be used to acidify the urine; orange juice should be avoided because it turns the urine alkaline.[14] One report indicates that ascorbic acid acts synergetically with the antipsychotic haloperidol to reverse psychotic symptoms from PCP.[15]

Receptor Interactions. There are at least two receptor types at which PCP is known to interact. The first is the σ receptor subclass, once (but no longer) thought to be a species of opioid receptor (see Chapter 2), and which perhaps mediates some of the quasi-psychedelic or dysphoric characteristics of some of the opioid drugs. More interesting is the interaction of PCP as an antagonist at a specific site (often called the PCP/NMDA site) on the second class of receptors, those responding to *N*-methyl-D-aspartate (NMDA).[16] There is considerable work in progress (see Chapter 1) to develop safe drugs antagonizing these receptors, since they are capable of minimizing the damage to the brain from stroke or oxygen deprivation; however, some of the drugs that have been developed (e.g. *selfotel*[17]) cause psychic effects much like PCP; and a study conducted at Washington University School of Medicine found that PCP and two NMDA receptor ligands developed as a neuroprotective agents, *MK-801* (**7-9**, Figure 7.2) and *tiletamine*, **7-8**, as well as *ketamine* **7-7**, may produce pathological changes in brain neurons when administered subcutaneously to adult rats in relatively low doses.[18]

Ketamine

After PCP was made a Schedule II drug, several novel "designer drug" variations of the PCP structure (such as **7-5** and **7-6**) appeared on the street. Meanwhile, legitimate pharmaceutical labs were trying to modify the structure in order to create a drug with the useful anesthetic properties of PCP but lacking its peculiar psychic effects. To date, the most successful replacement is ketamine (Ketalar, Ketaject, "Super-K," "K," **7-7**). The structure of this prescription anesthetic and many of its effects are similar to PCP, although it is only about one-tenth as active. It induces a state of sedation and amnesia during which the patient is still conscious, but is dissociated from the environment, immobile, and unresponsive to pain. Induction is rapid, and there is no respiratory depression, making it very useful for burn patients, diagnostic studies, and minor surgical procedures in young children. However, psychic disturbances like those from PCP (delirium, hallucinations) sometimes accompany recovery. These may last several hours, and are for unknown reasons more frequently observed in adults than in children and in

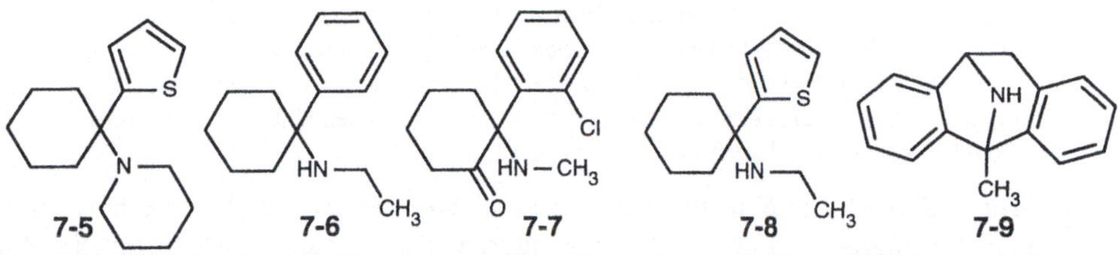

7-5 7-6 7-7 7-8 7-9

Figure 7.2 Drugs with activity similar to that of PCP.

women than in men; usually, in an effort to suppress these side effects, patients are premedicated with a benzodiazepine such as diazepam (Valium).

When the psychic effects of ketamine are sought out as such, they appear to be somewhere between LSD and PCP, with a high incidence of out-of-body experiences, and "the dissociative experiences often seem so genuine that afterward users are not sure that they have not actually left their bodies."[19]

> Where everyone who favored Ecstasy spoke of its mildness, the K people always led off by talking about its power. It was wild and strong—five thousand times stronger than LSD, one user told me after I pressed him for a comparison, although we both knew that in these realms numerical comparisons were meaningless. . . .
>
> One of the most diligent investigators of Vitamin K is neuroscientist John Lilly, who once took the drug every day for a hundred days. "On K I can look across the border into other realities," Lilly claims. "I can open my eyes in this reality and dimly see the alternate reality . . . I can experience the quantum reality. . . . On Vitamin K, I have experienced states in which I can contact the creators of the universe, as well as the local creative controllers."[20]
>
> Stan Grof, who is probably the most respected psychedelic researcher in America, calls K "an absolutely incredible substance. In some sense [it is] much more mysterious than LSD . . . . But I'm not quite convinced of its therapeutic power. I do, however, think it has incredible potential for coping with the problems of death. If you have a full-blown experience with K, you can never believe there is death, or that death can possibly influence who you are."[21]

Unlike PCP, ketamine is inactive orally; when used for its mind-altering effect, it is usually taken in about 100 mg doses as an IM injection. (Ketamine is packaged by Parke Davis in a multiuse sealed vial containing 500 mg in solution.) A less intense experience can be gained by evaporating the solution and insufflating the resulting crystalline powder. *Xochi Speaks* advises: "Because this experience induces unconsciousness or semi-consciousness, a partner *is required* to stand by for any physical needs of the subject. Likewise, fasting for several hours and voiding the bowels and bladder before dosing are highly recommended. . . . Ingestion of the liquid solution of this drug is best accomplished by administering half in each nostril while hanging the head upside down, off the edge of a bed."[22]

Peter Stafford relates the experience of a friend of his on a "low dose" (not otherwise specified) of ketamine:

> Then I reached a point at which I felt ready to die. . . . at the same time that I was having what in my normal state I would call the horror of death. . . . I just yielded, and then I entered a space in which there aren't any words. The words that have been used have been used a

> thousand times—starting with Buddha. I mean, at-one-with-the-universe, recognizing-your-godhead—all those words I later used to explore what I had experienced.
>
> The feeling was that I was "home." I didn't want to go anywhere, and I didn't need to go anywhere. It was a bliss state of a kind I never experienced before. I hung out there awhile, and then I came back. I didn't want to come back.[23]

The author of *The Essential Psychedelic Guide*, who says he has taken ketamine more than 100 times and finds it "the most intense, bizarre, and enjoyable psychedelic" he has tried,[24] cautions that the correct dosage is essential.[25] There are personality variables as well. A Russian psychiatrist, Igor Kungurtsev, has reported using ketamine to treat alcoholics:

> People who are very controlled and have difficulties letting go, or who have problems with relationships, often have negative experiences with ketamine. For them the dissolving of the individual self is horrible. For other patients who . . . have a deep capacity to love, the experience is usually blissful, even ecstatic. . . . Many people who never thought about spirituality or the meaning of life reported having experiences that one might read about only in spiritual texts or Eastern teachings . . . and that they now believe some part of them will continue to exist after death.[26]

THC and the Cannabinoids

Marijuana

History. Marijuana (marihuana, hemp, "grass," "pot," "weed," "reefer," "joint," "herb," "Acapulco gold," "dubie," "dagga," "kif") is a mix of the leaves and flowering tops of *Cannabis sativa* L, the common hemp plant; hashish is obtained from the clear resin secreted by the flowering tops of the plant and is four to eight times as potent as marijuana. Although it is originally of Eurasian origin, the plant grows well in almost all climates, and was once widely grown for its fiber, which was used for rope, clothing, and paper; and its seeds, which have been used for birdseed and as a source of oil for shellac. Escaping cultivation long ago, cannabis can now be found widely distributed throughout North America as "ditchweed" in pastures, waste ground, and along fences.[27] The resin responsible for the psychotropic effects is most abundant when the plant is grown in hot, dry climates; it is thought that the resin serves the purpose of protecting the plant from water loss. Both male and female plants have flowering leaves that secrete this resin, and most amateur and expert growers alike maintain that the female plants have far more of this resin and far greater potency than the male plants; it is traditional to weed out the male plants so that the female plants remain unpollinated and seedless (*sensemilla*).[28] However, some studies have shown that the content of Δ^9-THC (**7-10**, Figure 7.3), the chief component responsible for its psychotropic effects, is the same in both male and female plants.[29] Additionally, there is what Snyder calls the "confused sexual identity of the plant"—depending on the length of the daylight period in which it is planted, it can be hermaphroditic, or male and female; if planted in the winter, individual plants are likely to unpredictably change their sex as the year goes on.[30] Cannabis is also capable of assuming an astonishing variation of shapes and size; it is disputed whether these are responses to environmental factors such as soil and climate, to cultivation, or to the existence of distinct species such as *Cannabis indica, C. ruderalis, C. sativa*, and so forth.[31]

In India there are three cannabis preparations commonly available: *bhang* is the least potent and cheapest; it is made from dried leaves, seeds, and stems. *Ganja* is made from the flowering tops of the female plants and is about three times as powerful as bhang. *Charas* is the pure resin, which is known as *hashish* in the Middle East; it is four to eight times as potent as bhang. Any of these three types of marijuana can be smoked, eaten, or extracted into alcohol for use as a beverage; but since the most active component, Δ^9-THC, reaches the CNS much more rapidly on smoking, any given grade of marijuana will have about four times as immediately intoxicating an effect when smoked as when taken orally. However, the total final drug concentration in the brain and the length and intensity of the overall intoxication can be much higher (and the likelihood of a panic reaction proportionately greater) when the drug is taken orally than when smoked, for the simple reason that a much larger quantity of the material can be eaten than smoked, and what is eaten remains in the system whether the user likes it or not (usually they fall into a dreamlike stupor). Whether the psychotropic effects of cannabis are significantly different when it is smoked than when it is eaten has never been studied in a controlled manner; it is certainly demonstrable that one is in each case ingesting a somewhat different mix, but in both instances the predominant substance with known activity is Δ^9-THC.

Marijuana commonly available in the United States during the '60s and '70s was usually of the approximate quality of bhang, being mostly leaves. In the late '80s and '90s, the pressure of law enforcement has produced short, stocky varieties that are almost all "buds" with few leaves. These high-tech, inbred hybrids produce ganjalike material, and this has given rise to the fear that a much more dangerous type of marijuana is now being used than was prevalent in the '70s.[32] According to Grinspoon and Bakalar, the Δ^9-THC has gone from an average of 0.4% to 4% from 1970 to 1990, and a concentrated hashish oil sometimes called "the one" is now produced that contains as much as 60% Δ^9-THC.[33] The high-tech hybrids of the mid 1990s, can produce buds containing 10 to 14% THC.[34]

Some of the conflicting impressions of pot potency on the part of ordinary users may be due to a placebo effect based on an unconscious confounding of the aroma of marijuana with its potency (in some studies experienced users have been deceived by inert marijuana)[35]. In the Netherlands, after many years of quasi-legalization, some native-grown cannabis produced by hydroponic techniques in greenhouses has a far more intense aroma than conventional marijuana. This homegrown variety is colloquially referred to as "skunk," and considered more potent. But, of course, the many fragrant terpenes in marijuana, which are probably volatilized during shipping from, say North Africa to Holland, are not responsible for its psychotropic effect, since Δ^9-THC itself is odorless; and those who should know say that skunk is no more potent than the imported varieties.[36] Some similar phenomenon may be at work in the United States, where the pressures of prohibition have caused marijuana to be produced increasingly at the local level. A large number of clandestine "greenhouses" operate even in the middle of urban areas like New York City;[37] and the best marijuana is said to be a richly succulent gourmet variety grown indoors under pampered computerized temperature and humidity controls. But an enormous amount of good-quality marijuana is grown in the Midwest's Corn Belt, often between the rows of maize: it is seriously thought by some to be the nation's largest cash crop. A bushel of corn sells for roughly $2.50; a bushel of manicured marijuana for about $70,000.[38]

In any case, it can be argued that a variety of smokable marijuana more concentrated in Δ^9-THC would represent a step forward in health protection for the consumer, just as would a tobacco cigarette more concentrated in nicotine. In both cases, the health liability comes from the "inert" vegetable matter that is responsible for the carcinogens and solid particulates that damage the lungs. More potent material means fewer cigarettes, joints, or "bong hits" need to be inhaled for the same level of satisfaction, and hence a lowered overall health risk.

Etymology: The Assassins of the Old Man in the Mountains. There is a longstanding myth about the use of hashish by a group of Muslim terrorists at the time of the Crusades:

> **Assassins.** . . . Originally a sect of Moslem fanatics founded in Persia, about 1090, by Hassan ben Sabbah (better known as *the Old Man of the Mountain*), their terrorism was mainly directed against the Seljuk authority. . . . Their power was broken by 1273 through the attacks of the Mongols and Mameluke Sultan Bibars. Their name *hashishin* is derived from their reputed habit of dosing themselves with *hashish* or *bang* prior to their murderous assaults.[39]

The same story is told by such reputable authorities as *Webster's Third International Dictionary*, which defines an assassin as "one of a secret order of Muslims that at the time of the Crusades terrorized Christians and other enemies by secret murder committed under the influence of hashish."[40] The more recent (1992) edition of the *American Heritage Dictionary* is closer to the original myth: "Having been promised paradise in return for dying in action, the killers, it is said, were made to yearn for paradise by being given a life of pleasure that included the use of hashish."[41] The story is also found in chapter 31 of *The Count of Monte Cristo*, once an extremely popular novel, written by Alexander Dumas *père*, who was himself a member of "le club des Hachichins" (see next section).

In actuality, the word assassin may be derived from the proper name of the Old Man of the Mountain himself, Hasan Ben Sabáh. This is, at any rate, the view of Mirkhond:[42]

> After many mishaps and wanderings, Hasan became the head of the Persian sect of the *Ismailians*, a party of fanatics . . . . and it is yet disputed whether the word *Assassin*, which they have left in the language of modern Europe as their dark memorial, is derived from the *hashish*, or opiate of hemp-leaves (the Indian *bhang*, with which they maddened themselves to the sullen pitch of oriental desperation, or from the name of the founder of the dynasty. . . .[43]

Hashish is actually the Arabic word for grass, and another very ancient tradition cited by Grinspoon claims that the followers of the Old Man of the Mountain were so called because they despised all things of this world as so much grass. To add to the confusion, it seems that most of the stories in the West about the Old Man of the Mountain have their origins in various versions of Marco Polo's recollections, and in these there is no mention of cannabis at all, but of opium, and the opium is given not before but after the assassinations as a reward.[44]

In any case, from these stories it was a short if irresponsible step to the claim that marijuana caused violent behavior, a claim strongly endorsed by Anslinger, the first Commissioner of the Federal Bureau of Narcotics. This bureau was established in 1932 as (alcohol) Prohibition ended; and it took over from the defunct Alcohol Unit of the Treasury the enforcement of laws against opium and cocaine. They may not have had enough to do, and so Anslinger commendably took up a crusade against marijuana. In a 1937 article for the *American Magazine* entitled "Marihuana: Assassin of Youth" he wrote:

> In the year 1090, there was founded in Persia the religious and military order of the Assassins, whose history is one of cruelty, barbarity, and murder, and for good reason. The members were confirmed users of hashish, or marijuana, and it is from the Arabic "hashshashin" that we have the English word "assassin."[45]

The notion that marijuana lured youth into exotic oriental lubricities as well as homicidal violence was a fairly exciting idea, and made for good yellow journalism and government propaganda films like *Reefer Madness*.[46] As for sex, there is no evidence that marijuana leads to any more of it than would otherwise be customary (although some claim it enhances their appreciation of it and everything else); but as for marijuana affecting violent behavior, it definitely does. It inhibits it. In fact, a standard test that has been used to assay cannabimimetic activity is the rat muricidal inhibition protocol.[47] Rats will ordinarily kill a mouse introduced into their home cage within less than 5 seconds. But if the rats are given enough Δ^9-THC or another active cannabinoid, Lo! The lion lies down with the lamb; the child sleeps over the adder's nest.[48] Hasan Ben Sabáh's followers must have been savage rats indeed if they were able to kill while on cannabis.

As for the origins of the word cannabis, it is generally thought to be Indo-European, specifically Scythian, coming into Greek by way of Herodotus. In the 5th century B.C., Herodotus said that the Scythians, during funerals, would throw hemp onto hot stones, inhale the vapors, and "howl with joy."[49] But Sula Benet argues that the word is of Semitic origin and can be found in the Bible. She cites Yahweh's directive to Moses in Exodus 30:23 to make a holy oil consisting of "myrrh, sweet cinnamon, *kaneh bosm* and kassia." *Kaneh* is hemp, she says, and *bosm* means "aromatic." In the Mishna, the words have been fused into *kannabus*, and the same process could have occurred in Scythian.[50] Mechoulam finds Benet's hypothesis "not unreasonable," pointing out that *kaneh bosm* or *kneh-bosem* was one of the constituents of incense in Solomon's temple, and cannabis was definitely used as temple incense in Egypt. (However, neither in Egypt nor in Israel are any "howls of joy" recorded as attending its use.) But Mechoulam[51] thinks it more likely that the Hebrew *pannagh* mentioned in Ezekiel 28:17 was hemp; and Rabin claims that *pannagh* is cognate with the Sanscrit *bhanga*, and Persian *bang*, and was transformed in the Semitic languages by metathesis (the first and last consonants switching places) into Assyrian *qunnabu*, Arabic *kunnab*, and finally into Greek as *cannabis*.[52]

Religious, Social, and Recreational Uses. The innumerable historical and literary references to the use of marijuana and hashish are rivaled only by those of opium and alcohol;

and the vivid intricacy of the "dreams" experienced under high doses of hashish is equalled or exceeded only by the effects of the psychedelics. A sampling follows.

Shiva, Sādhus, Bhajans. Cannabis is mentioned as a "sacred grass" in the ancient Atharva Veda (2000-1400 BC).[53] Shiva, who with Brahma and Vishnu forms the primordial Hindu triad of deities, is frequently portrayed holding a bowl of herbs, including cannabis; and smoking marijuana is considered an act of worship by Shivites. Not only Shivites, but other Hindu mendicant holy men (*sādhus*) from a wide spectrum of Hindu sects also use cannabis, and it is considered a meritorious deed for layfolk to donate cannabis to them.

> One extreme example of ritual use is that of unusually austere *sādhus* called *aghoris*. Under the influence of cannabis, *aghoris* indulge in such ritual practices as eating excrement, urine, and the flesh of corpses. Professor Bharati[54] reports that cannabis functions as a deinhibiting agent in certain esoteric Tantric rituals. Hinduism is in many ways a puritanical religion, and cannabis helps to psychologically shore up adherents who partake of these somewhat exotic practices—dietary in the case of the *aghoris* and sexual in the case of Tantric rituals. . . .
>
> A second category of cannabis users consists of male devotees to sing at *bhajans*, Hindu devotional meetings. . . . At *bhajans*, the *gānjā* (marihuana) is passed around in a *chilam* (clay pipe) among the singers and musicians sitting on the floor. To partake of the *chilam* is not in any sense obligatory, but is clearly a way of symbolically stating the fact of devotional fellowship. As with *sādhus* it promotes good *bhakti*. . . . As with the *sādhus*, this devotional use of cannabis is publically known, but *bhajan* singers are not in any sense highly regarded because they use cannabis, however exemplary they may be as devotees. On the contrary, there is if anything a tendency to regard any layman *ganjari* (one who uses *gānjā* excessively) as slightly reprehensible, although not seriously objectionable.[55]

The British Colonial Government commissioned what is probably the most thorough study ever made of the actual use of cannabis in India, having in mind its regulation or prohibition should it prove to warrant this. The plant survived this intensive scrutiny, due in no small part to the enthusiastic testimony offered in its defense by native Indians who knew their culture best:

> To the Hindu the hemp plant is holy. . . . Bhang is the joy giver, the sky flier, the heavenly guide, the poor man's heaven, the soother of grief. . . . The students of the scriptures of Benares are given bhang before they sit to study. At Benares, Ujjain and other holy places, yogis take deep draughts of bhang that they may center their thoughts on the Eternal . . . . By the help of bhang ascetics pass days without food or drink. The supporting power of bhang has brought many a Hindu family safe through the miseries of famine.[56]

Jamaica. The use of cannabis was introduced to Jamaica by immigrants from India, as is evident from the Hindu terms used for the plant, *ganja*, the clay pipe, *chilam*, and so on. In Jamaica, smoking of marijuana by the middle and upper classes has a similar recreational significance as with middle-class Americans. But among males of the lower classes, it establishes a peer relationship with one of two groups: smokers, who use it regularly, smoked as "spliffs" (marijuana cigarettes, from one to eight or more daily), and nonsmokers. One is destined for life to be a smoker or a nonsmoker by a rite of passage to "manship." When a young man first tries a spliff, he either experiences nothing or he has a stereotyped dream of a little woman or man dancing happily; if he sees this, then he is a smoker for life. "The culturally standard vision of the "little dancing" person or creature, which characterizes the first *ganja* experience for some subjects, may be comparable to the vision quest for guardian spirits of the Plains Indians of North America."[57] Thereafter, those who use *ganja* do not experience dreams or hallucinations, nor do they seek them. Rather,

> The peasant takes *ganja* for energy when he works in his fields, the fisherman to ward off fatigue at sea. *Ganja* makes you "feel to work" and staves off hunger during work, but when taken before the evening meal, it enhances the appetite. If he smokes before an evening dance, the farmer can "win a contest"; if he takes *ganja* before bedtime, he can sleep restfully and wake refreshed and energized to start his day's work. Taken in congenial group settings, it can evoke religious meditation; taken in solitude, it is said to aid in problem solving. The subjective "mind altering effects" are thus selectively and conditionally experienced. *Ganja* teas and tonics are also used extensively in the working class—by males and females, adults and children, both as prophylactic and medication for a wide range of ailments. The *ganja* syndrome is characterized by situational determinants that reinforce working-class use and condition the range of reactions experienced, including the initial vision.[58]

Rastafarians. The Rastafarians are members of a Jamaican black political and religious group that advocates a return to Ethiopia as the promised land and worships Haile Selassie as their God and savior: the name of the sect derives from *Ras Tafari*, the former Amharic name of Haile Selassie (*ras* = prince; *tafari* = to be feared).

> Rastafarian metaphysics . . . emphasizes and brings into focus general concepts derived from working-class views of *ganja*. For them, it is "the wisdom weed," of divine origin, an *elixir vitae*, documented by Biblical chapter and verse which overrides man-made proscriptions. Religious authority thus validates and fortifies commitment to its use. . . the sacred source of *ganja* permits a sense of religious communion, marked by meditation and contemplation.[59]

Le Club des Hachichins. In the middle of the 19th century a group of famous Parisian writers, poets, painters, physicians, and other savants including Dr. Jacques-Joseph Moreau, Théophile Gautier, Charles Baudelaire, Gérard de Nerval, Honoré de Balzac, and Alexandre Dumas *père* would meet in the Hotel Lauzun (Pimodan) on the Isle St. Louis to eat *dawamesc*. This was a preparation of hashish imported from Egypt that certainly included cannabis resin but may also have contained any or all of the following, according to Dr. Moreau: "cinnamon, ginger, cloves, some aphrodisiacs, and perhaps also . . . powder of cantharides (Spanish fly) . . . . even opium, extract of *Datura*, and other narcotics." At the risk of some understatement, the good doctor concludes "the addition of these various substances to hashish assuredly modifies its effects to quite an extent."[60] It appears to have been Moreau who introduced the others to dawamesc; he had discovered its properties during extensive travels in Egypt. Moreau was interested in mental illness; anticipating the notions of psychiatrists a century later, he believed that the hashish experience provoked a model insanity in otherwise normal people, and he attempted to cure the mentally ill by dosing them with dawamesc.

The overall impression one gets from Moreau's extensive studies of hashish is that it produced effects remarkably similar to LSD and the classic psychedelics. Like the psychedelics, there were great differences in response from person to person:

> It does not have the same effect on everyone. The same dosage can produce extremely different results, at least in intensity. . . . I have met several people on whom hashish seemed

to have no effect. They resisted dosages that would have deeply affected others. I have become convinced that with some willpower one can stop or at least considerably diminish these effects, much as one controls a burst of anger. We shall see later how these reactions can be affected by external circumstances, by impressions that come from one's surroundings, and by one's state of mind.[61]

Like the psychedelics, partakers of dawamesc experienced distortions of perception verging on the hallucinatory, but that they remembered clearly afterwards and knew somehow at the time to be not really real. And there were clear instances of synesthesia and ego loss. Here are some excerpts from reports given to Moreau by two persons to whom he gave dawamesc; Moreau says they are the most complete he was able to observe in others.

On Thursday, December 5, I had taken hashish . . . and I quietly awaited the happy delirium that was supposed to seize me. I sat at the table, and I cannot add, as some people do, "after having relished this delicious paste," because to me it tasted horrible. I swallowed it with great effort. As I was eating oysters, I was seized with a fit of laughter, which ceased as soon as I directed my attention to two other people who, like me, had tasted the paste, and who were already seeing a lion's head on their plate. I was calm enough until the end of the dinner; then I took a spoon and stood *en garde* against a compotier of candied fruit with which I was preparing to duel.

Bursting into laughter, I left the dining room. Soon I felt a need to hear and make music; I sat down at the piano, and I began to play an air from *Black Domino*. I interrupted myself after several measures because a truly diabolic spectacle greeted my eyes: I thought I saw the image of my brother standing atop the piano. He stirred and presented a forked tail, all black and ending with three lanterns, one red, one green, and one white. This apparition presented itself to me several times during the course of the evening. . . . I cannot describe the thousand fantastic ideas that passed through my brain during the three hours that I was under the influence of the hashish. They seemed too bizarre to be credible. The people present questioned me from time to time and asked me if I wasn't making fun of them, since I possessed my reason in the midst of all that madness.[62]

The second report is from Théophile Gautier, describing his experiences after Moreau gave him "a few grams of dawamesc" at Gautier's request—although Gautier had protested his skepticism that anything could have the effects attributed to hashish. He was pleasantly disabused:

For many years, Orientals whose religion forbids the use of wine, have sought to satisfy by the use of various preparations their need for mental excitement common to all people, and that people of Western countries gratify by means of spirits and liquors. The aspiration toward an ideal is so strong in man that he tries to release the bonds that keep his soul within his body. Since ecstasy is not within the reach of all, he drinks his merriment, he smokes his oblivion, and eases his madness in the form of wine, tobacco, and hashish. . . .

After a few minutes a general sluggishness overcame me. It seemed that my body was dissolving and becoming transparent. I could clearly see in my chest the hashish I had eaten, in the form of an emerald glowing with a million sparkles. Billions of butterflies swarmed with wings fluttering like fans. Huge flowers with crystal calyces, enormous hollyhocks, streams of gold and silver flowed around me with a crackling like the explosion of fireworks. My hearing was fantastically sharpened. I heard the sound of colors: green, red, blue, and yellow sounds came to me in distinct waves. . . . I had never been so overwhelmed with bliss; I dissolved into

> nothingness; I was freed from my ego, that odious and everpresent witness; for the first time I conceived the existence of elemental spirits—angels and souls separate from bodies. I was like a sponge in the middle of the ocean. At every moment streams of happiness penetrated me, entering and leaving through my pores, since I had become porous. . . . By my calculations this state lasted approximately three hundred years, for the sensations followed one another in such numbers and so rapidly that an accurate appraisal of time was impossible. When the spell was over, I realized it had lasted a quarter hour.[63]

Another member of the club, poet and writer Baudelaire (1821-67) obtained his initial fame by the publication of *Les Fleurs du Mal* in 1857; as with James Joyce, D. H. Lawrence, and many others before and since, his reputation was assured when the French authorities banned six of the poems from *Les Fleurs* and fined its author for blasphemy and obscenity. Some of the wickedness we associate with the scent of marijuana today probably derives from the perfume of these evil flowers of the last century, for 3 years later, Baudelaire published *Les Paradis Artificiels*, which is a series of lurid "dreams," which he also calls "hallucinations," for many of which the *mise en scène* is the Club des Hachichins. As a clinical impression of the effects of hashish, *Les Paradis Artificiels* is probably less reliable than Gautier or any of the careful studies of Moreau; some of Baudelaire's dreams were under the influence of opium, many were recollected and amplified from long ago, and the book also contains translations into French from De Quincey's *Recollections of an English Opium Eater*, the genre of which Baudelaire was clearly emulating.[64]

Gertrude and Alice. When *The Alice B. Toklas Cook Book* was first published in 1954 "it caused a small sensation, partly because of the fuss about hashish fudge, but mostly because of what it was: a unique book by a unique person."[65] Indeed, the hashish fudge is a very minor page in a book of interesting recipes-cum-recollections of Picasso, Scott Fitzgerald, Sherwood Anderson, Matisse, and of course Gertrude Stein, with whom she lived in France from 1907 until Gertrude's death in 1946. And it was actually not Alice's own recipe, being one of many in a chapter of recipes contributed by her friends, in this case Brion Gysin, who also seems to be the author of the accompanying commentary on its psychotropic effects:[66]

> HASCHICH FUDGE
>
> This is the food of Paradise—of Baudelaire's Artificial Paradises: it might provide an entertaining refreshment for a Ladies' Bridge Club or a chapter meeting of the DAR. In Morocco it is thought to be good for warding off the common cold in damp winter weather and is, indeed, more effective if taken with large quantities of hot mint tea. Euphoria and brilliant storms of laughter; ecstatic reveries and extensions of one's personality on several simultaneous planes are to be complacently expected. Almost anything Saint Theresa did, you can do better if you can bear to be ravished by '*un évanouissement reveillé*.'[67]
>
> Take 1 teaspoon black peppercorns, 1 whole nutmeg, 4 average sticks of cinnamon, 1 teaspoon coriander. These should all be pulverised in a mortar. About a handful each of stoned dates, dried figs, shelled almonds and peanuts: chop these and mix them together. A bunch of *canibus* [sic] *sativa* can be pulverised. This along with the spices should be dusted over the mixed fruit and nuts, kneaded together. About a cup of sugar dissolved in a big pat of butter. Rolled into a cake and cut into pieces or made into balls about the size of a walnut, it should be eaten with care. Two pieces are quite sufficient.

> Obtaining the *canibus* may present certain difficulties, but the variety known as *canibus sativa* grows as a common weed, often unrecognised, everywhere in Europe, Asia and parts of Africa; besides being cultivated as a crop for the manufacture of rope. In the Americas, while often discouraged, its cousin, called *canibus indica*, has been observed even in city window boxes. It should be picked and dried as soon as it has gone to seed and while the plant is still green.[68]

The Beat Generation. The use of pot by the "beat generation" of the 1950s led to its explosive employment along with LSD and the psychedelics in the '60s. Beat poet Allen Ginsberg, author of *Howl* (1956), is firmly in the tradition of Gautier and Baudelaire when he argues that marijuana enhances artistic appreciation and catalyzes the creative process:

> [There is] much to be revealed about marijuana especially in this time and nation for the *general* public, for the actual experience of the smoked herb has been completely clouded by a fog of dirty language by the diminishing crowd of fakers who have not had the experience and yet insist on being centers of propaganda about the experience.[69]
>
> Although most scientific authors who present their reputable evidence for the harmlessness of marijuana make no claim for its surprising *usefulness*, I do make that claim: Marijuana is a useful catalyst for specific optical and aural aesthetic perceptions. I apprehended the structure of certain pieces of jazz and classical music in a new manner under the influence of marijuana, and these apprehensions have remained valid in years of normal consciousness.[70]

Some Sober Scientists. It is easy to dismiss the claims of Ginsberg. He is a beatnik, and he is a poet. But there are productive, intelligent, otherwise conventional Americans who use marijuana for essentially the same reasons Allen Ginsberg did. Of course, the more respectable their positions, the less likely they will want to tell you about their use of this Schedule I vegetation. But Lester Grinspoon, in *Marihuana Reconsidered*, has tracked down a highly successful scientist who teaches at a major university and who was willing to provide Grinspoon with a detailed and articulate description of his use of marijuana. A brief excerpt follows:

> My initial experiences [smoking marijuana] were entirely disappointing; there was no effect at all, and I began to entertain a variety of hypotheses about cannabis being a placebo which worked by expectation and hyperventilation rather than by chemistry. After about five or six unsuccessful attempts, however, it happened. I was lying on my back . . . idly examining the pattern of shadows on the ceiling. . . . I suddenly realized that I was examining an intricately detailed miniature Volkswagen, distinctly outlined by the shadows. I was very skeptical at this perception. . . . But it was all there, down to hubcaps, license plate, chrome, and even the small handle used for opening the trunk. . . .
>
> I want to explain that at no time did I think these things 'really' were out there. I knew there was no Volkswagen on the ceiling. . . . I don't feel any contradiction in these experiences. There's a part of me making, creating the perceptions which in everyday life would be bizarre; there's another part of me which is a kind of observer. About half of the pleasure comes from the observer-part appreciating the work of the creator-part. . . . In this sense, I suppose cannabis is psychotomimetic, but I find none of the panic or terror that accompanies some psychoses. Possibly this is because I know it's my own trip, and that I can come down rapidly any time I want to. . . .
>
> The cannabis experience has greatly improved my appreciation for art, a subject which I had never much appreciated before. . . . A very similar improvement in my appreciation of music

has occurred with cannabis. For the first time I have been able to hear the separate parts of a three-part harmony and the richness of the counterpoint. I have since discovered that professional musicians can quite easily keep many separate parts going simultaneously in their heads, but this was the first time for me. Again, the learning experience when high has at least to some extent carried over when I'm down. . . . I do not consider myself a religious person in the usual sense, but there is a religious aspect to some highs. The heightened sensitivity in all areas gives me a feeling of communion with my surroundings, both animate and inanimate. . . . There is a myth about such highs: the user has an illusion of great insight, but it does not survive scrutiny in the morning. I am convinced that this is an error, and that the devastating insights achieved when high are real insights; the main problem is putting these insights in a form acceptable to the quite different self that we are when we're down the next day.[71]

Dr. X had never read any descriptions of a marijuana high, and yet some of the effects are much too detailed to be coincidental: the ability to distinguish separate parts of a complex musical score is mentioned in almost the same words by Ludlow and Ginsberg.

The Active Constituents of Marijuana, the Cannabinoids. Chemically, cannabinoid structures are describable as a terpene joined to an alkyl-substituted resorcinol. But for purposes of nomenclature, the system is viewed by CAS as a benzopyran ring system rather than a substituted terpene, and this has led to two numbering systems, both widely employed (much like the spelling of the word marijuana/marihuana). In this chapter, the benzopyran numbering is used (Figure 7.3, **7-10** ***left***), but many workers use the terpenoid system (**7-10** ***right***), in which Δ^9-THC is Δ^1-THC, and Δ^8-THC is Δ^6-THC.

Extensive work by Adams at the University of Illinois in the early 1940s failed to isolate a psychoactive substance from cannabis, but simultaneous efforts by his group[72] and Todd's[73] in England to confirm the structure of cannabinol by synthesis led to the serendipitous synthesis of a non-natural Δ^9-THC isomer, $\Delta^{6a,10a}$-THC (commonly called Δ^3-THC by the alternate nomenclature), which proved to be quite psychoactive in both animal and human tests.[74] For many years, this was the only psychoactive cannabinoid known in more or less pure form. (Later work showed that it was actually a mixture of about 90% $\Delta^{6a,10a}$-THC and 10% of the $\Delta^{10,10a}$ isomer, which is also active.) From the significant activity of $\Delta^{6a,10a}$-THC, Adams was convinced that the actual natural component was a tetrahydrocannabinol isomer, but only in 1964 was Δ^9-THC isolated in pure form from cannabis by Gaoni and Mechoulam.[75]

There seems to be little controversy that Δ^9-THC is the principal psychoactive component of marijuana. The full CAS name for the substance, which occurs in cannabis as a single chiral isomer, is *(6a*R*,10a*R*) (−)-*Δ^9*-6a,10a-*trans*-tetrahydro-6,6,9-trimethyl-3-*

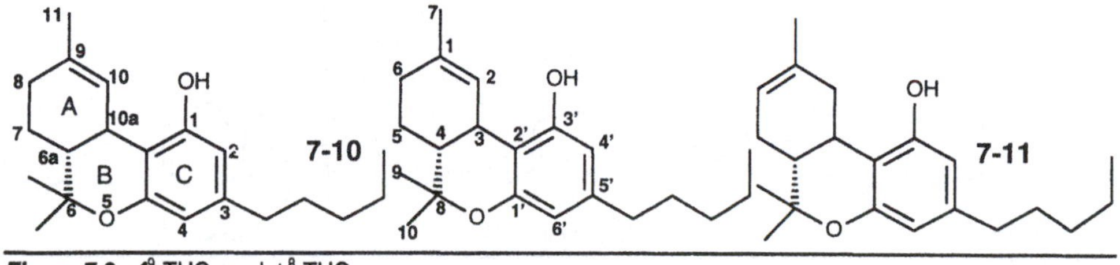

11 9 10 OH 8 A 10a 1 7 6a B C 2 6 5 O 4 3 **7-10** 7 1 6 2 OH 5 4 3 2' 3' 4' 9 8 O 1' 5' 10 6' OH **7-11**

Figure 7.3 Δ^9-THC and Δ^8-THC.

pentyl-6H-dibenzo[b,d]pyran-1-ol, **7-10**, but it is commonly called Δ^9-*tetrahydrocannabinol* or simply Δ^9-THC. There are many other closely related structures in the hemp plant, but in all probability the only structure that, by itself, will produce the distinct psychic "high" characteristic of the smoked or eaten plant material is Δ^9-THC. The Δ^8-THC (**7-11**) isomer is almost as active as Δ^9-THC, and it is occasionally isolated in very small amounts from cannabis; but it seems likely that its occurrence is actually an artifact of workup. There is continued controversy as to whether the natural mixture has a significantly different psychic effect than does pure Δ^9-THC, but any differences are almost certainly due to other cannabinoids, probably cannabidiol.[76]

The living cannabis plant actually appears to contain little or no Δ^9-THC; what Δ^9-THC is eventually present in the plant's leaves or resin is formed from carboxylic acids such as **7-12** and **7-13** (Figure 7.4). These β-enolic acids are easily decarboxylated, even under the fairly mild conditions of drying the plant material.[77]

Wallinsky[78] provides a typical GLC analysis of the cannabinoids (as a percentage of dry weight) in a 1971 street sample of marijuana: Δ^9-THC, 0.72%; *cannabidiol* (CBD, **7-14**), 0.04%; *cannabinol* (CBN, **7-15**, Figure 7.5) 0.26%; *cannabigerol* (**7-16**), 0.11%. For comparison, a 1-year-old sample of Lebanese hashish was found to contain 3.75% CBD; 3.30% Δ^9-THC; 1.30% CBN, 0.30% cannabigerol, and 0.19% cannabichromene (**7-17**). The amount of Δ^6-THC in either marijuana or hashish was so small as to escape detection.

Enthusiasts of hashish and marijuana have their preferences and subjective impressions of the respective potencies of materials grown in different regions and from different strains; any coffee shop in the Netherlands will produce a menu of offerings, and bystanders will provide ample advice and opinion. In a 1971 analysis of different strains of marijuana grown under identical conditions, Fetterman's group found a wide range of concentrations in the three major cannabinoid components, CBD, CBN, and Δ^9-THC. Plants grown from seeds originating in Mexico had Δ^9-THC concentrations with a range of 0.01-3.7%; from Turkey, 0.007-0.37%; from Minnesota 0.04-0.07%; from Thailand 1.3-2.2%. Thus, the Δ^9-THC ranged from a low of 0.007% to a high of nearly 4.0%.[79] However, there appear to be many factors affecting Δ^9-THC concentration other than the country of origin of the seeds. It appears that after a few generations in America, the plants become part of the great melting pot and have a Δ^9-THC content typical of other American plants; and the Δ^9-THC concentrations vary widely in a given climate depending on what time they are planted and harvested, and on the soil, light, temperature, and humidity at which they are grown. Plants that are "stressed" appear to produce more Δ^9-

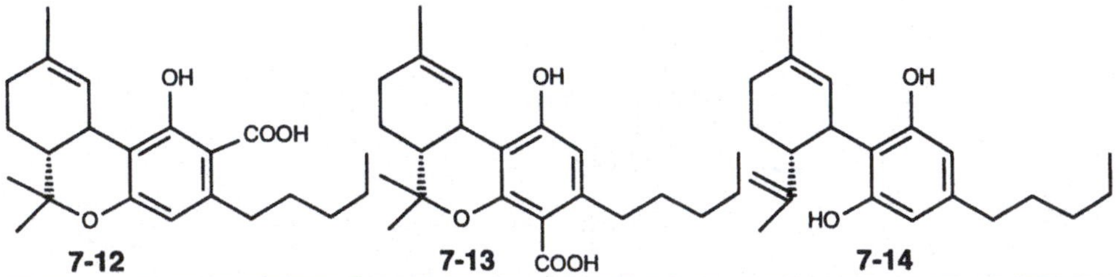

Figure 7.4 Δ^9-THC acids and cannabidiol.

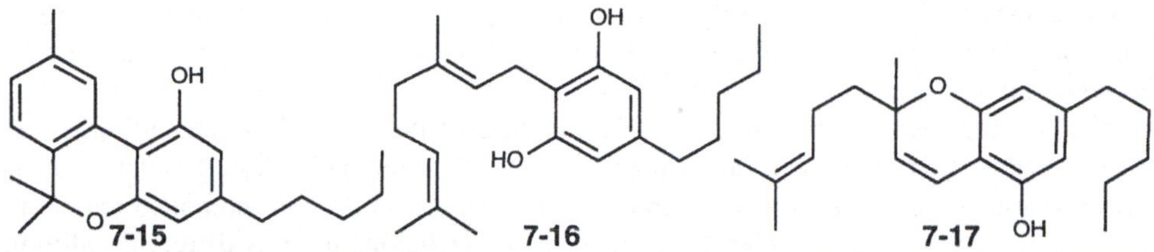

Figure 7.5 Cannabinol, cannabigerol, cannabichromene.

THC than those that are indulged.[80]

The acute psychic effects of cannabis are primarily due to Δ^9-THC, and CBD does *not* produce a psychological "high" in human subjects. But it has been shown that CBD is as effective or more effective than diazepam (Valium) in several measures of antianxiety.[81] The majority of those who have used (illicit) marijuana to self-medicate such conditions as chemotherapy nausea or paraplegic spasticity seem to prefer smoking cannabis to oral capsules of Δ^9-THC; while other factors (chiefly the bioavailability of smoked vs. oral material) certainly play a significant role in this, it is interesting that even those familiar with and fond of the effects of smoked cannabis often find oral Δ^9-THC unpleasant or anxiety-provoking. Thus, the CBD present with Δ^9-THC in the natural product may significantly contribute to the overall psychic effect of cannabis. None of the other cannabinoids present in cannabis have shown significant activity; however, the 11-hydroxy metabolites of Δ^9-THC and Δ^8-THC are probably active, contributing to the overall activity of the parent THCs.[82]

Some Inactive Constituents: Essential Oils. The aroma of smoked marijuana (which is something like burning leaves) is unmistakably characteristic to most people who have smelled it even once,[83] and may contribute significantly, by a Pavlovian mechanism, to many users' evaluation of the quality of their pot. But the active cannabinoid components are themselves tasteless and odorless. The aroma comes from the essential oils in marijuana, of which at least 25 have been identified: the most volatile of these are myrcene and *p*-cymene; and those most likely to contribute to the smell of the plant are thought to be caryophyllene (itself constituting nearly 50% of the essential oil by GLC), β-farnesene, α-selinene, β-phellandrene, limonene, and piperidine.[84] As for hashish, that sold in the Netherlands has a mild minty flavor that blends well with the chocolate used in the "space cakes" sold in some of the Amsterdam coffee shops.[85]

Acute Pharmacology. *Physical Effects.* The physical effects of smoked marijuana are quite minor: tachycardia (an increase in heart rate, but with no increase in blood pressure or respiratory rate) that is dose-related to the concentration of Δ^9-THC in the blood; a reddening of the conjunctivae (blood vessels in the eyes); and a decrease in intraocular pressure.[86] (But, contrary to myth, there is no change in pupil size.)[87] This belief, once common among law-enforcement personnel, has been refuted many times; it may have arisen from marijuana users looking at each other in dimly lit surroundings.)[88] Almost everyone using marijuana experiences an increased appetite ("the munchies"), and some find themselves eating ravenously. But there is no change in blood sugar induced by marijuana; the appetite enhancement presumably results from some central action on the hypothalamus.[89]

Dosage and Duration of Effect. Estimation of the actual dose an individual receives from either smoked or orally ingested cannabis is always difficult because of the considerable variation from several sources: in the amount of Δ^9-THC and other cannabinoids present in different samples of cannabis, in the technique used by different individuals when inhaling, and in the absorption of the highly water-insoluble Δ^9-THC from the gut. However, Weil's group of experienced marijuana smokers trained to inhale in a consistent manner and smoking a dose of 2.0 g of material assayed at 0.9% Δ^9-THC content (this would be two hefty joints of good-quality pot) showed maximum observable effects at 15 minutes after smoking; these effects diminished between 30 minutes and 1 hour after smoking; they were completely dissipated by 3 hours.[90] Orally ingested dronabinol (Marinol, capsules containing Δ^9-THC in sesame seed oil for use as an antiemetic) is said to be "slowly and erratically absorbed from the intestine," reaching peak plasma concentration in 2 to 3 hours; the manufacturers recommend adult dosages of about 10 mg every 3 to 4 hours beginning 4 to 12 hours before chemotherapy.[91] These data for smoked and oral cannabis are for its maximum effect: initial effects of smoked marijuana are felt within a few minutes; the first effects of oral ingestion occur in about 45 minutes.

Psychotropic Effects. The usual psychic effects of a few "hits" of the marijuana ordinarily available on college campuses in the United States consists of a subjective perception of relaxation and mild euphoria. The experience is often compared to the effects of one or two beers. But there are significant differences. The marijuana "high" is more subtle; it is typical for first-time users to say and believe that they feel nothing. The common myth is that this is because the Δ^9-THC must be absorbed in and "saturate" the body fat (lipid system) by repeated use until finally enough gets to the brain for the user to feel its psychological effects. This is almost certainly not true: objective physiological effects caused by the systemic presence of Δ^9-THC, such as reddening of the conjunctivae of the eyes and rapid heartbeat, can be shown to occur with first-time users. The more likely explanation is that the effects of small doses of Δ^9-THC (especially relative to known intoxicants such as alcohol) are so mild they can be fairly easily overridden or overlooked by the user, who must learn to appreciate them.

These are, at any rate, the conclusions of what was the first clinical study of the effects of smoked marijuana, which was conducted by Weil, Zinberg, and Nelson at Boston University School of Medicine in 1968.[92] The study compared the effects of marijuana on nine college students who had never smoked it before (in the heady days of the late '60s, it is amusing to note that the group had great difficulty finding nine such drug-naive students in the Boston metropolitan area) and nine who were regular users. Two of the results of the study are particularly interesting: Firstly, 8 of 9 naive subjects reported no subjective change (the other seemed to enjoy himself immensely). Weil later interpreted this as follows:

> Persons whose set toward the drug includes mucn anxiety (most first-time users, for example) can ignore the drug completely and pretend to themselves that nothing has happened. My Boston associates and I were repeatedly struck by this reaction in our marihuana-naive subjects. They would sit in the laboratory with red eyes and heart rates of 130 beats per minute

> (normal resting pulse is 70 or 80) after smoking two large joints and would have no subjective responses at all. After the sessions, some of them would ask, "Did I have a drug tonight?" (We, of course, did not know for sure until the experiments were finished and the double-blind code was broken.) Very few drugs that trigger altered states of consciousness can be ignored so completely.[93]

In contrast, all the marijuana-experienced subjects got thoroughly high. Secondly, Marijuana-naive subjects demonstrated some impairment on simple intellectual and psychomotor tests; in contrast, the marijuana-experienced users showed no impairment, and some showed a slight improvement in performance, even as they protested that they were quite "high." According to Weil, later studies[94] showed marijuana *did* impair performance of experienced marijuana users if they (1) used much higher doses than they are accustomed to, (2) used oral Δ^9-THC instead of smoking marijuana, or (3) are required to perform complex tasks unlike any they had performed previously when high. He explains:

> What pharmacologists cannot make sense of is that people who are high on marihuana cannot be shown, in objective terms, to be different from people who are not high. That is, if a marihuana user is allowed to smoke his usual doses and then to do things he has had a chance to practice while high, he does not appear to perform any differently from someone who is not high. Now, this pattern of users performing better than nonusers is a general phenomenon associated with all psychoactive drugs. For example, an alcoholic will vastly outperform a nondrinker on any test if the two are equally intoxicated; he has learned to compensate for the effects of the drug on his nervous system. But compensation can proceed only so far until it runs up against a ceiling imposed by the pharmacological action of the drug on lower brain centers. Again, since marihuana has no clinically significant action on lower brain centers, compensation can reach 100 percent with practice.[95]

This would seem a bit of an oversimplification. Whether "lower" or "higher" brain centers are involved, it is almost certain that *very* high doses of Δ^9-THC, such as can be gotten by high oral doses of hashish, would disorient a person to the extent that they could not compensate for the effects of the drug.[96] But perhaps this is a quibble: the point is that people who regularly use marijuana get quite skilled at performing their ordinary functions under their ordinary dosage; and that motor skills appear to be much less removed from the volitional control of the marijuana user than from that of the alcohol user.

Marijuana and Driving. What concerns most people, on a practical level, is the extent to which the motor and coordination skills needed in driving a car are affected by marijuana use. A study that measured exactly this was conducted in 1969 by Alfred Crancer, chief of research at the Department of Motor Vehicles in Washington State, using a standardized driving simulator apparatus that previously had been shown to provide an accurate correlation with the 5-year driving history of normal individuals. Chronic marijuana smokers were tested under three conditions: after smoking two marijuana cigarettes, after drinking enough alcohol to have a blood alcohol concentration of 0.10%, and drug-free. They performed in an essentially unchanged way after the marijuana as drug-free; but their skills significantly deteriorated under alcohol.[97]

As Snyder[98] points out, this was not a controlled or blinded study. The marijuana users would have a clear motivation to advertise the harmlessness of a marijuana high versus an alcoholic stupor. Nevertheless, with all the motivation in the world to enhance the reputation of alcohol, it is doubtful that a person who drank 6 ounces of 86-proof whiskey (the equivalent of the subjects' intake in this test) could perform well in this test; that they could, willy-nilly, do so well on two joints (a considerable amount of pot) says something. As Weil says, he would rather be in a car driven by a high marijuana smoker who had practiced driving while high than in one driven by a person with any amount of alcohol in them.[99]

The Dutch have allowed their citizens free access to high-quality marijuana and hashish for almost 20 years, and seem to have experienced few problems with marijuana use and driving.[100] In a Ph.D. dissertation from the Netherland's University of Limburg, occasional users of marijuana were given low, medium, and high doses of marijuana and their driving performance tested in rural and downtown settings. Surprisingly, the higher the dosage of marijuana, the more accurately the participants were able to maintain an exact distance away from the car in front of them (at 90 km/h). A group given alcohol until their blood registered at the legal limit was considerably more impaired than when sober. Perhaps even more significant for overall safety is than the degree of impairment of motor function may be the qualitative and quantitative effects of the two drugs on a user's judgment: alcohol inflates a false sense of confidence and aggressivity, while marijuana tends to inhibit aggression and accentuate caution. The results of the Limburg study bore this out, for those who took alcohol were unaware of the deterioration in their driving skills, while those who took marijuana felt they had driven more poorly than they actually had.[101]

In a 1991 critical review of more than 80 studies of the effect of marijuana on human performance, Chait and Pierri[102] found many of the results contradictory or inconclusive; but they maintain that "the most consistent finding was evidence that heavy marijuana smokers were relatively insensitive to marijuana-induced impairment on certain psychomotor tasks. . . suggest[ing] the development of tolerance to the behavioral effects of marijuana, a phenomenon that has frequently been observed in both laboratory animals and humans." Unfortunately, they also found that most of the studies reflected anything but the highest standards of scientific methodology:

> The most notable and frustrating aspect we encountered in conducting our analysis was the failure of many of the reviewed studies to provide the most basic methodological information that would be required to replicate the research. Many studies provided virtually no information as to the marijuana use history of the subject population, the instructions or incentives given to subjects regarding their performance, or pertinent facts concerning the exact nature of the behavioral tasks, the extent of practice subjects received, when the tasks were administered relative to drug treatment, or how long the tasks took to complete. On the other hand nearly all studies provided information on the amount of marijuana burned and the marijuana Δ^9-THC content (often to two decimal places), despite the fact that this information is largely useless for estimating the amount of THC actually delivered to the subject, given the uncertainties and variability inherent in the smoking process. . . . It is not surprising that we were unable to account for most of the incongruent findings.[103]

An article in the *New England Journal of Medicine* in August 1994 created a stir in the media: "Experiment in Memphis suggests many drive after using drugs,"[104] was a typical headline. Unfortunately, the "after" is quite critical in evaluating this study, wherein drivers observed driving recklessly were given drug tests by the Memphis Police

Department.[105] But the urine tests employed test for nonpsychoactive metabolites of cocaine and marijuana, metabolites that are present many hours and even days after any drug is taken. The tests do not prove that anyone was driving "under the influence" of either cocaine or marijuana. (This point is made by Dr. Petrie Rainey of Yale University School of Medicine in a letter to the editor some 8 months later.)[106] Since there were no controls (i.e., random samplings of the urine of those *not* driving recklessly), the study cannot even be used to "prove" what it most likely suggests: that people who drive recklessly are reckless in other aspects of their lives, and are thus more likely to take drugs. However believable this may be, it is not proved by the study; so widespread is the use of cocaine and marijuana in our society that the reckless drivers might have been under the average in their use of these drugs.[107]

Time Expansion. The remaining major psychic effect of Δ^9-THC is the very common phenomenon of time expansion. Users of cannabis (oral or smoked) asked to estimate elapsed time in a controlled setting will overestimate it, while users of alcohol by contrast will underestimate it.[108] Weil and Snyder suggest that most of the psychotropic effects produced by marijuana can be attributed to an intense focusing of consciousness on the immediate present—a state that Weil points out is the goal of many higher forms of religious meditation and awareness:

> Disturbance of immediate memory seems to be a common feature of all altered states of consciousness in which attention is focused on the present. . . . Is it a negative description of a condition that might just as well be looked at positively? I believe so. In fact, the ability to live entirely in the present, without paying attention to the immediate past or future, is precisely the goal of meditation and the exact aim of many religious disciplines. The rationale behind living in the present is stated in ancient Hindu writings and forms a prominent theme of Buddhist and Christian philosophy as well: to the extent that consciousness is diverted into the past and future—both of which are unreal—to that extent is it unavailable for use in the real here and now.[109]

Some quantitative measure of this focusing on present time may be given by a more recent study at the George Washington University Medical Center. In this study, "absorption," a trait capacity for "total attentional involvement," was reported to increase during episodes of marijuana intoxication; and the findings supported "the hypothesis that a specific type of alteration in consciousness that enhances capacity for total attentional involvement (absorption) characterizes marijuana intoxication, and that this enhancement may act as a reinforcer, possibly influencing future use."[110] "Depersonalization," an alteration in the experience of the self in which the usual sense of one's own reality is temporarily lost or changed, was shown by a group at Duke[111] in a double-blind study to be an effect of marijuana smoking strongly linked to a sense of temporal disintegration. Depersonalization had been previously reported in association with meditation,[112] as well as with several pathological conditions. However, in the case of marijuana and meditation, subjects regarded the state as a positive achievement. Again, the corollary with the familiar religious "cosmic sense" is obvious enough.

Metabolism: Urine Testing. Plasma concentrations of Δ^9-THC reach their highest levels quite soon after smoking marijuana begins; within an hour after the last puff, they are at 10% their maximum value. From then on, there is a slow leaching of Δ^9-THC into the plasma from fat tissues, in which it and some of the carboxylic acid metabolites are stored as glucuronides; hence the rate at which the last 10% declines to zero is much slower than the initial decline from the maximum. For this reason, a moderate user who smokes a single joint will have a positive urine sample for at most 5 days, even with the most sensitive detection method, which is gas chromatography-mass spectrophotometry (GC-MS); with the ordinary first-screening method, enzyme immunoassay test (EIA, also called EMIT),[113] the urine will be positive for at most 2 days. However, a very heavy user who stopped on 1 January could theoretically remain positive until 1 February, if tested by GC-MS[114] (but these are never used in a first-screen test because they are too expensive). EIA is neither as sensitive nor as reliable as GC-MS, by which it must always be confirmed; it is additionally susceptible to false positives from such common over-the-counter drugs as ibuprofen.[115] EIA tests are also easily confounded by addition of household bleach or vinegar, which alter the sensitive pH balance needed for proper enzyme activity; and all tests can of course be affected by dilution with water, either that added directly after an unwitnessed voiding or indirectly by deliberate intake of an unusually large quantity of fluids before testing.

However, regardless of the analytic method, the actual component for which the urine is tested is not Δ^9-THC itself. Although Δ^9-THC is the chief psychoactive component in marijuana and hashish (the concentration of Δ^6-THC is too small to be of concern), it is so lipophilic that analytic methods for it have proven intractable: it binds nonspecifically to tissue, protein, and even the glass surfaces of vessels used in workup.[116] In all cases, what is actually analyzed for is the presence of one or both of the major metabolites of Δ^9-THC, which are *11-hydroxy-Δ^9-THC* (**7-18**, Figure 7.6) and *11-nor-Δ^9-tetrahydrocannabinol-9-carboxylic acid* (THCA, **7-19**).[117] The terminal elimination half-life of 11-hydroxy-Δ^9-THC is approximately 15 to 18 hours.[118]

Chronic Pharmacology: Effects of Long-Term Use. *The "Amotivational Syndrome."* As long ago as 1843, Dr. Jacques-Joseph Moreau described long-term heavy users of hashish whom he observed while traveling in the Orient in terms that sound exactly like our amotivational syndrome—or what the '60s slang would have called a "pothead." However, he felt that the syndrome was relatively rare and that hashish was not as damaging as alcohol:

> But the misuse of hashish, I say, can in the long run bring about . . . . a state of constant drowsiness, hebetude [lethargy], mental apathy, and, as a consequence thereof, disappearance of spontaneity of action, willpower and the ability to make decisions. These psychic anomalies are visible by an expressionless physiognomy, a depressed, lax, and languid countenance, dull eyes . . . slow movements without energy, etc. Such are some of the symptoms characteristic

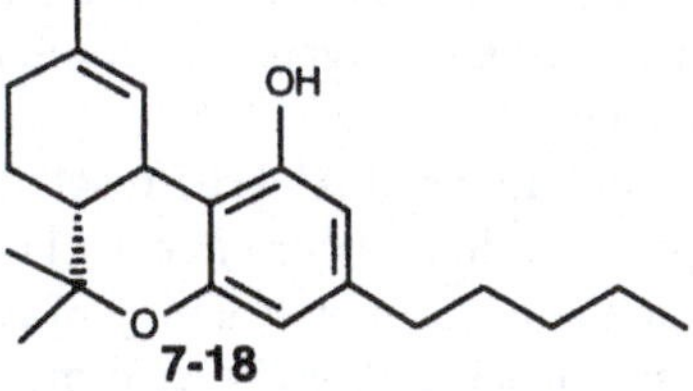

CH$_2$OH OH O **7-18**

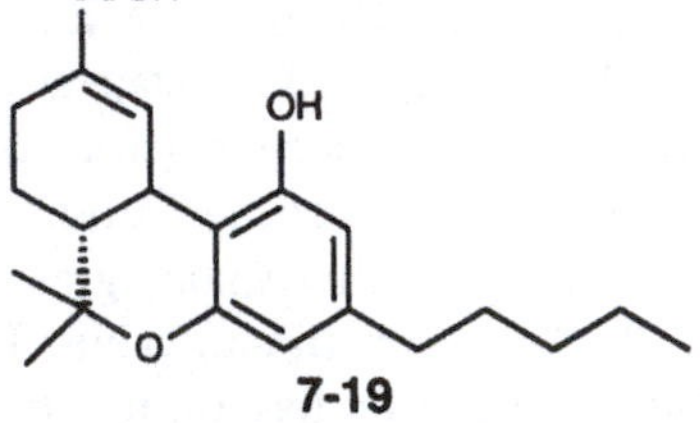

COOH OH O **7-19**

Figure 7.6 Major metabolites of Δ^9-THC.

> of the *excessive* use of hashish. We have had occasions to observe several examples of this.. . . . But, from what I have said before, one ought not to form an unfavorable idea about hashish. It is the same with hashish in Egypt as it is with wine and alcoholic beverages in Europe. The use of it is equally common. Almost all Moslems eat hashish, a very great number of them are addicted to it to an unbelievably high degree and, yet, *it is extremely seldom* that one encounters persons upon whom the hashish has had the disastrous effects we have spoken about here. If we disregard opium and other narcotics, wine and liquors are a thousand times more dangerous. Nevertheless, would it not be absurd to prohibit them and deprive us of their benefits because—by misuse of them—one runs the risk of harming one's health? The same can be said, and with a thousand times more reason, with regard to hashish, this marvelous substance to which the Orientals owe *undescribable* delights. In fact, it would be in vain to describe it to anybody who has not experienced it himself.[119]

In every culture there are those who for one reason or another have given up striving for the goals accepted and approved by the culture; such people often turn to drugs. In our culture one need only consider skid-row alcoholics, all of whom would register high on the amotivational scale to any unbiased observer. Does Thunderbird wine cause an amotivational syndrome? The idea that marijuana causes a specific and unique "amotivational syndrome" is just as unlikely.

Extensive studies of the effects of cannabis use in Jamaica (where use is widespread and the average daily consumption by users many times that used by citizens of the United States or Europe) show that even chronic heavy use does not result in any loss of motivation or striving after conventional goals. Actually, the Jamaican working class looks upon *ganja* smoking as a stimulus to hard work, somewhat like coca-chewing in the Andes.[120]

Respiratory Function. There has always been concern that chronic smoking of marijuana might lead to the same problems as chronic smoking of tobacco—among the most serious problems, emphysema and lung cancer. The way in which marijuana is smoked, by deliberately inhaling deeply and holding the smoke as long as possible, would seem to be likely to aggravate any damage that could be caused by smoke carcinogens; and it has been claimed that smoking marijuana introduces four times more tar into the lungs than does smoking tobacco.[121] Jaffe estimates (but gives no data supporting the prophecy) "that even one or two such [marijuana] cigarettes per day will increase the risk of lung cancer."[122] But most reviews have not found any evidence of chronic marijuana smoking causing lung cancer (although there are isolated reports of cancers of the upper aerodigestive tract; for example, an Australian group reported two cases of squamous cell carcinoma of the tongue in men who chronically smoked marijuana but had no other risk factors).[123] An obvious difference between users of marijuana and users of tobacco is that the former do not smoke a pack of joints a day, as is common with tobacco users. Unfortunately, until recently this has been mostly a speculative issue, since the overwhelming majority of those who used marijuana also smoked tobacco. Both in Jamaica, where use of ganja is almost universal among the working class, and in the Netherlands, where marijuana has been decriminalized and freely available for almost 20 years, marijuana or hashish is almost always mixed with tobacco when smoked; and in both cultures, tobacco is very widely smoked as well. It is almost impossible to find a control group that smokes only marijuana.

In 1994, a retrospective study was made by the Kaiser Permanente Medical Care Program in California of 450 self-reported "daily" users of marijuana who never used tobacco. The study correlated marijuana usage with hospital visits classified as respiratory, injury, or other. Somewhat unexpectedly, it found that respiratory visits were significantly increased for persons who had smoked marijuana for fewer than 10 years, but not for persons who had smoked longer; on the other hand, the reverse was true for injuries: those who had smoked longer had an increased number of hospital visits. Overall, the study concluded that "the relative risk for the marijuana-smoking group compared with the nonsmoking group was elevated but not statistically significant."[124]

Addiction Liability. As we saw in Chapter 1, when marijuana is compared with the five other most commonly used drugs—nicotine, cocaine, heroin, alcohol, and caffeine—it scores sixth or lowest in three of the five categories of addiction (withdrawal, reinforcement, dependence), and fourth in the categories of tolerance and intoxication. Of these five drugs, marijuana is certainly the least liable to induce addictive behavior overall.

In the Netherlands, there has been ample opportunity for anyone so inclined to develop an addictive relationship to marijuana or hashish for the last 17 years. Dr. Janhuib Blans, who holds dual appointments at the University of Amsterdam and at the Jellinek Center, which is the largest institute in the Netherlands specializing in addiction care and research, estimates that the total yearly number of referrals from the entire Amsterdam metropolitan area for what might be described as a problem of "cannabis addiction" is 60. Even of these 60, about half are persons who must take antipsychotic medications because of chronic schizophrenia, and attempt to self medicate themselves from the dysphoric effects of drugs like Haldol or chlorpromazine by intensive use of marijuana or hashish. "Marijuana addiction" is essentially not a problem: the main problems are alcohol, cocaine, heroin, and—to a surprisingly large extent—gambling.[125]

Other Long-Term Damage. Marijuana use is an emotional issue, and bias abounds, often from the best of motives. Parents who smoked reefers with heedless abandon in their teens suddenly find themselves painfully anxious to protect their own teenage children from experimenting with marijuana. Researchers are not immune from these influences (nor from less noble ones like the tempting largesse of grant funding from government agencies whose own *raison d'etre* obviously depends on the continued judgment that illicit drugs are dangerous). There is no substance—including water, salt, sugar, or Vitamin A—which if forced on an in vitro or in vivo biological system in massive enough overdose, will not cause some measurable damage, and of course Δ^9-THC is no exception. As Weil says:

> Aside from respiratory irritation, heavy marijuana use does not seem to cause other medical problems. Of course, warnings of the medical dangers of cannabis have been well publicized, with reports of everything from brain damage to injury of the immune and reproductive systems, but these are based on poor research, often conducted by passionate foes of the drug. Studies of populations that have smoked cannabis for many years do not reveal obvious illnesses that can be linked to marijuana.[126]

A recent review by Leo Hollister of the many studies of the effect of marijuana on immune function, most made in the 1970s concludes similarly:

> The evidence has been contradictory and is more supportive of some degree of immunosuppression only when one considers in vitro studies. These have been seriously flawed by the

> very high concentrations of drug used to produce immunosuppression and by the lack of comparisons with other membrane-active drugs. The closer that experimental studies have been to actual clinical situations, the less compelling has been the evidence. . . .
>
> The relationship between the use of social drugs and the development of clinical manifestations of AIDS has been of some interest, however. Persons infected with the virus but not diagnosed as AIDS have been told to avoid the use of marijuana and/or alcohol. This advice may be reasonable as a general health measure, but direct evidence that heeding this warning will prevent the ultimate damage to the immune system is totally lacking.[127]

A 1991 editorial in the *British Journal of Addiction* concedes that "since the Indian Hemp Commission's report in 1894 on the effects of cannabis, vigorous efforts by scientists have repeatedly failed to demonstrate long-term harmful effects associated with cannabis use and there has been little evidence to seriously challenge the conclusions of the Hemp commission that moderate use of cannabis produces no adverse effects."[128]

The best evidence for or against long-term damage obviously must come from study of that demographic group that has used marijuana more heavily than any other for the longest period of time, and that is arguably the Jamaican working class. An exhaustive study of this group, carried out from 1970 to 1972 under the auspices of the Center for Studies of Narcotic and Drug Abuse of the (U.S.) National Institute of Mental Health, concluded that there was no indication of an "amotivational syndrome"; a lowering of the incidence of alcoholism compared to similar populations that did not use marijuana; no evidence of cannabis being a "steppingstone" or "initiation" drug to heroin or cocaine (use of which were virtually unknown at that time in Jamaica); no correlation between marijuana use and crime; no indication of organic brain damage or chromosome damage; and *no* significant psychiatric or medical damage with one exception: some functional hypoxia among heavy, long-term smokers (which, because of these subjects' simultaneous extensive use of tobacco, could not be certainly attributed to marijuana use).[129]

Cannabinoid Structure-Activity Relationships. Mechoulam, Devane, and Glaser have recently authored a referenced review of cannabinoid SARs, and what follows is a brief synopsis of the general patterns.[130] Unless otherwise noted, by "activity" is meant specifically cannabimimetic activity (the marijuana "high").

- A dihydrobenzopyran-type structure (**7-10**), with an OH at the 1 position and an alkyl group at the 3 position of the *C* ring, is required for activity. Opening of the pyran ring to form cannabidiol (**7-14**) leads to complete loss of activity. But the pyran oxygen can be replaced by a nitrogen in compounds showing a high degree of activity, such as nantradol.
- Either esterification or etherification of the phenolic group lowers or abolishes activity in vitro; the esters show activity because of biological hydrolysis. Replacement of the phenolic group with a thiol eliminates activity.
- The *11-OH-*Δ^9*-THC* metabolite (**7-18**) is more active than its parent compound, as is also the case with the Δ^8 isomers. These compounds are rapidly metabolized by further oxidation to the corresponding aldehydes and carboxylic acids, which results in loss of cannabimimetic activity, although analgesic activity may be retained. The most potent cannabinoid substance thus far known is *11-OH-*Δ^8*-THC-DMH* (HU-210, **7-20**, Figure 7.7),[131] where the 11 position of Δ^8-THC has been hydroxyl-

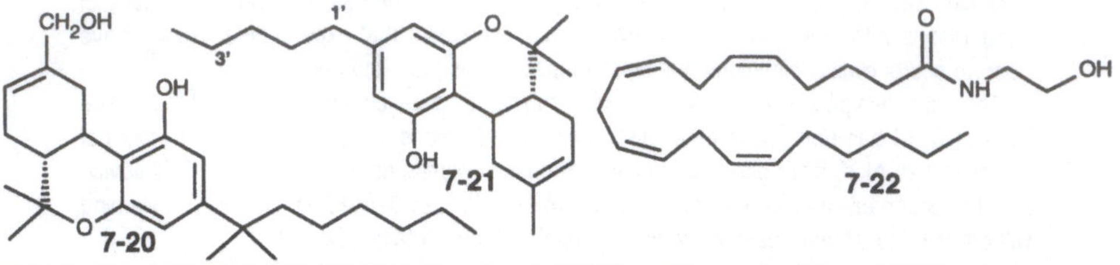

Figure 7.7 Structural variants of Δ^8-THC; anandamide.

ated and where the usual pentyl group in the 3 position of the *C* ring has been replaced with 1,1-dimethylheptyl. Both of these modifications enhance activity; unexpectedly, the corresponding modifications of Δ^9-THC result in a structure less active than the Δ^8 isomer **7-20**. In animal tests, the (−) enantiomer of **7-20** is almost 100 times as active as (−)-Δ^9-THC. Hydroxylation of the pentyl side chain occurs as one pathway of human metabolism: if the hydroxylation occurs at the benzylic 1′ position of **7-21**, activity is abolished; if it occurs at any other position in the side chain, activity is retained, and if it occurs at 3′, it is enhanced. Recent syntheses of 3′-hydroxy-Δ^9-THC and biological testing shows it is two to three times as active as the parent Δ^9-THC in animal behavior tests.[132]

- Double-bond THC isomers other than Δ^9-THC and Δ^8-THC have been synthesized and tested for activity. Only two show (diminished) activity: $\Delta^{6a,10a}$-THC and $\Delta^{10,10a}$-THC.
- A double bond in the *A* ring is not essential for activity; upon hydrogenation of this ring, two epimers are formed, and both are active. Indeed, if **7-20** is hydrogenated, the equatorial epimer is as active as the parent compound in animal discrimination tests, and is superior to it in binding to the cannabinoid receptor.
- The alkyl side chain at the 3 position of ring *C* is critical for activity. It must be at least five carbon atoms in length; elongation and branching enhance activity, with 1,1-dimethylheptyl and 1,2-dimethylheptyl showing the greatest activity.
- Despite the results of earlier work (which neglected to use starting materials of sufficient optical purity), a very high order of stereospecificity exists for the cannabimimetic activity of the natural (−)-Δ^9-THC. The unnatural (+) isomer is essentially inactive. Similarly, for the synthetic compound **7-20**, all psychotropic activity seems to lie in the (−) isomer; in pigeons, for example, the (+) isomer was inactive in doses 4,500 times higher than the ED_{50} for the (−) isomer. This is a degree of stereoselectivity on the order of that shown by the opiates, and is further evidence for a well-defined cannabinoid receptor system (and one that is common to mammals and birds).

Anandamide, an Endogenous Agonist at the Cannabinoid Receptor. In the late 1980s, Devane reported the results of his doctoral research at St. Louis University: the active constituent of cannabis, Δ^9-THC, binds to a specific G protein-coupled receptor in the brain.[133] In 1992, Devane and co-workers[134] found an endogenous agonist of this receptor by screening fractions of porcine brain extracts for their ability to displace a radiolabeled cannabinoid probe. They eventually isolated (0.6 mg from 4.5 kg of brain) a substance they named *anandamide*, **7-22**, from the Sanskrit word for bliss. High-resolution mass spectrometry gave a molecular formula of $C_{22}H_{37}NO_2$, and NOESY NMR indicated the compound was most probably the ethanolamide of arachidonic acid, i.e., *N*-(2-hydroxyethyl)-(*all-Z*)-5,8,11,14-eicosatetraenamide. Synthesis of this compound from arachidonyl chloride and ethanolamine confirmed that it was indeed anandamide. As is the case with the endogenous opioids and the morphine alkaloids, anandamide has a structure surprisingly different from the cannabinoids; and like the endorphins, anandamide is probably only one of a series of related natural compounds with cannabimimetic activity, since Devane's group was able to detect smaller amounts of other active substances in the brain extracts.

Shortly after the discovery of anandamide, Mechoulam's group at Hebrew University, Jerusalem, reported that when anandamide was administered in vivo to mice (intraperitoneally) it produced a positive response in a series of four tests that, taken together, are considered highly predictive of cannabimimetic activity. It is interesting to compare the activity of anandamide with that of Δ^8-THC in a classic test for analgesia: the animal is placed on a hot plate and the temperature is increased at a constant rate; the more effective the analgetic, the longer the time until the animal first licks its paw. Control animals showed this response at about 7 seconds; at a dose range of 20 mg/kg (when anandamide reaches maximal effect, it being no more active at 100 mg/kg), anandamide-treated mice respond in about 13 seconds; mice given Δ^8-THC in the same dose of 20 mg/kg did not respond until almost 36 seconds. In this test, the cannabinoid is almost three times as active as the natural ligand. Anandamide is presumably inactive orally, since it would be cleaved by gastric peptidases, but from these results it obviously crosses the blood-brain barrier.[135] Later work led to the characterization of two cannabinoid receptor subtypes, CB1 and CB2. The first type is found only in the CNS and the testes; the second in the periphery.[136]

Ancient and Modern Medical Uses. Cannabis has been used in folk medicine since before recorded history. Its use for medical conditions was recorded in Egypt as early as the 16th century BC; recently (1992), ashes from a 4th-century AD tomb near Jerusalem suggest that it may have been used in the ancient Middle East to help in childbirth—the ashes were from burned hemp, and were beside the skeletal remains of a teenage girl who died during labor. The smoke from the burning cannabis was presumably inhaled to ease the pain of childbirth.

Medical applications of cannabis are recorded in ancient Assyria, Egypt, Israel, Greece, and Rome; in medieval Christian and Moslem civilizations; in India, Persia, and China. Cannabis, usually in the form of an alcoholic tincture, was widely used in 19th-century Western medicine. In the summary of the historical record, Mechoulam[137] finds a total of 20 uses; of these exactly half have received some substantiation in the last two decades of research: analgetic, antiasthmatic, anticonvulsive, sedative, hypnotic, antirheumatic, antidiarrheal, antibiotic, appetite promoter, and antipyretic.

Between 1840 and 1900, more than 100 articles appeared in medical journals describing therapeutic uses of cannabis. It was claimed to be an appetite stimulant, muscle relaxant, analgesic, and soporific. In the first decade of the 20th century, the world-renowned physician, Sir William Osler of Johns Hopkins University considered it the best remedy for migraine headache.[138] Despite this rather impressive history—one equaled only by opium—application of cannabis products did not enter 20th-century medicine as did morphine and the opioids. One reason is probably the intrinsic variability of the cannabis extract, which could be very active or inert depending on when and where the hemp was harvested. By contrast, the appropriate time to excise the opium seed capsule is unmistakable. As chemistry and medicine progressed, cannabis fared no better; unlike morphine, Δ^9-THC is not an alkaloid, and the easy extraction procedure from plant material by alternate change from acid-aqueous to basic-organic solvent system could not be applied. As for the social opprobrium associated with marijuana products generally, these are obvious enough, and did their part to inhibit investigation of the medical uses of cannabis. The recent introduction of prescription Δ^9-THC (dronabinol) and at least one synthetic cannabinoid (nabilone) represents a first step toward the application of this ancient herb to modern medicine. The recent discovery that a whole series of endogenous cannabinoid agonists like anandamide probably interact with the arachidonic acid-prostaglandin system may finally allow hemp's rehabilitation.[139]

Nausea in Cancer Chemotherapy and AIDS. Many of the most widely used chemotherapeutic agents cause severe nausea. Dry heaves may last for hours or days after one treatment, and this may be followed by weeks of nausea. Spanish Inquisitor Tomás de Torquemada or the Nazi's Dr. Mengele would have enjoyed administering some these drugs: patients have been known to break bones or rupture the esophagus from the violence of their vomiting. Unable to eat and tormented by constant nausea, they often refuse to continue the chemotherapy or simply give up all will to live. Grinspoon and Bakalar (*Marihuana, the Forbidden Medicine*) have collected numerous case histories of individuals attempting to use marijuana for their medical problems and the enormous difficulties this entailed.[140] Here is the story told by Arnold and Mae Nutt about their oldest son, Keither, who developed testicular cancer in 1978 at age 22:

> In less than two months our son lost thirty pounds. He began to vomit bile. When there was nothing to vomit, he would simply retch and convulse. It was horrible for us to watch our child suffer such anguish from the disease and the treatment. . . . As a parent I was strongly opposed to marihuana and other illegal drugs. . . . It was hard to believe that an illegal drug could be of any help. If marihuana had medical value, we thought that the government would know and would make it legally available by prescription. But we were desperate, so we told Keither what we had read. . . . In desperation, we asked a close friend for help, an ordained Presbyterian minister who worked with local youth groups. Several days later he appeared at our door with some marihuana. It was the first time we had ever seen any.
>
> The next day we took the marihuana to Keither in the hospital. After he smoked it his vomiting abruptly stopped. The sudden change was amazing to see. Marihuana also put an end to his nausea. When he smoked it he was constantly hungry and actually began to put on weight. His mental outlook also underwent a startling improvement. Before he started to smoke marihuana, Keither would come home from chemotherapy, shut himself in his bedroom, stuff towels under the door to keep out the smell of dinner cooking, and remain in his room or the bathroom vomiting all evening. . . .
>
> Smoking marihuana dramatically changed his life. Immediately before the chemotherapy he smoked one marihuana cigarette, and afterward, if he felt queasy, all or part of a second. When we arrived home he stayed in the living room and talked with his brother and father. He joined the family for dinner and ate more than his share. He became outgoing and talkative,

> part of our family again. He never once experienced an adverse effect. Marihuana was the safest, most benign drug he received in the course of his battle against cancer. . . .
>
> My husband and I came to resent the fact that Keith's therapy was illegal. We felt like criminals. We are honest, simple people and hate having to sneak around. We were uncomfortable asking our closest friends, our minister, and our other son, Marc, to risk arrest in order to provide Keith with the medicine he so obviously needed.[141]

Keith was not alone: in the hospital they met Rev. and Mrs. Negen, the parents of a young girl receiving chemotherapy for leukemia; as pastor of a very conservative Dutch Reformed Church, Rev. Negen risked losing his pastorship for sending his sons into the streets of Grand Rapids to buy marijuana for their sister. These and other families testified before the Michigan House and Senate; Keith died on October 20, 1979, the night before the Michigan Congress voted by overwhelming margins in favor of the Michigan Marijuana as Medicine Bill.

A happier outcome, attributed in no small part to the benefits of smoked marijuana, was experienced by Stephen Jay Gould, Alexander Professor of Geology at Harvard University:

> I am a member of a very small, very fortunate, and very select group—the first survivors of the previously incurable cancer, abdominal mesothelioma. . . . I had surgery, followed by a month of radiation, chemotherapy, more surgery, and a subsequent year of additional chemotherapy. I found that I could control the less severe nausea of radiation by conventional medicines. But when I started intravenous chemotherapy (Adriamycin®), absolutely nothing in the available arsenal of antiemetics worked at all. I was miserable and came to dread the frequent treatments with an almost perverse intensity.
>
> I had heard that marihuana often worked well against nausea. I was reluctant to try it because I have never smoked any substance habitually (and didn't even know how to inhale). Moreover, I had tried marihuana twice (in the usual context of growing up in the sixties) and had hated it. (I am something of a Puritan on the subject of substances that, in any way, dull or alter mental states—for I value my rational mind with an academician's overweening arrogance. I do not drink alcohol at all, and have never used drugs in any "recreational" sense.) But anything to avoid nausea and the perverse wish it induces for an end of treatment.
>
> The rest of the story is short and sweet. Marihuana worked like a charm. I disliked the "side effect" of mental blurring (the "main effect" for recreational users), but the sheer bliss of not experiencing nausea—and then not having to fear it for all the days intervening between treatments—was the greatest boost I received in all my year of treatment, and surely had a most important effect upon my eventual cure. It is beyond my comprehension—and I fancy I am able to comprehend a lot, including much nonsense—that any humane person would withhold such a beneficial substance from people in such great need simply because others use it for different purposes.[142]

Similar anecdotal evidence abounds that smoked marijuana provides—for some people at any rate, in some medical conditions—a uniquely efficacious relief for nausea and vomiting. The two seemingly related phenomena are not well understood from a medical standpoint, however familiar they may seem to all of us. Like "pain," there may be many endogenous systems that elicit and control nausea; it is even quite likely that these are not the same systems that affect vomiting (emesis). Other species seem to have separated these two operations. Anyone who owns a cat knows that they are able to vomit with a cheerful efficiency not obviously associated with any signs of nausea. On the other hand, laboratory observation of rats seems to indicate the opposite paradigm: these creatures are physiologically unable to vomit, their esophagogastrointestinal system being an exclusively one-way street; but given the appropriate drugs, they seem to manifest all the symptoms of profound nausea.[143]

Perhaps this is one factor leading to the quite contradictory evidence surrounding the issue of whether smoked marijuana can be clinically demonstrated as efficacious against the nausea of chemotherapy. Other factors are that the nausea from one chemotherapy or radiation regimen is not necessarily equivalent to that from another, and the studies have rarely controlled for this. The people who benefit most from smoked marijuana seem to be those who have already experienced it as a pleasurable recreational drug. This in turn may be because they have become tolerant to the nausea induced by the very act of smoking, particularly if the cigarette also contains tobacco. In Weil's classic study of marijuana, he had to eliminate a surprising number of volunteers who "had described themselves in screening interviews as heavy cigarette smokers, 'inhaling' up to two packs of cigarettes a day,"[144] but when required to deeply inhale a tobacco cigarette (as one must do to get the full effect of the cannabinoid content of a marijuana cigarette), would become dizzy and nauseated from a dose of nicotine to which they were not tolerant. Finally, one person's "adverse" affects such as "mood changes, visual and time distortions," are another person's welcome distraction: at the complaint that oral Δ^9-THC was unsatisfactory in a 1979 study[145] because of "the tendency toward dysphoric effects by THC in elderly patients," the Woodstock hippies of '69 (now approaching the "elderly" designation themselves) would doubtless shrug their shoulders and dismiss the complainers as old fogies who didn't know how to relax and enjoy themselves.

Dronabinol. The principal antinauseant ingredient in marijuana is Δ^9-THC, which is now available under the generic name dronabinol (Marinol) for use as a Schedule II oral prescription drug to treat the nausea from cancer chemotherapy and the anorexia of AIDS. Dronabinol seems to be most effective against the nausea caused by methotrexate and cisplatin, and less effective against that caused by cyclophosphamide and doxorubicin. "Adverse reactions" are said to be mood changes, visual and time distortions, occasional dysphoria, and hallucinations. In a survey of oncologists conducted by a group at Harvard, it was found that nearly half of the 1,035 specialists responding would prescribe marijuana if it were legal; a slightly smaller percentage said that despite the illegality of the drug they had already recommended it to patients as a way of finding relief from nausea resulting from chemotherapy and for enhancing appetite.[146] It was felt by many that the mode of ingesting the drug (smoking) produced better results than an oral dose of dronabinol. This is probably in no small part due to the erratic absorption of the highly lipophilic drug from the intestinal tract—particularly the enteric system of a nauseated, vomiting patient. It may also be due to the coexistence of cannabidiol in marijuana smoke, which may reduce the anxiety provoked by Δ^9-THC taken alone.[147]

It seems obvious that it would be advantageous if there were a way of delivering pure Δ^9-THC (or a mixture of it and CBD) to the lungs without the admixture of vegetable tars, either as an aerosol (as many bronchodilating drugs are presently delivered) or by means of something like the smokeless cigarette introduced by R. J. Reynolds in 1988.

Nabilone. Nabilone (Cesamet, **7-23**, Figure 7.8) is a synthetic cannabinoid that has the advantage of greater water solubility than Δ^9-THC when used as an oral prescription

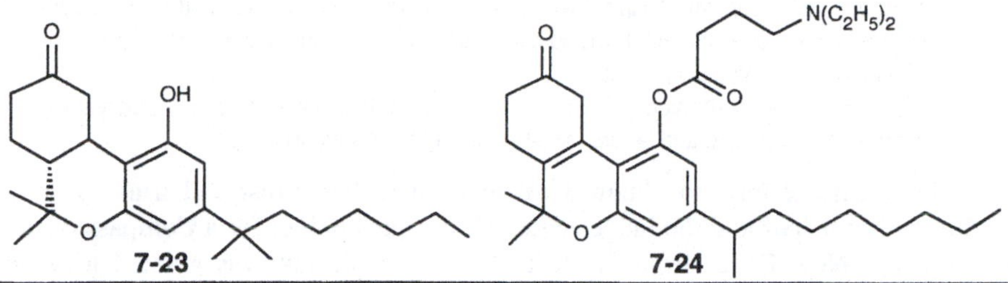

7-23 7-24

Figure 7.8 Nabilone and naboctate.

drug. It has an antiemetic effect much like Δ^9-THC; but it is possibly more effective against some anticancer agents such as cisplatin. It has been shown to be more effective than prochlorperazine (compazine), a phenothiazine antinauseant. The side effects are similar to those of dronabinol, but there is a lower incidence of tachycardia, and the psychotropic effects are less noticeable than with THC.[148]

Glaucoma. Glaucoma is a disorder resulting from an excess of intraocular fluid, which can lead to gradual loss of vision due to increasing intraocular pressure. "There is no doubt that marijuana and several cannabinoids, particularly Δ^9-THC, are able to lower the intraocular pressure, and this action certainly justifies their study as potential therapeutic agents for the treatment of glaucoma,"[149] writes Martin Adler, professor of pharmacology at Temple University School of Medicine. The fact that marijuana lowers intraocular pressure was discovered serendipitously in the course of an experiment at the UCLA designed to show whether—as the LPD then believed—cannabis dilated the pupils. Normal volunteers given government-grown cannabis were given complete ocular examinations. It was found that tearing was reduced and that intraocular pressures were reduced. Further experiments showed this was true for glaucoma patients as well; the effect lasted for 4 or 5 hours, with no indication of any deleterious effects on visual function or ocular structure.[150]

For the many sufferers from glaucoma who have discovered what seems to them the unique efficacy of marijuana in preserving their eyesight, the issue is much less theoretical. Elvy Musikka is a woman in her mid-40s who lives in Florida. In 1975 she was diagnosed with glaucoma; a friendly physician advised her to smoke marijuana. She was most reluctant to do so because she had been told it was as dangerous and as addictive as heroin. Since she had trouble inhaling, she used it in brownies. In 1977 she went to the University of Miami, hoping to be able to obtain legal marijuana. Instead, doctors there tried numerous drugs to no avail and finally scheduled her for surgery.

> At home that night I used a remaining bit of marihuana to bake some brownies, and ate two every twelve hours. The doctors were shocked when they checked my pressures as I arrived for surgery the next Monday morning—perfectly normal at 14 and 16! Regardless, they readied me for surgery . . . which turned out to be of no value. . . After this procedure I had less sight, more scar tissue, and higher pressures. . . .
>
> By 1980 I had little money and marihuana had gone up on price, so I started growing my own plants. I used the finest seeds, which produced small plants, hard to detect but productive. I only required three or four joints a day. My pressures became so close to normal that my

> doctors decided a corneal transplant was safe. It worked! I have never had such beautiful eyesight—it was so wonderful! I was so happy, until neighbors jumped the fence around my yard and stole my marihuana plants.
>
> My pressures went sky high. . . . So reluctantly and fearfully I went through surgery again. This time I hemorrhaged and before I knew it, my right eye was blind. . . .[151]

Her further efforts to obtain marijuana led to her arrest and trial. Her doctor testified on her behalf and the judge acquitted her; she applied for a Compassionate IND (Investigatory New Drug) from the federal government and was granted legal use of marijuana in October 1988. The sight in her right eye has returned; in her left eye, which was formally 20/400 but now 20/100, she has lost no peripheral vision.[152]

The problem with using marijuana itself or Δ^9-THC is twofold: the psychoactive high and the impossibility of overcoming this by topical application of Δ^9-THC directly to the eye because of its extreme water insolubility. Although nabilone reduces intraocular pressure, it is only effective if given orally, with some concomitant psychotropic effects.[153] There has been some progress in overcoming these difficulties; *naboctate*, **7-24**, lowered intraocular pressure in normal volunteers without side effects.

Other Ocular Phenomena. Fishermen in Kingston, Jamaica, have long claimed that their night vision is much improved by smoking hashish or drinking an extract of the plant in Jamaican white rum. The fishermen travel long distances at night in open canoes, often with no lights or compass. Pharmacologist M. E. West of the University of the West Indies, Kingston, tested this claim.[154] He traveled with a crew on a moonless night through a deep-water channel flanked by numerous coral reefs. West "sat in the boat and listened for the sound of us running aground . . . but heard nothing, only to be told a short while later that the boat was being docked. At daybreak it was impossible to believe that anyone could navigate a boat without compass and without light in such treacherous surroundings." Closer to home, frequent users report that they can chop onions without tears if they are stoned; on the other hand, contact-lens wearers lament this lack of lachrymation.

Neurological Disorders. *Muscle Spasticity in Spinal Cord Injury.* Many victims of spinal cord injuries such as paraplegia and quadriplegia have reported for years that the painful and debilitating involuntary muscle spasms of the legs from which many of them suffer—often the spasms are so severe as to throw them from their beds and make normal sleep impossible—are greatly alleviated after smoking marijuana. In a survey of such persons in 1982, 56% of respondents said that they had tried marijuana, and 88% of those reported that it helped reduce spasticity.[155]

G. Fred McBee was injured in a fall from a horse in 1964 when he was 16 years old. The fall crushed his fifth, sixth, and seventh cervical vertebrae and severed the spinal cord, leaving him with no sensation or movement in arms or legs. In 1981, the United Nations International Year of Disabled Persons, he and two others directed the "Continental Quest"—a coast-to-coast crossing of the United States by wheelchair. Nine years later, when *Idaho v. Hastings* was argued before the Idaho Supreme Court, he submitted the following testimony from Florida, where he was living with his wife of 24 years and his 11-year-old daughter:

A major obstacle to rehabilitation of paralyzed patients is muscle spasms. Mild spasms are good for you. They exercise paralyzed muscles. But . . . mine were so powerful that I had to be tied to my wheelchair or I would be thrown to the ground. And at night, my legs were tied to the bed rails so I wouldn't kick myself out of bed. . . . The intense spasming of my legs, back, and bladder was powerful enough to tear my muscles apart and to cause me excruciating pain above my level of injury. These spasms can be powerful enough to break bones. . . . To control them, my doctors prescribed heroic doses of muscle relaxants and tranquilizers, like Meprobamate, Valium, and Phenobarbital. They even experimented on me with Thorazine, an extremely powerful, mind-altering drug usually reserved for use by psychotic mental patients. And they offered to inject a chemical into my spinal cord which would dissolve my nerve tissue. They practically guaranteed me that this procedure would permanently end my spasms. The only side effect would be that it would also permanently end my ability to have an erection. I was seventeen years old, I had plans for my erections, and I decided I'd rather try to live with the spasms. . . . The first thing I did after leaving the hospital was to throw away all of my prescribed medications. I decided I'd rather live with spasms and pain than be "zoned out" on Valium or the host of other dangerous drugs that were prescribed for me. . . .

I don't know when I first heard rumors in the paralyzed community that marijuana significantly reduced spasms, aided in bladder control, and helped in getting a full night's sleep. But the rumors were pervasive. When I got together with the other "gimps," someone would eventually mention marijuana. I grew up on the Texas-Mexican border. I'd known about Mexicans smoking marijuana, and I'd been offered marijuana often during my early teens. But I never tried it. It was something an Anglo kid simply didn't do in South Texas.

By the late 1960s, however, the Viet Nam War was creating a lot of paralyzed veterans, many of whom were treated stateside at the Long Beach Veterans' Hospital where "Coming Home," an award winning film about a disabled vet, was later filmed. The paralyzed vets at the Long Beach VA were using marijuana to control their spasms and pain. My older cousin Jerry, himself a paralyzed vet in Long Beach, told me this. Of course, this marijuana use was surreptitious, hidden even from the doctors, because nobody was anxious to go to jail for twenty years. In 1971, while at the University of Oklahoma, I got together with five or six other paralyzed students. . . . We were curious to find out if the rumors about marijuana and spasms were true. That night each of us noticed, in his own way, that he "felt better" after smoking marijuana. None of us was spasming, which was remarkable. I'd never been in a room with this many para- and quadriplegics whose legs weren't dancing. And I got a full night's sleep—very unusual for me. . . .

I continued to smoke marijuana while attending college, competing as a world-class wheelchair athlete, writing articles and books, and representing the disabled for the United States and the United Nations. . . . For more than two decades I've been active in the disabled community. . . . To call marijuana's therapeutic use within the paralyzed community widespread would be an understatement. Marijuana is not simply the drug of choice among many paralyzed Americans, it is the only drug which provides us with effective, non-debilitating relief from spasms and pain, and permits us to function as competent individuals who work regular jobs and actively participate in society.

Paralyzed individuals live within a tightly knit community. We all know many others who smoke marijuana regularly, medically. . . . I resent the idea that so many thoroughly admirable people are considered to be criminals under the existing marijuana laws. I resent the prospect that, at any time, the police might kick down my door, put a gun in my face, handcuff me, take me to jail, place my daughter in state custody, revoke my wife's license to practice physical therapy, and confiscate my home and vehicles—and all because I choose to use the most medically effective medication for my condition. . . . I believe my therapeutic use of marijuana should be a medical issue between me and my doctor, not a legal issue between me and the criminal justice system.

Mr. William Bennet, our [1990] nation's "Drug Czar," is fond of stating, somewhat dogmatically, that he once taught Ethics at a university, and, therefore, he knows what is ethical and what is not. I have a degree in Philosophy. I, too, have studied Ethics, and I find Mr. Bennett's ethical arguments to be logically vapid, arrogant, and perverse.[156]

Multiple Sclerosis. In a case reported from the University of Göttingen, a 30-year-old MS patient showed acute improvement in his chronic motor handicaps when he smoked a marijuana cigarette. The effects were measured quantitatively by electromagnetic recordings, and the authors concluded that "cannabinoids may have powerful beneficial effects on both spasticity and ataxia."[157] But in a double-blind, randomized, placebo-controlled study of 10 MS patients and 10 normal subjects, with a computerized video camera measuring posture and balance, the motor skills of both normal and MS patients deteriorated when smoking marijuana; and the deterioration was greater for the MS patients than for the normal subjects.[158] However, the benefit many MS patients claim and seek from marijuana is rarely improved posture; it is reduced spasticity and pain. In 1981, Petro and Ellenberger reported on 9 patients, all naive to marijuana, given Δ^9-THC in 5 or 10 mg doses or placebo under a double-blind crossover regimen. Compared with placebo, the 10-mg Δ^9-THC significantly reduced spasticity as quantified by clinical and electromyograph measurements.[159]

Analgesia. Extensive studies at Pfizer research laboratories by M. Ross Johnson[160] and others led to the development of a considerably modified cannabinoid structure with potent analgetic activity. This compound, with a nitrogen replacing the oxygen of the dihydropyran ring, a 5-phenyl-2-pentoxy group replacing the pentyl group, and other alterations of Δ^9-THC, was called *nantradol.* As originally synthesized it was a 50:50 mixture of two pairs of racemic diastereoisomers, **7-25/7-26** (Figure 7.9) and the pair with the opposite chirality at 2′. When nantradol was compared in such standard measures of analgesia as the mouse writhing and rat tail pinch tests, it showed from two to seven times greater potency than morphine. Like morphine, nantradol also had considerable antidiarrheal and antitussive activity; but it did not bind to opiate receptors and its effects were not blocked by naloxone. Unlike morphine and Δ^9-THC, nantradol did not induce tolerance to its analgesic effects. When the enantiomers were separated for further study, it was found that *levonantradol*, **7-26**, was the most active component, being more than 100 times as active as its enantiomer, *dextronantradol*, **7-25**, and four times as active as the original nantradol mixture. Unfortunately, despite this promising pharmacological profile,

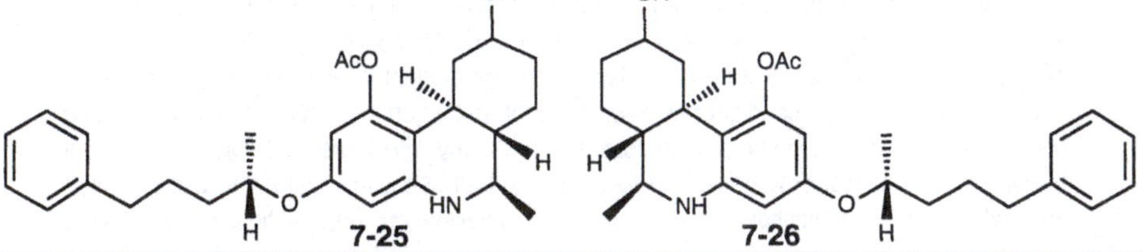

OH AcO H H HN H O 7-25 — OH H OAc H NH O H 7-26

Figure 7.9 Dextronantradol and levonantradol.

levonantradol is not free from the disturbing charms of its distant forbear, Δ^9-THC. It is a potent cannabimimetic, that is, it provides a decided psychotropic "high," which may be regarded as dysphoric or euphoric as your taste may so incline, but in any case has ruled out its acceptance as a prescription drug.

Legal Aspects of the Medical Use of Marijuana

The issue of whether marijuana can be prescribed by a physician for use by a patient for the treatment of a medical disorder is considerably narrower than the issue of the legalization of marijuana. It amounts to moving marijuana from the Schedule I status it now enjoys to the Schedule II category wherein are found cocaine, morphine, and Δ^9-THC itself. As a matter of fact, the DEA was ordered to do this by Francis L. Young, Chief Administrative Law Judge, USDEA, on 6 September 1988. After nearly 2 years of extensive testimony from patients, physicians, and pharmacologists, Judge Young ruled:

> The administrative law judge concludes that the provisions of the CSA [Controlled Substances Act] permit and require the transfer of marijuana from Schedule I to Schedule II. The judge realizes that strong emotions are aroused on both sides of any discussion concerning the use of marijuana. . . . Marijuana can be harmful. Marijuana is abused. But the same is true of dozens of drugs or substances that are listed in Schedule II so that they can be employed in treatment by physicians in proper cases, despite their abuse potential. . . .
>
> There are those who, in all sincerity, argue that the transfer of marijuana to Schedule II will "send a signal" that marijuana is "OK" generally for recreational use. This argument is specious. It presents no valid reason for refraining from taking an action required by law in light of the evidence. . . . The fear of sending such a signal cannot be permitted to override the legitimate need, amply demonstrated in this record, of countless sufferers for the relief marijuana can provide when prescribed by a physician in a legitimate case.
>
> The evidence in this record clearly shows that marijuana has been accepted as capable of relieving the distress of great numbers of very ill people and of doing so with safety under medical supervision. It would be unreasonable, arbitrary, and capricious for the DEA to continue to stand between those sufferers and the benefit of this substance in light of the evidence in this record.[161]

Although as of this writing [1996], marijuana remains a Schedule I substance, there are successful efforts to allow its use for the sick and dying. There are thought to be several dozen groups which are engaged in this, most prominently the Cannabis Buyers' Club of San Francisco—the "Taj Mahal of Cannabis Buyers' Clubs," according to Allen St. Pierre, deputy director of the National Organization for the Reform of marijuana Laws (NORML).[162] This is located near City Hall and openly sells marijuana banana bread and brownies as well as eight grades of loose plant material. A series of compassionate use resolutions by the city authorities allows its continued operation. But to enter this Taj Mahal, where hundreds of smokers chat at cafe tables and bars as they smoke the pungent herb, you must have a doctor's note to prove you have AIDS or cancer or one of the other conditions marijuana is claimed to alleviate:

> Dr. Marcus Conant, a leading AIDS doctor in San Francisco, said he frequently tells patients that they can have a prescription of Marinol, which costs several dollars a pill, or they can go to the Cannabis Buyers' Club, smoke a joint and probably will find it helps more.

"I'm not advocating everybody growing it in their flower box," Dr. Conant said, "but we have physicians with their hands tied because they can't use a drug that we have 40 years' experience showing is effective and has a reasonable place in the medical armamentarium."[163]

The political winds may be shifting, at least in California: a 1994 survey of voters there found that they backed legalizing medical marijuana by almost three to one. Three years earlier, 80% of San Francisco voters expressed the same view.[164]

Syntheses of Δ⁹-THC

The first successful synthesis of Δ^9-THC was reported by Mechoulam and Gaoni within a year of their isolation of the substance as the active principle in cannabis. It was modeled on the presumed biogenetic pathway: *citral* (**7-27**, Figure 7.10) was reacted with the lithium derivative of *olivetol* (**7-28**, protected as its dimethyl ether); tosylation and demethylation formed (±)-cannabidiol, which under acid conditions cyclized to (±)-Δ^9-THC.[165] The yield was only about 2%, and almost simultaneously a simpler and more effective route from essentially the same starting materials was published by Taylor, Lenard, and Shvo.[166] Both the products and the yields are very dependent on the solvent and the Lewis catalyst employed (e.g., 1% BF_3 in methylene chloride produces (±)-Δ^9-THC in the last step in about 20% yield, while 20% BF_3 produces (±)-Δ^8-THC); but with the incorporation of improvements later suggested by Mechoulam[167] the overall pathway (Figure 7.10) represents a reasonably efficient synthesis of racemic (±)-Δ^9-THC or (±)-Δ^6-THC. As for intermediates **7-29** and **7-31**, they represent one of several plausible mechanisms suggested for the conversion.[168]

Several syntheses of the natural (–)-Δ^9-THC have been developed, starting with chiral terpenoids. The first practical synthesis started with *cis-verbenol* (**7-33**, Figure 7.11), which can be readily synthesized from pinene. Condensation with olivetol under the proper conditions forms (–)-Δ^8-THC, **7-35**, which can be transformed in high yield into

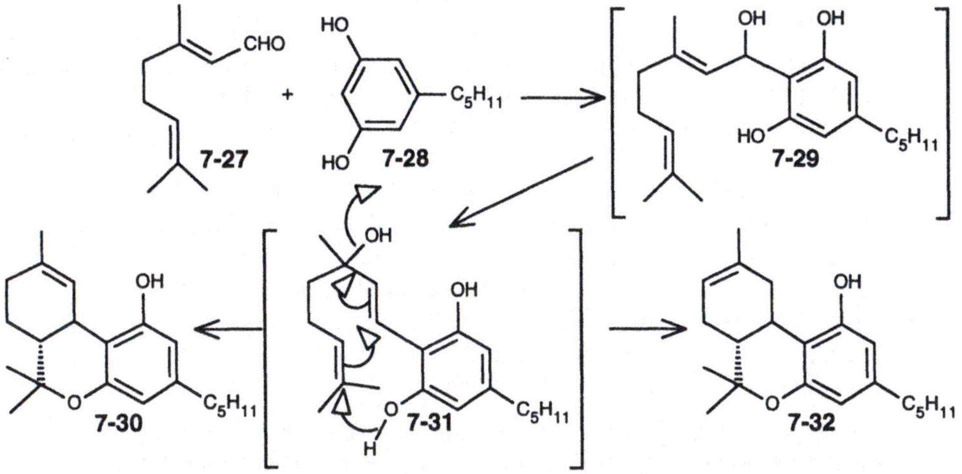

CHO
+
HO
C_5H_{11}
HO
7-27
7-28
OH OH
HO
C_5H_{11}
7-29
OH
OH
OH
OH
O
O
O
H
C_5H_{11}
C_5H_{11}
C_5H_{11}
7-30
7-31
7-32

Figure 7.10 Synthesis of (±)-Δ^9-THC and (±)-Δ^8-THC from citral and olivetol.

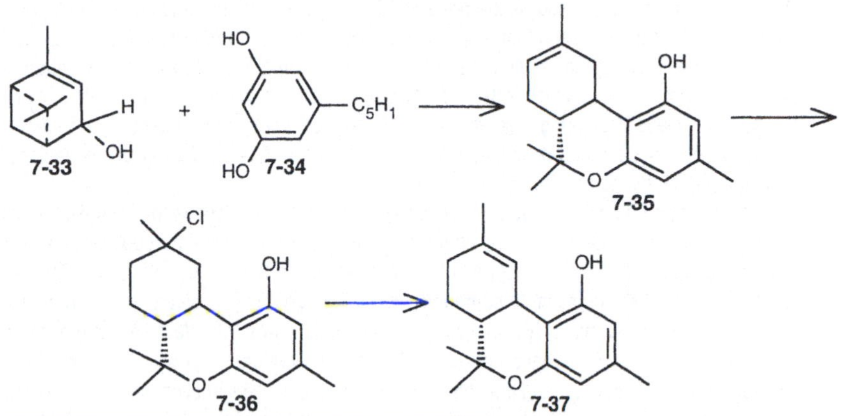

Figure 7.11 Synthesis of (–)-Δ^9-THC (7-37) from *cis*-verbenol and olivetol.

(–)-Δ^9-THC, **7-37**, by successive addition of HCl (**7-36**) followed by dehydrochlorination with potassium *tert*-amylate.[169]

More recently, Razdan, Handrick, and Dalzell have developed two one-step syntheses of (–)-Δ^8-THC; from *p*-mentha-2,8-dien-1-ol and olivetol,[170] and from *cis*-chrysanthenol and olivetol.[171]

ETCs

Nitrous Oxide

Sir Humphrey Davy incurred not only the "odium of discovering sodium," but that of the first synthesis of nitrous oxide (N_2O, "laughing gas," "whippits") in 1776. He laughed uproariously upon inhaling it, and soon had his friends over for parties to share the fun (it was a gas). They were high-quality friends: poet and opium addict Samuel Coleridge, poet laureate Robert Southey, potter Josiah Wedgwood, and physician-lexicographer Peter Roget.[172]

William James, brother of novelist Henry James, teacher of Gertrude Stein, and author of *The Varieties of Religious Experience*, in a footnote to an essay attacking the popular Hegelian philosophy of the day, described his own experiments with nitrous oxide:

> Since the preceding article was written, some observations on the effects of nitrous-oxide-gas-intoxication which I was prompted to make by reading the pamphlet called *The Anaesthetic Revelation and the Gist of Philosophy*, by Benjamin Paul Blood, Amsterdam, N.Y., 1874, have made me understand better than ever before both the strength and the weakness of Hegel's philosophy. I strongly urge others to repeat the experiment, which with pure gas is short and harmless enough. The effects will of course vary with the individual, just as they vary in the same individual from time to time; but it is probable that in the former case, as in the latter, a generic resemblance will obtain. With me, as with every other person of whom I have heard,

the keynote of the experience is the tremendously exciting sense of an in tense metaphysical illumination. Truth lies open to the view in depth beneath depth of almost blinding evidence. The mind sees all the logical relations of being with an apparent subtlety and instantaneity to which its normal consciousness offers no parallel; only as sobriety returns, the feeling of insight fades, and one is left staring vacantly at a few disjointed words and phrases, as one stares at a cadaverous-looking snowpeak from which the sunset glow has just fled, or at the black cinder left by an extinguished brand.

The immense emotional sense of *reconciliation* which characterizes the "maudlin" stage of alcoholic drunkenness—a stage which seems silly to lookers-on, but the subjective rapture of which probably constitutes a chief part of the temptation to the vice—is well known. The centre and periphery of things seem to come together. The ego and its objects, the *meum* and the *tuum*, are one. Now this, only a thousand-fold enhanced, was the effect upon me of the gas: and its first result was to make peal through me with unutterable power the conviction that Hegelism was true after all, and that the deepest convictions of my intellect hitherto were wrong. Whatever idea or representation occurred to the mind was seized by the same logical forceps, and served to illustrate the same truth; and that truth was that every opposition, among whatsoever things, vanishes in a higher unity in which it is based; that all contradictions, so called, are but differences; that all differences are of degree; that all degrees are of a common kind; that unbroken continuity is of the essence of being; and that we are literally in the midst of *an infinite*, to perceive the existence of which is the utmost we can attain. Without the *same* as a basis, how could strife occur? . . . .

It is impossible to convey an idea of the torrential character of the identification of opposites as it streams through the mind in this experience. I have sheet after sheet of phrases dictated or written during the intoxication, which to the sober reader seem meaningless drivel, but which at the moment of transcribing were fused in the fire of infinite rationality. . . . Let me transcribe a few sentences:

"What's a mistake but a kind of take? . . . .

Emphasis, emphasis; there must be some emphasis in order for there to be a phasis. . . ."

The most coherent and articulate sentence which came was this:—

"There are no differences but differences of degree between different degrees of difference and no difference."

This phrase has the true Hegelian ring, being in fact a regular *sich als sich auf sich selbst beziehende Negativität.* And true Hegelians will *überhaupt* be able to read between the lines. . . . But for the assurance of a certain amount of respect from them, I should hardly have ventured to print what must be such caviare to the general. . . .

My conclusion is that the togetherness of things in a common world, the law of sharing, may, when perceived, engender a very powerful emotion; that Hegel was so unusually susceptible to this emotion throughout his life that its gratification became his supreme end. . . .[173]

In the *Psychological Review* of 1898, James printed a letter sent to him from an unidentified British correspondent, who describes an experience he underwent under nitrous oxide at a dentist's office while an undergraduate at Oxford:

The next experience I became aware of, who shall relate! My God! *I knew everything!* A vast inrush of obvious and absolutely satisfying solutions to all possible problems overwhelmed by entire being, and an all-embracing unification of hitherto contending and apparently diverse aspects of truth took possession of my soul by force. . . .

Then, in a flash, this state of intellectual ecstasy was succeeded by one that I shall never forget, because it was still more novel to me than the other —I mean a state of moral ecstasy. I was seized with an immense yearning to take back this truth to the feeble, sorrowing, strug-

> gling world in which I had lived. . . . I saw that previous prophets had been rejected only because the truths they brought were partial and on that account not convincing. . . . But I thought I was dying and should not be able to tell them. I had never cared much for life, but it was then that I prayed and strove to live for the world's sake, as I had never prayed and striven before.[174]

Unfortunately, the vision was interrupted at this point by a "little pink man with a kindly face" who turned out to be the dentist, recommending that the student go away for a change of air since he was obviously suffering from a sluggish liver.

Many high school and college students buy grocery-store whipped cream or small canisters of nitrous oxide from kitchen supply stores and inhale the gas at parties. Presumably, if they knew, their parents would be aghast. Balloons of the gas are sold on Bourbon Street in New Orleans during Mardi Gras.

Epidemiology on the recreational use of nitrous oxide is sparse. But studies of dentists, anesthesiologists, and nurses exposed to daily low-level concentrations of nitrous oxide have shown that there are long-term health hazards, including nerve, liver, and blood disorders, and infertility.[175]

Volatile Nitrites

Amyl nitrite ("Amyls," "Poppers," "Snappers," "Pearls," Aspirol, Vaporal) is an older chemical name for a technical grade of material that is actually a mixture of isomers, of which the principal one is isoamyl (isopentyl) nitrite, $Me_2CHCH_2CH_2ONO$. A volatile liquid, its vapors have been inhaled for over a century to relieve the pain of angina pectoris. Amyl nitrite causes a rapid peripheral vasodilation, as does nitroglycerin, by the same mechanism of interaction with NO receptors. But unlike nitroglycerin, it can be inhaled and hence provide relief in as little as 30 seconds. Amyl nitrite is also used as an antidote for cyanide poisoning; the exact mechanism for this effect is disputed, but nitrite ion is known to convert hemoglobin to methemoglobin. It is packaged for medicinal purposes in thin glass ampules covered by cloth: breaking the cloth (with a popping sound, hence the street name) releases the volatile liquid. Inhaling causes flushing of the face, a pounding of blood rushing to the head (to compensate for lowered blood pressure consequent upon vasodilation). It is an experience, although not one which everyone enjoys. Too much of the experience results in a throbbing headache.

Amyl nitrite was at one time a prescription drug; it was so harmless (the body very rapidly metabolizes nitrites) that the FDA allowed it to be sold OTC. But many young people seemed to relish the rush. After all, as Lincoln said, "Those who like this sort of thing will find this the sort of thing they like." The FDA was concerned (being concerned is their job), and so they made it a prescription product once again. The market responded by packaging *butyl* and *isobutyl nitrite*, with equivalent pharmacologic effects, under such trade names as Rush and Locker Room. At this point, the FDA either lost its will to fight or gained a sense of humor, and the products thus far remain legal. Of course, they are (ostensibly) not to be used for human consumption, but apparently as room deodorizers: the label on the Rush bottle says: "NOT FOR SALE TO MINORS. Remove cap, allow to stand, aroma will develop. Contains Isobutyl Nitrite."

Locker Room? Well, locker rooms often smell pretty bad. In fact, butyl nitrite smells pretty bad; some speculate that it is called this "because of its musty smell."[176] Much more likely is the (homo)sexual connotations of a locker room. These products are sold with sexual paraphernalia; they are one of the many Made in America products displayed in the windows of Amsterdam's Sex Shops. Many believe that the inhalation of nitrites prolongs and intensifies orgasm, and this belief is particularly strong in the gay community, despite any hard scientific evidence to support it.[177]

Inhalants: Chloroform, Ether, Butane

When the anesthetic properties of two volatile liquids, chloroform, $CHCl_3$, and (diethyl) ether, $(CH_3CH_2)_2O$, were first discovered in the 19th century, they became somewhat popular as the centerpiece (literally—often chloroform was placed in a large shallow punch bowl to be sniffed) of otherwise rather elegant get-togethers. How many would-be socialites came to an untimely end by the delayed liver toxicity that can follow heavy use of chloroform is not known. In the case of ether, the greatest danger is that of violent explosions, sometimes triggered merely by static electricity.

Toward the end of the 19th century, the British authorities attempted to reform the Irish drinking habits by heavy taxation of hard liquor. The wily Irish responded by smuggling ether from hospitals. They found that if you drank a shot of ether it gave a quick 15 to 30-minute high much like alcohol, followed by a return to hangover-free sobriety—and the production of elephantine flatus, in this case highly combustible, which made the simultaneous use of another popular psychoactive substance, tobacco, very dangerous. The British authorities realized they had a prohibition-created drug epidemic—and a major fire hazard—on their hands: "etheromania" was sweeping Ulster! The common sense that made them so successful an imperial power soon reasserted itself, and the Irish were allowed their alcohol again.[178]

One of the most amusing accounts of the use of inhaled ether to provide a transcendental experience is that of Oliver Wendell Holmes, poet, author of *The Autocrat of the Breakfast-Table* (1858), father of the U.S. Supreme Court Justice, and quondam dean of Harvard Medical School:

> I once inhaled a pretty full dose of ether, with the determination to put on record, at the earliest moment of regaining consciousness, the thought I should find uppermost in my mind. The mighty music of the triumphal march into nothingness reverberated through my brain, and filled me with a sense of infinite possibilities, which made me an archangel for a moment. The veil of eternity was lifted. The one great truth which underlies all human experience and is the key to all the mysteries that philosophy has sought in vain to solve, flashed upon me in a sudden revelation. Henceforth all was clear: a few words had lifted my intelligence to the level of the knowledge of the cherubim. As my natural condition returned, I remembered my resolution; and, staggering to my desk, I wrote, in ill-shaped, straggling characters, the all-embracing truth still glimmering in my consciousness. The words were these (children may smile; the wise will ponder): "A strong smell of turpentine prevails throughout."[179]

Less amusing, and more unfortunate, is the prevalence of inhaling various organic solvents and gases for the "narcosis" it may produce. This is most common among young male teenagers and pre-teenagers.[180] The lower molecular weight hydrocarbons like propane and butane, found in backyard grills and cigarette lighters, do not cause any direct effects on the CNS, they simply deprive the brain of oxygen (if inhaled in concentrated quantities). Naturally, this results in stupor or unconsciousness; sometimes, if derring-do be too daring done, in death. Cincinnati Children's Hospital reported two deaths in 1990 from sniffing hydrocarbons.[181] An 11-year-old boy was found dead in the bathroom of a movie theater; a container of butane lighter fuel and a plastic bag were nearby, and postmortem examination showed butane in his blood and lung tissues. A 15-year-old boy was found unconscious in his back yard. His companions said they had inhaled propane from plastic bags filled from the tank of a gas grill. They also torched the gas while exhaling. However, the boy—who died on the way to the hospital—had no burns; propane was found on postmortem examination of his blood and lung tissues. Friends and school officials said sniffing these gases was common among children in this upper-middle-class neighborhood.

Hexane is a known neurotoxin (its biometabolite, hexane-2,5-dione is believed to be the causative agent), but is not usually found in current commercial solvents. Benzene, C_6H_6, upon prolonged exposure, can induce leukemia or aplastic anemia (bone marrow failure) in susceptible individuals. It is found in significant amounts in unleaded gasoline, but it is no longer used as the solvent for markers and glues. Benzene has been replaced by toluene (methyl benzene, $C_6H_5CH_3$), which is not carcinogenic or toxic to bone marrow (toluene can be oxidized by liver cytochrome systems to benzoic acid and excreted; benzene presents a more difficult challenge). The sort of stupor induced by benzene and toluene resembles that from the general anesthetics and probably has the same origin.[182]

Absinthe: Van Gogh's Yellow Eyes

To visit the Rijksmuseum Vincent van Gogh in Amsterdam is to be overwhelmed by the indefinably passionate intensity of this strange man who sold during his lifetime only a token handful of the paintings he gave his life to create. And to be overwhelmed by the images—a yellow sun in a green sky haloing a writhing earthworm sower; midnight stars whorling in a rainbow firewheel dance over an incandescent and exultant cypress; a phosphorescent skull grinning eerily as it smokes a clumsily rolled cigarette. Is the man's art a reflection of his life, or vice versa? Why did he cut off his left earlobe? He himself couldn't remember, by one account. What were the strange seizures that seemed to drive him to his death; were they florid psychosis, epilepsy, or something else? Why, above all, did he commit suicide—and why by the finally effective, but horribly slow and agonizing means of a bullet wound to the stomach?

Many suggestions have been made—epilepsy, alcoholism, manic depression, schizophrenia, sunstroke, neurosyphilis, and Ménière's disease, among others—but the most thorough study of van Gogh's physical and psychological condition was done by the University of Kansas Medical Center's Professor of Biochemistry and Molecular Biology Wilfred Arnold, who with Dr. Loretta S. Loftus suggested that van Gogh suffered from acute intermittent porphyria. This is a diagnosis still frequently missed today and one that was unknown in van Gogh's day. It is a disease whose episodic neuropsychiatric and

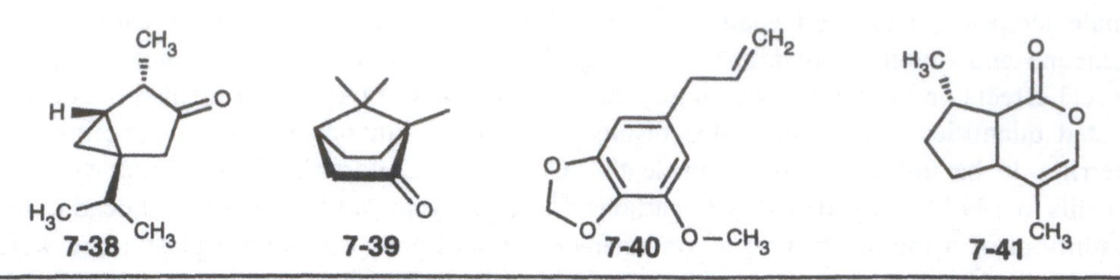

Figure 7.12 β-Thujone, camphor, myristicin, nepetalactone.

gastrointestinal complaints can be to a considerable degree suppressed by regimens such as a high-carbohydrate diet. Unfortunately, it is also powerfully activated by malnutrition, alcohol, and certain terpenes such as *thujone* (**7-38**, Figure 7.12) and *camphor*, **7-39**.[183] And Vincent was strangely wont to tipple his turpentine paint-thinner, treat his insomnia with double-doses of camphor, and toast his all-too-often absent suppers with absinthe. Some of these peculiar dietary quirks were as much the symptoms as the aggravation of the disease, argues Arnold; they may also have been caused by a form of pica (a symptom of malnutrition characterized by compulsive ingestion of nonnutrient materials such as clay or even paint).[184]

But what about absinthe? A gloriously green liqueur, when diluted with water it was transformed into a gentle opalescent yellow *impression* of itself. No wonder Lautrec and Gaugin used it to enhance the excitement of Montmartre! No wonder Poe used it to bypass the boredom of Baltimore! But everybody who was chi-chi liked it in the late 19th century: "artists painted and poets personified; men and women embraced the ritual of presentation as well as the appearance, taste, and excitement of this alcoholic drink."[185] Why was it so popular? Probably simply because it was the mode. Whether any of the herbs in absinthe contribute psychotropic or addictive substances to the mix has never definitely been shown. But that the concoction could cause seizures and brain damage in experimental animals, and was statistically associated with neurological damage in humans (including hallucinations and seizures like van Gogh's), was shown early on by Valentin Magnan.[186] Absinthe was an alcoholic extract of wormwood, anise, fennel, hyssop, and lemon balm; and Magnan showed that it was the extract of wormwood, *Artemisia absinthium*, that caused the problem. The constituents of wormwood are thujone, several isomeric thujyl alcohols and their esters, limonene, pinenes, and lesser ingredients; and the toxic ingredient appears to be thujone, which occurs as an enantiomeric mixture of about 33% (−) or α-thujone and 67% (+) or β-thujone (the isomers differ only in the stereochemistry at the methyl group). Later workers showed that the two isomers differ significantly in their convulsive dose in mice,[187] but there has been no systematic study to date of the consequences of long-term chronic administration in either animals or humans. In 1912, absinthe was banned in the United States; in 1915, it was prohibited in France.[188]

The English word "vermouth" comes from the Dutch word for wormwood, and there is a small quantity of wormwood still used as a flavoring in sweet vermouths. However, since the 1973 guidelines of the Council of Europe, *Artemisia absinthium* cannot be added to any foodstuff in amounts that will permit the thujone content of the final product to exceed 10 ppm. Still, if one must go the way of alcoholic excess, it might be better to go with vodka than vermouth.[189]

Nutmeg

William S. Burroughs[190] reports that the common household spice nutmeg, if swallowed in sufficiently large quantity, produces an effect "vaguely similar to marijuana" with the undesirable bonus of headache and nausea. Few users, including Burroughs, show any interest in repeating the experience. It is said to be used by prison inmates and sailors, smuggled from the kitchen, because they can get nothing better. According to Jaffe, 1 to 2 grated nutmegs, or about 1 to 2 tablespoons of the powdered spice, produces, "after a latency of several hours, leaden feelings in the extremities . . . feelings of depersonalization and unreality . . . dry mouth, thirst, rapid heart rate, and red, flushed face."[191]

Weil quotes a 22-year-old man who "choked down a whole can" of nutmeg one evening, over an ounce, and reports "it was a little like a light dose of pot. However, the next morning, I had a splitting headache, my heart was racing, and my mouth was really dry. . . . It lasted all day. I would never eat nutmeg again. I don't even like the taste of it in food anymore."[192] Shulgin has shown that a chief constituent of nutmeg oil is *myristicin*, which is 3-methoxy-4,5-methylenedioxyallylbenzene, **7-40**; by the formal addition of ammonia or an amine to the allyl double bond of myristicin, substances could be produced that would bear a close resemblance to several known psychedelic drugs; whether this process occurs *in vivo* and is the origin of nutmeg's psychotropic effects is not known.[193]

A Purrfect Drug

What would concerned parents say were they to hear of a drug that caused those individuals who snorted it to respond by sniffing, chin and cheek rubbing, licking and chewing with head shaking, and headover roll and body rubbing?! Yet this is just what some mammals will do in animal tests when exposed to even a few whiffs of *nepetalactone*, **7-41**, a monoterpenoid constituting up to 99% of the essential oil of *Nepeta cataria* L. Still more alarming is the appearance of a designer variant of the natural substance, *dihydronepetalactone*, which is even more powerfully attractive to addicts.[194]

Fortunately for those concerned parents, only the domestic cat, *Felis domestica*, and other members of the cat family (panthers, etc.) appear to be affected by this plant. And not all of them, either; response depends on an trait inherited by the cat as an autosomal dominant.[195] We do not know what the cat feels, but those possessing the trait in its full manifestation seem, when given catnip—insofar as we may judge or cats may express—happy. Euphoric, the pharmacologists would probably say. That nepetalactone is a psychotropic drug for only one branch of the mammalian family, the *Felidae*, is quite unusual; most other psychotropic drugs show broad activity across the various mammalian species. (Nepetalactone may confuse insects, too; or at least the aphids: it is both an insect repellant[196] and an aphid sex pheromone.[197]) Just throw a bag of catnip into a cage in the lion house at the zoo, and you will see why, were cats kings, nepetalactone would be a Schedule I substance. But only when sniffed! When cats were fed mammoth doses of nepetalactone in a gelatin capsule, they were totally unresponsive.[198]

With this rather hodgepodge assortment of etceteras, our tour of psychoactive drugs comes to a close. There are many reactions one might have to all of these strange stories of strange substances and the human involvement with them—an interaction variously enriching, amusing, futile, or tragic. Some will respond in repugnance: "Drugs are not the answer!" A second, more reflective response may be: "Drugs are not the problem."

At the end of a book it is appropriate to remember Last Words. A famous case of Last Words was that of Gertrude Stein. As some may recall, she lay dying of cancer in a French hospital just at the close of World War II. When it had become obvious that nothing more could be done, she turned to Alice B. Toklas, sitting at her bedside, and murmured, "What is the answer?" Alice said nothing. "In that case," said Gertrude, "what is the question?"[199] And the rest was silence. Ending life on an interrogative may at first seem an admission of defeat or denial, but Gertrude's simplicities often concealed a quiet wisdom.

If drugs are not the answer, what is the question?

References and Notes

The following abbreviations are used for frequent sources in these References/Notes:

BMC4: *Burger's Medicinal Chemistry*, 4th ed., Wolff, M. E., Ed., Wiley: New York, 1981. The **bold** number following *BMC4* indicates the volume number (1-3); this is followed by the page number.

C&C: *Cannabis and Culture* [*World Anthropology*, Tax, S., Ed.,], Rubin, V., Ed., Mouton: The Hague [Aldine: Chicago], 1975.

CCPT: *The Cannabinoids: Chemical, Pharmacologic, and Therapeutic Aspects*, Agurell, S.; Dewey, W. L.; Willette, R. E., Eds., Academic Press: Orlando, FL, 1984.

CTA: *Cannabinoids as Therapeutic Agents*, Mechoulam, R., Ed., CRC Press: Boca Raton, FL, 1986.

DE95: *Drug Evaluations: Annual 1995*; American Medical Association Division of Drugs and Toxicology, Department of Drugs; Bennett, D. R., Ed.; AMA: Chicago, 1995.

JJM: Moreau, J-J., *Hashish and Mental Illness*, Peters, H.; Nahas, G. G., Eds., Barnett, G. J., Trans., Raven Press: New York, 1973. Originally published in 1845 as *Du Hachisch et de l'Aliénation Mentale; Études Psychologiques.*

G&G8: *Goodman and Gilman's Pharmacological Basis of Therapeutics*, 8th ed., Gilman, A. G.; Rall, T. W.; Nies, A. S.; Taylor, P., Eds., Pergamon: New York, 1990.

G&G9: *Goodman and Gilman's The Pharmacological Basis of Therapeutics*, 9th ed., Hardman, J. G.; Limbird, L. E., et al., Eds.; McGraw-Hill: New York, 1996.

GJ: Rubin, V.; Comitas, L, *Ganja in Jamaica* [Vol. 26 of *New Babylon: Studies in the Social Sciences*], Mouton: The Hague, 1975.

GRIN: Grinspoon, L., *Marihuana Reconsidered*, 2nd ed., Harvard University Press: Cambridge, MA, 1977.

L&ID: *Licit and Illicit Drugs: The Consumers Union Report on Narcotics, Stimulants, Depressants, Inhalants, Hallucinogens, and Marijuana—Including Caffeine,*

Nicotine and Alcohol, Brecher, E. W. and the Editors of Consumer Reports, Eds., Little, Brown: Boston, 1972.

MI11: *The Merck Index*, 11th ed.; Budavari, S.; O'Neil, M. J., et al., Eds.; Merck: Rahway, NJ, 1989. References are given by monograph number.

MI12: *The Merck Index*, 12th ed.; Budavari, S.; O'Neil, M. J., et al., Eds.; Merck: Rahway, NJ, 1996. References are given by monograph number.

M/C: *Marijuana/Cannabinoids: Neurobiology and Neurophysiology*, Murphy, L.; Bartke, A., Eds., CRC Press: Boca Raton, FL, 1992.

MECH: *Marijuana: Chemistry, Pharmacology, Metabolism, and Clinical Effects*, Mechoulam, R., Ed., Academic Press: New York, 1973.

MFM: *Marihuana, the Forbidden Medicine*, Grinspoon, L.; Bakalar, J. B., Yale University Press: New Haven, CT, 1993.

PsyDR: *Psychedelic Drugs Reconsidered*, Grinspoon, L., Bakalar, J. B.; Basic Books: New York, 1979.

SA: *Substance Abuse: A Comprehensive Textbook,* 2nd ed.; Lowinson, J. H., Ruiz, P., Millman, R. B., Langrod, J. G., Eds., Williams & Wilkins: Baltimore, 1992.

1. From an essay of 1844. Cited in *The Hashish Club: An Anthology of Drug Literature. Volume One: The Founding of the Modern Tradition: From Coleridge to Crowley,* Haining, P., Ed., Peter Owen Limited: London, 1975, p. 84. Contrasting cultural bias is shown by A. E. Fossier, who opines: "The dominant race and most enlightened countries are alcoholic, whilst the races and nations addicted to hemp and opium, some of which once attained to heights of culture and civilization have deteriorated both mentally and physically" ("The marihuana menace," *New Orleans Med. Surg. J.*, **1931**, *84*, 249).
2. Dumas, A., *The Count of Monte Cristo*, Heritage Press: New York, 1941, Vol. 2, Chapter 31, p. 27.
3. Alcott, L. M., "Perilous Play." In: *Louisa May Alcott Unmasked: Collected Thrillers*, Stern, M. ed, Northeastern University Press: Boston, 1995, p. 694. (This is a collection of short stories that have been discovered only recently. Alcott wrote them pseudonymously for mass appeal to help subsidize her family. "Perilous Play" shows that she was well acquainted with the mental effects of hashish—whether through reading, observation, or firsthand experience is not known.)
4. Zukin, S. R.; Zukin, R. S, "Phencyclidine," *SA*, pp. 290-302.
5. There are, or course, some serious conceptual and even constitutional ambiguities built into such a law, as Shulgin has pointed out: Shulgin, A. T., "How similar is substantially similar?" *J. Forensic Sci.*, **1990**, *35*, 10-12. Cited in Morgan, J. P., "Controlled substance analogues: Current clinical and social issues," *SA*, 328-333.

6. *The New York Times*, 19 March 1994; 28 March 1994; 9 April 1994. The news reports were remarkably sympathetic to this white employed mother. "Wrestling Private Demons: Driver in Crash that Killed Priests Kept Problems to Herself," is the headline of the March 28 article.

7. Thombs, D. L., "A review of PCP abuse trends and perceptions," *Public Health Reports*, **1989**, *104*, 325-328.

8. *MM16*, p. 1566.

9. Griefenstein et al., *Anesth. Analg.* (N.Y.) **1958**, *37*, 283.

10. Johnstone, M; Evans, V.; Baigel, S. "Sernyl (Cl-395) in clinical anesthesia." *Br. J. Anesthesia*, **31**, *1959*, 433-439.

11. Luby et al., *Am. Med. Assoc. Arch. Neurology Psych.*, **81:** 113/363-119/369.

12. *PsyDR*, p. 33.

13. Balster, R. L.; Johanson, C. E.; et al., "Phencyclidine self-administration in the rhesus monkey," *Pharmacol. Biochem. Behav.*, **1973**, *1*, 167-172.

14. Pearlson, G. D., "Psychiatric and medical syndromes associated with phencyclidine (PCP) abuse." *Johns Hopkins Med. J.*, **148**, *1981*, 25-33.

15. Giannini, A. J.; Loiselle, R. H.; DiMarzio, L. R.; Giannini, M. C., "Augmentation of haloperidol by ascorbic acid in phencyclidine intoxication," *Am. J. Psychiatry*, **1987**, *144*, 1207-1209.

16. (a) Pechnick, R. N., "Effects of opioids on the hypothalamo-pituitary-adrenal axis." In: *Annu. Rev. Pharmacol. Toxicol.*, **1993**, *32*, 353-82 (pp 362f). (b) Quirion, R.; Chicheportiche, R.; et al., "Classification and nomenclature of phencyclidine and sigma receptor sites," *Trends Neurosci.*, **1988**, *10*, 444-446.

17. *The New York Times*, 19 February 1994.

18. Olney, J. W.; Labruyere, J.; Price, M. T., "Pathological changes induced in cerebrocortical neurons by phencyclidine and related drugs," *Science*, **1989**, *244*, 1360-1362.

19. *PsyDR*, p. 34. On the other hand, as with the traditional psychedelic drugs, there are workers who are convinced that the ketamine state is a model psychosis. At the National Institute of Mental Health, research psychiatrist Alan Breier is using positron emission tomography (PET) to study normal human volunteers while they are experiencing a ketamine-induced "psychosis." He is hoping to correlate the ketamine state with dopamine functioning. Brennan, M. B., "Positron emission tomography merges chemistry with biological imaging," *C&EN*, 19 February 1996, 26-33.

20. Stevens, J., *Storming Heaven: LSD and the American Dream*, Atlantic Monthly Press, New York, 1987, pp. 364-365.

21. *Ibid.*, p. 365. Elizabeth Kübler-Ross, author of *On Death and Dying*, describes an "iatrogenically" (presumably ketamine?) induced out-of-body experience that, she says, led her to "a rebirth beyond human description." Kübler-Ross, E., "Life, Death, and Life after Death." Originally a lecture recorded on tape in 1980. In: Kübler-Ross, E., *On Life After Death*, Celestial Arts: Berkeley, CA, 1991, pp. 41-69. Cited material pp. 64, 67. See also: Rogo, D. S., *The Return From Silence: A Study of Near-Death Experiences*, Aquarian Press: Wellingborough, Northamptonshire, England, 1989 [in particular Chapter 6, "Experiencing death through drugs?"].

22. *Xochi Speaks*, a poster describing the proper approach to be taken to the dozen most popular psychedelics. Fraser, L., "Xochipilli: a context for ecstasy," *The Whole Earth Review*, **1991**, *75*, 38-43, or by writing LordNose! PO Box 170473X, SF, CA 94117-0473.

23. Stafford, P., *Psychedelics Encyclopedia,* 3rd ed., Bigwood, J., Ed., Ronin: Berkeley, CA, 1992, p. 394.

24. Turner, D. M., *The Essential Psychedelic Guide,* Panther Press: San Francisco, CA, 1994, p. 64. The author of this book has studied the effects of numerous psychedelic drugs on himself (herself?) hundreds of times, and ranks the drugs on a 1 to 10 intensity scale. Thus the harmala alkaloids rate a 1 to 2; ecstasy rates a 2 to 4; psilocybin mushrooms a 3 to 7; LSD and peyote a 3 to 8; and DMT a 9 to 10. Ketamine, "the ultimate psychedelic journey," gets a 10 to infinity rating. While ketamine is different in some ways from the traditional psychedelics, it is nonetheless quite similar: Turner reports "spectacular and hallucinatory" visuals with eyes opened on coming out of a ketamine experience, which he compares to those of DMT. *Ibid.* p. 66 *et passim.*

25. This, in addition to the obvious differences in set and setting, may account for the rather undramatic results of two clinical trials of low-dose (35-58-mg infusion over a period of 40 minutes to an hour): (a) Malhotra, A. K.; Pinals, D. A.; et al., "NMDA receptor function and human cognition: The effects of ketamine in healthy volunteers," *Neuropsychopharmacology,* **1996**, *14*, 301-307. (b) Krsytal, J. H.; Karper, L. P.; et al., "Subanesthetic effects of the noncompetitive NMDA antagonist, ketamine, in humans," *Arch. Gen. Psychiatry,* **1994**, *51*, 199-214.

26. Kungurtsev, I., *The Albert Hofmann Foundation Bulletin*, Fall, 1991 (no page given). Cited in Turner, D. M., op. cit., pp. 67-68.

27. It has been noted growing wild, blooming about mid-August, in all but one of the 22 counties of Wisconsin, Illinois, Indiana, and Michigan surveyed in the Morton Arboretum's definitive work: Swink, F.; Wilhelm, G., *Plants of the Chicago Region: A Checklist of the Vascular Flora of the Chicago Region, with Keys; Notes on Local Distribution, Ecology, and Taxonomy; and a System for Evaluation of Plant Communities*, The Morton Arboretum: Lisle, IL, 1979, p. 121. Much of the wild weed is said to be the result of a failed federal WWII program, "Hemp for Victory," which encouraged farmers in the Corn Belt to grow hemp for fiber. See Schlosser, E., "Reefer madness," *Atlantic Monthly,* **1994**, *274*, (August), p. 45-63.

28. Grinspoon, L.; Bakalar, J. B., "Marihuana," *SA*, pp. 236-246.

29. Valle, J. R.; Lapa, A. J.; Barros, G. G., *J. Pharm. Pharmacol.*, **1968**, *20*, 798. Cited in Snyder, S. H., *Uses of Marijuana*, Oxford University Press: New York, 1971, p. 5. See also Mechoulam, R., "Cannabinoid chemistry," *MECH*, pp. 1-99, particularly pp. 11-13 and references cited therein.

30. Schaffner, J. H., *Ecology*, **1923**, *4*, 323. Cited in Snyder, S. H., *Uses of Marijuana*, Oxford University Press: New York, 1971, p. 97.

31. For a thorough discussion, see Schultes, R. E.; Klein, W. M.; Plowman, T.; Lockwood, T. E., "Cannabis: An example of taxonomic neglect," *C&C*, pp. 21-38.

32. For some stunning color photographs of these plants and an intriguing article on their horticultural development, see: Pollan, M., "Marijuana in the '90s: high tech, high crime, high stakes," *The New York Times Magazine*, Sunday 19 February 1995, pp. 30-57. As Milton Friedman has pointed out, the development of more concentrated drug forms is the inevitable economic consequence of prohibition, and Pollan comes to the same conclu-

sion: "It was all a little bit mad, and yet a gardener couldn't help but be impressed; . . . only later . . . did I fix on what may be the maddest part of all: that the credit for this most dubious of achievements belonged not only to the gifted, obsessed gardener and his willing plants but to the obsessions of a Government as well" (p. 57).

33. Grinspoon, L.; Bakalar, J. B., "Marihuana," *SA*, pp. 236-246. No reference is given to support the analyses of Δ^9-THC content given. In Holland, where ordinary marijuana and hashish are freely available, hashish oil is considered a hard drug and banned along with heroin and cocaine.

34. According to Pollan, the "Potomac Indica" variety has some strains with 14%; the DEA estimates the usual sensemilla presently available ranges from 8 to 10% THC: Pollan, M., "Marijuana in the 90s: high tech, high crime, high stakes," *The New York Times Magazine*, 19 February 1995, p. 35.

35. Jones, R. T.; Stone, G. C., *Psychopharmacologia*, **1970**, *18*, 108. Cited in Snyder, S. H., *Uses of Marijuana*, Oxford University Press: New York, 1971, p. 57.

36. So I was told in July 1994 during a conversation in Amsterdam with Reijer Elzinga, who is the chairman of the Bond van Cannabis Detallisten (the Union of Cannabis Retailers), representing the owners of the soft-drug "coffee-shops" that sell marijuana and hashish.

    For a fascinating portrayal of the economic and sociological history of the quasi-legalization of marijuana and cannabis in Amsterdam: Jansen, A. C. M., *Cannabis in Amsterdam: A Geography of Hashish and Marijuana*, Dick Coutinho: Muiderberg, 1991. The book is hard to find in the United States; I was able to obtain a copy at the Jellinek Center in Amsterdam, at the recommendation of Dr. Janhuib Blans. A book review in the *Journal of Psychoactive Drugs* appeared in **1993**, *25*, pp. 271-272.

37. Treaster, J. B., "Growing like, uh, weeds: Do it yourselfers find niche in urban marijuana market," *The New York Times*, 5 August 1994, p. B1.

38. Schlosser, E., "Reefer madness," *Atlantic Monthly*, **1994**, *274*, (August), p. 45-63. And there are such extensive fields of marijuana in Morocco that in June 1995, marijuana pollen was detected in record-breaking quantities along a 250-mile stretch of the Spanish coast, 25 miles across the Mediterranean. "This is exceptional," said Eugenio Domínguez, coordinator for the Spanish Network for Aerobiology, which monitors pollen counts as part of a European warning system for people with allergies. "We've never measured marijuana pollen in so many places." A bumper crop is predicted for Europe's pot consumers. (Simons, M., "Signs in wind of Morocco drug crop," *The New York Times,* Sunday 18 June 1995, p. 14.

39. Evans, I. H., *Brewer's Dictionary of Phrase and Fable*, Harper & Row: New York, 1970, p. 54.

40. *Webster's Third New International Dictionary of the English Language Unabridged*, G & C Merriam: Springfield, MA, 1967, p. 130.

41. *The American Heritage Dictionary of the English Language*, 3rd ed., Houghton Mifflin: Boston, 1992, p. 110.

42. In his *History of the Assassins*, as quoted in the *Calcutta Review*, as cited by Edward Fitzgerald in the preface to the 1872 edition of his translation of the *Rubaiyat* of Omar Khayyam. Small world! It turns out that when the Old Man of the Mountain was but a young lad learning his Koran, the then-also-young Omar was his classmate and school-buddy.

43. Fitzgerald, E., "Omar Khayyam: the astronomer-poet of Persia." In: *Rubáiyát of Omar Khayyám*, Dolphin Doubleday: Garden City, NY.

44. For much of the above, I am indebted to Grinspoon: *GRIN*, pp. 291-300.

45. Anslinger, H.; Cooper, C. R., "Marihuana: Assassin of Youth," *American Magazine*, **1937**, (July), p. 150. Cited in *GRIN* p. 291.

46. *Reefer Madness* is a 1936 film which "depicts a group of high-school students trying marijuana with murder, rape, prostitution and madness as the swift result of their folly." (Anderson, P., "The pot lobby," *The New York Times Magazine*, 21 January 1973, pp. 8-9; cited in Ray, O., *Drugs, Society, and Human Behavior,* 2nd ed., C. V. Mosby: Saint Louis, MO, 1978, p. 413.) The movie is so silly that it has the direct opposite of its intended effect, and has been distributed widely by the National Organization for the Reform of Marijuana Laws (NORML) as an argument for legalization. The movie was created as part of Federal Bureau of Narcotics Commissioner Harry J. Anslinger's anti-marijuana campaign in which "the public was led to believe that marijuana was addictive and caused violent crimes, psychosis, and mental deterioration. The film . . . may be a joke to the sophisticated today, but it was once regarded as a serious attempt to address a social problem, and the atmosphere and attitudes it exemplified and promoted continue to influence American culture today." (*MFM*, p. 8.)

47. Matsumoto, K.; Stark, P.; Meister, R. G., *J. Med. Chem.*, **1977**, *20*, 17-24. The muricidal rat test is described on p. 22.

48. On the other hand, there *is* one drug that, given to rats in large doses, will cause them to attack each other and even mutilate themselves in an enraged frenzy. The drug is caffeine—see Chapter 4.

49. Herodotus, *History of the Persian War*, iv, 142.

50. Benet, S., "Early diffusion and folk uses of hemp," *C&C*, p. 39-49.

51. Mechoulam, R., "The pharmacohistory of *Cannabis sativa*," *CTA* 1-19.

52. Rabin, C., "Rice in the Bible," *J. Semitic Studies*, **1966**, *11*, 2.

53. Chopra, G. S., "Man and marijuana," *The International Journal of the Addictions*, **1969**, *4*, 215-247.

54. Bharati, A., *The Tantric Tradition*, Rider: London, 1965. [Cited in *C&C*.]

55. Fisher, J., "Cannabis in Nepal: an overview," *C&C*, pp. 245-255; quoted material p. 250.

56. *Indian Hemp Drug Commission Report* (1893). Reprint, Thomas Jefferson Press: Silver Spring, MD, 1969, p. 492. Cited by Snyder, S., *Uses of Marijuana*, Oxford University Press: New York, 1971, p. 20.

57. Rubin, V., "The 'Ganja Vision' in Jamaica," *C&C*, pp. 262-263.

58. *Ibid.,* pp. 264-265.

59. *GJ*, p. 146.

60. *JJM*, p. 3.

61. *JJM*, p. 4

62. *JJM*, pp. 7-10.

63. *JJM*, pp. 11-13. *Le Club des Hachichins*, cautions Richard B. Grant [*Théophile Gautier*, Twayne: Boston, 1975, p. 131] was "written to be published and read. How much did Gautier invent or borrow from other accounts (like De Quincey's) that in their turn may

have been in part invented or copied? How much did Gautier falsify his narratives to make them more interesting or readable?" Professor Grant seems infected with Gautier's own initial skepticism. Just because something is "written to be published and read" hardly ipso facto makes the author out to be a liar. The internal evidence seems to authenticate the work: as Grant himself comments, the lack of thematic integrity or structured plot is what makes these descriptions poor fiction—but this is exactly what makes them credible *reportage*. To the doubting Thomas in us all, Moreau must be given the last word: "In fact, it would be in vain to describe it to anybody who has not experienced it himself."

64. For more on Gautier and Baudelaire, as well as Fitz Hugh Ludlow (author of *The Hasheesh Eater: Being Passages from the Life of a Pythagorean*) see Grinspoon's thorough analysis, with extensive quotations from Gautier: *GRIN*, 55-116. Also Solomon, S. H., *Uses of Marijuana*, Oxford University Press: New York, 1971, pp. 26-30.

65. Bessie, S. M., "A happy publisher's note to the new edition." In: *The Alice B. Toklas Cook Book*, Harper & Row Perennial: New York, 1984, p. viii.

66. Alice seems to have been quite surprised at the clamor roused by the inclusion of this recipe in her book. Apparently, she had forgotten it was there, since she wrote from Paris to Donald Gallup in October of 1954: "I hope you werent [sic] as shocked as I was by the notice in *Time* of the hashish fudge. I was also furious until I discovered it was really in the cook book! Contributed by one of Carl's [Carl van Vechten] most enchanting friends—Brion Gysin—so that the laugh was on me. Thornton [Wilder] said that no one would believe in my innocence as I had pulled the best publicity stunt of the year—that Harper had telegraphed from London to the Attorney General to see if there would be any trouble in printing it!" *Staying on Alone: Letters of Alice B. Toklas*, Burns, E., Ed., Random House Vintage: New York, 1973, p. 310.

    Brion Gysin was a painter and writer who had met Gertrude Stein and Alice briefly in the 1930s in Paris, but who lived many years in Morocco, where marijuana and hashish would have been readily available. (*Staying on Alone*, p. 253.) He also figured prominently in the life of one of the most accomplished drug users of all time, William Burroughs. (Burroughs, W., *My Education: A Book of Dreams*, Viking: New York, 1995.) Gysin died of cancer in the mid-1980s.

    Perhaps the notion of serving marijuana fudge at a DAR social is less outrageous than it at first seems. This author (DMP) recalls a story told him by an older Jesuit priest who had been the superior of the Jesuit community in Sri Lanka (then Ceylon) after WWII. The use of marijuana was technically illegal on the island (cultivation of real tea being preferred), and fairly large quantities were often confiscated by the police. It was then the custom in Jesuit communities to have quite elaborate many-course meals on major religious feasts, preceded and followed by wines, liquors, etc. After the superior had received many unsolicited expressions of grateful enthusiasm from the community members on the success of several of these events (a rare phenomenon in Jesuit communities), the superior felt he should congratulate the kitchen help, all local employees. There was a reason, he found, that everyone felt such a warm glow of good feeling and companionship after these meals: the main course was always liberally seasoned with marijuana provided gratis from the local constabulary, who felt their abundant supply should be put to a benevolent use. The practice continued with the encouragement of the superior, who expressed his regret that the grumpy postprandial tone so prevalent in most United States Jesuit communities could not be similarly benefited.

    In 1914, the *New York Post*, reviewing Gertrude Stein's *Tender Buttons*, speculated that Gertrude had developed her inimitable style by eating hashish. (See Souhami, Diana,

Gertrude and Alice, Pandora: London, 1991, p. 120.) But the fact that the recipe was not Alice's, and she had even forgotten she had included it, makes it likely that she had never tasted the fudge herself, let alone served it to Gertrude. A more likely drug etiology, one long prior to her meeting Alice, might be the babblings transcribed by her Radcliffe mentor, William James, under the influence of nitrous oxide. This was an experience he prized and encouraged (see below in this chapter), and one that she perhaps indulged in, thereby discovering for herself an experienced stream of consciousness.

67. *un évanouissement reveillé*: literally, something like "a wide-awake faint," but the French, even to one but feebly sensitive to its charms, seems much more evocative. Perhaps Gysin is quoting Baudelaire. The mention of St. Theresa is an allusion to Stein's 1934 opera, *Three Saints in Four Acts*, for which Virgil Thompson wrote the music.

68. Toklas, A., *The Alice B. Toklas Cook Book*, Harper & Row Perennial: New York, 1984, pp. 259-260.

69. Ginsberg, A., "The great marijuana hoax: first manifesto to end the bringdown," *Atlantic Monthly*, **1966**, November. A longer version printed in: *Marihuana Papers*, Solomon, D., Ed., Bobbs-Merrill: Indianapolis, 1966, pp. 184-185. Cited in *GRIN*, pp. 103-104.

70. *Ibid.*, p. 196 (*GRIN*, p. 104).

71. *GRIN*, pp. 110-113.

72. Adams, R.; Baker, B. R., *J. Am. Chem. Soc.*, **1940**, *62*, 2401, 2405.

73. Ghosh, R.; Todd, A. R.; Wilkinson, S., *J. Chem. Soc.*, **1940**, 1121, 1393.

74. To test the cannabimimetic effects on a naive human being, Adams sent some capsules containing $\Delta^{6a,10a}$-THC to a fellow chemist "of high standing in university circles" and asked him to record their effects. The effects were powerful indeed, and caused near panic on the part of the volunteer, who perhaps unfortunately chose the unfamiliar circumstances of a long-distance train ride to try it out. Other volunteers using natural THC and a few synthetic variants reported effects generally similar to those they associated with mild doses of alcohol, but with less motor incoordination. "The feeling was very different from that of being at one or another stage of intoxication," reported the high-stander in university circles; "for I looked perfectly clear and normal and I could stand erect without swaying and execute motions with considerable precision." Nonetheless, he "greatly resented" the unexpectedly strong "loss of control." Another chemist, who was not very responsive to alcohol generally, wishing to experience the "distorted time and space" characteristic of marijuana intoxication, finally took a triple dose. During a dinner at which he ate with "ravenous" hunger, he reported that "he was able to comprehend a question but by the time the answer was given, which was immediately, he couldn't remember the question." Nonetheless, "he was able to hold his own and then some in a poker game composed of expert players." The one consistent difference all subjects noted between alcohol and the THC was the development of an "enormous appetite" and "uncontrollable bursts of laughter or giggles." Adams reported these and other research results in the Harvey Lecture of 19 February 1942: Adams, R., "Marihuana," *Bull. N. Y. Acad. Med.*, **1942**, *18*, 705-730.

75. Gaoni, Y.; Mechoulam, R., *J. Am. Chem. Soc.*, **1964**, *86*, 1646; **1971**, *93*, 217.

76. A study of the interaction of pure CBN and CBD with Δ^9-THC showed no effect on the pharmacokinetics of Δ^9-THC, ruling out the possibility that the presence of one cannabinoid alters the liver enzymes responsible for metabolizing the others: Agurell, Stig, et al., "Pharmacokinetics of [Δ^9-THC] in man," *CCPT*, pp. 165-185.

77. Turner, J. C.; Mahlberg, P. G., "Separation of acid and neutral cannabinoids in *Cannabis sativa* L. using HPLC," *CCPT*, pp. 79-89.

78. Willinsky, M. D., "Analytical aspects of cannabis chemistry," *MECH*, pp. 137-165. See table on p. 142.

79. Fetterman, P. S.; Doornbos, N. J.; Keither, E. S., et al., *Experientia*, **1971**, *27*, 988. Cited in Willinsky, M. D., "Analytical aspects of cannabis chemistry," *MECH*, pp. 137-165. See table on p. 142.

80. See Mechoulam, R., "Cannabinoid chemistry," *MECH*, pp. 1-99, particularly pp. 11-13 and 88-89.

81. In animals: (a) Musty, R. E., "Possible anxiolytic effects of cannabidiol," *CCPT* pp. 795-813. (b) Guimaraes, F.; Chiaretti, T. M.; Graeff, F. G.; Zuardi, A. W., "Antianxiety effect of cannabidiol in the elevated plus-maze," *Psychopharmacologia*, **1990**, *100*, 558. (c) Onaivi, E. S.; Green, M. R.; Martin, B. R., "Pharmacological characterizations of cannabinoids in the elevated plus maze," *J. Pharmacol. Exp. Thr.*, **1990**, *252*, 1002. In humans: (d) Consroe, P.; Snider, S. R., "Therapeutic potential of cannabinoids in neurological disorders," *CTA*, 21-50.

82. Mechoulam, R.; Edery, H., "Structure-activity relationships in the cannabinoid series," *MECH*, pp. 101-136.

83. In letters to *High Times* (August, 1994, *228*, p. 79), collegiate folk wisdom advises clandestine smokers of pot to blow the smoke through a toilet-paper roll stuffed with fabric-softener sheets (Bounce, etc.) to avoid detection in the dorms; both the smoke and the odor are claimed to be totally absorbed. Perhaps restaurants with a no-smoking policy could provide these devices to their (tobacco) addicted patrons.

84. Willinsky, M. D., "Analytical aspects of cannabis chemistry," *MECH*, pp. 137-165. Leading references on pp. 139-140.

85. "Coffee shops" in Rotterdam did not seem to sell "space cakes," and only a few in Amsterdam did. Reijer Elzinga, owner of several of the more toney marijuana outlets (Kadinsky, Cum Laude, International Front Page) and the chairperson for the Union of Cannabis Retailers, said that he did not sell these cakes in his stores and discouraged their sale by other members of the Union. Oral hashish was unpredictable in its effects: customers would consume the cake, feel cheated because there was no effect in 30 minutes, eat more, then suddenly be overwhelmed in another 15 minutes by much more drug effect than they had bargained for. Unlike smoked cannabis, one is locked into the dosage one has swallowed, and panic reactions are much more common. Hashish itself, as small brown sticks, is sold in all the coffee shops for use as smoking material, usually by inserting it into tobacco cigarettes. Used in this way, it is completely equivalent to smoked marijuana, and the dose can be "titrated" by the smoker.

    I nibbled on a stick of hashish and found the flavor as described above. I had expected something much worse, relying on the descriptions from French 19th century literature. "The most common hashish preparation . . ." writes Moreau (*JJM*, p. 3), "is a greasy extract. The leaves and the flowers of the plant are boiled in water to which fresh butter has been added. When the mixture has been reduced by evaporation to the consistency of syrup, it is strained through a cloth. This extract contains the active ingredient and has a very nauseating taste." The revolting taste of the hashish preparations used by Moreau and the other members of the *Club des Hachichins* (see also Chapter 31 of Dumas' *The Count of Monte Cristo*, where Franz tastes some hashish given him by Sinbad the Sailor, a.k.a. the Count of MC, and finds it tastes terrible) may have been due to the labor-intensive method of harvesting the resin employed a century ago: naked men would jog through a field of blossoming cannabis, and the resin that clung to their sweaty skin would be scraped off and collected for processing. Although the clerk at one coffee shop opined that the hashish was still collected in this manner, I am very doubtful that it is. There are certainly no sweaty joggers in the Dutch hydroponic greenhouses.

86. The reddening of the eyes is a physiological effect from the cannabinoid constituents, not an irritation from the smoke: it also occurs when hashish is taken orally, according to Miras, C. J., *Hashish: Its Chemistry and Pharmacology*, Little, Brown: Boston, 1965, pp. 37-47.

87. Jaffe, J. H., "Drug addiction and drug abuse," *G&G8*, p. 551.

88. There are other studies that showed some pupil enlargement; others a fluctuating miosis/mydriasis; for a summary with references, see Adler, M. W.; Geller, E. B., "Ocular effects of cannabinoids," *CTA*, pp. 51-70, in particular pp. 65-66.

89. Some have opined that the appetite is not directly stimulated, but that eating is just one of many activities whose enjoyment is generally enhanced by THC: Weil, A. T.; Zinberg, N. E.; Nelson, J. M., "Clinical and psychological effects of marijuana in man," *Science*, **1968**, *162*, 1234-1242. Supportive of this view is the testimony of natives of India that bhang has traditionally been used by the poor during famine to stave off the pains of hunger. But in 1993, the FDA approved dronabinol (Marinol), which is pure Δ^9-THC, for a new indication: the treatment of anorexia associated with weight loss in patients with AIDS (*Drug Therapy*, March 1993, p. 4) after objective studies showed significant gains in weight when cachetic patients were given dronabinol. One would expect that people who use marijuana regularly would be chubby, but this does not seem to be the case. I have discussed this matter with at least one (skinny) paraplegic who uses marijuana on a daily basis because of his medical condition, and there seems not to be an acquired tolerance to this effect—he said he often finds himself raiding the refrigerator after a few hits on his evening bong.

90. Weil, A. T.; op. cit. (*Science*, **1968**), p. 1238.

91. *Drug Evaluations: Annual 1992*, American Medical Association Division of Drugs and Toxicology, Department of Drugs; Bennett, D. R., Ed.; AMA; Chicago, 1992, p. 444.

92. Weil, A. T.; op. cit. (*Science*, **1968**).

93. Weil, A., *The Natural Mind: A New Way of Looking at Drugs and the Higher Consciousness*, Houghton Mifflin: Boston, 1972, pp. 83-84.

94. See Clark, L. D., et al., *Am. J. Psychiat.*, **1968**, *125*, 379-384; *Arch. Gen. Psychiatry*, **1970**, *23*, 192-198; Manno, J. S.; Kiplinger, G. F., et al., *Clin. Pharm. Ther.*, **1970**, *11*, 808. Discussed in Snyder, S., loc. cit. (*Uses of Marijuana*), pp. 60-65.

95. Weil, A., op. cit. (*The Natural Mind . . .*), p. 86-87.

96. "At a high enough dose, CNS function would probably be affected to the extent that performance on virtually any laboratory behavioral task would be disrupted. (Such a dose might well be so aversive, however, that it would rarely be self-administered intentionally by users in the natural environment.)" Chait, L. D.; Pierri, J., "Effects of smoked marijuana on human performance: a critical review," *M/C*, p. 411.

97. Crancer, Jr., A.; Dille, J. M.; et al., *Science*, **1969**, *164*, 851.

98. Snyder, S., loc. cit. (*Uses of Marijuana*), pp. 62-63.

99. Weil, A., op. cit. (*The Natural Mind . . .*), p. 87-88.

100. In an informal poll taken by the author during the summer of 1994, several health officials and police officers in the cities of Rotterdam and Amsterdam were asked whether they had a problem with motor vehicle accidents occurring with drivers operating under the influence of cannabis. They seemed surprised at the question, but universally replied that they were unaware of it ever happening. Several volunteered, however, that alcohol remained a very serious problem.

101. Kerssemakers, R., (review of Robbe, H. W. J., *Influence of Marijuana on Driving*, University of Limburg: Maastricht, 1994) in: *Jellinek Quarterly* [News bulletin of the Jellinek Institute and the Amsterdam Institute for Addiction Research], **1995**, *2*, (5), 9-10.

102. Chait, L. D.; Pierri, J.; loc. cit., *M/C*, pp. 387-424. Cited material on p. 410.

103. Chait, L. D.; Pierri, J., "Effects of smoked marijuana on human performance: a critical review," *M/C*, pp. 387-424. Cited material on p. 411.

104. *The New York Times*, 28 August 1994, p. 30.

105. Brookoff, D.; Cook, C. S.; et al., "Testing reckless drivers for cocaine and marijuana," *NEJM*, **1994**, *331*, 518-522.

106. Rainey, P. M., *NEJM*, **1995**, *332*, 892-893. In rebuttal, Dr. Brookoff replies that "the majority of subjects who tested positive for these metabolites admitted to using the parent drug within 12 hours." But, of course, neither marijuana or cocaine, in however large a dose, is behaviorally active for 12 hours.

107. Perhaps this is an instance of laudable empathy making questionable science: in the *New York Times* article cited above, it is said that Dr. Brookoff, the principle investigator, "had become involved in the issue of drugged driving after a man hit a friend's two daughters while driving on the wrong side of the road. The man was never tested for drugs, but Dr. Brookoff *believed* he was high on marijuana" [emphasis added].

108. Jones, R. T.; Stone, G. C., *Psychopharmacologia*, **1970**, *18*, 108.

109. Weil, A., ibid (*The Natural* . . .) pp. 90-91. See also Snyder, *Uses of Marijuana*, Oxford: New York, 1971, pp. 68-70, analyzing the work of Melges et al., *Arch. Gen. Psychiat.*, **1970**, *23*, 204-210; *Science*, **1970**, *168*, 1118.

Aldous Huxley quotes Jalal-uddin Rumi: "the Sufi is the son of time present." (Huxley, A, *The Perennial Philosophy*, Harper: New York, 1945, p. 187. The Sufis are often accused of using hashish to enhance their communion with Allah; some of them may have, and it seems consonant with their emphasis on the ecstatic experience of God and their use of dance and *dhikr* (chanting the Divine names) to effect this, but the accusation is just as likely to be simply another detraction by the conventional enemies of this nonconformist group. Not only Sufis are the sons of time present. What is more characteristic of Jesus (or the hippies) than Matthew V: "Consider the lilies of the field: they neither toil nor spin. Yet I tell you that not even Solomon in all his glory was arrayed as one of these. . . . Therefore do not worry about tomorrow, for tomorrow will worry about itself. Sufficient for the day is the evil thereof." And what is more central to the wisdom of Siddhartha Gautama than the Silence of the Buddha, his refusal to answer questions about the future of the soul, "the meaning of life"? The meaning of life, as any Zen master will tell you, pouring his tea, is pouring his tea. A rose is . . . a rose. *Id cuius maior cogitari nequit . . . cogitari nequit.* When Oliver Wendell Holmes, like many others, saw the Answer to every question while passing out under ether, but found later that it was no longer the answer . . . it no longer was. The goalpost had moved, and he hadn't noticed. In the absolute attention to the absolute present one glimpses Boethius' definition of Eternity: *interminabilis Vitae tota simul et perfecta possessio.*

110. Fabian, W. D., Jr.; Fishkin, S. M., "Psychological absorption. Affect investment in marijuana intoxication," *J. Nerv. Ment. Dis.*, **1991**, *179*, 39-43.

111. Mathew, R. J.; Wilson, W. H.; et al., "Depersonalization after marijuana smoking," *Bio. Psychiatry*, **1993**, *33*, 431-441.

112. Castillo, R. J., "Depersonalization and meditation," *Psychiatry*, **1990**, *53*, 158-168.

113. The THC molecule, and the molecules of any presently used illicit drugs, are much too tiny to elicit an antibody response. EIA test materials are prepared by covalently linking the nonantigenic drug molecule to larger protein molecules such as bovine serum albumin. Rabbits are then injected with this modified albumin, and the antibodies that develop collected for the EIA assay. But only part of the small drug molecule sticks out from the protein-drug complex; and in any case the antibody is responding for the most part to the overwhelmingly more numerous surface features of the protein; consequently, other small molecules with some similarity to the drug molecule can bind almost indistinguishably to the protein, and result in a false positive. In the case of ibuprofen (4′-*iso*-butylphenyl-2-propionic acid), the *iso*-butylphenyl moiety is fairly similar to the *n*-pentylphenyl group of 11-*nor*-Δ^9-tetrahydrocannabinol-9-carboxylic acid (THCA); in each case, the carboxylic acid functionality has been covalently bonded to the enzyme.

114. Dackis, C. A.; Pottash, A. L. C.; Annitto, W.; et al., "Persistence of urinary marijuana levels after supervised abstinence," *Am. J. Psychiatry*, **1982**, *139*, 1196-1198.

115. Gold, M. S., *Cocaine* [Vol. 3 of *Drugs of Abuse: A Comprehensive Series for Clinicians*], Plenum Medical Book Company: New York, 1993, p. 177.

116. Hawks, R. L., "Developments in cannabinoid analysis of body fluids: implications for forensic applications." In: *CCPT*, p. 123.

117. Schwartz, R. H.; Hawks, R. L., "Laboratory detection of marijuana use," *JAMA*, **1985**, *254*, 788-792.

118. Anderson, P. O.; McGuire, G. G., "Delta-9-tetrahydrocannabinol as antiemetic," *Am. J. Hosp. Pharm.* **1981**, *38*, 639-646. Cited in *Drug Evaluations: Annual 1992*, American Medical Association Division of Drugs and Toxicology, Department of Drugs; Bennett, D. R., Ed.; AMA; Chicago, 1992, p. 444.

119. Moreau, J-J, "Recherches sur les aliénés en orient," *Ann. Méd. Pscy.* **1843**, *I*. Quoted in *JJM*, pp. xviii-xix.

120. (a) *GJ*, pp. 146-151. (b) Beaubrun, M. H., "Cannabis or alcohol: the Jamaican experience," *C&C*, pp. 485-494.

121. Wu, T.-C.; Tashkin, D. P.; Djahed, B.; Rose, J. E., "Pulmonary hazards of smoking marijuana as compared with tobacco," *N. Engl. J. Med.*, **1988**, *318*, 347-351.

122. Jaffe, J. H., "Drug addiction and drug abuse." In: *G&G8*, p. 552

123. Caplan, G. A.; Brigham, B. A., "Marijuana smoking and carcinoma of the tongue. Is there an association?" *Cancer*, **1990**, *66*, 1005-1006.

124. Polen, M. R.; Sidney, S.; Tekawa, I. S.; et al., "Health care use by frequent marijuana smokers who do not smoke tobacco," *Western J. Med.*, **1993**, *158*, 596-601. See also *The Drug Policy Letter*, **1993**, *20*, 26.

125. From a personal interview of the author with Dr. Blans in the summer of 1994. Gambling is available in the Netherlands in the form of the state lotto, as in the United States, but also in the form of computerized gaming machines, which can be found in most bars and are concentrated in game parlors in the tourist areas. These are something like the

"one-armed bandit" slot machines of Reno, Atlantic City, etc. There has been a great deal of study in the Netherlands on what sort of (programmed) reward/winning strategy in these machines provokes less harmful or addictive behavior on the part of those who use them. Like everything else, tastes differ: the author watched in friendly neighborhood bars as some people dumped guilders obsessively into these machines while others showed not the remotest interest. But, according to Dr. Blans and others, there are a sizeable number of people who incur severe financial damage from the habit.

126. Weil, A.; Rosen, W., *From Chocolate to Morphine: Everything You Need to Know About Mind-Altering Drugs*, revised and updated, Houghton-Mifflin: Boston, 1993, p. 118. "I have smoked marijuana every day for the past 26 years," says Richard Cowan, director of the National Organization for the Reform of Marijuana Laws (NORML), "and that's for the record. According to the marijuana critics, I should be dead or vegetative at least, but I'm in good health, and I seem to be reasonably coherent." Indeed, his brain functioning seems to be OK; William Buckley, certainly no vegetable himself, has described Cowan's arguments for the legalization of marijuana presented in two *National Review* articles as "tough-minded and analytically resourceful."

127. Hollister, L. E., "Marijuana and immunity," *Journal of Psychoactive Drugs*, **1992**, *24*, 159-164. So anxious are many about the imagined horrors of marijuana smoking that one frequently hears from the educators of our youth on college campuses the admonition that marijuana lowers immune function, followed by the reminder of how careful one must be in an era of AIDS—leaving the impression, wholly undocumented, that those who smoke marijuana are more likely to get AIDS.

    A further irony is the fact that the oral prescription drug dronabinol is officially indicated by the FDA for use as an antinauseant in the treatment of AIDS—but of course dronabinol is nothing other than Δ^9-THC, the very substance that was involved in the megadose in vitro tests of immune function that are used to substantiate the warnings that AIDS victims ought not to smoke pot.

128. Deahl, M., "Cannabis and memory loss," *British Journal of Addiction*, **1991**, *86*, 249-252. The author goes on to caution how any new evidence implicating cannabis with persistent harmful effects should be "subject to critical scrutiny . . . if accusations of prejudice and moral bias are to be avoided." But then, in a forthright rejection of this caution, the author wrings his hands about the "strong evidence" presented in a study of short-term memory impairment of "10 cannabis-dependent North American middle-class adolescents' [sic] participants of a community drug treatment program." This is a very small cohort to constitute strong evidence of anything, but more damaging is the supposed control: 10 healthy adolescents matched for IQ, etc., *but not in a community drug program* and who, of course, had never smoked marijuana. Ought it not be obvious that participants in such a program have more or less conceded that their drug use has done them harm, and will they not consequently be likely to unconsciously perform less ably on a test of short-term memory, inasmuch as they have been told since their earliest youth that marijuana will damage this function?

129. *GJ*, pp. vi; 164-166. See also the extensive tabular compilation of data in the Appendices.

130. Mechoulam, R.; Devane, W. A.; Glaser, R., "Cannabinoid geometry and biological activity," *M/C*, pp. 1-34.

131. But see the interesting new class of aminoalkyl*indoles* that show cannabinoid activity: Eissenstat, M. A.; Bell, M. R.; et al., "Aminoalkylindoles: structure-activity relationships of novel cannabinoid mimetics," *J. Med. Chem.*, **1995**, *38*, 3094-3105. A cannabinoid an-

tagonist which is active in vivo has also been reported: Rinaldi-Carmona, M.; Barth, F.; et al., "FEBS Lett.", **1994**, *350*, 240-244. And a 1-pentyl-3-acyl pyrrole has shown antinociceptive activity comparable to tetrahydrocannabinol: (a) Reitz, A. B.; Jetter, M. C.; et al., "Centrally acting analgesics," *ARMC*, **30**, 11-20. (b) Lainton, J. A.; Huffman, J. W.; et al., *Tet. Lett.,*, **1995**, *36*, 1401.

132. (a) Handrick, G. R.; Duffley, R. P.; Lambert, G.; et al., *J. Med. Chem.*, **1982**, *25*, 1447-1450. See also (b) Singer, M.; Siegel, C.; Gordon, P. M.; Dutta, A. K.; Razdan, R. K., *Synthesis*, **1994**, 486-488.

133. Devane, W. A.; Dysarz, F. A.; Johnson, M. R.; Melvin, L. S.; Howlett, A. C. *Mol. Pharmacol.*, **1988**, *34*, 605. Devane, W. A., thesis, 1989, St. Louis University, St. Louis, MO.

134. Devane, W. A.; Hanuš, L.; Mechoulam, R.; et al., "Isolation and structure of a brain constituent that binds to the cannabinoid receptor," *Science*, **1992**, *258*, 1946-1949.

135. Fride, E.; Mechoulam, R., "Pharmacological activity of the cannabinoid receptor agonist, anandamide, a brain constituent," *Eur. J. Pharmacol.*, **1993**, *231*, 313-314.

136. Howlett, A. C., "Pharmacology of cannabinoid receptors," *Annu. Rev. Pharmacol. Toxicol.*, **1995**, *35*, 607-634.

137. Mechoulam, R., "Preface," and "The pharmacohistory of *Cannabis sativa,*" *CTA*, pp. i-19. A full history of the historical medical uses of cannabis can be found in this excellent essay.

138. Grinspoon, L; Bakalar, J. B., "Marihuana as medicine," *JAMA*, **1995**, *23*, 1875-1876.

139. Devane, W. A., "New dawn of cannabinoid pharmacology," *TIPS*, **1994**, *15*, 40-41.

140. Grinspoon's impatience with the consequences of criminalizing marijuana and other drugs is not based on a frivolous whim; it originated when he saw the effects of chemotherapy—and marijuana—on his own young son, who eventually died of leukemia. See *GRIN.*

141. *MFM*, 30-31.

142. *MFM*, 36-38.

143. Levitt, M., "Cannabinoids as antiemetics in cancer chemotherapy," *CTA*, pp. 71-83; for a discussion of the phenomena of nausea/emesis, with leading references, see pp. 76-77.

144. Weil, A. T.; op. cit. (*Science*, **1968**), p. 1237.

145. Frytak, S.; Moertel, C.; et al., "Delta-9-tetrahydrocannabinol as an antiemetic for patients receiving cancer chemotherapy, a comparison with prochlorperazine and a placebo," *Ann. Int. Med.*, **1979**, *91*, 819. Cited in Levitt, M., op. cit., p. 75.

146. Treaster, J. B., "Agency says marijuana is not proven medicine," *The New York Times*, 19 March 1992.

147. (a) Chang, A. E., et al., "Delta-9-tetrahydrocannabinol as an antiemetic in cancer patients receiving high-dose methotrexate: a prospective, randomized evaluation." *Ann. Int. Med.*, **91**, *1979*, 819-824. (b) Zuardi, A. W.; Shirakawa, I.; Finkelbarb, E.; Karniiol, I. G., "Action of cannabidiol on the anxiety and other effects produced by delta-9-THC in normal subjects," *Psychopharmacology*, **76**, *1976*, 245-250.

148. Archer, R. A.; Stark, P.; Lemberger, L, "Nabilone," *CTA*, pp. 85-103.

149. Adler, M. W.; Geller, E. B., "Ocular effects of cannabinoids," *CTA*, pp. 51-70. Quoted material p. 66.

150. Hepler, R. S.; Frank, I. M., "Marihuana smoking and intraocular pressure," *JAMA*, **217**, *1971*, 1392.

151. *MFM*, 54-56.

152. But in March 1992, the Public Health Service said it would not provide marijuana for medical purposes to any more patients because "it might make them sicker." They were primarily concerned that the already weakened immune systems of AIDS patients might be further compromised. Furthermore, said spokesman Bill Grigg, "we know that marijuana contains substances which can cause lung problems; and AIDS patients are prone to pneumonia and other lung infections." *The New York Times*, 21 March 1992, p. A21.

Substances are Schedule I rather than II only in that they "have no accepted medical use." The United States Court of Appeals for the District of Columbia ruled in April 1992 that the Drug Enforcement Administration had based its scheduling of marijuana on factors like the drug's general availability and its use by a substantial number of doctors. Because the drug is illegal, the court said, meeting these criteria would be all but impossible. But Robert C. Bonner, chief of the DEA, rejected this argument, saying that "the general availability of a drug is irrelevant to whether it has a currently accepted use." *The New York Times*, 19 March 1992.

153. Archer, R. A.; Stark, P.; Lemberger, L, "Nabilone," *CTA*, pp. 85-103; see pp. 96-97.

154. *Nature*, **1991**, *351*, 703. Cited by K. M. Reese, *C&EN*, 5 August 1991, p. 44.

155. Malec, J.; Harvey, R. F.; Cayner, J. J., "Cannabis effect on spasticity in spinal cord injury," *Arch. Phys. Med. Rehabil.*, **1982**, *21*, 81. Cited in *CTA*, p. 41.

156. McBee, G. F., submission in *Idaho v. Hastings.* In: *Muscle Spasm, Pain and Marijuana Therapy: Testimony from Federal and State Court Proceedings on Marijuana's Medical Use in the Treatment of Multiple Sclerosis, Paralysis, and Chronic Pain*, Randall, R. C., Ed., Galen Press: Washington, DC, 1991, pp. 89-96.

157. Meinck, H. M.; Schonle, P. W.; Conrad, B., "Effect of cannabinoids on spasticity and ataxia in multiple sclerosis," *J. Neurol.* **1989**, *236*, 120-122.

158. Greenberg, H. S.; Werness, S. A.; Pugh, J. E.; et al., "Short-term effects of smoking marijuana on balance in patients with multiple sclerosis and normal volunteers," *Clin. Pharmacol. Ther.*, **1994**, *55*, 324-328.

159. Petro, D. J.; Ellenberger, C., "Treatment of human spasticity with Δ^9-tetrahydrocannabinol," *J. Clin. Pharmacol.*, **1981**, *21*, 413S.

160. Johnson, M. R.; Melvin, L. S., "The discovery of nonclassical cannabinoid analgetics," *CTA*, pp. 121-145.

161. Young, F. L. In: *Muscle Spasm, Pain and Marijuana Therapy: Testimony from Federal and State Court Proceedings on Marijuana's Medical Use in the Treatment of Multiple Sclerosis, Paralysis, and Chronic Pain*, Randall, R. C., Ed., Galen Press: Washington, DC, 1991, pp. 207-208.

162. Cited in Goldberg, C., "Marijuana club helps those in pain: Menu offers various forms of the drug to sick or dying people," *The New York Times*, 25 February 1996, 16.

163. *Ibid.* However, even as these pages were going to press, the Cannabis Buyer's Club was raided by agents of the California Bureau of Narcotics Enforcement at the instigation of the state's Attorney General, Dan Lungren: Golden, T., "Agents crack down on marijuana buying club," *The New York Times*, 5 August 1996, A8.

164. *Ibid.* In Germany, the Federal Constitutional Court ruled in April 1994 that possession of small amounts of marijuana or hashish for personal use could no longer be subject to criminal prosecution. In declaring such criminal laws unconstitutional, the court relied on its judgment that the dangers resulting from the use of cannabis products cited in the original Narcotic Substances Act [Betäubungsmittelgesetz] of 1971 had proved over time to be greatly exaggerated. It denied that any citizen had a "right to intoxication," but it admitted that nicotine and alcohol caused as much or greater harm than did cannabis. But it pointed out that nicotine does not cause intoxication, and that alcohol had other effects and uses than causing intoxication. In any case, the court somewhat desperately concluded, it would be futile for the states to attempt banning alcohol, so widespread is its customary usage: "Zudem könne der Gesetzgeber den Genuß von Alkohol wegen der herkömmlichen Konsumgewohnheiten nicht effektiv unterbinden." (*Süddeutsche Zeitung*, 29 April 1994.)

Since the court decision, according to an article in *The New York Times* of 18 May 1994, many prominent German users have felt free to come forward and discuss their habits: Joshka Fischer, Minister of Environmental Affairs in the German state of Hesse and the leading figure in Germany's Green Party, said in a published interview that he had "smoked hashish with quite positive effects." While police officials have said they will not tolerate the public use of drugs, Christian Kerls, a dentist in Munich, has started to provide nervous patients with marijuana cigarettes and a special waiting room in which to smoke them before they are treated; other dentists are expected to follow suit.

165. Mechoulam, R.; Gaoni, Y., *J. Am. Chem. Soc.*, **1965**, *87*, 3273.

166. Taylor, E. C.; Lenard, K.; Shvo, Y., *J. Am. Chem. Soc.* **1966**, *88*, 367.

167. Mechoulam, R.; Braun, P.; Gaoni, Y., *J. Am. Chem. Soc.* **1972**, *94*, 6159.

168. For a discussion and leading references, as well as a more complete history of the early synthetic approaches to Δ^9-THC, see Mechoulam, R., *MECH*, pp. 38-50.

169. (a) Mechoulam, R.; Braun, P.; Gaoni, Y., *J. Am. Chem. Soc.* **1972**, *94*, 6159. (b) Mechoulam, R.; Gaoni, Y., *Tetrahedron Lett.*, **1967**, 5349. (c) Petrzilka, T.; Haefliger, W.; Sikemeier, C., *Helv. Chim. Acta*, **1969**, *52*, 1102.

170. Razdan, R. K.; Handrick, G. R.; Dalzell, H. C., *J. Am. Chem. Soc.*, **1974**, *96*, 5860.

171. Razdan, R. K.; Handrick, G. R.; Dalzell, H. C., *Experientia*, **1975**, *31*, 16-17.

172. Nagle, D. R., "Anesthetic addiction and drunkenness," *Int. J. Addictions*, **1968**, *3*, 33. See *L&ID*, pp. 312-314.

173. James, W., "On Some Hegelisms." In: *The Will to Believe and Other Essays in Popular Philosophy*, Burkhardt, F. H.; Bowers, F.; Skrupskelis, I. K., Eds., Harvard University Press, Cambridge, 1979, pp. 219-221. Originally published in *Mind* (April, 1882), then revised for *The Will to Believe*, published in 1897.

174. *The Will to Believe and Other Essays in Popular Philosophy*, Burkhardt, F. H.; Bowers, F.; Skrupskelis, I. K., Eds., Harvard University Press, Cambridge, 1979, Appendix IV, pp. 439.

175. Baird, P. A., "Occupational exposure to nitrous oxide—not a laughing matter," *NEJM*, **1992**, *327*, 1026-1027.

176. Winick, C., "Substances of use and abuse and sexual behavior," *SA*, pp. 722-732, quoted material p. 729. [What is a "substance of use"?]

177. Weil, A.; Rosen, W.; op. cit. (*From Chocolate to . . .*), pp. 129-131.

178. Connell, K. H., "Ether drinking in Ulster," *Q. J. Stud. Alcohol*, **1965**, *26*, 629-653. Cited in Blum, et al., "Drugs, behavior, and crime," *Society and Drugs: Social and Cultural Observations*, Jossey-Bass: San Francisco, 1969, Vol. I, pp. 35-36. Cited in *GRIN*, p. 345.

179. Holmes, O. W., *Mechanism in Thought and Morals* (Phi Beta Kappa address, Harvard University, 29 June 1870), J. R. Osgood: Boston, 1871. Cited in *L&LD*, p. 316n.

180. Some speculations as to why kids do this: Caputo, R. A., "Volatile substance misuse in children and youth: a consideration of theories," *Int. J. Addictions*, **1993**, *28*, 1015-1032.

181. (a) *NEJM,* **1990**, *323*, 1638. A case of partial paralysis in a 15-year-old-boy from butane inhalation: Gray, M. Y.; Lazarus, J. H., "Butane inhalation and hemiparesis," *Clinical Tox.* **1993**, *31*, 483-485.

182. *G&G8*, pp. 1621-1626.

183. Was van Gogh a chemical genius as well as an artistic one? Did he know that these two substances were *isomers*, each having the formula $C_{10}H_{16}O$?

184. Arnold, W. N., *Vincent van Gogh: Chemicals, Crises, and Creativity*, Boston: Birkhäuser, 1992.

185. Arnold, *Vincent*, p. 103.

186. Magnan, V., "On the comparative action of alcohol and absinthe," *The Lancet*, **1874**, *2*, 410-412. For complete references see Arnold, *Vincent*, references to chapter 4, pp. 134-137.

187. Rice, K. C.; Wilson, R. S., "(-)-3-Isothujone, a small nonnitrogenous molecule with antinociceptive activity in mice," *J. Med. Chem*, **1976**, *19*, 1054-1057.

188. Jean-Claude Michel, in a passionate letter from Argenteuil, France, to *C&EN* (6 Sept 1993, p. 58), protests that the date was 1915, not 1922 as stated erroneously in *C&EN* 9 November 1992, p. 56. Even earlier, he says, when WWI broke out in August 1914, the use of the beverage was already severely restricted. "However, it is possible that certain stocks were disposed of clandestinely after the war, especially in the Parisian artistic and literary milieux." For the straight dope, Michel recommends Marie-Claude Delahaye-Dutrillaux's "admirable monograph" *L'Absinthe: Histoire de la Fée Verté*, [history of the green fairy] Berger-Levrault: Paris, 1983.

189. As to whether the excessively yellow pigmentation of some of Van Gogh's paintings reflected a chromatopsia, specifically a xanthopsia, see Arnold, W. N.; Dalton, T. P.; et al., "A search for santonin in *Artemisia pontica*, the Other Wormwood of Old Absinthe," *J. Chem. Ed.*, **1991**, *68*, 27-28.

190. Burroughs, W. S., "Letter from a master addict to dangerous drugs." Appendix to Burroughs, W. S., *Naked Lunch*, Grove Weidenfeld: New York, 1959, p. 231. [Originally printed in *The British Journal of Addiction*, *53* (2).]

191. Jaffe, J. H., *G&G8,* p. 559.

192. Weil, A., op. cit. (*From Chocolate to . . .*), pp. 135-136.

193. Shulgin, A. T., "Possible implication of myristicin as a psychotropic substance," *Nature*, **1966**, *210*, 380-384.

194. Wolinsky, J.; Eustace, E. J., "Syntheses of the dihydronepetalactones," *J. Org. Chem.*, **1972**, *37*, 3376-3378.

195. Todd, N. B., *J. Heredity*, **1962**, *53*, 54.

196. Regnier, F. E.; Waller, G. R.; et al., "The biosynthesis of methylcyclopentane monoterpenoids II," *Phytochemistry*, **1968**, *7*, pp. 221-230.

197. Suemune, H.; Oda, K.; et al., *Chem. Pharm. Bull.*, **1988**, *36*, 172-177.

198. Waller, G. R.; Price, G. H.; Mitchell, E. D., "Feline attractant, cis, trans-nepetalactone: metabolism in the domestic cat," *Science*, **1969**, *164*, 1281-1282.

199. Cited in Mellow, J. R., *Charmed Circle: Gertrude Stein & Company,* Praeger: New York, 1974, p. 468. The story comes from Toklas, A. B., *What Is Remembered,* Holt, Rinehart & Winston: New York, 1963, p. 173.

Appendix I
HONC: The Four Key Elements

Introduction

The purpose of these four appendices is to provide a brief introduction or refresher on the basic rules of organic chemical bonding—that is, how the structures of all drugs are formed. The basic ideas are quite simple and easily modeled by the well-worn device of a set of wooden balls (the atoms) held together by sticks (the bonds). About 60% of these atoms, it will turn out, are the simplest possible atom there is, hydrogen (so simple it can only manage one solitary stick bond). Most of the other atoms are carbon, with four bonds. A few key atoms in almost every drug are nitrogen, with three bonds. There is usually a sprinkling of oxygen atoms, commanding two sticks each. And with that rather meager assortment of possibilities, almost every drug in this book can be built. Of course, if you read through this book's main chapters with no previous acquaintance with organic chemistry, you will see no "balls" or "sticks"—just a series of hexagons, pentagons, and odd line drawings. These are chemical shorthand for three-dimensional drug structures, and deciphering this shorthand is what these appendices are all about.

Chemical Structures

We will start with the molecular structure of a very familiar chemical substance, one that has the honor of being the simplest possible compound of carbon and hydrogen, namely *methane*. Methane (CH_4) is the "natural" gas that is piped into most urban areas for cooking or heating. It is odorless, colorless, and of course quite flammable. (The penetrating stink of leaking gas is due to a small amount of ethyl mercaptan, a sulfur-containing substance that is added to the methane as a safety precaution. Ethyl mercaptan is itself closely related in structure to the chemical used by skunks as a defense; the human nose happens to be terrifically sensitive to this class of compound.)

Methane is explosive when mixed with air; it is the gas responsible for many mine disasters. It is also a natural by-product of organic decomposition and is responsible for the eerie "will 'o the wisp" blue light of marsh gas. It is a major constituent of the *flatus* (gas expelled from the intestines) of ruminants such as sheep, goats, and cattle. Ruminants have a second stomach in which they can digest cellulose (plant fiber) using a symbiotic bacterium also found in the stomachs of termites and cockroaches. The expansion of human population has led to (1) deforestation, (2) the recycling of trees used for paper pulp, and (3) expanded grazing land. This results in a great growth in the tree-trunk-chomping termite and flatulating ruminant populations, with a concomitant increase in the atmospheric methane, which contributes to the greenhouse effect and the destruction of the ozone layer.

Nonruminant mammals, such as humans, generally produce noncombustible carbon dioxide (CO_2) flatus. However, hydrogen gas (H_2) "is produced in large quantities after eating certain fruits and vegetables (e.g., baked beans)."[1] Since H_2 is violently explosive when mixed in nearly any proportion with air (it was the gas responsible for the *Hindenburg* disaster), smoking after such meals should be avoided.[2] On the other hand, in some individuals the intestines do produce methane, but this is independent of dietary intake: "some people consistently excrete large quantities of CH_4; others, little or none. Apparently familial, this trait appears during infancy and persists for life."[3] Here, the risk of conflagration appears less severe, but because it is genetically determined, it is unavoidable.

Methane contains four hydrogen atoms and one carbon atom; the carbon atom (C) is in the center with the four hydrogen (H) atoms symmetrically radiating from it. Carbon atoms are about 12 times as massive as hydrogen atoms, and so are drawn as larger spheres; we can imagine each hydrogen atom as a smaller sphere, with rods connecting them to the carbon atoms. Figure A1.1 shows three different perspectives of the structure of methane using a computer modeling program.

The four hydrogen atoms in methane are at the corners of a tetrahedron—a pyramid having a triangle for its base. This type of carbon bonding is consequently called *tetrahedral*, and it is the most common way for carbon to form molecules; later we will see two other types of carbon bonding. Figure A1.2 gives us a fourth view of methane on the left, a tetrahedron on the right, and in the center the methane molecule with a tetrahedron superimposed on it.

Models like these for methane are called "Tinkertoy" or "ball and stick" models. Although such models are obviously quite simple, they still represent a great deal of understanding of chemical structure and activity; and the essential notion embodied in these models—that *distinct chemical substances* (which term encompasses all vitamins,

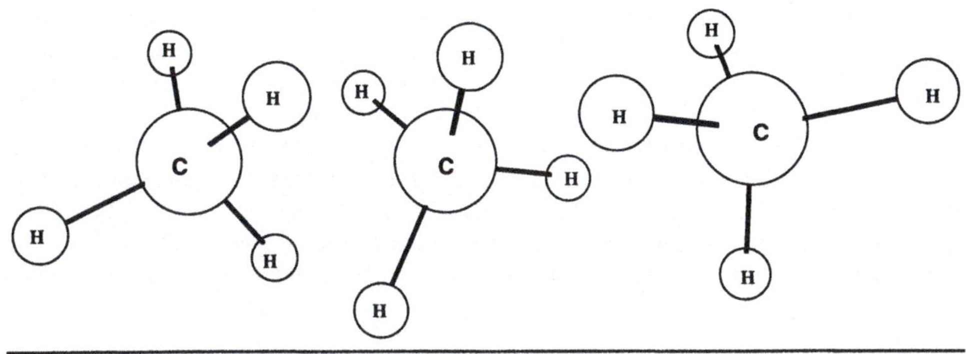

Figure A1.1 Three views of methane.

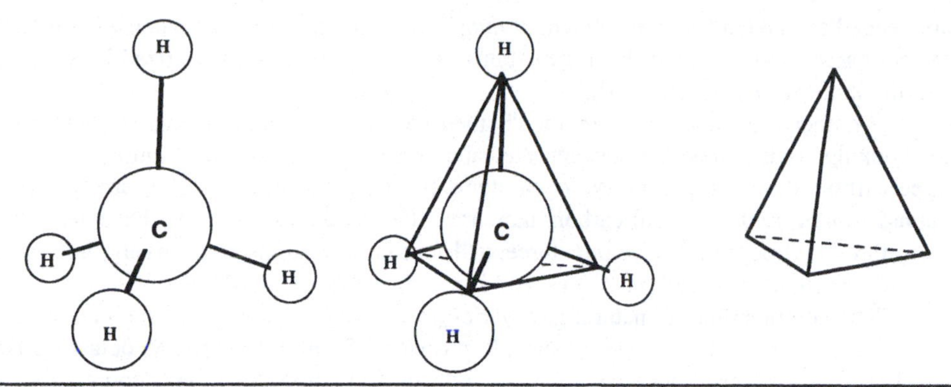

Figure A1.2 The tetrahedral structure of methane.

all drugs, and many foods) consist of atoms (of carbon, hydrogen, nitrogen, etc., the "balls" of these models) rigidly held together by bonds (the "sticks")—represents an enormous intellectual insight achieved by the collective work of many scientists from about 1860-1920.

Nowadays we know a great deal more about what constitutes a bond: a bond is a pair of electrons shared by the two atoms it holds together; each electron interacts as a Fermi particle described by some complex quantum mechanical equations proposed in the early 20th century by several Nobel laureate physicists (Schroedinger, Dirac, and Heisenberg).

Let us consider another commonly encountered substance, *butane* (Figure A1.3). Butane is ordinarily a gas; when it is pumped into a cigarette lighter under pressure (just as air is pumped into an automobile tire), it turns into a liquid (condenses). This feature allows for the compact storage of a large amount of the fuel in a small container: methane will not liquefy under these conditions and only enough gas for a few lights could be contained in the lighter under similar pressure.

Butane looks like four methanes holding hands; it has four carbon atoms bonded together by three carbon-to-carbon bonds; two views of butane are shown in Figure A1.3. Each carbon atom then has two (or three) hydrogen atoms attached to it—however many

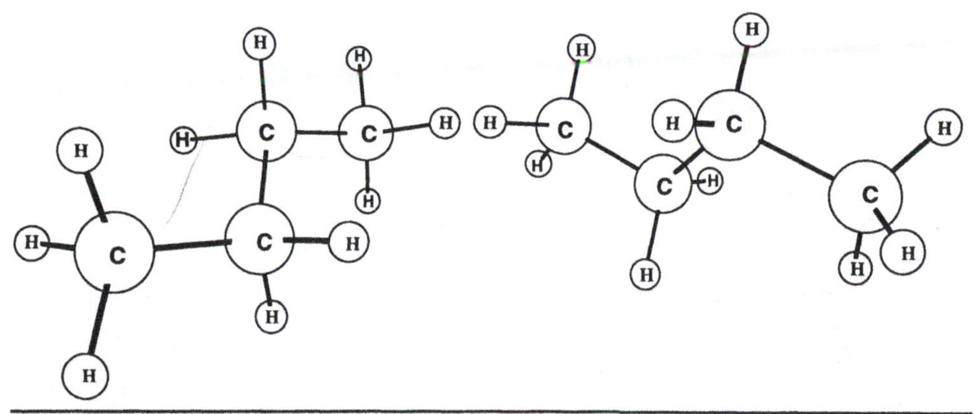

Figure A1.3 Two views of butane.

are needed to make a full complement of *four bonds per carbon atom.* Butane has a total of four carbon atoms and 10 hydrogen atoms; this atom count is summarized in what we call the *molecular formula*: C_4H_{10}.

Another molecule, *propane*, the "bottled gas" used in camping and in rural areas for cooking, is intermediate between methane and butane. Propane (Figure A1.4) has three carbon atoms and eight hydrogen atoms (C_3H_8). Because methane, propane, and butane contain hydrogen and carbon, they are called *hydrocarbons.* We have seen that these three hydrocarbons have one, three, and four carbon atoms respectively; is there a hydrocarbon with two carbons? Yes, and it is called *ethane* (Figure A1.5).

The composition of "natural gas" varies depending on its origin; American natural gas is about 85% methane, 9% ethane, 3% propane, 2% nitrogen, and 1% butane. Like gasoline, the combustion of natural gas produces CO_2; however, the amount of CO_2 produced per unit of energy obtained is smaller for natural gas. Many buses and trucks have begun using natural gas to reduce total CO_2 production with its consequent effect on global warming. If you have ever passed an oil refinery, you may have seen large flames at the top of some of the towers; this is ethane (which is a commercially unimportant by-product of the distillation of crude petroleum) being "flamed off." "Bottled gas" or "suburban propane" contains about 90% propane, 5% ethane, and 5% butane. All of these hydrocarbons are more or less narcotic if inhaled in high concentrations.

In addition to methane, ethane, propane, and butane, chemists have characterized hundreds of other hydrocarbons over the years; we will see some of these later.

The HONC Rule

Hydrogen, oxygen, nitrogen, and carbon are the four most abundant elements in living organisms, and their bonding together can be easily described by the "HONC" rule, a useful mnemonic that summarizes the bonding properties of these four most important elements: *1-2-3-4, H-O-N-C.* That is: H forms one bond; O forms two; N three; and C four.

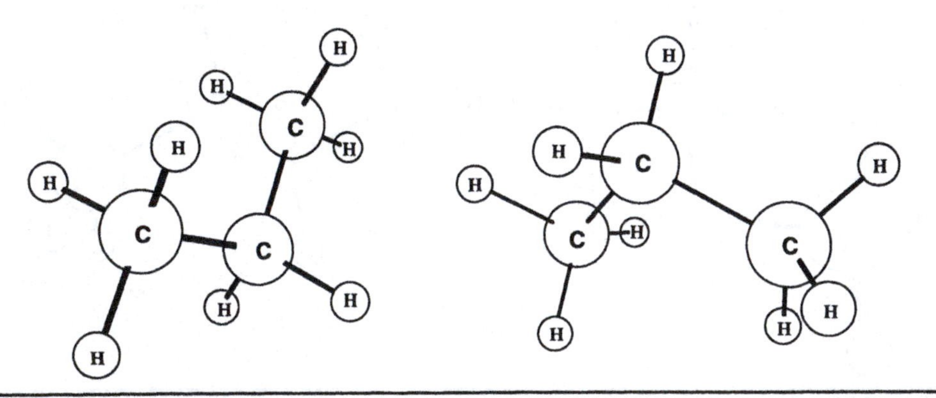

Figure A1.4 Two views of propane.

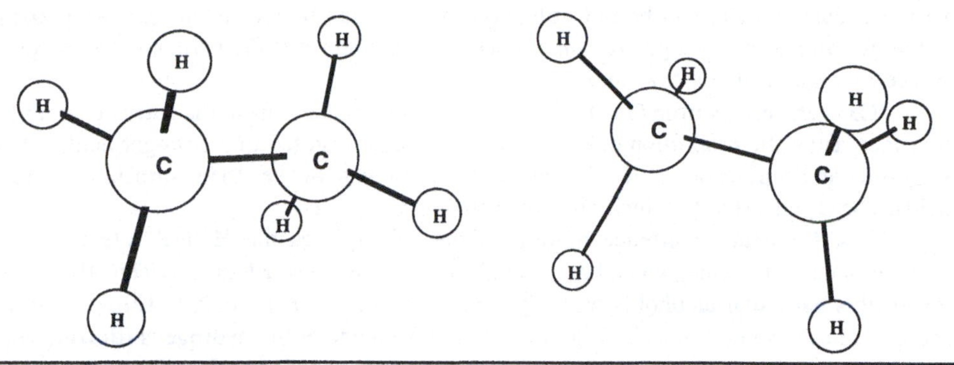

Figure A1.5 Two views of ethane.

Hydrogen Gets One

Hydrogen, one of the two elements found in hydrocarbons, is the most abundant element in the universe. Hydrogen characteristically forms *one* bond with other atoms, as it does in the four hydrocarbons we have considered.

Hydrogen, when it is not bonded to any other element, is a colorless, odorless gas, H_2. (Each of the two hydrogen atoms still forms one bond, with the other atom in the pair: H–H.) It is the least dense, lightest gas there is, and it was formerly used to fill dirigible balloons like the "Hindenburg." Because of hydrogen's explosive reaction with the oxygen in the air, modern dirigibles like the Goodyear blimp are filled with the inert element helium He.

Hydrogen is the most abundant element in the universe not only by number of atoms but by total mass (weight): much of the material of interstellar space and of the stars themselves is hydrogen. Hydrogen is the simplest atom, and the first formed, in enormous quantity, after the Big Bang. The nuclear fusion of hydrogen atoms within stars and our sun is what provides most of their energy and what eventually leads to the production of the heavier elements. Essentially, we and the rest of the universe as we know it have evolved from hydrogen. More immediately, life as we know it on this planet has evolved from hydrogen, thanks to its most important compound, water: the very word hydrogen is from the Greek ὑδρογενη, "the water-maker," the product of the combustion of H_2 with air or oxygen being water, H_2O.

Oxygen Gets Two

From the familiar formula for water, H_2O, you can deduce by now that oxygen always forms two bonds. Oxygen occurs as about 20% of the air we breathe and is necessary for life. Oxygen is slightly soluble in water (more soluble in cold than in hot water), and fish "breathe" the oxygen dissolved in water. Excessive phosphates from older detergent formulations fertilize the overgrowth of algae; eventually this process can reduce the amount of oxygen in the water below the point at which fish and other aquatic life can survive. Just as there can be too little oxygen, there can be too much: oxygen is toxic when not diluted by nitrogen or some other inert gas, and it is dangerous to inhale pure oxygen for more than a few hours.

Oxygen derives from Greek ὀξυγενη, "the acid-maker," from the early observation that increasing the proportion of oxygen in a molecule resulted in a stronger acid. Thus sulfurous and nitrous acids, H_2SO_3 and HNO_2, are much weaker than sulfuric and nitric acids, H_2SO_4 and HNO_3, which contain more oxygen.

One frequent occurrence of oxygen that exemplifies the HONC rule is in the alcohol in beer and wine, which has a chemical structure intermediate between ethane and water; this particular alcohol is properly called *ethanol* or *ethyl alcohol*. Ethanol can be thought of as a water molecule (Figure A1.6) in which one of the hydrogens (the one with the arrow pointing to it) has been replaced by the ethyl group, CH_3CH_2. Ethanol can also be thought of as a modified molecule of ethane: in the ethane molecule of Figure A1.6, imagine that we take away the H atom that has an arrow pointing to it and substitute for it an O–H (an oxygen atom with a hydrogen attached to it; this group is called the *hydroxy* group). If we do this, we will have the molecule ethanol (Figure A1.6 center); consequently, we say that ethanol is *ethane with a hydroxy substituent,* and ethanol can also be called *hydroxyethane*, although this is not the usual name. This substitution changes the properties of ethane drastically: although ethane is a colorless, odor less gas, ethanol is a clear liquid much like water and soluble in it.

Similarly, *methanol* or *methyl alcohol* (Figure A1.7) is structurally intermediate between water and methane, and can be called *hydroxymethane*. Methanol (center) can be thought of as a water molecule (left) in which one of the two hydrogen atoms (the one with the arrow pointing to it) has been replaced by a CH_3 (the methyl group). Or it can be thought of as a methane molecule (right), in which one of the H atoms has been replaced by a hydroxy group, OH. (It is easiest to see this if you imagine that the H with the arrow has been replaced by an OH, but any of the four H atoms could be replaced

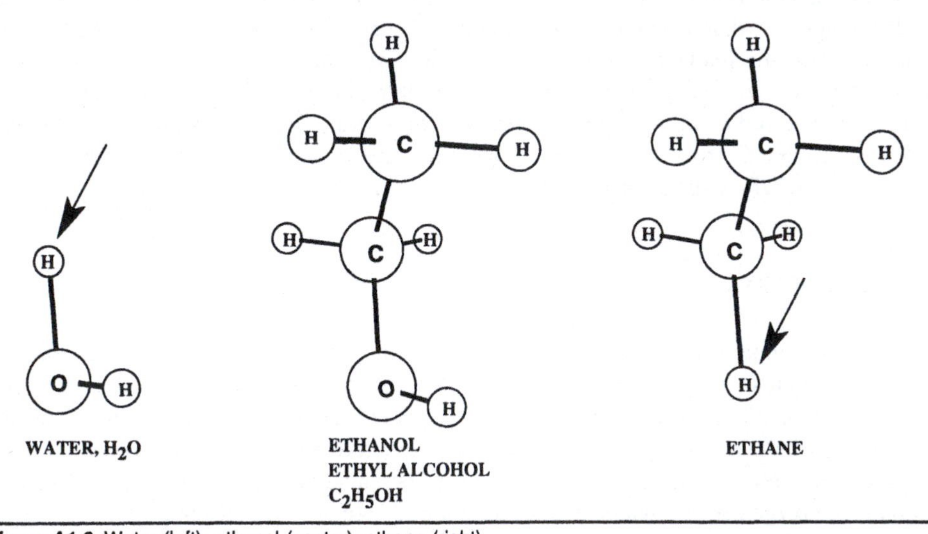

Figure A1.6 Water (left); ethanol (center); ethane (right).

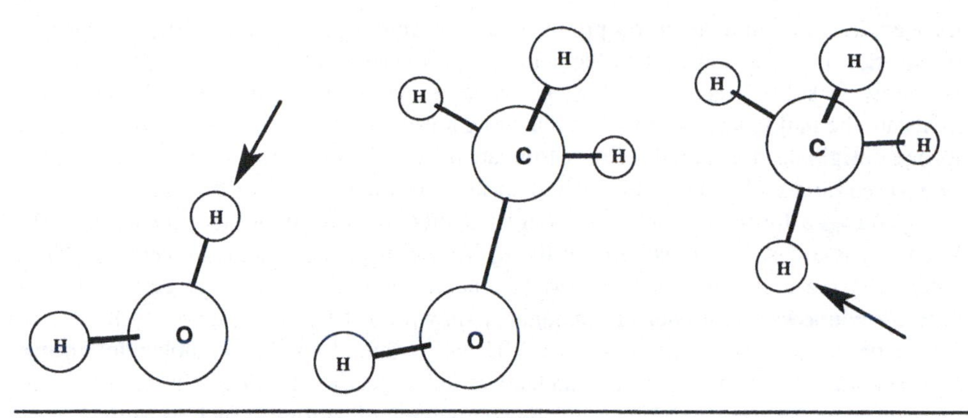

Figure A1.7 Water (left); Methanol (center); Methane (right).

with the same result; the molecule of methanol that results would simply be differently oriented in space.) Although methane is a gas, methanol, like ethanol, is a liquid that is very soluble in water.

Methyl alcohol was originally made by a process called the "destructive distillation of wood": wood was heated in a metal tank without any exposure to air; the vapors when condensed contained methanol. The word comes from the Greek word μεθυς [methus] meaning *wine* and ὑλη [hule] meaning *wood*; methane is so named from its structural similarity to methyl alcohol.

Like ethyl alcohol, methyl alcohol is an intoxicant, but unlike ethyl alcohol it is a severe poison causing blindness, paralysis, or death. Surprisingly, ethanol can act as a partial antidote to methanol. Some liqueurs naturally contain significant amounts of methanol, but the relatively large concentration of ethanol counteracts its toxicity. How this happens is explained in Chapter 3.

Nitrogen Gets Three

We haven't seen this element yet. Nitrogen usually forms three bonds. (We say "usually" because in certain conditions it has an extra bond and an electric charge, and this often has a very important consequences—for example, as shown in Chapter 4, this extra bond makes the difference between the cocaine that is snorted and "crack" cocaine, which is smoked).

The element nitrogen occurs as about 70% of the air we breathe. It is quite inert, since the two nitrogen atoms in the gas N_2 are very strongly bonded to each other. In fact, as the HONC rule predicts, there are *three* bonds between the atoms: N≡N. Nitrogen is essential for all living things. Proteins, RNA, DNA, and most vitamins are just some of the essentials of life that contain nitrogen. However, despite its abundance in the atmosphere, nitrogen is difficult to "fix" (i.e., it is difficult to break the N_2 bond and bring the nitrogen into combination with other elements). Lightning is sufficiently energetic to break the N_2 bond, combine the nitrogen atoms with oxygen from the air, and produce nitrogenous acids (thunderstorms produce a certain amount of acid rain, although this type of acid rain is not as harmful as the acid rain produced from burning coal, which owes its acidity to sulfuric acid). Lightning and some symbiotic bacteria found in tuberous plants are the major natural processes that bring nitrogen into the organic world. The word nitrogen originates through the Greek from an ancient Semitic root, *nitro*, for the mineral soda, or sodium carbonate (which paradoxically does not contain nitrogen).

Nitrogen forms a simple and familiar compound with hydrogen: *ammonia*, NH_3. Ammonia is a gas, but unlike any of the hydrocarbon gases, this gas is very soluble in water, and the solution is sold as the "ammonia" you can buy in grocery stores for cleaning windows. A model of ammonia is shown on the left in Figure A1.8. Notice the shape of the ammonia molecule. Usually, the atoms in a molecule arrange themselves as far apart as possible, and we would expect the hydrogen atoms in ammonia to do this. However, in many of the compounds of nitrogen, such as ammonia, an unshared electron pair is present that takes up as much space as a fourth atom would, pushing the other three atoms away from it. In ammonia, this results in a molecule with an overall shape much like methane's. It is conventional to indicate this by drawing an "electron cloud" containing two electrons (the dots), as in the model in the center of Figure A1.8. This unshared electron pair is important because it allows nitrogen to sometimes form a fourth bond. When the unshared electrons are taken into account, it is evident that the ammonia molecule has a tetrahedral shape, as shown in the model on the right of Figure A1.8.

An "electron cloud" may sound like an odd concept; it comes from an effort to portray with images from our ordinary experience something that cannot really be imagined. Electrons are so small that, according to quantum physics, they have no exact location in space, as do objects like baseballs in our macroscopic world. We can only assign a region of probability to their location, and so we speak of an electron cloud, having no distinct boundaries.

Ammonia, like butane, can be compressed, though not as easily. When a gas is compressed, it gives off heat; when a compressed gas is allowed to rapidly decompress (expand), it conversely becomes cold and takes in heat. If the gas is compressed to a liquid or the liquid expands back to a gas, there is a still greater liberation or absorption of heat. This phenomenon is what makes possible the construction of refrigerators, air-conditioners, and heat pumps. Before WWII, the only gas available for refrigeration

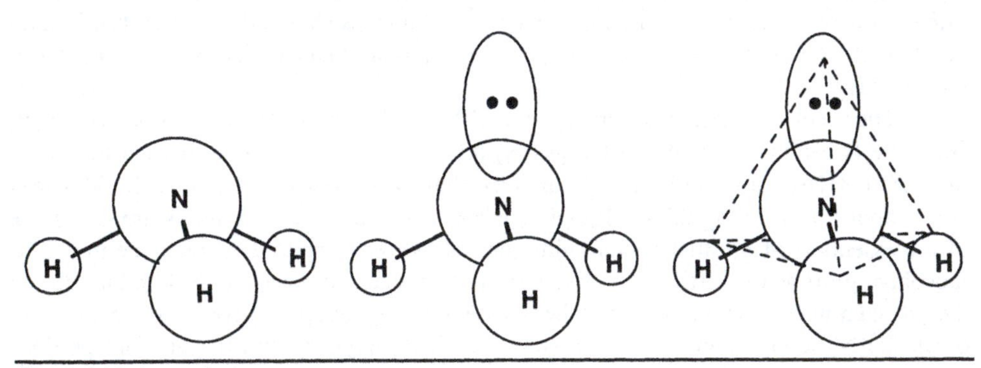

Figure A1.8 Ammonia.

was ammonia. To compress ammonia requires fairly high pressures and heavy machinery, so ice was made in local factories and distributed to homes, where it was used in iceboxes (a large block of ice melted slowly on top of the insulated food compartment). Since then the Freon chlorofluorocarbon refrigerants have been developed, making possible home freezers, refrigerators, and air conditioners. However, the leakage into the atmosphere of these substances, which biodegrade very slowly, is now known to catalyze the destruction of the ozone layer.

Ammonia is also used as a fertilizer, the gas being injected directly into the moist ground, where it dissolves and can be absorbed by plant roots. This ammonia is usually made by combining the nitrogen in the air with hydrogen gas, using a platinum catalyst and high pressure—a process of nitrogen fixation that earned its inventor, Fritz Haber, the Nobel prize.

Carbon Gets Four

Because of the atomic structure of carbon, it is uniquely stable when each carbon atom has *four* bonds attaching it to other atoms; this is the case with methane, ethane, propane, and butane.

What is the element carbon like when it is by itself and not bonded to hydrogen? Perhaps you know there are two common forms of carbon: *diamond* and *graphite*. In diamonds, carbon atoms are bonded to each other in an endless latticework, each having the same tetrahedral configuration as methane: the structure is very rigid and very stable (though at high temperatures diamonds burn in air to form CO_2). Graphite (the writing material in "lead" pencils, which have no actual lead) has an arrangement of flat sheets of carbon atoms joined together in hexagonal rings like some bathroom floor tiles; we will see this type of carbon bonding later when we come to benzene. These flat sheets slide over each other easily, which is why graphite is soft and can be used as a lubricant. Graphite is softer than paper; diamond is harder than steel. They are both pure carbon; the only difference is the type of bonds between the carbon atoms. In the last few years, a new form of carbon has been detected in outer space and synthesized on our planet; it consists of a fish net of carbon atoms in hexagonal and pentagonal rings enclosing a spheroidal cavity. If you examine a soccer ball you will see that it is made by sewing leather hexagons and pentagons together; the corresponding carbon structure is called *soccerballane*.

Using the HONC Rule

Just by inspecting the model labeled **A** in Figure A1.9 we can deduce that it contains one carbon (the atom with four bonds), one oxygen (the atom with two bonds), and four hydrogens (those with one bond: three attached to the carbon atom and one attached to the oxygen atom). The molecular formula for **A** must be CH_4O. Similarly, **B** has one N, one O, two C, and 7 H; its molecular formula is C_2H_7NO; C is C_2H_6O.

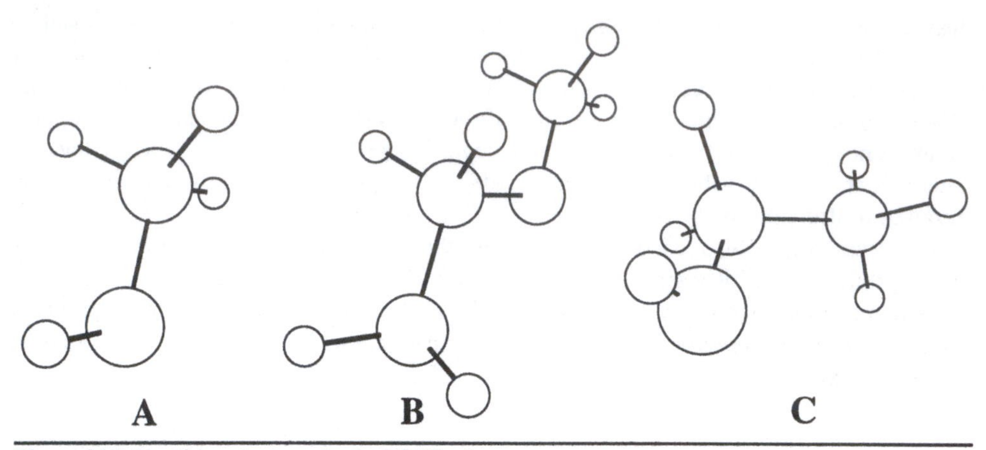

Figure A1.9 Identifying structures by the HONC rule.

Isomers

Butane and Isobutane

In Figure A1.10 are two very similar chemical structures. **D** is the structure for butane, while **E**, though very similar to butane, is actually a different substance with slightly different physical and chemical properties. **E** is called *iso*butane.

What is the difference between butane and isobutane? Both butane and isobutane have the same number of carbon atoms and hydrogen atoms, so they have the same molecular formula, C_4H_{10}. But D has the sequence CH_3–CH_2–CH_2–CH_3 , while **E** has the sequence CH_3–$CH(CH_3)$–CH_3. There is a branch in the carbon chain backbone at the second carbon in **E**. Another way to see the difference is to count the number of methyl (CH_3) groups in each structure. **D** has only two; but **E** has three. The center carbon atom in **E** has three other carbon atoms bonded to it, while no carbon atom in **D** is bonded to any more than two others. **D** is said to be a straight-chain hydrocarbon

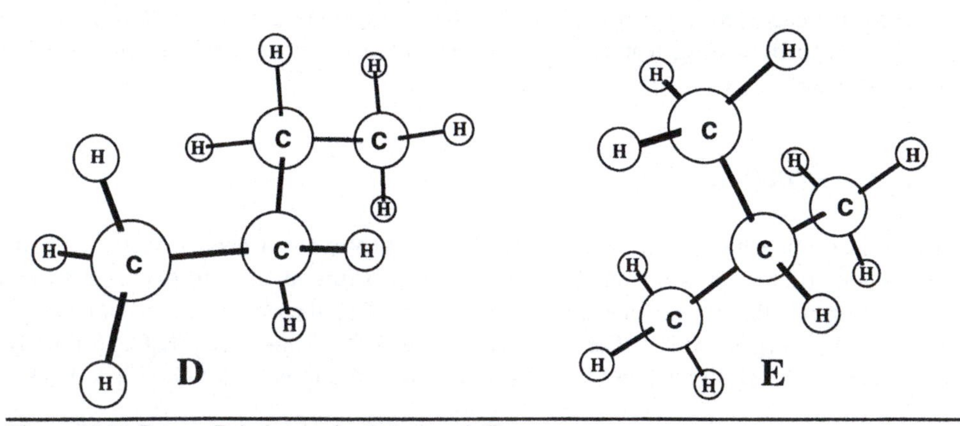

Figure A1.10 Butane, **D**; isobutane (methylpropane), **E**.

and **E** a branched hydrocarbon. In straight-chain hydrocarbons, no carbon is attached to more than two other carbons; in branched hydrocarbons, at least one carbon is bonded to three (or four) other carbon atoms.

Two substances like these that have the same molecular formula but different structural formulas are called *isomers*, and this is why **E** is called *iso*butane. This is an older naming system, in which the straight-chain isomer **D** is called *normal*-butane. **E** is more properly called *methylpropane* because it is a propane molecule with a methyl (CH_3) group attached to the middle carbon. ("Straight" is of course misleading; a road as zigzagged as a butane molecule would never be called straight—but at least there would be no forks in the road, as there would be with isobutane).

Ethanol and Dimethylether

Butane and isobutane are very similar substances. In fact, they are quite difficult to separate and there is probably a considerable quantity of isobutane mixed with the butane in lighter fuel. However, isomer pairs sometimes have very different properties; this is the case with ethanol, CH_3CH_2OH, and its isomer, *dimethylether*, CH_3–O–CH_3, as shown in Figure A1.11. Dimethylether is a very volatile, extremely flammable gas, quite different from the familiar liquid, ethanol. Notice that ethanol and dimethylether have the same total number of carbon, hydrogen, and oxygen atoms; both have the molecular formula C_2H_6O. But the connections between the atoms are different. Ethanol can be transformed into dimethylether by switching the positions of the circled hydrogen atom and the circled methyl group. The reverse process converts dimethylether back into ethanol.

Conformers

Figure A1.12 displays a model of butane, **F**, and of methylpropane or isobutane, **G**.

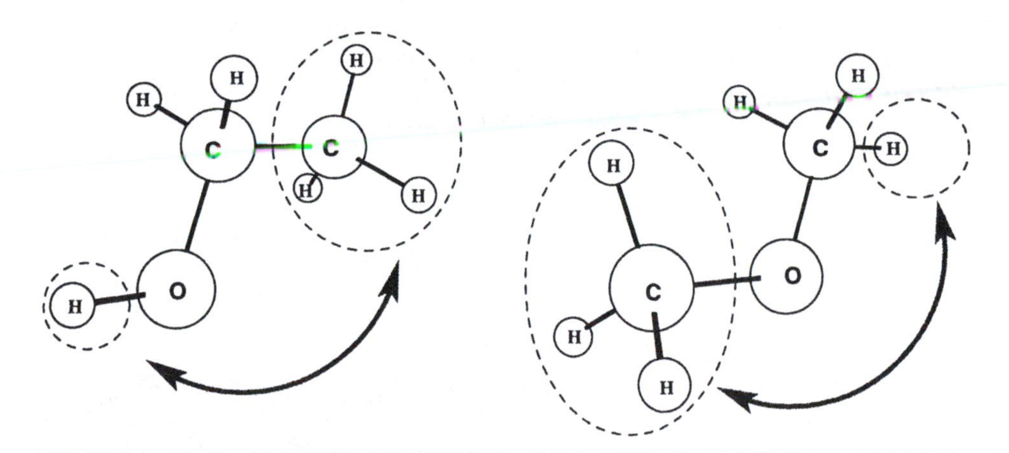

Figure A1.11 Ethanol (left) and its isomer, dimethylether (right).

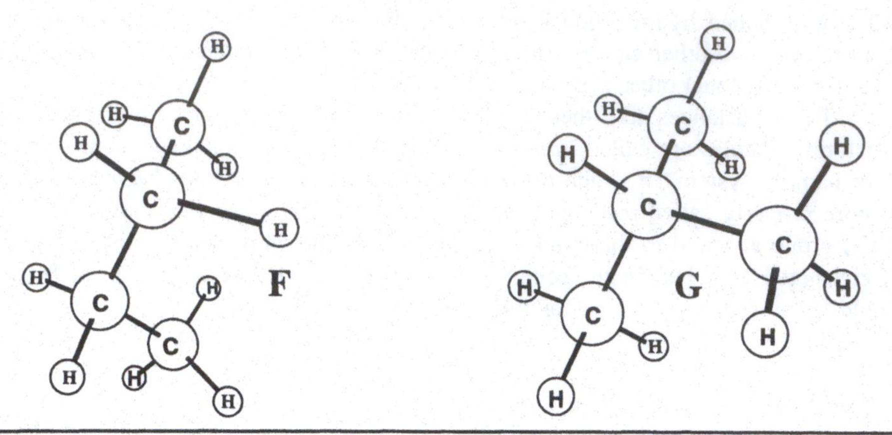

Figure A1.12 Two new conformers of butane, **F**, and methylpropane, **G**.

These models are slightly different from models **D** and **E** of the same substances which we saw in Figure A1.10. **F** and **G** are different *conformers* of **D** and **E**; **F** can be transformed into **D**, or **G** into **E**, simply by twisting or rotating the carbon atoms on each side of the single bonds joining them; but **D** cannot be changed into **E** or **G**, or **F** into **E** or **G**, without breaking one of the bonds joining a carbon atom to another carbon atom and reconnecting them.

Figure A1.13 shows three different views of the ethane molecule, **H**, **I**, and **J**; in each perspective it is the same conformer of ethane and there has been no rotation around the carbon–carbon bond. The particular conformer of ethane shown in Figure A1.13 is called *eclipsed*, because (viewed as in **J** along the carbon–carbon bond) the three hydrogens of the first carbon eclipse the three hydrogens of the second carbon; because this crowds the molecule, it is the least stable conformation of ethane. In Figure A1.14 is shown how **K** (which is the same as **J** in Figure A1.13) can be transformed into a different conformer, **L**, by rotating the back methyl group 60° as the arrows in **K** indicate. Similarly, **L** can be transformed into conformer **M** by rotating the front methyl bond of **L** as its arrows indicate. **K** and **M** are both the eclipsed conformation of ethane. Conformer **L**, where the hydrogens of the front and back methyl groups are as far apart as possible, is the most preferred (stable) conformation of ethane and is called *staggered*. Most ethane molecules spend most of the time comfortably

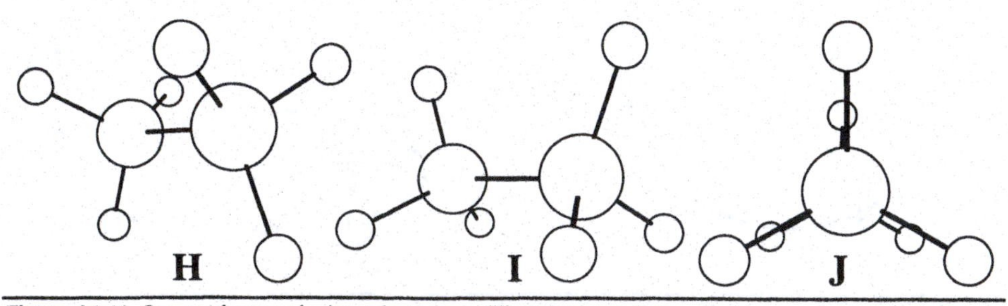

Figure A1.13 One conformer of ethane from three different perspectives.

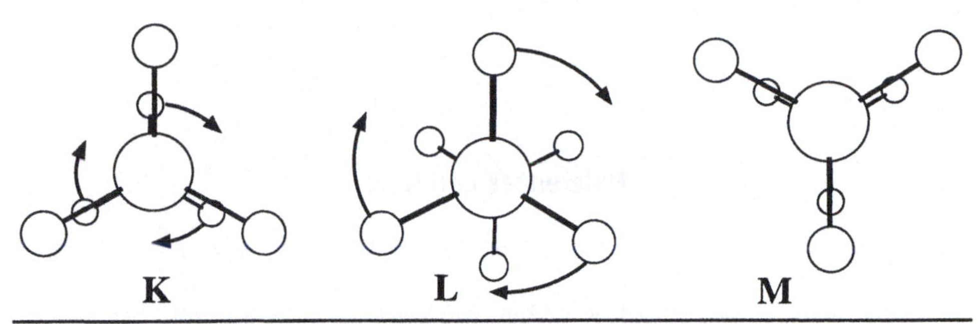

Figure A1.14 Rotation of back CH_3 (**K→L**) and front CH_3 (**L→M**) of ethane.

staggered; but the constant jostling they undergo bounces all of them some of the time into the eclipsed position. It may help to visualize the eclipsed/staggered conformations of ethane if we imagine the molecule as HONC, the molecular dog of Figure A1.15. **N** shows HONC standing on all fours; his feet eclipse each other and his head eclipses his tail; in **O**, however, he has assumed a staggered, comfort-station conformation.

The differences between conformers of the same molecule are often minor, as they are with ethane; the carbons at each end of a single bond are twirling and rotating at very great speeds so that any two conformers are interconverting all of the time. With more complicated molecules, some conformers may not be as easily interconvertible, and it has been shown that many drugs exert their pharmacological activity through one specific conformation of the many available to them. Isomers, on the other hand, are always distinct substances that often have significantly different properties. It is impossible to transform one isomer into another without breaking the existing bonds and making new ones, a process that requires either the introduction of a great deal of energy or (in living systems) a very specific catalytic enzyme.

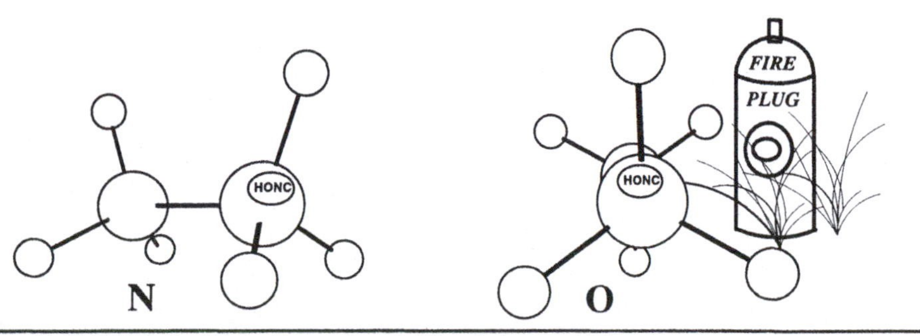

Figure A1.15 HONC, the Ethane Dog, *eclipsed* (**N**) and *staggered* (**O**).

References and Notes

1. *The Merck Manual,* 15th ed., R. Berkow, Ed., Merck: Rahway, NJ, 1987, p 811.
2. Well, this is overly cautious. But smoking is never good for you.
3. *The Merck Manual,* 15th ed., R. Berkow, Ed., Merck: Rahway, NJ, 1987, p 811. One can only speculate on the infernally combustible comestibles of Dante's demon Barbariccia, who summoned his legions with a colonic trumpet: "*ed elli avea del cul fatto trombetta*" (*Inferno*, canto xxi, 139). But the gas expelled from these Gargantuan regions was doubtless the foul-smelling hydrogen sulfide, H_2S. The odor frequently associated with mammalian flatus derives from traces of sulfides (methane and hydrogen are odorless). Peter Brimblecombe of the University of East Anglia warns against visiting art galleries in wet woolen clothes or after having eaten beans: "Wet woolens and flatulence are key sources of gases called sulfides, whose corrosive powers threaten artworks." [*Chemical & Engineering News*, February 1991.] This is problematic: East Anglia has been densely populated with soggy sheep and woolens ever since the Romans persuaded the populace to wear clothes instead of blue paint. No study has yet been able to support humorist Dave Barry's claim that flatulence is caused by riding in crowded elevators.

Appendix II
Return to the Second Dimension

Planar Projections

The ball-and-stick models we have used so far have the advantage of portraying quite accurately the geometry of organic molecules. But they become cumbersome when used to represent more than a few atoms, and it requires considerable artistry (or a sophisticated computer program) to draw them. Consequently, two-dimensional, planar representations of organic structures are commonly used.

Figure A2.1 shows two representations of diphenhydramine, which is the active agent in Benadryl, Sominex, Nytol, Dramamine, and several other common over-the-counter drugs (the same drug is active either as an antihistamine or a soporific). The left model is actually closest to portraying the amount of space taken up by the atoms, but the bonds are hidden; the right model minimizes the space filled by the atoms and emphasizes the underlying structure of the chemical bonds holding the molecule together. In both models there are at least three hydrogen atoms that are hidden behind other atoms. There are just too many hydrogen atoms in organic compounds to make an uncluttered drawing, and the abstract portrayals discussed here are really a necessity for clear communication.

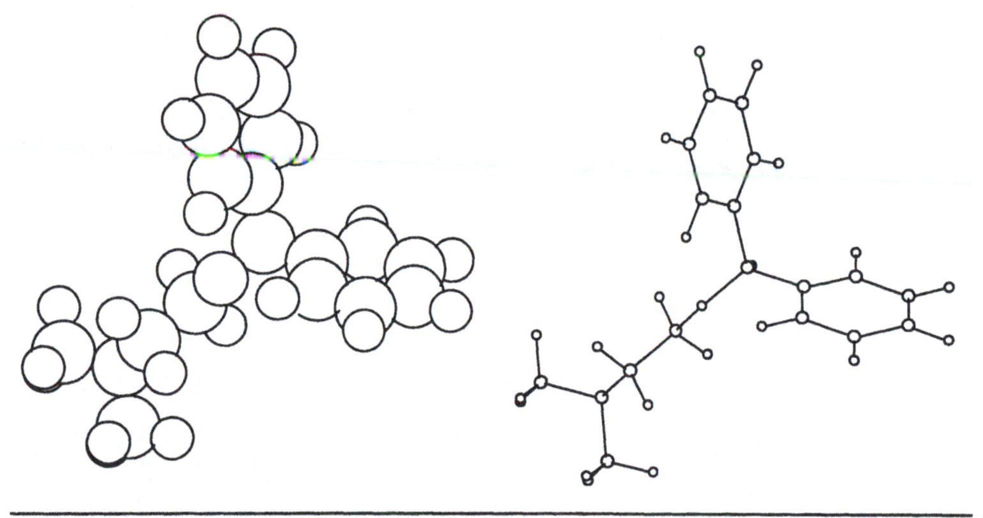

Figure A2.1 Two representations of diphenhydramine.

Like the various planar projections of our spherical earth that have been developed over the years by cartographers, planar projections of three-dimensional molecules have their virtues and limitations. The projection closest to the ball-and-stick models we have used so far replaces the balls with the letter abbreviations for the atoms involved (H, O, N, and C), replaces the sticks with a line, and simply squashes the molecule down onto the flat paper surface. However, we soon discover that when we make a plane projection of, for example, butane using this "letter-and-line" procedure, the results vary considerably depending on which perspective or conformation we start with. In Figure A2.2, to the right of each ball and stick image of butane, there is a letter-and-line projection that attempts to portray the same conformation and perspective. Because perspectives are relative, and conformations continually interchanging, there is no real difference between what all four of these structures represent. And because it is so much easier to draw the typographer's tidy right angles as in the bottom right, this sort of projection tends to predominate in chemical and medical publications. However, it is important to remember that the geometry of the real molecule is anything but flat or rectangular, as this projection seems to suggest, and that it is the same substance represented by either of the seemingly quite different planar projections. The same problem occurs in comparing the apparent size of equatorial and arctic land masses on a Mercator projection or on a globe.

In Figure A2.3, the hydrocarbons we considered in Appendix I are shown using this traditional letter-and-line projection. Notice that the hydrogens in branched molecules like isobutane begin to get in the way, and one of the carbon–carbon bonds has to be drawn unrealistically longer than the others. Some other confusions can easily result from using these projections; the four planar projections in Figure A2.4 are all perfectly equivalent and represent the same molecule, butane; but they look quite different. We can see they are really the same by ignoring the different shapes they suggest (the true shape of the

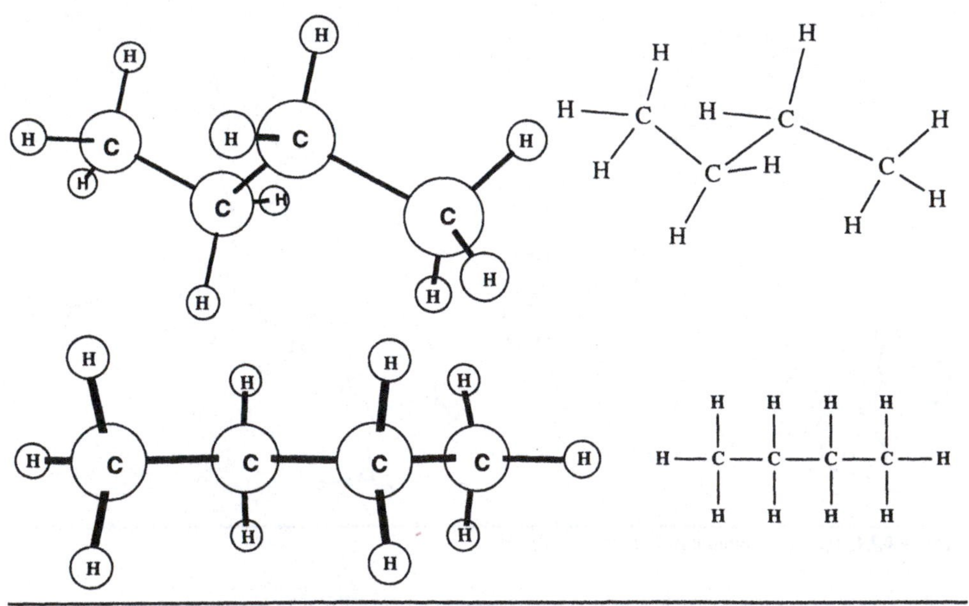

Figure A2.2 Four representations of butane. *Left*, ball and stick; *right*, letter and line.

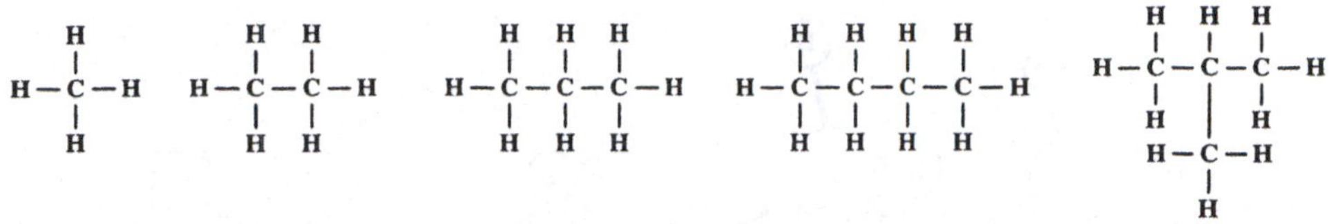

Figure A2.3 Methane, ethane, propane, butane, and isobutane.

butane molecule is given more or less by the ball-and-stick model) and simply counting the hydrogens attached to each carbon. If we start at either of the two circled CH_3 ends of any of the four projections, we find that this terminal methyl group is attached to a CH_2, which is attached to a second CH_2, which is in turn attached to the second terminal CH_3 methyl group. The sequence is always $CH_3CH_2CH_2CH_3$.

However, *iso*butane has the different sequence $CH_3CH(CH_3)CH_3$; it is a structurally distinct isomer, as we found in Appendix I. Unlike butane, isobutane has *three* CH_3 groups, *one* CH group, and *no* CH_2 group. Isomers are very important because even the very slight variation in structure that they represent usually has dramatic consequences in physiological activity. Two examples: adding a single methyl group to the opioid fentanyl makes it about a thousand times as potent; adding a methyl group to carcinogenic benzene changes it to toluene, which has no known cancer-causing activity.

With still larger molecules, the hydrogen atoms tend to get more and more in the way of each other; consequently, an even more condensed two-dimensional projection has been developed that observes the following rules:

- A line is used to represent a bond between atoms. If no atoms are indicated at either end of the line, it is assumed that the atoms are carbon atoms; the symbol for the carbon atom can be (and usually is) omitted.
- Hydrogen atoms attached to carbon atoms, and the bonds attaching them, can be (and usually are) omitted.

Structures drawn according to these rules are called *line drawings* because they often consist of simple lines or geometric figures: the end of each line and the intersection of any two, three, or four lines represents a carbon atom. But, as Figure A2.5 shows—where the ball-and-stick butane at the left is the equivalent of any of the four simple line drawings at the right—line drawings can create surprisingly (perhaps deceptively) simple representations of molecular structures.

We can become more familiar with the use of this type of line drawing by carefully

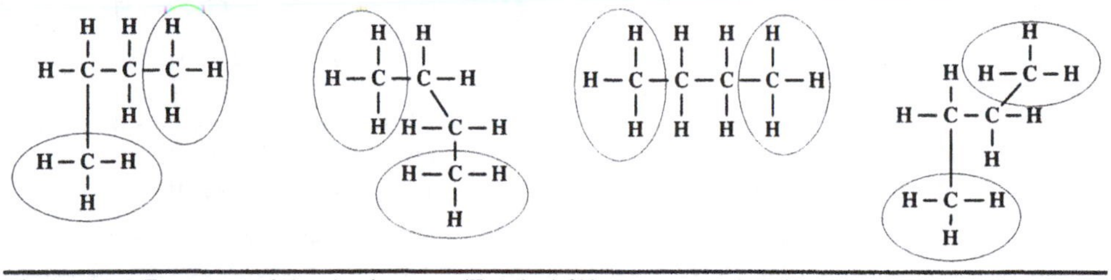

Figure A2.4 Four ways of drawing butane. (Terminal CH_3 groups are circled.)

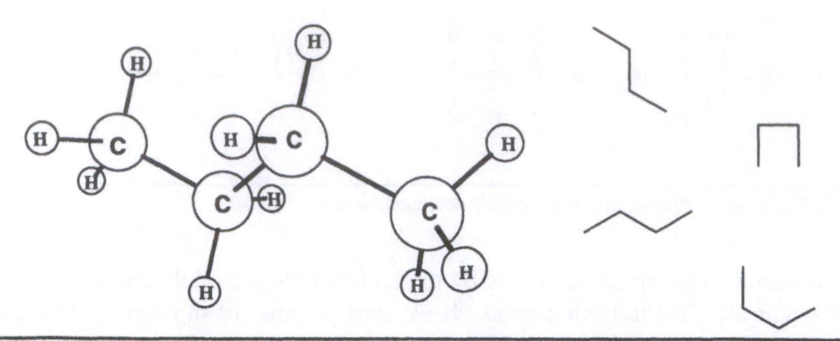

Figure A2.5 Five representations of butane, C_4H_{10}. *Left,* ball and stick; *right,* four line drawings.

considering the different ways it can be used to represent the same molecule, *isooctane* (pronounced to rhyme with "I sew socks, Jane": *eye-zo-AHK-tane*). Isooctane is an old-fashioned industrial term for the substance chemists now call 2,2,4-trimethylpentane. (We will see later in this appendix how this modern naming system works.) Isooctane is of interest, however, because it forms the basis of the octane rating system used for gasoline. In this system, the efficiency and "knocking" of an engine running on pure isooctane is given a relative rating of 100, while the much poorer performance of the same engine running on pure heptane is given a rating of 0. Heptane is a straight-chain hydrocarbon containing seven carbon atoms, while isooctane is a branched isomer of octane, which contains eight carbon atoms. The smoother combustion of isooctane is due to the more extensive branching and the larger number of methyl groups in its structure.

Figure A2.6 shows several acceptable ways in which isooctane can be drawn. **A** is the style preferred by chemists most of the time; it has a certain elegant simplicity, and to those not initiated into the mysteries, it reveals as little as possible. (Since the days of alchemy, chemists have liked their cabalistic symbols.) Other implicit rules are:

- If the (alphabetical) symbol for a carbon atom is written, any hydrogens attached to it must also be explicitly indicated.

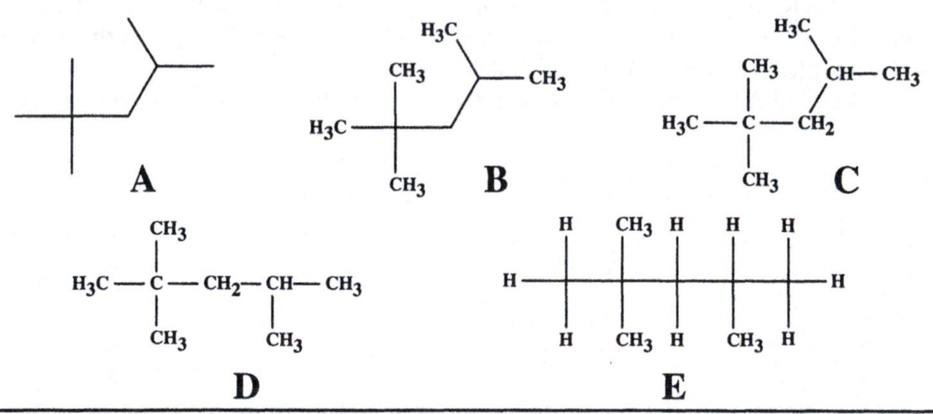

Figure A2.6 Five ways of drawing isooctane, C_8H_{18}.

- Any atoms other than carbon, and all hydrogens bonded to atoms other than carbon, must be explicitly written.

While the hydrogen atoms bonded to an explicitly written carbon must be written, the bonds (lines) joining these hydrogens to the carbon atom need not be drawn and are usually omitted, as in **C** and **D**. In **E**, some of the bonds to hydrogen are drawn. Figure A2.7, shows some examples of acceptable (OK) and unacceptable (not-OK) usages. Notice the structure in the top right: it is customary to indicate the geometry around a carbon atom by using a solid wedge for a bond coming out of the plane of the page toward the reader, and a hatched wedge for a bond going back behind the page.

Counting the Carbons

Here is a doodling game for practicing line formulas. Draw some geometric figures with no intersection point having more than four lines coming from it. If these figures were hydrocarbons, how many carbon and hydrogen atoms would they have? Here are some hints:

- An angle or corner with four lines radiating from it is a carbon atom with no implicit hydrogen atoms.
- An angle or corner with three lines radiating from it is a carbon atom with one implicit hydrogen, CH.
- An angle or corner with two lines radiating from it is a carbon atom with two implicit hydrogens, CH_2.
- The dead-end point of a radiating line is a terminal methyl group, CH_3.

Figure A2.8 shows some doodlings to which we have given whimsical names having the systematic hydrocarbon ending "ane". (Actually, most of these structures have been synthesized or are found in nature; the chemical names are of course not like the ones we have given them—except for our "windowpane," one chemical name for which is [4.4.4.4]fenestrane, from the Latin, *fenestra*, meaning window; our "housane" is [3.4.4.4]fenestrane in this system.)

What is the molecular formula for "trianglane"? There are three corners representing carbon atoms each of which has two bonds going to other carbon atoms; there must be two implicit hydrogens at each corner, so the molecular formula is C_3H_6.

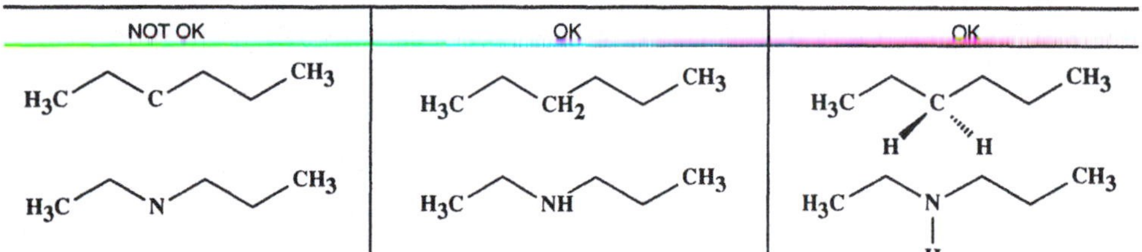

NOT OK	OK	OK
H_3C, C, CH_3	H_3C, CH_2, CH_3	H_3C, C, H, H, CH_3
H_3C, N, CH_3	H_3C, NH, CH_3	H_3C, N, H, CH_3
C, C	H_3C, CH_3	

Figure A2.7 Conventions used in drawing line formulas.

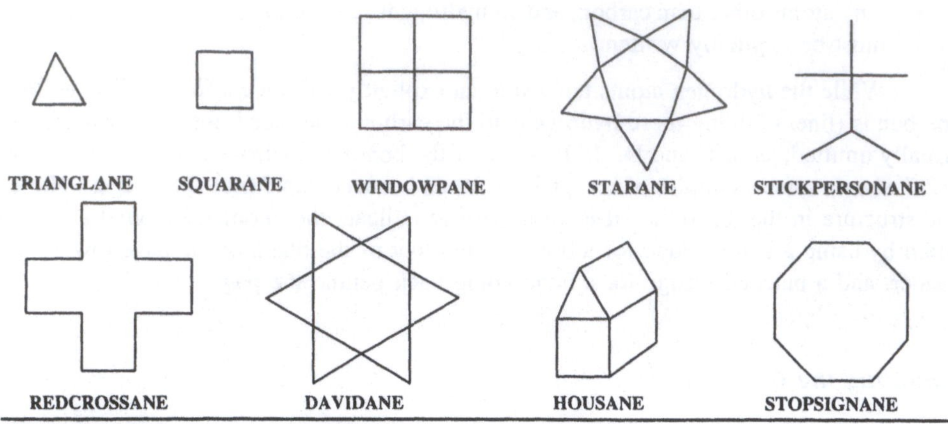

Figure A2.8 Some imaginary hydrocarbons.

That was easy. Larger molecules like "housane" and "davidane" are more difficult. It is easier if we first use the rules just discussed to figure out how many carbons and hydrogens are at each corner. Then we can redraw the molecules making the carbons and hydrogens explicit as in Figure A2.9. When we add up the total number of atoms of each element, we see that davidane is $C_{12}H_{12}$ and housane is C_8H_{10}.

We will leave to the reader the challenge of determining the molecular formula for the other molecules in Figure A2.8. Only the structures that we have called trianglane, squarane, stickpersonane, redcrossane, and stopsignane are known, stable substances. The chief reason for nature's unwillingness to accommodate our imagination here was enunciated in the 19th century by Adolf von Baeyer as the "strain theory": the tetrahedral angle of carbon is unnaturally constrained when it is forced to form rings of four or three atoms (squares and triangles); while one or two of these configurations per molecule can be tolerated, compounding them—as in housane and windowpane—leads to very unstable structures.

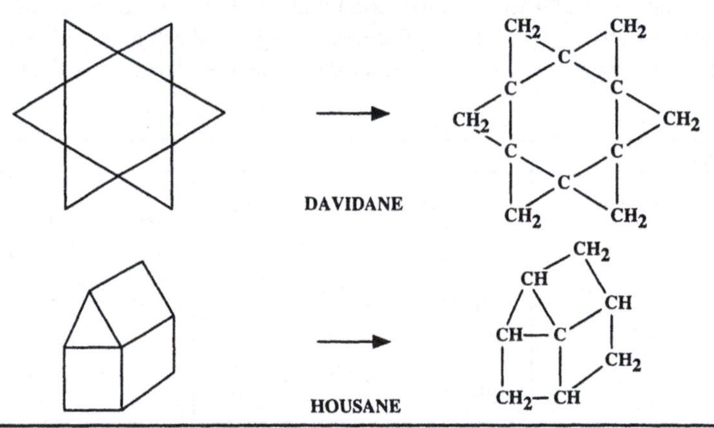

Figure A2.9 Counting the atoms in "housane" and "davidane."

Finding the isomers

Here is another way to develop an intuition for line structures. How many distinct structures/isomer sets are there in Figure A2.10? We can approach this systematically. First, count the carbons in each structure; there are 7 in **F**; 8 in **J**, **K**, **L**, and **M**; 9 in **I**; 12 in **G** and **H**. Therefore, **F** and **I** are distinct substances, both from each other and from all the remaining molecules. We look now at the set of four 8-carbon molecules **J**, **K**, **L**, and **M**. All four molecules have the same molecular formula, C_8H_{18}, and are at least isomers. Are any of them identical? To be identical, they must have the same structure as well as the same number of molecules of carbon, hydrogen, etc. The first criterion to examine is, what is the *longest unbranched carbon chain* in each molecule? We can be sure that a chain is unbranched if we can trace a pencil over its entire length without ever raising the pencil from the paper. In Figure A2.11 we have traced the longest unbranched carbon chain in each molecule **J–M** and numbered each carbon on the way. From this we can see that **J** and **M** have six carbons in their longest unbranched chain, while **K** and **L** have seven; **J** has a two-carbon branch at carbon 3, while **M** has two one-carbon branches—one at carbon 3, one at carbon 4. Therefore **J** and **M** are distinct substances, both from each other and from **K** and **L**. (**J** is 3-ethylhexane; **M** is 3,4-dimethylhexane.) When we examine **K** and **L** in the same way, we see that both of them are really identical; each has a 7-carbon chain as its longest unbranched chain, and each has a one-carbon (methyl) branch on the same carbon, 4. (Each is really 4-methylheptane.)

We turn finally to the set of two 12-carbon molecules, **G** and **H**, and we can see that these are really identical, even though one looks like the red cross and the other like the star of David. Each structure has 12 corners, and every corner must be a CH_2 group, since they all have two bonds coming from them. Both drawings represent $C_{12}H_{24}$, the systematic name for which is *cyclododecane*.

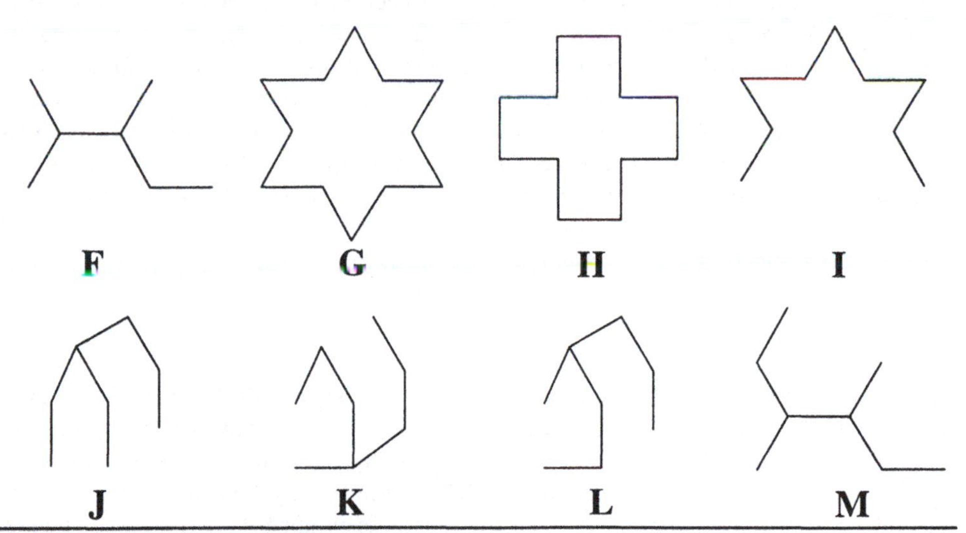

Figure A2.10 Which are isomers/identical?

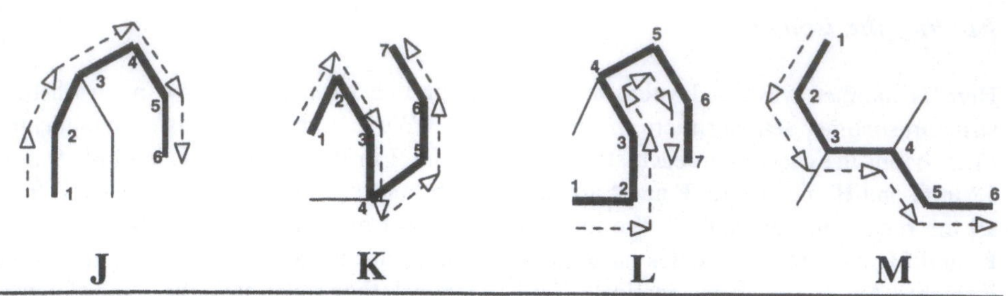

Figure A2.11 Finding the longest unbranched chain.

Systematic Nomenclature (Naming System) for the Alkanes

Elaborate and detailed conventions have been developed over the years by the International Union of Pure and Applied Chemistry (IUPAC) and by Chemical Abstracts to determine an unequivocal name for all but the most complicated organic structure. The system is really a language, with prefixes, suffixes, and declensions; the rule-making bodies, like grammarians the world over, often come in after the fact and codify the names the chemists have given their creations or discoveries. And like most languages, it is easier to listen to a native speaker than study a grammar book. So we will start by naming some compounds and then most of the rules will seem fairly obvious. But first some basic vocabulary.

The Straight-Chain Alkanes

In Table A2.1 are the names and line formulas (except for methane) for the first 10 *alkanes*. Many of these names are easy to remember if you associate them with more familiar words that have a similar origin in the Greek or Latin counting system. Thus pentane and *pentagon*, hexane and *hexagon*, heptane and *heptagon*, octane and *octopus*, nonane and *nonagon*, decane and *decade* are cognates. It is a bit more difficult to remember the names of the first four alkanes. *Methane* derives from methyl alcohol, whose etymology we have seen previously. *Ethane* comes from the Greek "to burn"; *propane* from the Greek word for fat, because *propanoic/propionic acid* is found in milk and butterfat. *Butane* derives from *butyric acid*, found in butter (Latin *butyrum*, "butter"—from the Greek for "cow-cheese").

The Cycloalkanes

In addition to the straight-chain alkanes there are *cycloalkanes*, some of which are shown in Table A2.2. These are carbon chains joined head to tail in a ring, and they are named by prefixing *cyclo-* to the name for the straight-chain hydrocarbon with the same number of carbon atoms. It is impossible, of course, for there to be a cyclomethane or a cycloethane, but starting with cyclopropane, there are an endless number of possible cycloalkanes.

Table A2.1 The Alkanes

METHANE	CH_4	
ETHANE	CH_3CH_3	
PROPANE	$CH_3CH_2CH_3$	
BUTANE	$CH_3CH_2CH_2CH_3$	
PENTANE	$CH_3CH_2CH_2CH_2CH_3$	
HEXANE	$CH_3CH_2CH_2CH_2CH_2CH_3$	
HEPTANE	$CH_3CH_2CH_2CH_2CH_2CH_2CH_3$	
OCTANE	$CH_3CH_2CH_2CH_2CH_2CH_2CH_2CH_3$	
NONANE	$CH_3CH_2CH_2CH_2CH_2CH_2CH_2CH_2CH_3$	
DECANE	$CH_3CH_2CH_2CH_2CH_2CH_2CH_2CH_2CH_2CH_3$	

Hundreds of these structures have been found in nature and hundreds more have been synthesized.

The simplest of these cycloalkanes is cyclopropane, which we called "trianglane" earlier. It is a gas which has been used to induce surgical anesthesia. Notice that our "squarane" in Figure A2.8 is cyclobutane, and "stopsignane" is cyclooctane. Cyclohexane and cyclopentane are particularly important as components of the structures of many of the chemicals which occur in nature; the steroids, for example, contain three cyclohexane rings and a cyclopentane ring. In Table A2.2, we have drawn some of the larger cycloalkanes in more than one form: these large rings have numerous possible conformations accessible to them, but those having hexagonal or pentagonal angles are usually more stable.

Finding the Longest Chain or Ring

The first step in naming any compound is to find the longest straight-chain alkane chain, or—if there is a ring—the largest cycloalkane. In **N** of Figure A2.12, the longest unbranched carbon chain contains eight atoms; these eight atoms have been numbered. Since the longest chain is eight, we see from Table A2.1 that the parent name for **N** is octane. A few points should be noted. Could we have started our numbering at the carbon labeled *a*? Yes, it would have made no difference, because we would still have a chain of eight atoms; similarly, we could have ended our count on the atom labeled *d* instead of the one labeled eight. (Depending on the substituent pattern, we will soon see

Table A2.2 The Cycloalkanes

CYCLOPROPANE

CYCLOBUTANE

CYCLOPENTANE

CYCLOHEXANE or or or

CYCLOHEPTANE

CYCLOOCTANE or

CYCLONONANE or

CYCLODECANE or

that sometimes one way of counting the longest chain is preferable, but here it is immaterial.) But to have started with the atom labeled *b* would have been a mistake, because we would have had a chain of only seven atoms, and not the longest possible chain.

Naming the Branches

Our name has to convey the fact that there are four branchings off this longest chain: *a, b, c,* and *d*. Each of these is a one-carbon branch, so we call them *methyl* groups. Methane, CH_4, is the name for a one-carbon alkane, so *methyl* is the name for a one-carbon substituent, CH_3; similarly, *ethyl, propyl, butyl, pentyl, hexyl, heptyl, octyl, nonyl,*

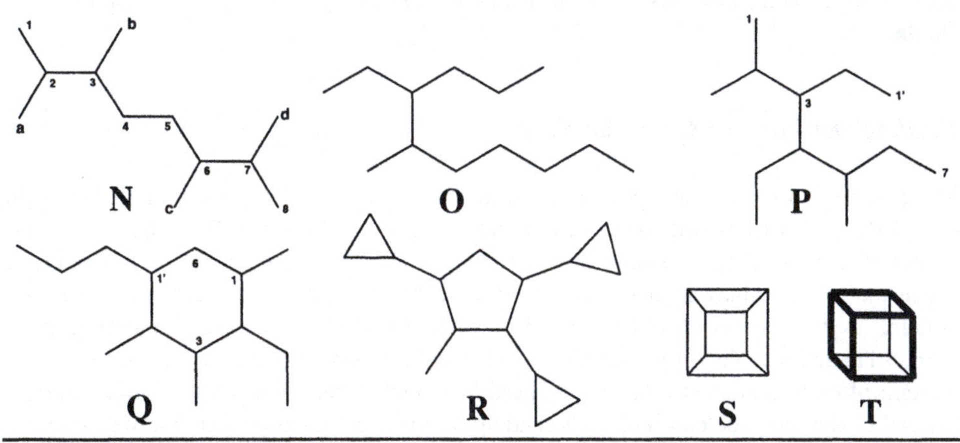

Figure A2.12 Some structures to be named.

decyl are the names for substituents of chain length 2–10. These substituent names function as adjectival prefixes modifying the name for the longest unbranched chain in the molecule, which is a noun with the suffix *-ane*.

Counting the Branches: The Multiplying Prefixes

There are four methyl groups, and we indicate this by saying *tetra*methyl. The multiplying prefixes are, from 1 to 10: *mono-, di-, tri-, tetra-, penta-, hexa-, hepta-, octa-, nona-, deca-* (but *mono-* is almost always omitted). So far, our name for **N** is *tetramethyloctane*. Notice that we write the name as one uninterrupted word; this is probably due to the profound influence German scientists and their language had on the early development of organic chemistry in the late 19th and early 20th century.

Locating the Branches

Tetramethyloctane is still not specific enough to unambiguously indicate the structure for **N**; in Figure A2.13 are shown six structures, and all have a longest chain of eight carbons and four methyl-group branches; that is, they are *all* tetramethyloctanes. We will see that by carefully locating and naming where these four methyl branches occur along the eight-carbon parent chain we will be able to decide which of these six tetramethyloctanes are genuinely distinct isomers and which are actually different ways of drawing the same molecule.

Let's look at **N** and **U**; the methyl group *c* that was attached to carbon 6 in **N** has been moved to carbon 4 in **U**; the other three methyl groups are at carbons 2, 3, and 7 in both molecules. So the name for **N** is *2,3,6,7-tetramethyloctane*, and the name for **U** is *2,3,4,7-tetramethyloctane*. (Notice that every substituent gets its own number; there is

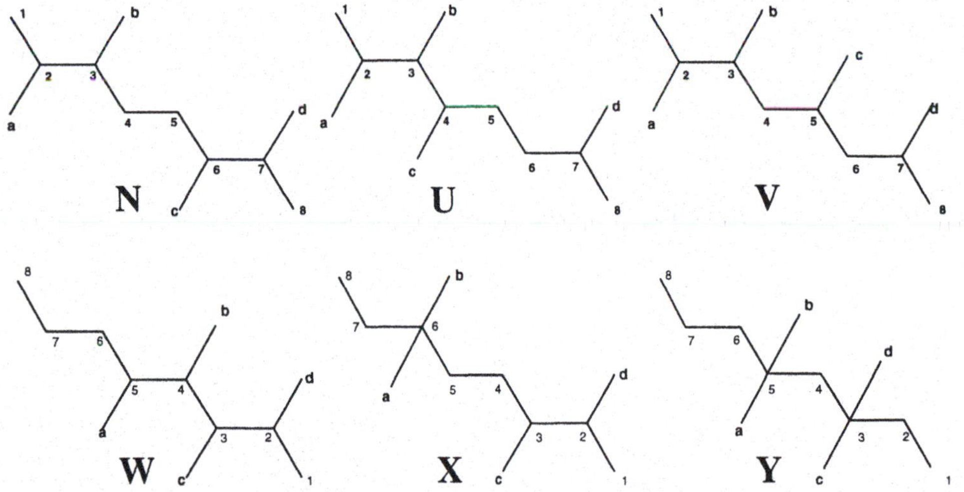

Figure A2.13 Some tetramethyloctanes.

also a somewhat arbitrary convention whereby numbers are separated from each other by commas, while numbers are separated from letters by hyphens.)

Going on to **V**, it is easy to see that this is *2,3,5,7-tetramethyloctane*. We tentatively give **W** the name 4,5,6,7-tetramethyloctane, but actually this is less than correct. Less than correct, but more than incorrect: anyone asked to draw the structure for 4,5,6,7-tetramethyloctane would come up with the same structure **W**, so this name fulfills the primary purpose of describing the structure. But there is a better name: *2,3,4,5-tetramethyloctane*, which we get by numbering our octane chain starting at the other end than we have used so far. We choose the second name as the more correct systematic name because the first number in this name, 2, is lower than 4, the first number in the previous name. (The importance of this rule is mainly for indexing, where we want to be able to assign one and only one name for each structure.)

In **X**, we see that we should number the chain in the same direction as we used for **W**. A further rule: we give a number for every substituent, even though we have two identical substituents on the same carbon. Thus **X** is *2,3,6,6-tetramethyloctane*, and **Y** is *3,3,5,5-tetramethyloctane*.

Now let's look at **O** of Figure A2.12. From the way it is drawn in the figure, it is a little surprising to find that the longest chain has not six, not nine, but 10 carbon atoms. This is because the first thing one naturally does is follow the more or less direct path from left to right, a habit ingrained from reading European languages, and hard to overcome. In Figure A2.14 we have indicated the longest chain in **O** and circled the two substituent branches Then we have made a new conformer by rotating the top of the molecule 180° around the vertical bond (process **a**); we have transformed the perspective of this conformer by flipping it over like a pancake top-to-bottom (process **b**); finally, we have redrawn the molecule (process **c**) so that in the last conformer the longest chain goes in a more or less straight line from left to right, to match our intuition. Now we count along this longest chain and we see that branching from the fourth carbon is a two-carbon substituent chain (ethyl); from the fifth carbon is a one-carbon substituent (methyl): the proper name for **O** is therefore *4-ethyl-5-methyldecane*.

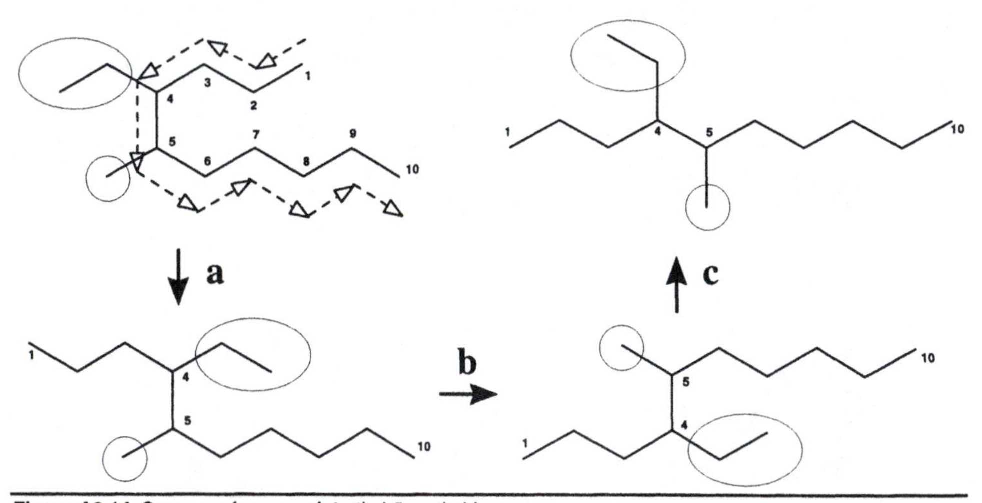

Figure A2.14 Some conformers of 4-ethyl-5-methyldecane.

In **P** of Figure A2.12, the longest chain is seven carbon atoms, but we have to decide whether to count 1→7 or 1′→7. We choose 1→7 for two reasons: the first substituent comes earlier along this path, giving it a lower number; and there are more substituents (4 to 3) along this path than along 1′→7. The best name is *3,4-diethyl-2,5-dimethylheptane*; substituents are listed in alphabetical order, ignoring the multiplicative prefixes *di*, *tri*, etc. (The use of two numbers and the multiplicative prefixes is somewhat redundant, but such is the convention.)

In **Q** of Figure A2.12 we have our first cycloalkane. Should we choose to number the carbon atoms in the cyclohexane ring starting at 1′ and going counterclockwise to 6 or at 1 and going clockwise? (Any path but one of these gives us higher numbers for our substituents.) Either path will give us 1 as the number for the first substituent; either path will give us 5 substituents. We decide the tie by giving precedence to a 1-*m*ethyl rather than a 1-*p*ropyl substituent (hence 1 rather than 1′) because *m* precedes *p* in the alphabet. **Q** is *2-ethyl-1,3,4-trimethyl-5-propylcyclohexane*. For the same reason, because in **R** tri*c*yclopropyl will come before *m*ethyl, we choose the name *1,2,4-tricyclopropyl-3-methylcyclopentane* rather than counting in a counterclockwise manner.

Here is a summary of all of the rules we have used:

- Find and name the longest straight-chain hydrocarbon (the parent chain) or the largest ring within the molecule; if there is a tie, pick the chain or ring having the greatest number of substituents (if there is still a tie, pick the chain having the lowest-numbered first substituent).
- Carbon-chain branchings (substituents) off this longest chain are named by changing the "ane" ending to "yl" of their corresponding alkane; that is, a three-carbon branch chain is called "propyl"; a five-carbon substituent "pentyl," and so forth.
- Two identical substituents are prefixed with "di-" (that is, "3,3-dimethyl," "4,5-dipropyl"); similarly, tri-, tetra-, penta-, hexa-, hepta-, octa-, nona-, deca.
- A number is given each and every substituent to indicate where the branch occurs along the parent chain; the numbering starts at that end of the chain or that corner of the ring that gives the lower numbers, or, in a tie, the prior alphabetical letter to the first substituent.
- The substituents are listed in alphabetical order, ignoring the multiplicative prefixes.

Finally, we have **S** and **T** in Figure A2.12. These are really the same structure; the eight carbon atoms are arranged in a cube, as in **T**; and a cube is (considered simply as to its topological connectivity) two squares joined to each other by four lines from each corner, as in **S**. Chemists spent many years trying to synthesize this interesting molecule, and when Philip Eaton of the University of Chicago succeeded in 1964, he naturally called it *cubane*. This was not good enough for the purists of nomenclature, of course; its proper name is *pentacyclo*$[4.2.0.0^{2,5}.0^{3,8}.0^{4,7}]$*octane*. While such names are necessary for the alphabetization of databases, they are almost useless for conversation between chemists, who continue to call it cubane.

How does the name for cubane work? *Pentacyclo-* means that there are five rings in all; if you take a scissors and snip one bond of cubane at a time until there are no rings left, you will have made five cuts. *Octane* means there are eight carbons. The numbers in brackets are really a set of instructions. The interested reader can understand the nature

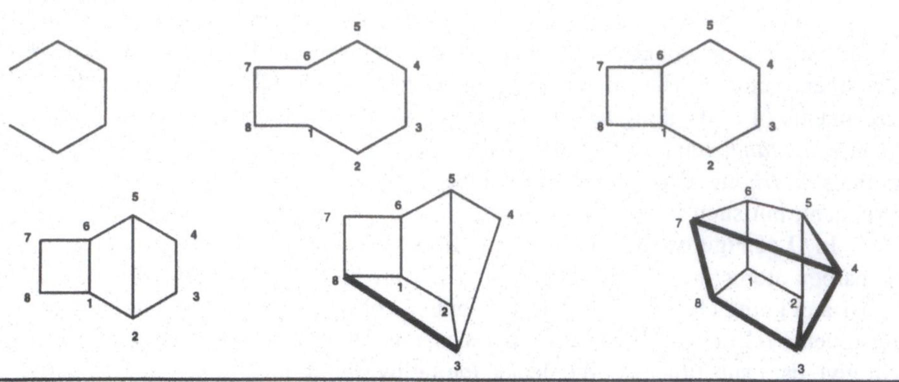

Figure A2.15 Cubane, or Pentacyclo[4.2.0.$0^{2,5}$.$0^{3,8}$.$0^{4,7}$]octane.

of these instructions by following the steps outlined in Figure A2.15. The **4** means take any two carbon atoms and connect them with a four-carbon bridge, as in the top left structure of Figure A2.15. The **2** means connect these same two carbons with a second two-carbon bridge, as in the top center structure. Since we now have all eight carbons of the octane molecule, we number them, starting with one of the original two carbons and going over the longest bridge first, then over the next longest. The **0** instructs us to connect the same original two carbons with a third, zero-carbon bridge (i.e. draw a bond between them, as in the top right structure). The $\mathbf{0^{2,5}}$ tells us to draw a zero-carbon bridge (i.e. a bond) between the second and fifth carbons (the meaning of the superscripted 2,5), as in the bottom left structure. Similarly, $\mathbf{0^{3,8}}$ tells us to draw a zero-carbon bridge between carbons three and eight, giving the bottom center structure. Finally, $\mathbf{0^{4,7}}$ tells us to make a zero-carbon bridge between carbons four and seven, which produces the bottom right structure. We can see that this is indeed a distorted cube; of course the true molecule has equivalent bonds and is a true cube.

There are a large number of other rules to cover even more complex situations, and new nomenclature is being formulated all the time. It is easy to see one limitation to the rules we have given: what if a branch chain is itself branched? There are of course rules to cover this, but some branched substituents occur so often that they have been given special names themselves: *isopropyl, isobutyl, secondary* butyl (abbreviated sec-butyl) and *tertiary* butyl (tert-butyl). These are displayed in Table A2.3, where each is used as a substituent of cyclopentane.

Two common pharmacy items that owe their name to these prefixes are also found in Table A2.3. Rubbing alcohol sold in drug stores is either ethanol that has been "denatured" (a poison added to curtail its use as a beverage), or it is *isopropyl alcohol,* whose structure is shown in Table A2.3. If ingested, isopropyl alcohol causes nausea, headache, anesthesia, coma, or—if 100 mL (about 3 oz) or more is consumed—death.

Ibuprofen is the active ingredient in Advil, Medipren, Motrin, and Nuprin. Its structure contains an *isobu*tyl group attached to a *phen*yl (benzene) ring that is in turn attached to a *pro*pionic acid group; the generic name ibuprofen is composed of pieces of the full chemical name, with phenyl spelled phonetically as *fen.* One feature of the ibuprofen structure in Table A2.3 is new to us: the three carbon–carbon double bonds in

Table A2.3 Branched Substituent Names

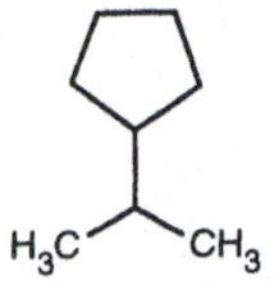

Isopropylcyclopentane	Isobutylcyclopentane	sec-butylcyclopentane	tert-butylcyclopentane
Isopropyl alcohol	Ibuprofen [2-(4-isobutylphenyl)propionic acid]		

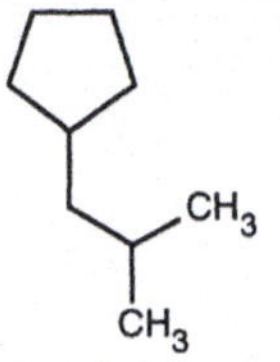

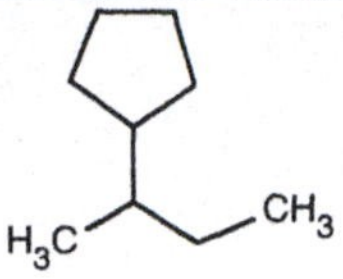

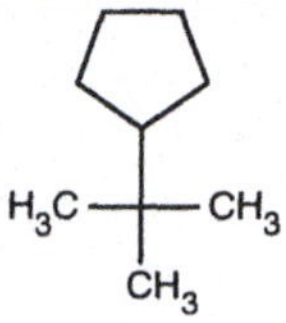

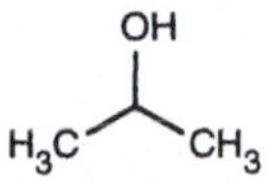

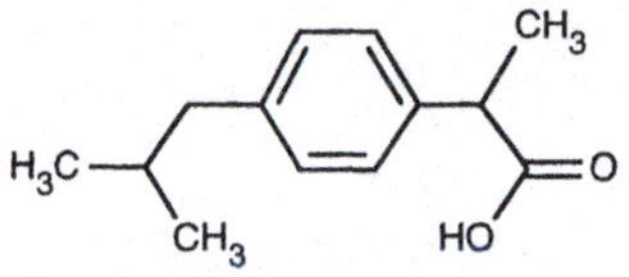

the phenyl ring and the carbon–oxygen double bond in the propionic acid part of the molecule. We will look at double and triple bonds in Appendix III.

Appendix III
How Shall I Bind Thee? Let Me Count the Ways

Carbon–Carbon Double Bonds: Alkenes

At the end of Appendix II we saw the structure of ibuprofen, which has several instances of something new to us: double bonds. On the left in Figure A3.1 is a line drawing of the simplest structure containing a carbon–carbon bond: *ethylene*, as it is commonly called, or *ethene* (the more correct name). Since ethene has the molecular formula C_2H_4 it is not an isomer of ethane, which is C_2H_6. The geometry of the two molecules is also quite different. All six atoms in the molecule of ethene are in the same plane: the ball-and-stick model in the middle of Figure A3.1 shows ethene as a flat molecule in the plane of the paper; on the right is the same molecule viewed in perspective. Because the three bonds extending from each carbon atom are directed to the three points of an equilateral triangle, double-bonded carbons are said to be trigonal in contrast to single-bonded carbons which are tetrahedral.

"Stop!" you are perhaps saying at this point. Aren't these two ball-and-stick structures wrong? There should be two "sticks" for a carbon–carbon double bond, but there's only one! Unfortunately, and confusing as this is, computer-generated molecular modeling systems give only the relative positions of the atoms in space and a single vector line to indicate whatever bond(s) join(s) them. To lessen the confusion, ball-and-stick models are often accompanied by a line drawing: the line drawing gives the exact number of atoms and shows each type of bond, while the ball-and-stick model is used to display the geometry and orientation of the molecule in space.

A small quantity of ethene is present with methane in natural gas; it has been used as an inhalation anesthetic. Ethene is synthesized by maturing fruit, where it functions as the natural hormone that triggers the ripening process. Commercially, fruit is often picked green and then induced to ripen after transport by "gassing" it with ethylene. Unfortunately, the ethene comes from outside the fruit, not from its own internal production: the outside is transformed into the beautiful color of the ripened fruit, but the inside still has all the rich flavor of wet cardstock.

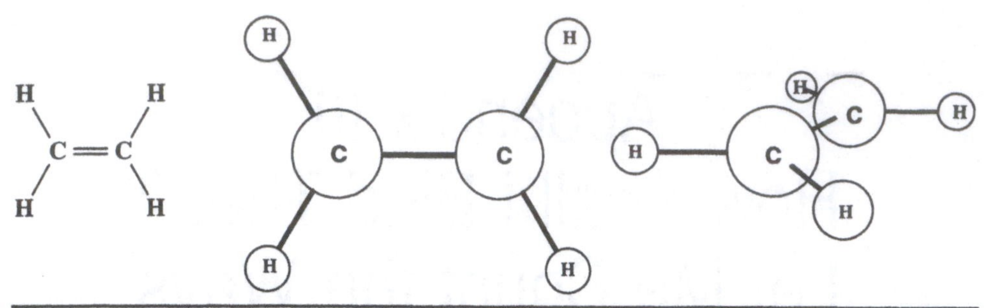

Figure A3.1 Ethylene (ethene).

Hydrocarbons that have a double bond are named like the alkanes and cycloalkanes with these additional rules:

- The parent chain is chosen to be the longest chain which contains the double bond.
- The *-ane* ending is changed to *-ene* (pronounced to rhyme with "bean").
- The parent chain is numbered from the end which is closest to the double bond, and a number is used to indicate at which carbon the double bond starts.

Cis-trans Isomerism

A further difference between alkenes and alkanes is this: whereas the carbon atoms at each end of a single bond like that in ethane can spin 360°, resulting in different conformers (such as the staggered and eclipsed versions of HONC, the ethane dog), the carbon atoms at each end of a double bond cannot rotate relative to each other—to do so would break the second bond. This characteristic of carbon–carbon double bonds results in the phenomenon of *cis-trans* isomerism. In Figure A3.2, **A** and **D** are identical (flipping either one like a pancake around the *y*-axis without changing any bonds gives you the other one); these are two drawings of *trans*-2-butene. The molecule is *trans* because the CH_3 groups are on opposite sides of the double bond. **B** and **C** are also identical (via a pancake-flip around the *x*-axis) and *cis* because the two CH_3 groups are on the same side of the molecule. Note that if there are two identical groups attached to one of the carbons in a double bond, *cis-trans* isomerism is no longer possible; the one side of the molecule is indistinguishable from the other.

Further examples of alkene nomenclature are given in Figure A3.3. Some points

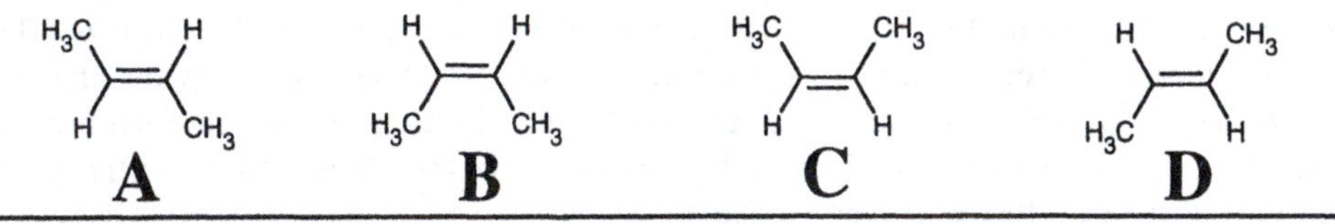

Figure A3.2 *trans*-2-butene (**A** and **D**); *cis*-2-butene (**B** and **C**).

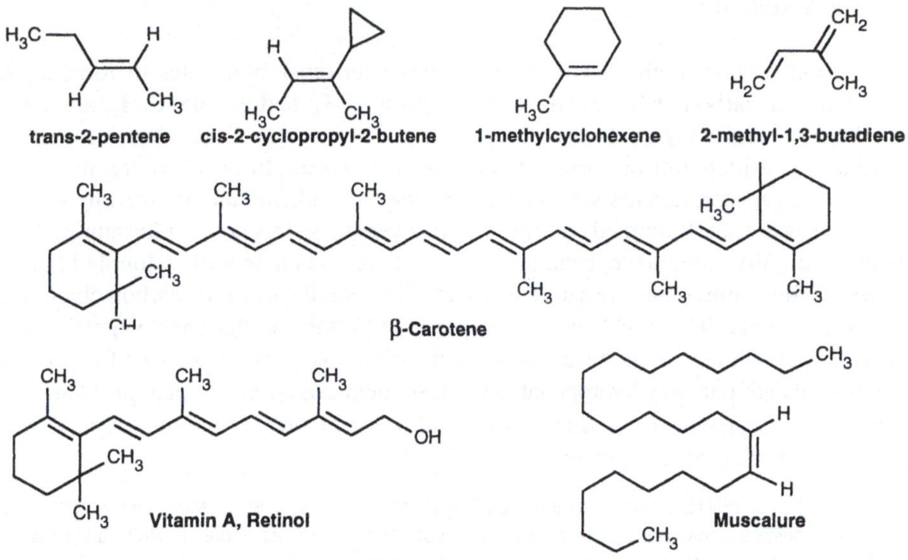

trans-2-pentene cis-2-cyclopropyl-2-butene 1-methylcyclohexene 2-methyl-1,3-butadiene

β-Carotene

Vitamin A, Retinol

Muscalure

Figure A3.3 Alkenes and cycloalkenes.

worth mentioning: (1) The *cis* prefix is not considered necessary for 1-methylcyclohexene, since it is geometrically impossible to have a *trans* double bond inside so small a ring; (2) when there are two double bonds, as in 2-methyl-1,3-butadiene, the alkane name is suffixed with *diene* (similarly *triene,* for three double bonds, etc.) and the *a* is retained for euphony.

The structure of *β-carotene* is interesting because it has so many double bonds alternating with single bonds: these are called *conjugated* double bonds. The electron pairs that constitute conjugated double bonds interact with each other in several important ways: one consequence is the absorption of light of a particular frequency. The β-carotene molecule is a deep reddish-orange and, as its name suggests, is responsible for the color of carrots. (It and other carotenes of similar structure contribute to the color of tomatoes, spinach, etc.; they are also used to add a more appetizing color to margarine, which is otherwise the color of Crisco.) It is transformed in the human body to *Vitamin A*, which becomes part of the complex structure rhodopsin in the retina of the eye. Vitamin A is one-half of the β-carotene molecule with an OH added. All the double bonds in these molecules are *trans*, and the phenomenon of vision depends critically on a *trans*- to *cis*-flip which occurs when light strikes the retina. It was long thought that the antioxidant properties of β-carotene inhibited the development of cancer; but large controlled studies of smokers given dietary supplements of β-carotene have been disappointing.[1] *Muscalure* is one of many insect pheromones (intraspecies chemical signaling agents) that have been discovered; it is the sex pheromone of the female common house fly, *Musca domestica.*

Alkene Reactivity

The carbon–carbon double bond usually reacts with other molecules to form the more stable carbon–carbon single bond. If hydrogen gas, H_2, and ethene, C_2H_4, are squeezed together (put under high pressure), they react to form ethane, C_2H_6. Much less pressure is needed if a platinum or nickel metal catalyst is used. In an older terminology that reflects this process, alkanes were called *saturated* and alkenes *unsaturated* hydrocarbons (the latter were not "saturated" with H_2). Most vegetable fats contain unsaturated carbon chains (e.g., olive oil and soybean oil have an 18-carbon chain with a double bond in the middle), while animal fats are saturated (beef tallow has the same 18-carbon chain without the double bond). The double bond is responsible for the fact that most vegetable fats are oils (liquids), while animal fats are waxy semi-solids. To make margarine from vegetable oils, the oils are partially hydrogenated to make them semi-solids (some problems arising from this process will be discussed in the following chapter). Here is a typical label on the side of a stick of margarine:

INGREDIENTS: Liquid soybean oil, partially hydrogenated soybean oil, water, salt, vegetable monodiglycerides, whey, vegetable lecithin, sodium benzoate as a preservative, artificially flavored, colored with beta carotene, vitamin A palmitate added.

NUTRITION INFORMATION PER SERVING:

Serving size (1 Tbsp) (14g)

Fat (100% of calories)	11g	Calories	100
Polyunsaturates	3g	Saturates	2g

Alkenes also can react with themselves to produce molecules with more stable single bonds. Although ethylene is a gas, at high temperatures it can be made to polymerize (combine with itself many times) to form *polyethylene,* the material used to make plastic bags, Rubbermaid housewares, television and computer cabinets, and most of the numberless plastic materials we use each day. The process of polymer formation can be visualized as shown in Figure A3.4, where one of the double bonds in each ethene

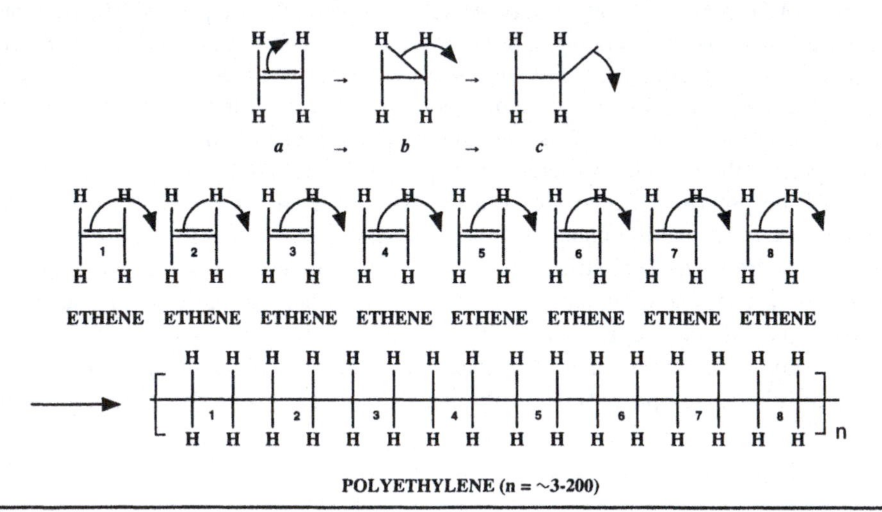

Figure A3.4 Formation of poly(ethylene).

molecule swings out from between the two carbons it had joined (as in ***a→b→c***) and becomes a new (single) bond joining two previously separate ethene molecules. The eight separate ethene molecules of Figure A3.4 thereby become part of one enormous polyethylene molecule. This process goes on until the supply of ethene is exhausted; in the final plastic, some polymer chains come from 50, some 3,500 ethene units, with all the possibilities in between; the polymer chains are intertwined and tangled like a bowl of spaghetti. The final plastic has no double bonds in it except a few residues at the unreacted end of the chains: it is a mixture of many straight-chain alkanes having about 800–1000 carbon atoms each. Because all bonds are saturated, it is quite stable and inert—and nonbiodegradable.

There are many other types of polymers or plastics that are based on the same process as that which forms polyethylene. For instance, *polypropylene* is polymerized propene: there are three forms, *isotactic* (Greek: same arrangement), *syndyotactic* (Greek: two-at-a-time arrangement), and *atactic* (Greek: no arrangement), depending on how the methyl groups are arranged relative to each other (Figure A3.5). The isotactic polymer forms fibers that are used for fishing gear, ropes, insulated underwear and camping equipment (Amerfil, Herculon, and Tuff-Lite). We will see later that PVC and Styrofoam have a similar structure.

Carbon–Carbon Triple Bonds: The Alkynes

Figure A3.6 displays three ball-and-stick models of acetylene viewed from different perspectives. As you can see, the acetylene molecule is linear: all four atoms lie in a straight line. Hydrocarbons that have a triple bond are named like the alkenes and cycloalkenes:

- The parent chain is chosen to be the longest chain that contains the triple bond (if there is a double bond as well as a triple bond, the double bond takes precedence).
- The *-ane* ending is changed to *-yne* (pronounced to rhyme with "fine").
- The parent chain is numbered from the end which is closest to the triple bond, and a number is used to indicate at which carbon the triple bond starts.

Alkynes are not at all as common as alkenes—although there is a large number of

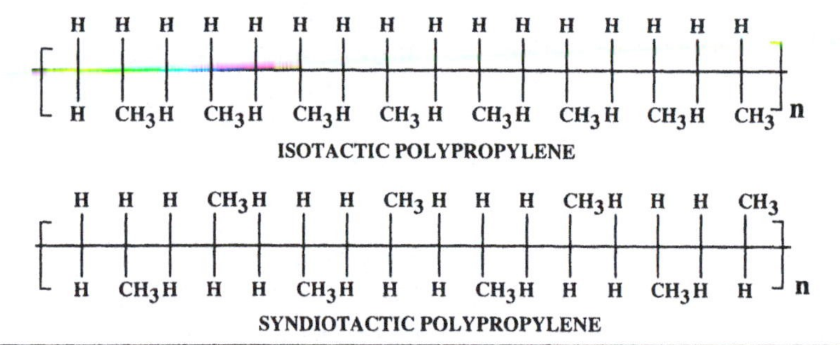

Figure A3.5 Poly(propene): syndiotactic and isotactic forms.

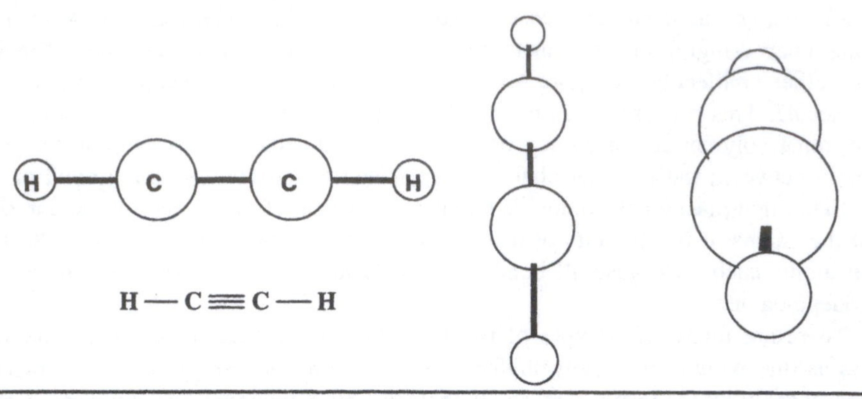

Figure A3.6 Three perspectives of acetylene (ethyne).

very unusual compounds containing triply bonded carbon atoms produced by flowers of the aster family (*Compositae*); an ethyne group also constitutes the critical modification of the natural female sex hormones that allows steroids to be taken orally for birth control (or male hormones for body-building).

Acetylene itself is an important industrial commodity: it is sold in tanks as a compressed gas for use in oxyacetylene welding and as fuel for motor boats. The high energy of the carbon–carbon alkyne bond contributes to the unusually intense heat of combustion of acetylene. Acetylene will explode and decompose into its elements if put under greater pressure than about 2 atmospheres. Commercial tanks of compressed acetylene contain enough of the liquid solvent *acetone,* in which acetylene is quite soluble, to maintain the pressure below the explosive point.

Acetylene can be polymerized in a manner analogous to ethylene: the product contains alternating double bonds (Figure A3.7). Because of the residual double bonds, polyacetylene conducts electricity and has either a silver (the form in which all the double bonds are *trans*) or copper (all *cis*) color; polyethylene, on the other hand, with no double bonds, is used to insulate electrical wires and is transparent. The all-*trans* polyacetylene finds a use in making plastic bags for wrapping computer chips: the chips are very sensitive to static electricity, and the conductive plastic allows any developing charge to be dissipated.

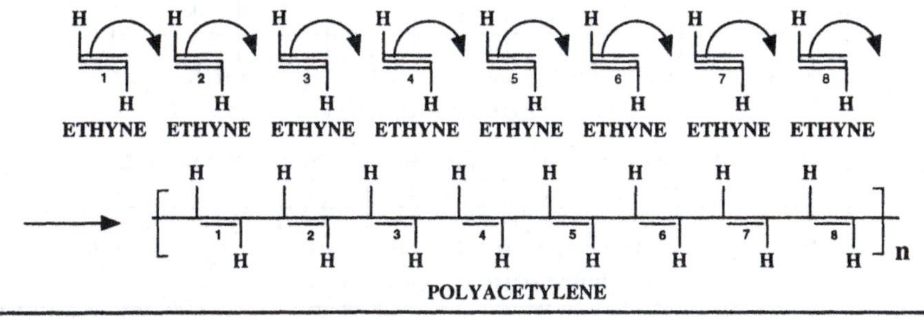

Figure A3.7 Polymerization of acetylene.

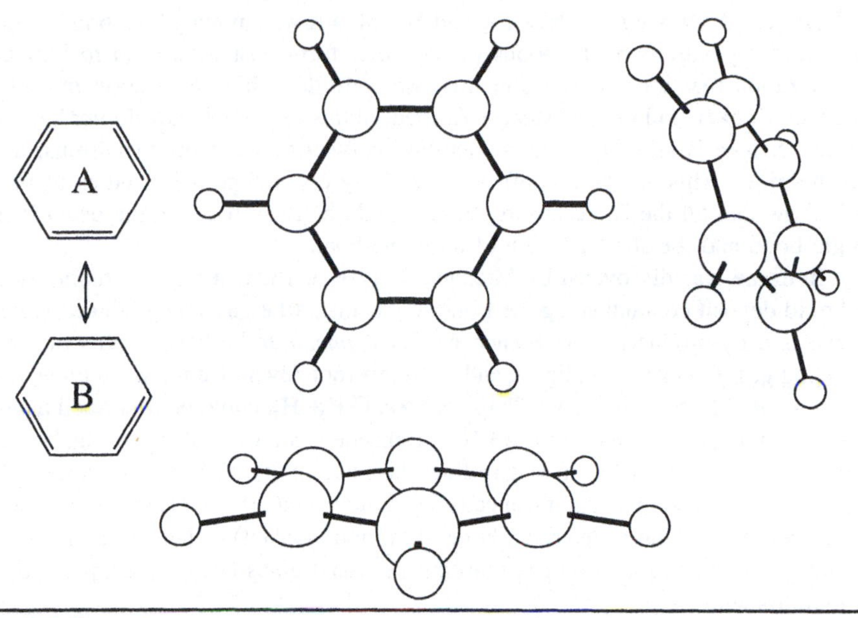

Figure A3.8 Benzene.

Triple Carbon–Carbon Double Bonds: The Aromatics

Figure A3.8 shows three conjugated double bonds introduced into a cyclohexane ring. The resulting structure, "1,3,5-cyclohexatriene," has some unusual properties. There is no Baeyer ring strain, because the internal angles of a planar hexagon have the same 120° angle of a trigonal carbon. Indeed, it turns out that this is an extraordinarily stable structure, and the paradigm of a whole class of important organic structures. This special ring is called *benzene:* in Figure A3.8 three perspectives of the benzene ring are drawn showing that it is a completely planar system. Benzene has other remarkable features; for instance, the two forms **A** and **B** connected by a double-headed arrow are actually identi-

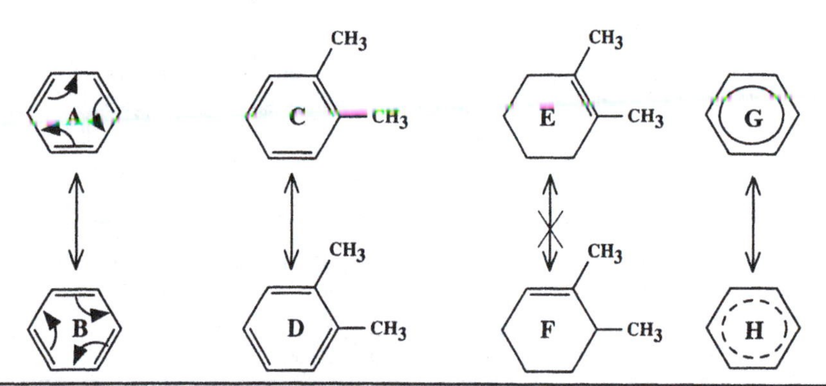

Figure A3.9 Resonance.

cal. In Figure A3.9, we have drawn **A** and **B** with arrows "moving" the bonds from one form into the positions of the bonds of the other form, just as we did to indicate the polymerization reactions above. But then we were describing a reaction in which old bonds were broken and new substances formed; in this case the bonds do *not* "move" but the true benzene is all of the time something halfway between the two forms, a "resonance hybrid." This has been confirmed by X-ray crystallographic studies of benzene, which show that all the bonds are the same length, halfway between the usual length of a single bond and the shorter length of a double bond.

Benzene was discovered by Michael Faraday in 1825: it was the main ingredient of a liquid deposit contaminating the London gas lines (the gas was produced from coal; benzene is a by-product). The French named it *phenique* because of this origin in illuminating gas (Greek to shed light), and from this root developed the word *phenyl*, which is the adjectival form of benzene. Thus *toluene*, $C_6H_5CH_3$, could be described as *methylbenzene* or *phenylmethane* (Figure A3.10). Benzene is a clear, colorless, highly flammable liquid: it is used in lead-free gasolines because of its high octane rating. It is a common industrial solvent, but it can cause *aplastic anemia* (bone marrow depression that can be irreversibly fatal or require a bone marrow transplant) or leukemia in susceptible individuals. Its use as a solvent in marking pens and glues has been supplanted by the less toxic toluene.

Styrene, or phenylethene, can be polymerized by heating to 200° forming *polystyrene*. The process is essentially the same one we have seen before for polyethylene. When polystyrene is mixed under pressure with a volatile solvent like Freon (Appendix IV), and the pressure suddenly released, the solvent turns to a gas: the resulting *styrofoam* is used for insulation, food packaging, and coffee cups. (However, the chlorofluorocarbon foaming agent contributes to deterioration of the ozone layer; more recent processes use carbon dioxide.)

The first indications of the phenomenon of resonance in benzene were discovered in the 19th century, when chemists tried to find two different 1,2-dimethylbenzenes, **C** and **D** (Figure A3.9). They expected to be able to find two different isomers because the two dimethylcyclohexenes, the 1,2-isomer, **E**, and the 2,3-isomer, **F**, were distinguishable

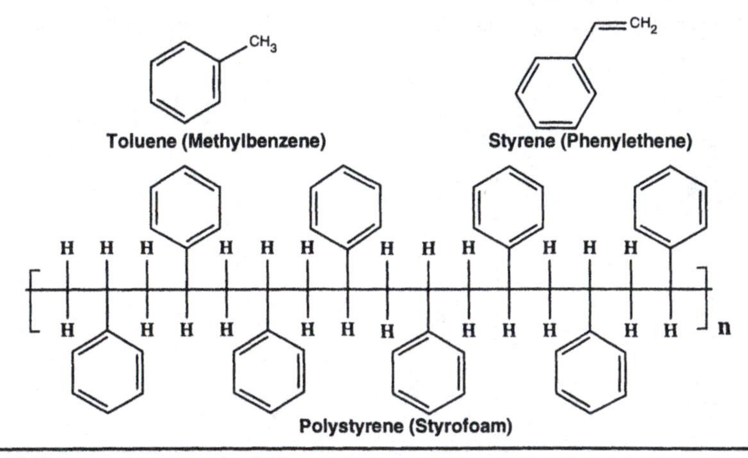

Figure A3.10 Toluene, styrene, and polystyrene.

isomers with different physical properties—for instance, different boiling points. (Let the obsessive-compulsive reader note that there are exactly two and only two further 1,2-dimethylcyclohexene isomers which we have spared the other readers by leaving unmentioned.) But there was then, is now, and presumably ever shall be one and only one 1,2-dimethylbenzene (of course there are also the 1,3- and 1,4- isomers). To emphasize the unique character of the double bonds in benzene, some chemists like to draw them in a special way: as **G** or **H**.

For many years, chemists tried in vain to synthesize *1,3-cyclobutadiene* (Figure A3.11). Neither this compound nor *1,3,5,7-cyclooctatetraene* are aromatic; instead of being stabilized by their alternating double bonds, they are *destabilized*: the systems are *antiaromatic.*

An analogy: blowing a stream of air over a beer bottle will sound a note of only those frequencies that are harmonics of the size of the bottle. In a roughly similar way, the quantum-mechanical wave properties of the electron pairs constituting the double bonds in these cyclic systems will be destabilizing if the number of double bonds is even and stabilizing if it is odd. In *pyrrole*, the unshared electron pair on nitrogen provides a pair of electrons equivalent to a third double bond.

When two benzene rings have two carbon atoms and the bond joining them shared in common, the rings are said to be *fused.* As more and more benzene rings are fused together the structures are referred to as *polycyclic aromatic hydrocarbons* (PAHs). Coal and oil contain very complex PAHs, and graphite, one of the allotropes of elemental carbon, is essentially an endless sheet of fused benzene rings, like hexagonal floor tiles. Aromatic systems with an element other than carbon in the ring are called *heterocyclic.*

Naphthalene (pronounced to rhyme with "LAUGH the lean"; cf. *diphthong* and *ophthalmology*) is one of several chemicals formerly used to repel moths from woolen clothing ("moth balls"); it was also taken orally to expel intestinal worms, that is as an *anthelmintic* (pronounced "Aunt-hell-MIN-tick"—from Latin *anti* + the Greek *helmintos* for worm). Both uses should have been abandoned because of naphthalene's toxicity; however, as late as 1991, incidents continue to be reported of children developing hemo-

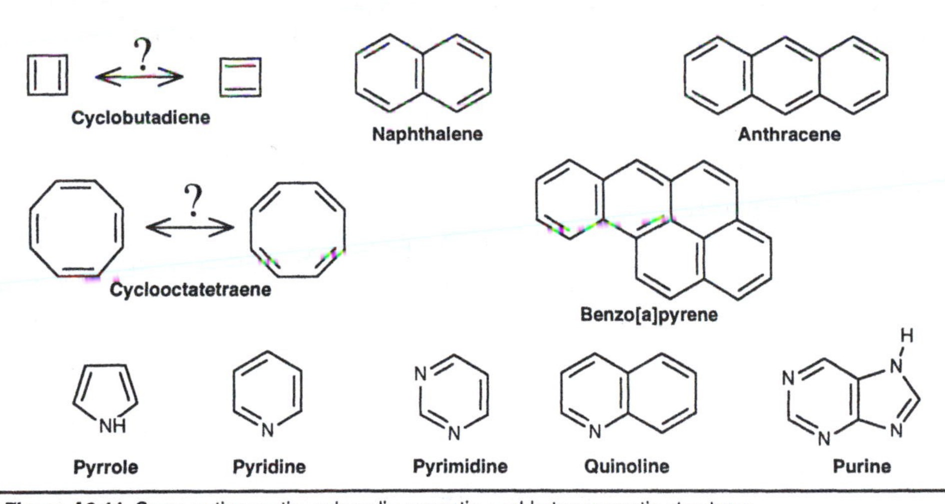

Figure A3.11 Some antiaromatic, polycyclic aromatic, and heteroaromatic structures.

lytic anemia from eating older moth ball formulations. *Benzo[a]pyrene*, a PAH found in the tar from cigarette smoke (see Chapter 4), is one of the most powerful carcinogens known. *Quinoline* forms part of the molecular structure of *quinine*, a complex antimalarial alkaloid used to flavor tonic water (see Appendix IV, Figure A4.11); *pyrimidine* and *purine* form the ring systems of the nucleic acid base pairs in DNA and RNA.

Carbon–Oxygen Double Bonds: Aldehydes

Carbon can form a double bond not only with a second carbon atom, but with an oxygen atom; this C=O group is called a *carbonyl*. Following the HONC rule, the double bond to carbon leaves oxygen with no further possibility of bonding to any other atom; the carbon atom, however, can form two further bonds, and depending on what these bonds are, the compound is either an aldehyde, a ketone, an acid, an ester, or an amide. If one of the bonds from the carbonyl group is to a hydrogen atom, the molecule is an *aldehyde* (the word derives from *al*[cohol] *dehyd*[rogen]: if a hydrogen atom is taken from both the OH and the CH of an alcohol like methanol, the product is formaldehyde). Figure A3.12 shows two perspectives of the simplest possible aldehyde, *formaldehyde*, in which has the carbonyl is bonded to two hydrogens. Notice that the molecule is flat, and that the three bonds from the doubly bonded carbon extend to the corners of an equilateral triangle: carbon is *trigonal* in forming the carbonyl group, just as it is in the alkenes. The IUPAC rules for naming aldehydes direct that the ending of the longest chain containing the CHO group should be modified: the final *-ane* is changed to *-anal* (only the *-an* is retentive). No number is needed to indicate the position of the carbonyl group: aldehydes are by definition at the terminal carbon of the chain. By these rules, formaldehyde should be called *methanal* and acetaldehyde *ethanal*, but almost no one does this. The ordinary names for aldehydes derive from the names of the corresponding carboxylic acids, which we will see in Appendix IV. This naming system developed because aldehydes are easily converted to carboxylic acids, often just by the oxygen of the air.

Formaldehyde is a gas with a somewhat evil reputation: it is very soluble in water (the solution is called *formalin*), and this solution is used to preserve biological speci-

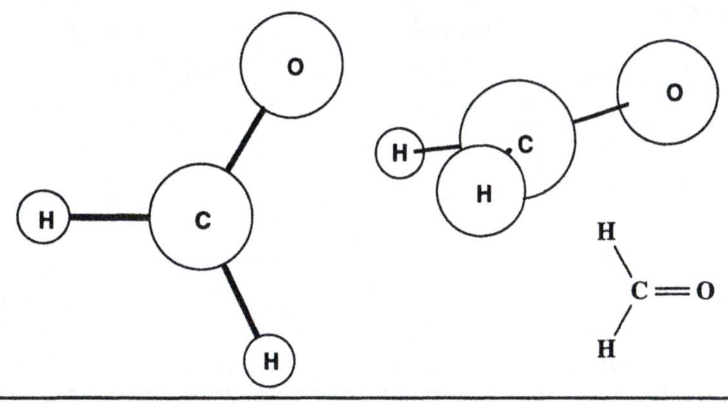

Figure A3.12 Two perspectives of formaldehyde.

mens—and to embalm the human body. Significant quantities of formaldehyde have been found to be given off by new carpets, contributing to the "sick building syndrome." *Acetaldehyde* (Figure A3.13) is the next simplest aldehyde: it is a volatile liquid that readily polymerizes to a trimer, *paraldehyde*, which is an addictive hypnotic. *Benzaldehyde* is found in the ripe seed of the bitter almond, *Prunus amygdalus*, and is responsible for much of the aroma and flavor of almonds; complexed with hydrogen cyanide (the gas of Hitler's gas chambers) in the glycoside *amygdalin (Laetrile)* it was once touted as a cure for cancer (and death often is). *Perillaldehyde* is found in mandarin peel oil from *Citrus reticulata.* Perillaldehyde forms an *oxime* that is 2000 times as sweet as table sugar (sucrose) and is used as a sweetening agent in Japan ("perilla sugar," perillartine). Note the carbon–nitrogen double bond; there is a carbon–nitrogen triple bond in hydrogen cyanide, $HC \equiv N$. *Vanillin* occurs naturally in the seed pods of the vanilla plant (and in potato parings), but it is cheaply synthesized from the waste lignin of the wood pulp industry. There is about 1 g of vanillin in 400 g of vanilla pods.

Why does the sensitive palate find something wanting in the flavor of vanillin compared to the flavor of vanilla? Is it because the former is synthetic, the latter natural? No, it is because there is, indeed, something wanting: the other 398 g of vanilla, which contains thousands of other more subtle contrasting and harmonizing flavors from thousands of other chemicals. The chemical vanillin can be extracted from vanilla pods and purified thoroughly so that it is chemically indistinguishable from the synthetic substance: it also tastes just as vapid. The same factors are at play when good cooks prefer sea salt (a mixture of a dozen or more salts) to ordinary salt (which is >99.9% pure NaCl); or real coffee to the decaffeinated (decaffeination deflavors too). The phenomenon of taste (really smell) is amazingly sensitive and complex: for example, often the aroma of A + B is quite different from either component alone—the whole is greater than the sum of its parts.

Another item: are vanilla pods poisonous? Like everything, even water (yes, if you drink too much water you can suffer electrolyte depletion, heart failure, and death), it is a matter of dosage. There is 1 g of vanillin in 400 g of vanilla; the LD_{50} of vanillin

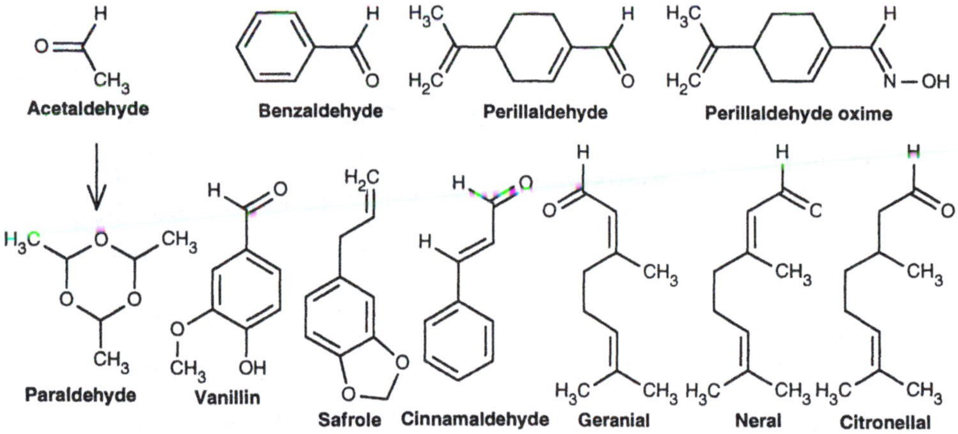

Figure A3.13 Some common aldehydes.

administered orally to rats (the dosage needed to cause 50% of the population to die; as for the other 50%, the record is silent) is 1.580 g/kg (1.580 g vanillin for each kg of rat body weight). Assuming that a population of 154-lb (70-kg) humans is essentially the same as a population of giant rats, feeding each person 1.580 x 70 = 110.6 g (0.24 lb) of vanillin, or 400 x 0.24 = 97 lb of vanilla pods, will kill half of them. Since genuine vanilla ice cream conservatively contains no more than 10% vanilla you have a 50% chance of survival if you limit your consumption of vanilla ice cream to no more than 1000 lb daily. It might seem better to stick to chocolate, but chocolate contains caffeine, which causes savage self-mutilative behavior in laboratory animals (see Chapter 4).

Still again: is sassafras tea good for your health? According to the *Peterson Field Guide to Medicinal Plants: Eastern & Central N. America* (Houghton Miflin: Boston, 1990, p 278) it is "a folk remedy for stomach aches, gout, arthritis, high blood pressure, rheumatism, kidney ailments, colds, fevers, skin eruptions." Yet the oil of sassafras is about 75% *safrole*, which is listed by the EPA as a carcinogen. The Field Guide notes "Safrole . . . reportedly is carcinogenic. Banned by FDA. Yet the safrole in a 12-ounce can of old-fashioned root beer is not as carcinogenic as the alcohol (ethanol) in a can of beer."

Cinnamaldehyde, as you would expect, was first isolated from Ceylonese and Chinese cinnamon oils. It is a colorless oil with a strong odor of Big Red or Dentine chewing gum. *Geranial*, *neral*, and *citronellal* are three of about a dozen structurally similar compounds that occur in geranium, rose, citronella, and lemon oils and contribute to their aromas. They are used in perfumes and soaps; citronellal is also used as an insect repellant.

Carbon–Oxygen Double Bonds: Ketones

When the carbonyl group occurs anywhere but at the end of a carbon chain, that is when both bonds to the carbonyl group are attached to carbon atoms, the structure is said to be a *ketone*. The simplest possible ketone is *acetone* (Figure A3.14), which according to IUPAC rules should be called *propanone* (the suffix of the longest chain containing the

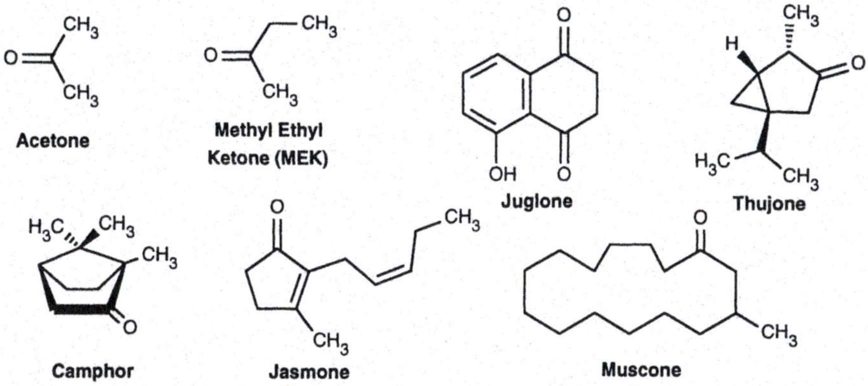

Acetone; Methyl Ethyl Ketone (MEK); Juglone; Thujone; Camphor; Jasmone; Muscone

Figure A3.14 Some common ketones.

carbonyl group is changed from *-ane* to *-anone*; if necessary, a number precedes the name to show the position of the carbon bond). It is inefficient to use two names for the same functional group, but the early experimenters were very impressed by the much greater reactivity of aldehydes, which unlike ketones are easily oxidized to carboxylic acids.

Acetone is a common industrial and household solvent; it is found in many nail-polish removers. Hardware stores sell gallon cans of MEK (methyl ethyl ketone, butanone), which is used as a paint remover and smells like shlivovitz, a Central European liqueur distilled from fermented plums (шливо, plum), which tastes like high-octane lighter fluid. *Juglone* is an interesting example of an allelopath (a chemical given off by one plant species which affects another). It is found in black walnut (*Juglones nigra*) shells and in pecans; it inhibits the growth of other plants (e.g., tomatoes and raspberries will not grow near walnut trees). *Thujone* is the chief constituent of the oil derived from the dried leaves and flowering tops of *Artemisia absinthium* (wormwood), used to make absinthe, a now-outlawed addictive liqueur that probably contributed significantly to van Gogh's seizures and suicide (see Chapter 7). Small amounts of wormwood are still used to make vermouth (vermouth and wormwood derive from the same Old High German root). The same painter may have compounded his problems by attempts to medicate himself with massive doses of *camphor*, a fragrant solid originally obtained by steam distillation of 50-year-old *Cinnamomum camphora* trees; most of the camphor used in the United States is now synthesized from pinene, which is obtained from the rosin of pine trees. Ingestion or injection of significant quantities of camphor causes nausea, convulsions, or death: it is also used as a plasticizer in the manufacture of celluloid, lacquers, and varnishes; as a moth repellant; in embalming fluid; in ointments for topical application; and as an additive to paregoric to prevent its use by opium addicts. *Jasmone* is the chief fragrant principle in jasmine flowers: the natural substance is the *cis* ketone, while the synthetic variety is usually *trans*: both have similar odors. The jasmone molecule is also incorporated into a larger structure, jasmolin, one of the insecticidal pyrethrins produced by flowers of *Chrysanthemum (Pyrethrum) cinerariaefolium* and used in domestic bug sprays. *Muscone* is one of the odorous principles of musk, originally obtained from the musk glands of the male musk deer, *Moschus moschiferus*, of the civet cat, *Viverra civetta*, or of Louisiana muskrats. Animal lovers will be glad to know that much of it is now synthesized.

References and Notes

1. A Finnish trial showed that beta carotene supplementation was associated with an *increase* in lung cancer among smokers; two later randomized trials showed either no effect or confirmed a negative one: (a) Hennekens, C. H.; et al., "Lack of effect of long-term supplementation with beta carotene on the incidence of malignant neoplasms and cardiovascular disease," *N. Engl. J. Med.,* **1996**, *334*, 1145-1149. (b) Omenn, G. S.; et al., "Effects of a combination of beta carotene and vitamin A on lung cancer and cardiovasular disease," *N. Engl. J. Med.,* **1996**, *334*, 1150-1155.

Appendix IV
Through the Looking-Glass

The Halogens

Of the 100 or so elements making up our universe, only a few other than H, O, N, and C make up the structures of most drugs. The most important are the family of four elements called the halogens: fluorine, F; chlorine, Cl; bromine, Br; and iodine, I. The word halogen comes from the Greek and means salt-producer; the most common halogen is chlorine, which is found in enormous quantities in the ocean as sodium chloride, NaCl, table salt.

Like hydrogen, oxygen, and nitrogen, the halogens exist in their elemental state as diatomic molecules: F_2, Cl_2, Br_2, I_2. There is a progression of properties corresponding to the increase in atomic weight from fluorine to iodine: fluorine is a very pale green gas, chlorine a yellow gas, bromine a red-brown liquid, and iodine a purple-black solid. Fluorine is the most reactive element known; consequently, it is very dangerous and rarely used as the free element. Chlorine is an industrial chemical made from the electrical decomposition of salt water and shipped as a liquefied gas. It is used to purify water for city water supplies and swimming pools. Elemental bromine is probably only found in chemistry labs; but a solution of I_2 in alcohol, known as tincture of iodine, has long been familiar as a topical antiseptic.

The Alkyl Halides (Haloalkanes)

Halogens HONC like Hydrogen; that is, they all form one and only one bond. Almost any hydrogen atom of an organic molecule can be replaced by a halogen atom to make a new substance, usually with considerably different properties. If one or more of the hydrogen atoms of an alkane is replaced by a halogen, the resulting compound is called an *alkyl halide*. Thus CH_3X is called *methyl fluoride, methyl chloride, methyl bromide, methyl iodide*, where X is the respective halogen. These compounds are also named by prefixing *fluoro-, chloro-, bromo-, iodo-* to the name of the parent compound, with the expected numerical prefixes. Some of this nomenclature is illustrated in Figure A4.1.

Chloroform has been known since the 19th century, when it was used for general anesthesia. A volatile, dense, colorless liquid with a sweetish aroma, it rapidly produces

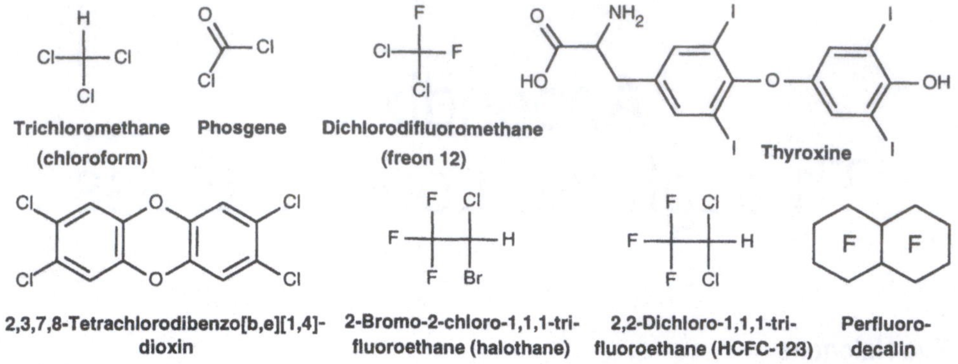

Trichloromethane (chloroform) | Phosgene | Dichlorodifluoromethane (freon 12) | Thyroxine

2,3,7,8-Tetrachlorodibenzo[b,e][1,4]-dioxin | 2-Bromo-2-chloro-1,1,1-trifluoroethane (halothane) | 2,2-Dichloro-1,1,1-trifluoroethane (HCFC-123) | Perfluorodecalin

Figure A4.1 Some alkyl halides.

unconsciousness when its vapors are inhaled. A chloroform-soaked sponge is the classic device used by kidnappers and muggers to knock out their victims; in Victorian England, the upper gentry were fond of chloroform parties, where guests would sit around a large punch bowl filled with chloroform and sniff themselves into giddiness or stupor. Unfortunately, it easily causes a *cris de foie*—damage to the liver that is sometimes fatal.

Phosgene is one of the products formed in small amounts when chloroform is exposed to light and oxygen in the air (Gk light-generated). Some of the toxicity of chloroform may be due to traces of phosgene or conversion of chloroform to phosgene in the liver: phosgene is very toxic and was used in WWI as a war gas.

Freon 12 has been the most common refrigerant in refrigerators and freezers, and in home and car air conditioners until recently. It is no longer used in aerosol sprays as a propellant, being replaced by carbon dioxide or butane, and it is being phased out of use altogether by international agreement. It is virtually indestructible and contributes to the deterioration of the ozone layer. There are other freons with similar uses (freon 14 is tetrafluoromethane; freon 114 is 1,2-dichloro-1,1,2,2-tetrafluoroethane); most freons were originally patented by DuPont. All of these compounds are collectively referred to as *chlorofluorocarbons*, or CFCs, and various new, more biodegradable (and more expensive) substitutes have been developed by DuPont and other companies to replace them.

Halothane is a modern inhalation anesthetic that lacks the toxicity of chloroform but possesses its advantage of being nonflammable (many anesthetics such as cyclopropane and ether are very flammable and form explosive mixtures with air easily ignited by a static electric spark). Halothane sometimes causes a rare but serious form of hepatitis, probably via an oxygenated metabolite, F_3CCOCl, that binds irreversibly to protein.[1]

HCFC-123 is one candidate to replace CFCs such as the freons. It is a *hydrohalocarbon*; the residual unhalogenated hydrogens make these compounds susceptible to oxidation in the lower atmosphere, where they react before reaching the ozone layer. Unfortunately, this makes them more biologically active than the inert CFCs. HCFC-123 showed no significant toxicity in short-term testing (its structure is based on that of halothane); however, on 2-year lifetime inhalation by male rats, benign tumors developed in their pancreas and testes. Paradoxically, the animals exposed to HCFC-123 did not die from the tumors; in fact, those receiving higher doses lived longer than controls.[2]

Thyroxine is an amino acid hormone of the thyroid gland that has a stimulating effect on metabolism. *Grave's disease*, so prevalent amongst the first family during the Bush presidency, is caused by excessive production of this hormone by the thyroid. Ingested iodine (as sodium iodide or iodate) is used by the body to form thyroxine in the thyroid: if the iodine is made radioactive, the cells of the thyroid are preferentially killed and an even presidentially overactive thyroid can be thereby stabilized. Most regions of the world contain enough iodine in the soil and water for proper thyroid functioning. If there is insufficient iodine in the diet (as in the Midwestern United States), it must be added ("iodized" salt is NaCl with about 0.1% added NaI) to prevent *goiter*, which is the overgrowth of the thyroid attempting to compensate for iodine deficiency. Lack of dietary iodine can cause other problems. In China, the Public Health Ministry estimates that more than 10 million cases of mental retardation in almost every province of the country are a result of iodine shortages during brain development; it is likely that tens of millions more suffer from lowered intelligence. "With the lack of iodine, the brain just does not wire correctly in early development," says Glen Maberly, chairman of Emeroy University's department of international health.[3] Only the L form of thyroxine chiral amino acid (see below) is active physiologically as a thyroid hormone; however, the D form has been used as an antihyperlipoproteinemic (cholesterol-lowering) drug.

Dioxin is not itself a commercial product but a by-product of the industrial preparation of several other chemicals (e.g., the fungicide 2,4,5-trichlorohydroxybenzene or the herbicide *Agent Orange*); it is found in trace amounts in paper that has been bleached with chlorine (hence it contaminates water runoff from paper pulp mills and can be detected in probably harmlessly minute quantities in coffee filters). It is one of the most potent carcinogens known in some animal studies (particularly rodents); however, recent evidence indicates it may not be as severe a carcinogen in humans as was formerly thought.[4]

Perfluorodecalin is a structure consisting of two fused cylcohexane rings in which *all* the hydrogen atoms have been replaced by fluorine atoms (this is the meaning of the F symbol inside each ring). When all hydrogens of a hydrocarbon are replaced by fluorine, the resulting compound is usually very unreactive. In this case the colorless liquid that results dissolves oxygen gas to a remarkable degree: a mouse can be placed at the bottom of a tank of perfluorodecalin and will continue to live, breathing the liquid as though it were air. The substance is being tested for use as an emergency blood substitute: animals whose entire bloodstream has been replaced with perfluorodecalin seem to show no ill effects. Unlike transfused blood, perfluorodecalin needs no refrigeration, has an indefinite shelf life, and cannot transmit hepatitis, HIV, or other pathogens.[5]

Polymerized Haloalkenes

In 1928, B. F. Goodrich research chemist Waldo Lonsbury Semon concocted poly(chloroethene), as it is most properly designated, a substance that would profoundly alter

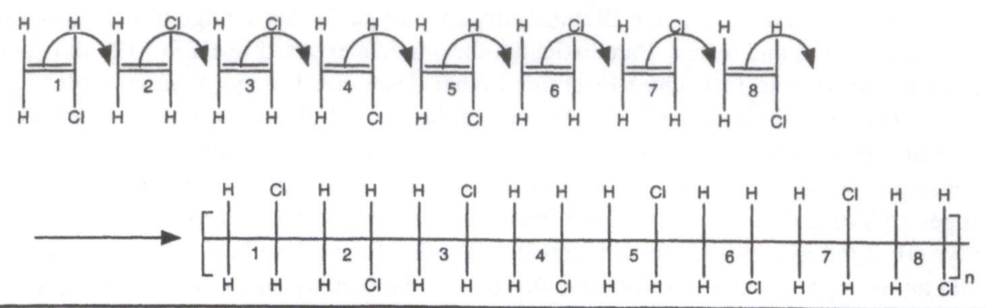

Figure A4.2 Formation of polyvinyl chloride (PVC).

the texture and smell of everyday life. The polymer is more commonly called polyvinyl chloride, or PVC (Figure A4.2), using an older system of nomenclature. Most people just call it "vinyl." It has been used to make phonograph records, go-go boots, raincoats, garden hoses, floor tiles, dresses, and umbrellas. It is what gives new cars their new-car smell. It is the second most widely employed plastic (after polyethylene) and constitutes a $20-billion-a-year industry, with 44 billion pounds produced each year.[6] PVC is also used for electrical wire coverings (it is nonflammable because of the chlorine), and for shoe soles. Pipes made of PVC are cheap and more easily installed than lead-soldered copper; many cities allow PVC piping only for drainage and sewage because small amounts of plasticizers and chloroalkanes may leach into drinking water if it is used for pressurized piping. The trade-off is the lead that leaches from older pipes and solder. A similar halogenated polymer is made from 1,1-dichloroethene (vinylidene chloride); the copolymer of this substance and vinyl chloride is marketed as Saran wrap. Still a third is poly(tetrafluoroethene) or Teflon (Figure A4.3). Teflon is familiar as the nonstick coating for pots and pans. You may have wondered: if nothing sticks to Teflon, how is it fastened to the metal pan? This is a very good question.

Chiral Molecules, Optical Isomers

Most of the chemical structures produced by the living world are *chiral*. This word comes from the Greek word for *hand*, and refers to the property whereby the right hand

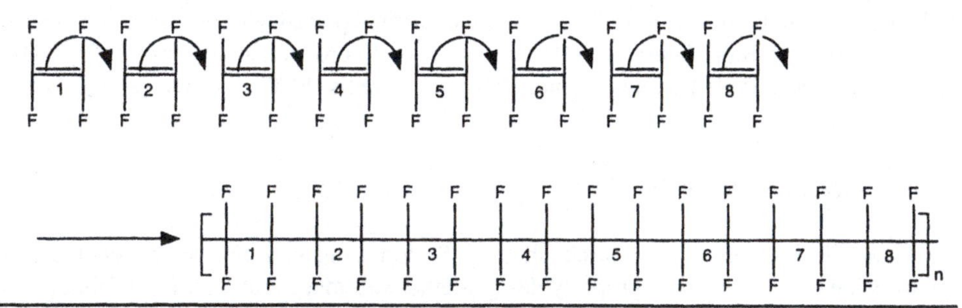

Figure A4.3 Formation of poly(tetrafluoroethene) (Teflon).

is the mirror image of the left. If you hold your hands together with the fingers just touching, each hand can be imagined as the mirror reflection of the other; if you place your right hand against a mirror, the reflection in the mirror is that of a left hand.

Many of the objects around us are nonchiral; that is, they are the same as their mirror image. Drinking glasses are nonchiral; so are cups and saucers. Telephones and scissors are chiral. A car made for the United States is chiral; it is the mirror image of one made for use in Great Britain.

There are several ways in which an organic structure can be chiral, but by far the most common is if there are one or more carbon atoms in the molecule that have *four different* groups attached (thus doubly and triply bonded carbon atoms are always nonchiral). In Figure A4.4 is an imaginary molecule, a carbon atom with four distinct substituents: ***A***, ***B***, ***C***, and ***D***. The broken line represents a mirror. By convention, the bold bond going from the carbon atom to substituent ***C*** represents a bond going forward out of the plane of the paper toward the reader, while the hatched line from the carbon atom to substituent ***D*** represents a bond going backward behind the plane of the paper away from the reader; the bonds to substituents ***A*** and ***B*** are in the plane of the paper. Imagine that *you* are the molecule on the right side of Figure A4.4 and that ***D*** is your right hand, ***C*** is your left hand, and ***A*** is your head. If you get up and, like Alice in Wonderland, walk around and through the mirror, and turn around and face where you came from, you will discover that you are not the same as your original reflection (the molecule on the left of Figure A4.4). Your reflection has a right hand labeled ***C*** and a left hand labeled ***D***. If, on the other hand (?), both your right and left hands had been ***C***, when you entered the looking-glass world, you would have found that you were identical with your mirror-image. A carbon atom with two or more identical substituents is superimposable on (identical with) its mirror image; one with four distinct substituents is not. Any molecule not superimposable upon its mirror image is said to be chiral.

Thus, of all the alkyl halides in Figure A4.1, only halothane and thyroxine contain chiral carbon atoms; each has exactly one. Some examples of molecules containing chiral carbons are drawn in Figure A4.5; carbons that are chiral have been boxed. Carbon 4 in 1,1-dichloro-4-methylhexane is chiral because there are four different groups attached to this carbon (remember that there is an implicit H atom); carbon 1 is not chiral because there are two identical Cl substituents; carbons 2, 3, 5, and 6 each have two identical H substituents. The carbon of BCFM is obviously chiral; in thyroxine, all carbons are

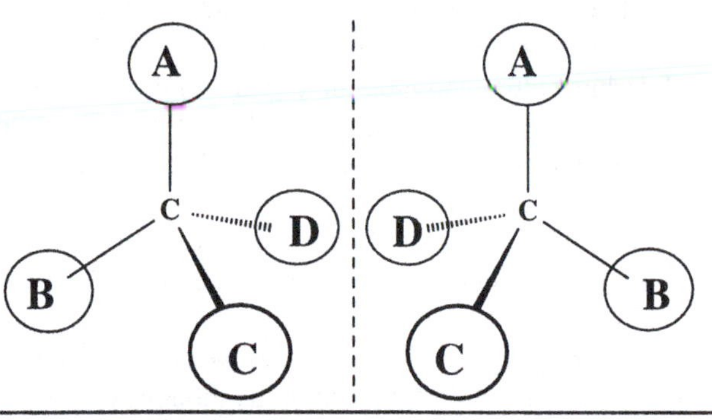

Figure A4.4 A chiral carbon and its mirror image.

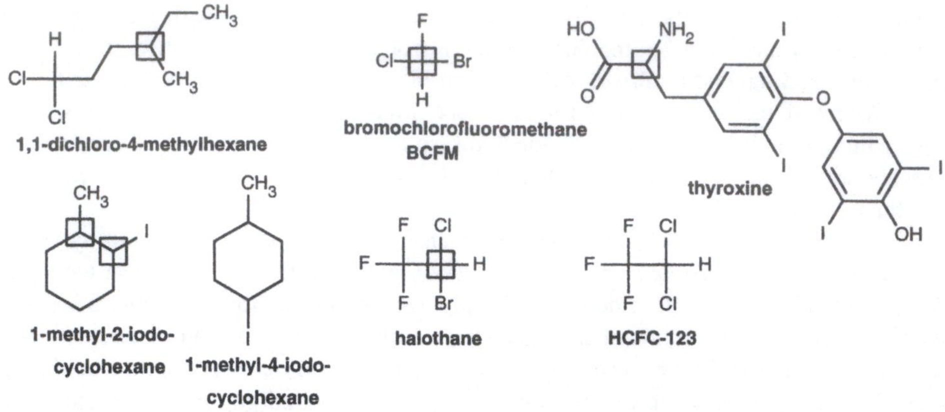

1,1-dichloro-4-methylhexane

bromochlorofluoromethane BCFM

thyroxine

1-methyl-2-iodo-cyclohexane

1-methyl-4-iodo-cyclohexane

halothane

HCFC-123

Figure A4.5 Squared carbons are chiral.

double bonded (hence planar and nonchiral) except for the boxed one, which is chiral, and the adjacent carbon, which is not because it has two H substituents. The 1,2- and 1,4- isomers of methyliodocyclohexane are more difficult: the reason the 1,4- isomer is not chiral is that it has a plane of symmetry: a mirror plane perpendicular to the paper and intersecting the methyl and iodo groups would reflect the left half of the molecule onto the right. This is called a mirror plane of symmetry and any molecule having such a mirror plane cannot be chiral. Another way of seeing this is to realize that either the carbon of the cyclohexane ring to which the methyl group is substituted or that to which the iodo group is substituted has two identical branches: the set of atoms encountered going around the ring from these carbons clockwise is the same as that encountered going counterclockwise.

Naming Chiral Molecules

The two molecules related as nonsuperimposable mirror images of each other are called *enantiomers*; each is said to be the enantiomer of the other. In Figure A4.6, enantiomeric pairs of halothane and BCFM are drawn as three-dimensional projections reflecting their mirror images. In the top left enantiomer of halothane, the hydrogen atom is shown projecting back behind the plane of the page; now imagine that you were to grasp this molecule by the hydrogen atom and look directly down on the carbon–hydrogen axis, so that the carbon atom eclipses the hydrogen atom below it. You would see the molecule from the perspective shown below the arrow, the bottom left structure of halothane marked ***S***. The same process carried out on the mirror image of this molecule, the top right enantiomer of halothane, results in the perspective below the arrow marked ***R***. Notice the direction of the sequence Br→Cl→CF_3 is counterclockwise in ***S*** whereas it is clockwise in ***R***.

With a bit more imaginative effort, try to think of grasping the top left enantiomer of BCFM in Figure A4.6 by the hydrogen atom as though it were a bouquet of flowers,

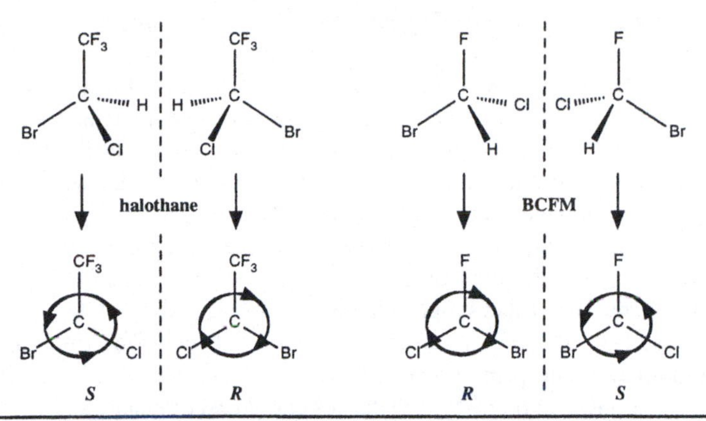

Figure A4.6 Three-dimensional projections of two chiral molecules, halothane and BCFM.

the C–H bond being the stems of the flowers. Now you want to smell the flowers, so you put your nose right over the C–H axis, C side up, and you see the perspective displayed in the bottom left structure of BCFM, marked ***R***, where the sequence of Br→Cl→F is clockwise. The same process repeated for the top right enantiomer of BCFM results in the bottom right perspective, marked ***S***.

If you have been able to follow this bit of imaginary geometric legerdemain (note the chiral chauvinism implicit in this word!), you understand the essentials of what is called the Kahn-Ingold-Prelog nomenclature for chiral carbons. Somewhat simplified, the rules are:

- Prioritize the four different atoms attached to the chiral carbon according to their atomic size (the bigger the better). The elements we have had so far thus prioritized are I>Br>Cl>S>F>O>N>C>H; hence Br>Cl>CF_3 is the correct priority for halothane (the C of CF_3 is all that is needed to establish priority; if it were CI_3 it would still fall below Br or Cl), and Br>Cl>F is the correct priority for BCFM.
- If there is a tie, resolve it by going to the next atom down the chain, then the next one, until the tie is resolved.
- A carbon doubly bonded to an oxygen counts as a carbon attached to two oxygens; a carbon triply bonded to a nitrogen counts as a carbon attached to three nitrogens, and so forth.
- View the chiral carbon by grasping it like a bouquet of flowers with your hand around the atom of *lowest* priority, and note whether the direction of top>middle>bottom priority of the remaining three atoms is clockwise or counterclockwise.
- If it is clockwise, the atom has ***R*** chirality; if counterclockwise it has ***S***.

R stands for Latin *rectus*, or right, and means clockwise; *S* stands for the Latin *sinistrus*, or left, and means counterclockwise. (Again, we ask any left-handers, whose life expectancy is already shorter than the rest of us, to excuse the chiral chauvinism implicit in all these terms.)

Physical and Pharmacological Differences Between Enantiomers

Is all this more than an exercise in geometry? Yes. There are a few physical differences between a bottle of (*R*)-halothane and a bottle of (*S*)-halothane; one enantiomer will rotate plane polarized light a given number of degrees clockwise, while the other will rotate plane polarized light the same number of degrees counterclockwise. What is plane polarized light? If you have ever used polaroid/ sunglasses, you have seen it: light going through these glasses is limited to vibrations in one plane; the plane is altered by some number of degrees when the light passes through an enantiomerically pure chiral substance. If an enantiomer rotates light clockwise, this is indicated by prefixing (+); if counterclockwise (–). There is no direct correlation between *R*/*S* structure and (+)/(–): some *R* enantiomers rotate plane polarized light clockwise, some counterclockwise. In almost every other physical property, substances composed of enantiomeric molecules are indistinguishable from each other.

More important are the pharmacological and physiological differences between the otherwise very similar chemical substances that differ only in being enantiomers. Many drugs interact with the body by fitting into its natural *receptors* (discussed more extensively in Chapter 1). These receptors are nearly always chiral: the chiral receptor is like a glove into which the chiral drug must fit, and as we all know, left hands do not fit well into right gloves. Because it is usually difficult to synthesize exclusively one of a pair of enantiomers or to separate them from each other, many prescription drugs consist of a 50/50 mixture of *R* and *S* forms. Such a mixture is called *racemic.*

The FDA has recently begun to require more detailed studies, when possible (some substances are difficult to impossible to separate into their constituent enantiomeric pairs), of the activity of each member of the chiral pair constituting a drug. Shown in Figure A4.7 are some examples of drugs, the *R* and *S* isomers of which have varying activity.

All the pharmacological activity may reside on one isomer. An example is *methyldopa*, where all antihypertensive activity is found in the *S* isomer. The activity of ibuprofen lies also in the *S* isomer; however, it is marketed as the racemate because the *R*

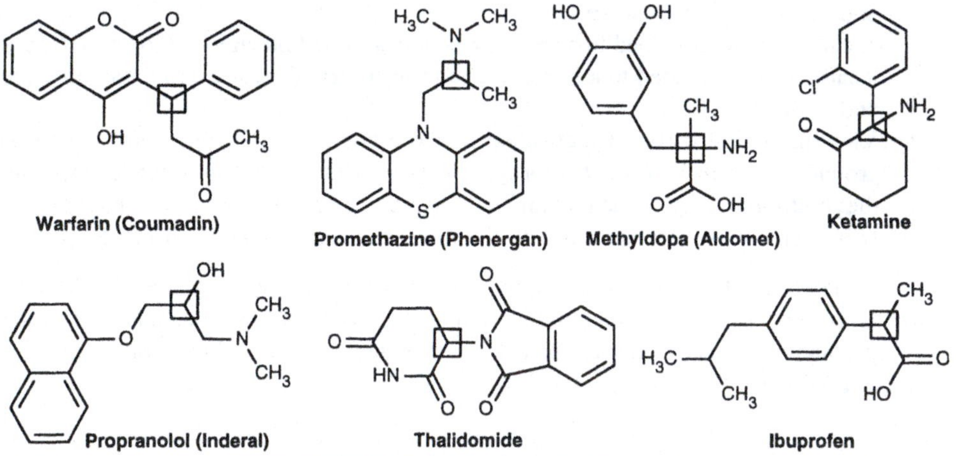

Warfarin (Coumadin) | Promethazine (Phenergan) | Methyldopa (Aldomet) | Ketamine

Propranolol (Inderal) | Thalidomide | Ibuprofen

Figure A4.7 Some drug structures containing chiral carbon atoms.

isomer is converted in the body to the active *S* form within about 30 minutes of ingestion. On a mg/mg basis, the pure *S*-ibuprofen would act more quickly than the racemate, and *dexibuprofen*, the pure *S*-(+)-isomer, has been marketed since 1994 in Austria as Seractil because of its improved efficacy at lower dosage in the treatment of rheumatoid arthritis.[7]

The isomers may have nearly identical pharmacological activity. An example is the antihistamine *promethazine*, both enantiomers of which appear equipotent and have similar side effects and toxicity. (Notice that promethazine contains an element we have not seen yet: sulfur, S, which is in the same family as oxygen and like oxygen forms two bonds.)

The isomers have similar activity but one is more potent. Most drugs used as racemic mixtures fall into this category. Examples are *warfarin* (an anticlotting agent), *thyroxine*, and *propranolol* (a widely used antihypertensive β-blocker sold under the trade name Inderal). While racemic drugs are administered in exactly equal amounts, it can often be shown that isomer ratios in plasma after absorption vary markedly from one drug to another.

The isomers have entirely different pharmacological activities. The dissociative anesthetic *ketamine* has a structure and activity related to that of the street drug PCP ("angel dust," see Chapter 7). There are occasional side effects in the use of ketamine and the side effects resemble the psychic excitation from PCP. It turns out that these undesired effects are primarily found in the *R*-(-) enantiomer, while the "good" properties come from the *S*-(+) isomer.

In the early '60s, a popular sedative was marketed in Europe and widely used by women in their first trimester of pregnancy. Unfortunately, inadequate testing had allowed a teratogenic drug to be marketed: the so-called "thalidomide babies" born to many of these women suffered from *phocomelia*, stunted or missing arms and legs. Since then, tests for teratogenicity have been required of all drugs. The drug was of course withdrawn; but later research indicates only the (+) isomer of this chiral molecule was teratogenic. Thalidomide is now being used experimentally with promising results by researchers at Johns Hopkins to treat (a) a form of leprosy, *erythema nodosum leprosum*; (b) rheumatoid arthritis; (c) *actinic prurigo* and *discoid lupus erythematosus*; (d) *graft-versus-host disease*, in which white cells from transplanted bone marrow attack host tissues.[8]

Most of the opiate class of drugs (Chapter 2) are chiral, and usually the addictive and narcotic properties lie in only one of the isomers. For instance *Darvon* (dextropropoxyphene) is an analgesic liable to abuse and is listed as a controlled substance; its isomer, levopropoxyphene or *Novrad* (Darvon spelled backward) is not narcotic and has been used as an antitussive (see Chapter 2).

Molecules with More than One Chiral Carbon

It is common for large molecules to have two or more, even dozens, of distinct chiral carbons. An analogy that exemplifies some of the possibilities when there are two chiral carbons in a molecule is found in Figure A4.8. Seven packages contain sets of grey and white gloves. **A** is the "enantiomer" of **C** in the sense that each glove in **A** has its mate, or mirror image, in **C**. Is **A** the enantiomer of **B**? No, nor is **C** the mirror image of **B**.

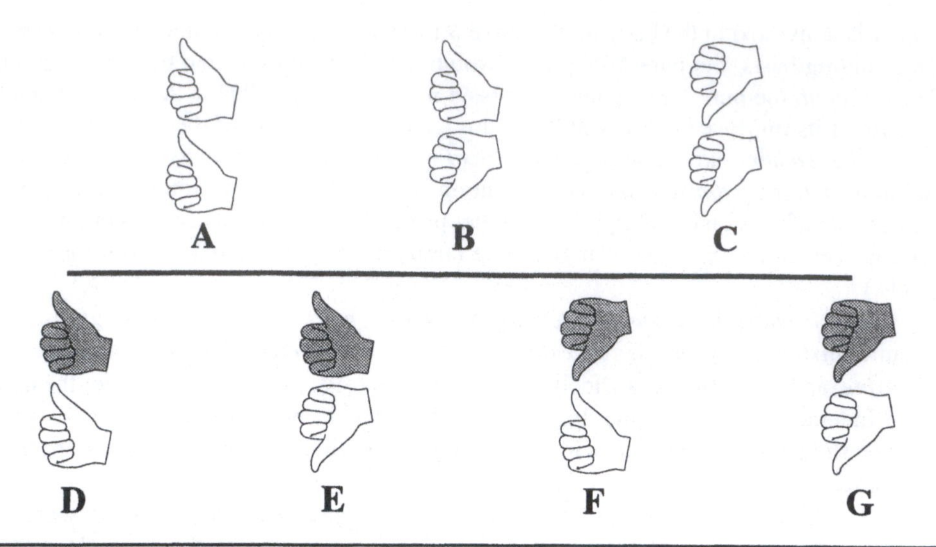

Figure A4.8 Two 2-glove assortments.

B is actually its own mirror image; it contains a normal pair of matching gloves. Similarly, **D** and **G** are enantiomeric pairs, as are **E** and **F**. But **D** is not the mirror image of **E** or **F**. A pair like **D–E** or **D–F** or **A–B** is called a *diastereomeric* pair; a molecule one half of which is the mirror image of the other, like **B**, is said to be *meso*.

Figure A4.9 shows some molecules corresponding to the sets of gloves in Figure A4.8. Notice that chiral isomers are more subtly different than structural isomers or *cis-trans* isomers: all three structures, **A**, **B**, and **C**, are 1,2-dibromo-1,2-dichloroethane just as all four structures, **D–G** are 1-bromo-1,2-dichloro-2-fluoroethane. The difference lies only in the chirality around each carbon; the complete name for E would be 1-*(R)*-2-*(S)*-1-bromo-1,2-dichloro-2-fluoroethane. As with the sets of gloves, **A** and **C** are each chiral

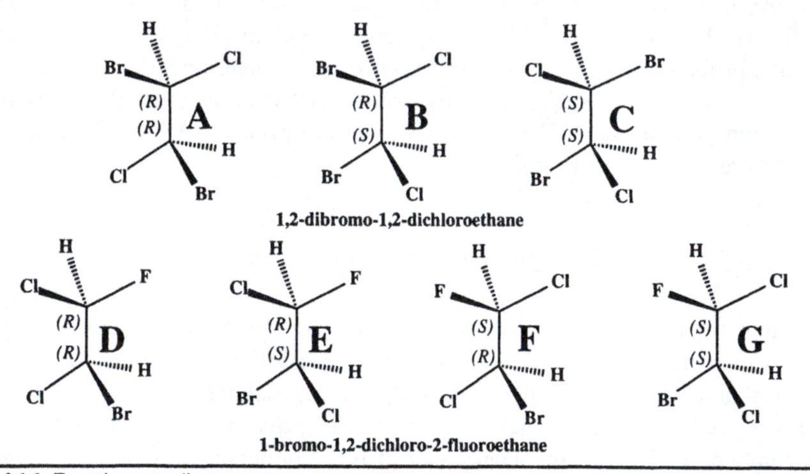

Figure A4.9 Enantiomers, diastereomers, meso structures.

molecules and constitute a pair of enantiomers; **B** is *not* a chiral molecule because it is identical with its own mirror image and hence superimposable upon it. In terms of physical properties, **A** and **C** will have the same melting point, boiling point, and so forth, and will differ only in their equal and opposite effect upon the plane of rotation of polarized light. **B** will differ from either of these enantiomers (it is their diastereomer) in melting point and boiling point, and it will not affect polarized light. It is likely to have considerably different properties physiologically than either **A** or **C**. All molecules **D–G** are chiral; **D** and **G**, and **E** and **F** are two pairs of enantiomers. **D** will have the same melting point as **G**; **E** the same as **F**; but **D** will melt at a different temperature than its diastereomer **E** or **F**. While the members of an enantiomeric pair like **E** and **F** are quite difficult to separate, diastereomers like **D** and **E** are relatively easy to separate. Physiologically, they are likely to be very different in their activity.

Some drugs with two chiral centers are shown in Figure A4.10. *Labetalol* (Normodyne) reduces blood pressure in patients with hypertension. It was first marketed as a racemic mixture of all four stereoisomers: β-blocking activity comes from the (*R,R*) isomer, α-blocking activity from the (1*S*,1′*R*) isomer; the other isomers are inactive. In 1989, the all-*R* isomer, *dilevalol* (Dilevalon)[9] by Schering Plough. It has more β-blocking and less α-blocking activity than labetalol, and lowers blood pressure with less change in heart rate.[10]

A much older pair of drugs is found in nature: quinidine and quinine, which differ only in their chirality at positions 9 and 2. Quinine is the 2*S*,9*R* isomer, whereas quin-

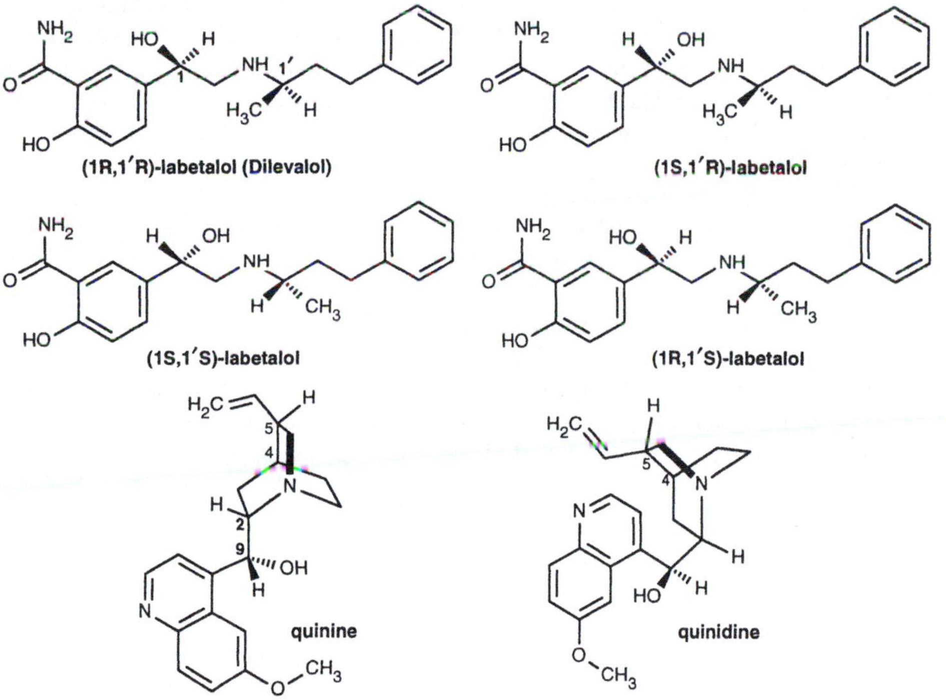

(1R,1′R)-labetalol (Dilevalol)

(1S,1′R)-labetalol

(1S,1′S)-labetalol

(1R,1′S)-labetalol

quinine

quinidine

Figure A4.10 Some drugs with two chiral centers.

idine is the 2*R*,9*S*. Quinine and quinidine are *not* enantiomers but diastereomers, because carbons 4 and 5 in each molecule are also chiral, and the chirality at these centers is identical for both molecules. These alkaloids and several closely related ones (cinchonide, cinchonine, epiquinine, and epiquinidine) occur in the bark of the cinchona tree ("Jesuit bark"), and both are effective antimalarial drugs. However, quinidine is twice as effective as quinine in reducing electrical conduction in the atrium of the heart; it is prescribed to maintain proper heart rhythm and prevent tachycardia. Quinine is still used as an antimalarial (and as a flavoring in tonic water).

Carboxylic Acids

There is another important class of molecules that we should be aware of since it occurs so frequently: carboxylic acids (Figure A4.11). Carboxylic acids are compounds containing the COOH group, and the simplest one of all is formic acid, the acid with which ants and bees season their bites and stings (L *formica*, ant hill). Acetic acid contains a two-carbon chain and is the acid found (about 5%) in vinegar (L *acetum*, vinegar; Old French *vin aigre*, sick wine). As you can see, the carboxylic acid function, COOH, consists of a hydroxy group attached to the carbon of a carbonyl group. You may know that acids are substances that produce *hydrogen ions* in water; water removes the hydrogen (as a +1 ion) of the OH of carboxylic acids, leaving behind a negatively charged ion; in Figure A4.11 we see that this negative ion is stabilized by resonance between two equivalent forms. The hydrogen ion, sometimes called a proton (really it is a solvated proton better represented by H_3O^+), is what is responsible for the tart or sour flavor of vinegar, lemon juice, dry white wine, and many other foods. Carboxylic acids are not at all as strong or caustic as inorganic acids like hydrochloric acid (muriatic, HCl, the acid found in dilute form in mammalian stomachs that is essential for digestion), nitric acid (HNO_3), or sulfuric acid (H_2SO_4).

The names we have used for the acids above are traditional; of course, there is an IUPAC approved systematic method. However, the traditional names are very commonly used. For the first six straight chain carboxylic acids, the traditional names are: *formic* (methanoic); *acetic* (ethanoic); *propionic* (propanoic); *butyric* (butanoic); *valeric* (pentanoic); *caproic* (hexanoic). Figure A4.12 shows the structures of some common carboxylic acids.

Lactic acid is widely distributed in nature. It occurs as the racemic form in sour milk as a result of natural *lactobacillus*, the family of bacteria that gives us yogurt, cheese, and sauerkraut; it is also found in molasses, apples, beer, and wines. The (*S*) form occurs in small quantities in the blood and muscle; the concentration increases

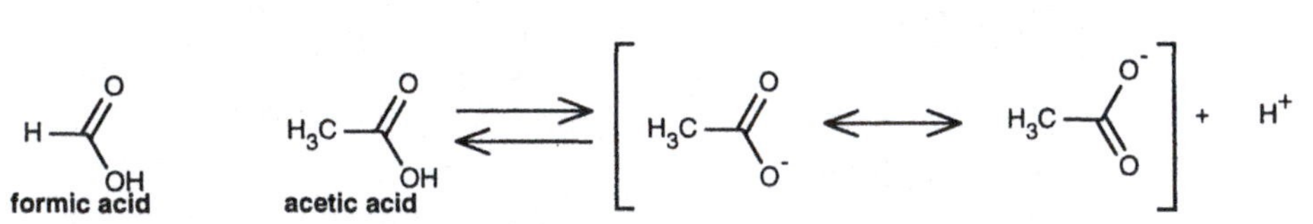

formic acid acetic acid

Figure A4.11 Acetic and formic acids; ionization of acetic acid.

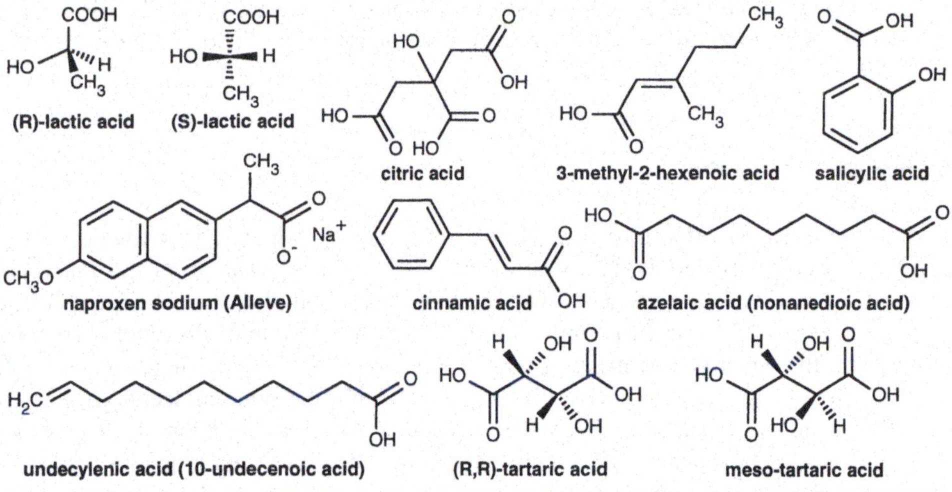

(R)-lactic acid (S)-lactic acid citric acid 3-methyl-2-hexenoic acid salicylic acid

naproxen sodium (Alleve) cinnamic acid azelaic acid (nonanedioic acid)

undecylenic acid (10-undecenoic acid) (R,R)-tartaric acid meso-tartaric acid

Figure A4.12 Some common carboxylic acids.

significantly after strenuous exercise, contributing to the pain of sore muscles.

Citric acid is widely distributed in plants and animals (it is a key component of the fundamental Krebs cycle). The tartness of lemons and limes is due to citric acid, which is added to many candies and beverages for flavoring and as a preservative.

3-Methyl-2-hexenoic acid (both *E* and *Z* forms) has been recently shown to be the chief constituent responsible for human underarm odor.[11] This was long thought to be caused by isovaleric (3-methylbutanoic) acid, but research scientist Dr. George Preti remained unconvinced. Isovaleric acid was "not what I smell after cutting the lawn on a hot Saturday," he maintained. This discovery is important not only for the deodorant industry: the stew of chemicals found in the underarm has earlier been shown to influence the length and timing of the menstrual cycle. (There is promise here for an all-natural contraceptive perfume!)

Salicylic acid is found in willow tree (*salix*) bark, which has been known for centuries to reduce fever. It is a topical keratolytic and antifungal but is quite irritating to the stomach; aspirin is a semisynthetic modification that is more easily ingested. *Cinnamic acid* occurs in cinnamon and cocoa leaves; the esters are used in perfumes. The methyl ester (see the next section) has an odor reminiscent of strawberries. *Azelaic acid* is a dicarboxylic acid that was recently introduced (Skinoren) as a drug for topical application to control acne.[12] *Undecylenic acid* occurs naturally in sweat, and its odor, *pace* Dr. Preti, is to many reminiscent thereof; combined with its zinc salt (Desenex), it has long been used as an effective topical antifungal to control tinea pedis (athlete's foot).

Tartaric acid is widely distributed in nature as the (*R*,*R*) form, usually in fruits and fruit juices. It was observed in antiquity that a fine crystalline crust often developed in wine as it aged; the Romans called this *faecula.* We now know that this crystalline material is actually the acid potassium salt of tartaric acid. Some California chardonnays and many German rieslings will throw a crystalline deposit of potassium bitartrate ("cream of tartar," German *Weinstein*) when chilled. The potassium salt of tartaric acid is tasteless, but the acid itself is used as a flavoring agent in many soft drinks, and the diethyl and dibutyl esters (below) are used for lacquers. The *meso* and (*S*,*S*) forms are made from the (*R*,*R*) form.

Esters and Lactones

Carboxylic acids are often found in nature as their esters. An ester is a compound that can be formally regarded as the result of the elimination of water from the functional group of an acid and an alcohol, as with the formation of ethyl acetate in Figure A4.13. Esters are named by using two words: the first is the alkyl group of the alcohol (methyl, ethyl, etc.); the second is the name of the acid with the ending *-oic* changed to *-oate*. In certain compounds, a hydroxy group is positioned within the same molecule as the acid at a distance allowing an intramolecular esterification; the cyclic compounds thus formed are called *lactones* (nepetalactone, ascorbic acid). Lactones with very large rings (erythromycin A) are called *macrolides*.

Ethyl acetate is a common industrial solvent for varnishes and lacquers. *Ethyl 2-methylbutanoate* along with hexanal and 2-hexenal are the compounds chiefly responsible for the characteristic aroma and taste of apples. *Ethyl (methyl) trans:2-cis:4-decadienoate* are the esters that are chiefly responsible for the flavor and aroma of pears.[13]

Mevalonic acid is a key precursor in the endogenous synthesis of cholesterol; formation of mevalonic acid is the rate-determining (the slowest, bottle-neck) step in the process. Some recent drugs (lovastatin, simvastatin) function as *hypocholesterolemics* by blocking the enzyme controlling this key step. Mevalonic acid exists as a rapidly equilibrating mixture of its open-chain and lactone forms.

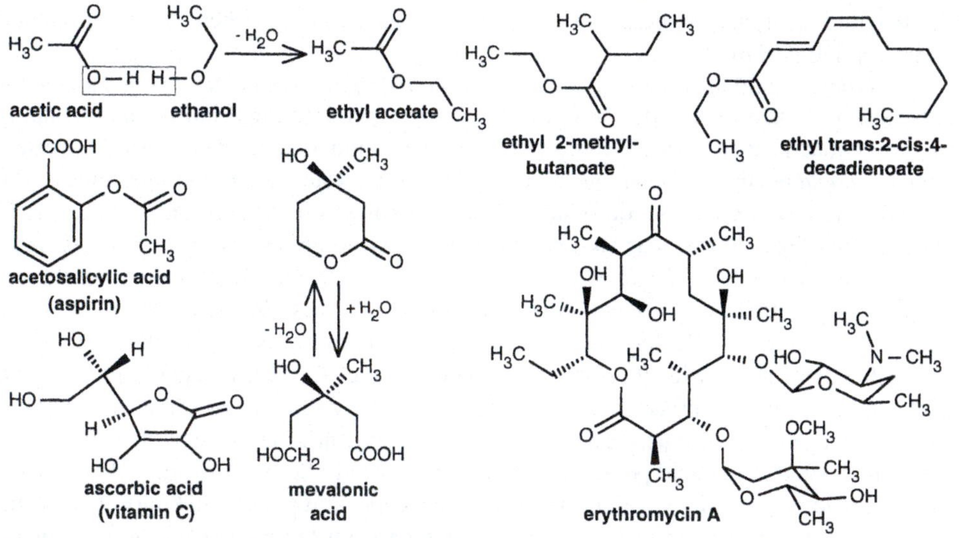

acetic acid, ethanol, $-H_2O$, ethyl acetate, ethyl 2-methylbutanoate, ethyl trans:2-cis:4-decadienoate, acetosalicylic acid (aspirin), mevalonic acid ($+H_2O$ / $-H_2O$), ascorbic acid (vitamin C), erythromycin A

Figure A4.13 Some common esters and lactones.

Ascorbic acid (Vitamin C) is an essential component in the human diet; it is needed in the biosynthesis of collagen, which is the structural protein that provides mechanical strength to bone, tendon, cartilage, and skin. Persons deprived of vitamin C, as were sailors on long voyages until the XVIII century, develop *scurvy* (Latin *scorbutus*; hence, with the Greek alpha privative, *ascorbic*) from faulty collagen metabolism. It is characterized by skin lesions, fragile blood vessels, and bleeding gums. The British Navy began requiring daily consumption of limes, rich in vitamin C, to prevent scurvy; consequently, to this day, the British are called—by those not wishing to honor them—*limeys*.

Erythromycin A and *troleandomycin* are antibiotics derived from *Streptomyces erythreus S.* and *antibioticus*, respectively, and are the only macrolide antibiotics presently marketed in the United States. Erythromycin is the preferred alternative for the treatment of many infections in penicillin-allergic individuals. It is considered among the safest antibiotics in use today. Despite the intimidating number of chiral centers, it has been synthesized by several groups.[14]

Fats, Oils, Fatty Acids

Figure A4.14 shows the structure of *glycerol* (also known as glycerin or 1,2,3-propanetriol). This oily triol is used in hand lotions as a moisturizer and is added to tobacco products to prevent them from drying out: glycerol has such an affinity for water that it absorbs water from the air. It is an inexpensive by-product of the soap manufacturing industry. Soaps are made by breaking up fats with lye, NaOH, forming the sodium salts of long fatty acids (soaps) and glycerol.

The distinction between a fat and an oil is simply whether the material is a liquid or a solid at room temperature. Both fats and oils consist of *triglycerides*, that is, they are triple esters of glycerol with three molecules of long-chain fatty acids. A triglyceride can be *simple* (with all three acid molecules the same) or *mixed*. With mixed triglycerides, many different triglycerides can be formed; for instance, with only four different fatty acids in a fat, 40 chemically distinct triglycerides are possible. And natural fats are nearly always mixed: butter has more than 28 different fatty acids. Because of this complexity, fats are characterized not by their individual triglyceride components but by the overall percentage of each fatty acid present. Fatty acids themselves are also characterized, as in Figure A4.14, by the number of carbon atoms in the chain (which is almost always unbranched) and the number of double bonds. Since oleic acid has 18 carbon atoms and one double bond, it is said to be an (18:1) fatty acid, while lauric is (12:0). Triglycerides containing no double bonds are said to be "saturated"; those with one or more double bonds are "unsaturated"; as a general rule, the saturated triglycerides are found in animal fats, and the unsaturated in vegetable oils.

In the popular media we are presently being treated to a chorus of well-intentioned advice as to what fats in what foods are "good" or "bad." Even the venerable McDonald's has made concessions (ridiculed by Wendy's) to the seaweed and tofu constituency. The best current thinking is that Americans should replace much of the red meat and dairy products (any food containing "land-mammal fat") in their diet with vegetable oils (olive, corn, etc. but *not* coconut or palm oils), lean fowl, and fish.[15] Much of the evidence for

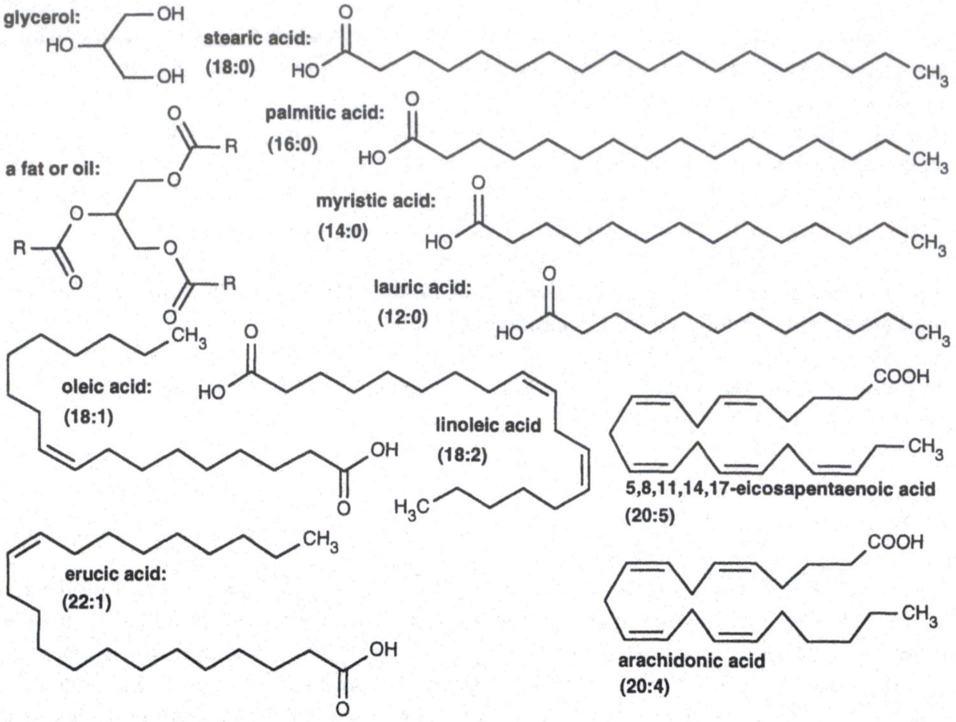

glycerol:
stearic acid: (18:0)
a fat or oil:
palmitic acid: (16:0)
myristic acid: (14:0)
lauric acid: (12:0)
oleic acid: (18:1)
linoleic acid (18:2)
5,8,11,14,17-eicosapentaenoic acid (20:5)
erucic acid: (22:1)
arachidonic acid (20:4)

Figure A4.14 Fats and fatty acids.

this lies in the extremely low incidence of cancer and heart disease (relative to the U.S. population) to be found among the Japanese and Eskimos (whose fat intake is primarily from fish oils) and the Cretans (who eat mostly pasta, olive oil, and garlic).

Stearic [octadecanoic] acid is widely found in mostly animal fats. Despite being a saturated acid, it does *not* increase LDL ("bad") cholesterol or atherosclerosis, as do the following three saturated acids. *Palmitic (hexadecanoic), myristic (tetradecanoic),* and *lauric (dodecanoic) acids* are all found predominantly in land-mammal fats and tropical oils. They have been shown to increase LDL cholesterol levels and cause atherosclerosis in extensive tests on animals. They should be reduced to low levels in the human diet.

Oleic [(Z)-9-octadecenoic] acid is the most widely distributed fatty acid in nature. It makes up 20 to 50% of corn oil and 83% of olive oil (whence its name). Olive oil is the chief source of fats among native Greeks, who have an extraordinarily low rate of coronary heart disease and most cancers. *Linoleic [(Z,Z)-9,12-octadecadienoic] acid* is a constituent of many vegetable oils. It and several similar fatty acids (linolenic, arachidonic) are essential nutrients in small laboratory animals; *essential fatty acid deficiency* is rare in humans, but can occur in infants fed nonfat milk.[16] Some recent, if preliminary, research at the Boston University School of Medicine claims that the problems in the American diet that lead to atherosclerosis are really due to inadequate amounts of linoleic and linolenic acids: they claim that people who have reduced their fat intake excessively have the same plaque buildups as those who eat saturated fats or high-calorie diets. Supporting the claim is the lowered levels of these two fatty acids in the plasma of those with confirmed coronary heart disease as compared with controls.[17] The Boston group says we should eat soybeans, nuts, seeds, canola oil, purslane. And fish, chimes in Dr. William Castelli, director of the Framingham Heart Study, who has been researching heart disease and diet for years: "If you can't be a fish," he says, "the next best thing is to eat one."[18]

(Z,Z,Z,Z,Z)-5,8,11,14,17-Eicosapentaenoic acid is an important constituent of the fats of the marine food chain, and is found in cod-liver and other fish oils in the Eskimo diet. As *omega-3* and *eskima*, it has been touted as a health food supplement. It is a precursor of the prostaglandins (as is *arachidonic* acid, which differs from it only in lacking the double bond at the 17 position).

Erucic acid (13-docosenoic acid). Erucic acid makes up to 50% of the fat in the seeds of the rapeseed plant (canola oil), and about 80% of the oil from nasturtium seeds. It and oleic oil are the chief components of "Lorenzo's oil," which is a rapeseed oil particularly high in erucic acid. An eponymous movie in 1993 popularized the use of the oil in the treatment of Lorenzo Odone's adrenoleukodystrophy. This is an inherited neurological disease that affects almost exclusively young males, is characterized by a buildup of long-chain fatty acids in the bloodstream, and eventually leads to paralysis and death. Some evidence that a restricted diet supplemented by Lorenzo's oil does slow progression of the disease, although only slightly, has been presented in studies at Johns Hopkins University directed by Dr. Hugo Moser.[19]

It should be noted that all the double bonds in the natural fats of Figure A4.14 are *cis (Z)*. As we learned in Appendix III, margarines are made by hydrogenating unsaturated vegetable oils. Since most vegetable oils are composed chiefly of oleic and linoleic acids, hydrogenation results in transforming these acids to stearic acid (18:0). However, there is another effect: the nickel or platinum catalyst that allows hydrogen to be added to these double bonds can also catalyze the transformation of the double bonds in natural oleic and linoleic acids from *cis* to *trans*. These "unnatural" fatty acids can make up as much as 35% of the fat in some margarines; they may compete with the natural acids and, like animal fats, adversely affect LDL and HDL cholesterol ratios.[20]

As a matter of perspective, it should be said that smoking and lack of exercise contribute much more significantly to cardiovascular mortality and morbidity in the U.S. population than fatty diets (the problem is fatty asses, not fatty acids). And what is "the only dietary factor consistently associated with [reducing] the risk of coronary heart disease"? Alcohol, of course.[21] Hence we should "enjoy an occasional meatball when the urge becomes irresistible, preferably with a glass of red zinfandel," while considering that "the world's vast array of vegetarian dishes ... provides an eating adventure, between the occasional meatballs, that Americans are only beginning to explore."[22]

Amides, Peptides, and Lactams

Another important class of chemical structures is introduced in Figure A4.15. *Amides* are formed from carboxylic acids and amines in the same way that esters are formed from

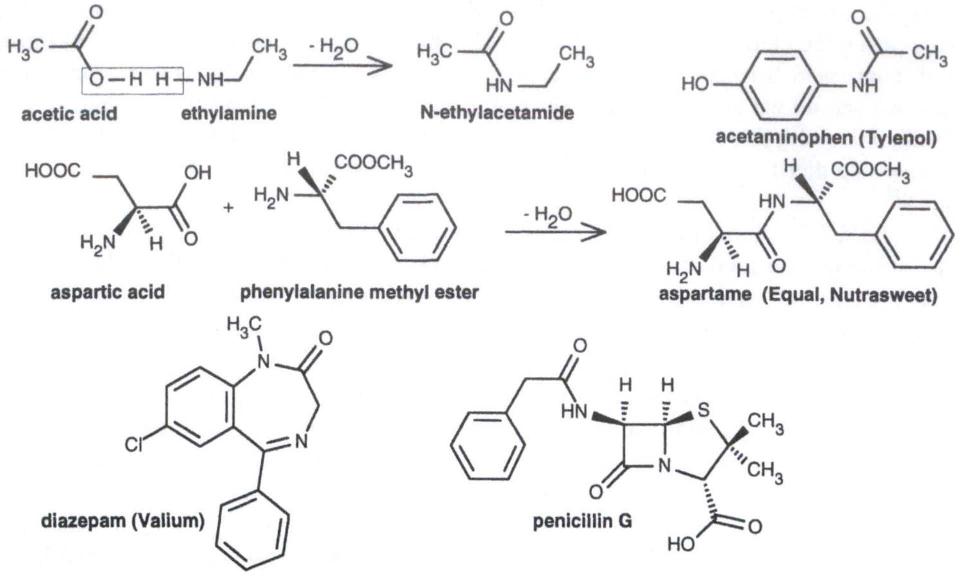

Figure A4.15 Amides, peptides, and lactams.

acids and alcohols. Shown in the figure is the formation of *N-ethylacetamide* from acetic acid and ethylamine. Amides are named by replacing the *-ic* suffix of the acid with *-amide*; if there is an alkyl group affixed to the nitrogen, this is indicated by an italicized capital *N*- followed by the name of the group. *Acetaminophen,* sold under the brand name of Tylenol, is a familiar example of a simple amide; it is the amide formed from 4-hydroxyphenylamine and acetic acid.

The most important occurrence of the amide linkage is between *amino acids.* There are some 20 amino acids found in the biosphere of this planet. Every one of these amino acids has a chiral carbon atom from which there are four groups: an $-NH_2$, a -COOH, an -H, and an alkyl chain. The amino acids differ only in the structure of the alkyl chain: the configuration around the chiral alpha carbon is always *S* (commonly called L from an older system, in which D [dextro] and L [levo] more or less correspond to *R* and *S*). The fact that almost all organisms, from bacteria to bulldogs, from frogs to Frenchmen, have the same *S*-only sort of amino acids, constitutes one of the strongest pieces of evidence supporting the evolutionary theory.

These 20 or so amino acids function as monomers in a polyamide or poly*peptide* chain. One of the major types of polypeptide are *proteins.* For instance collagen, which we mentioned previously in the context of Vitamin C, is the most abundant protein in many vertebrates; it is constructed from a triple-stranded polypeptide fiber of about 1000 units each, and has a tensile strength equal to hard-drawn copper wire. An equally important occurrence of the polypeptide structure is found in *enzymes*, catalysts that regulate nearly every process in living organisms. Enzymes can select molecules having the proper stereochemistry and structure from a sea of thousands of competing molecular species, and transform them with a minimum energy expenditure into a specific product.

Most drugs act by inhibiting or enhancing enzymatic reactions. Still other proteins function as hormones (insulin, corticotropin). Hair, wool, and silk are all polypeptides.

Organisms vary in their ability to synthesize amino acids; human beings (and albino rats) can make 10 of the 20 they need endogenously, while the remaining 10, called *essential amino acids*, must be present in the diet (usually as part of the protein we or the rats eat: our stomach enzymes [proteases] chop these proteins up into their constitutive amino acids, which we then recycle).

Aspartame (Nutrasweet) is a simple dipeptide formed from the methyl ester of *phenylalanine*, one of the essential amino acids, and *aspartic acid*, one of the nonessential amino acids. Aspartame is 160 times as sweet as the *sucrose* of ordinary table sugar (which is produced by the sugar cane and sugar beet plants).

Just as an acid with an OH group properly positioned can form a lactone, an acid with a properly positioned NH_2 can form a *lactam*. *Diazepam (Valium)* is one of the most widely prescribed minor tranquilizers; it is one of a large class of drugs called *benzodiazepines* (see Chapter 3). The seven-atom ring fused to the benzene ring in these compounds is a lactam. *Penicillin G* and all the other penicillins contain a four-atom ring called a β-lactam. There is also an amide linkage in the side chain. The more recent class of similar antibiotics called the *cephalosporins* also has a β-lactam ring, but the five-membered ring containing the nitrogen and sulfur of the penicillins is expanded in the cephalosporins to six atoms. Several cephalosporins are produced naturally by the busy internal chemistry of fungal life, but ingenious synthetic chemists have developed a fairly straightforward method of converting a given penicillin antibiotic to its analogous cephalosporin, and this seems to fool the bacteria which have developed resistance to the older penicillin analogue.

Polyamides and Polyesters

We should not leave the subject of esters and amides without mentioning some important synthetic polymers that we encounter every day: *polyesters* and *polyamides*, which are shown in Figure A4.16.

Polyethylene terephthalate is the most common polyester: it is the polymer formed by esterifying terephthalic acid (1,4-benzenedicarboxylic acid) with ethylene glycol (ethanediol). This form of the polymer is used in polyester clothing and polyester blends under the trade name Dacron. The terminal unreacted COOH groups are often methylated, and this variant is called Mylar and Fortrel.

Nylons were the among the first useful synthetic polymers. The nylons and neoprene (a chlorinated synthetic rubber) were both the discoveries of Wallace Hume Carothers of DuPont. Carothers was a brilliant experimentalist who was tortured by bouts of severe depression: in 1938, shortly before the patent was filed for the first nylon, he committed suicide by drinking potassium cyanide (KCN) in lemonade. Today's long-lasting automobile tires are constructed with Carothers's nylon and neoprene.

The word nylon is a variation of "no run" spelled backwards: one of the first, and enormously profitable, uses of nylon was in women's stockings, where it replaced the more fragile silk. *Nylon 6* is polymerized *6-aminohexanoic acid;* however, it is made by

polyethylene terephthalate [R = H, Dacron, Amilar; R = CH_3, Mylar]

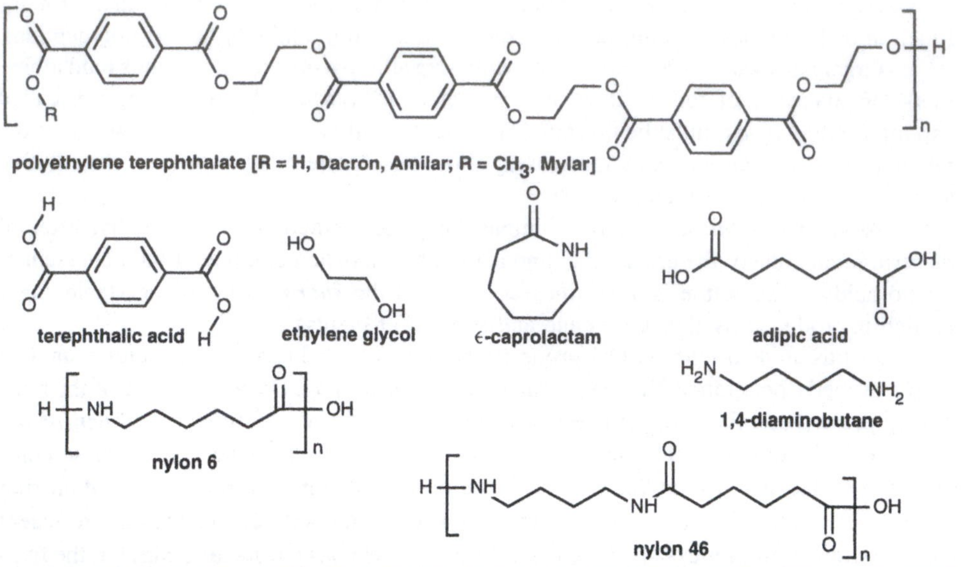

terephthalic acid   ethylene glycol   ϵ-caprolactam   adipic acid

1,4-diaminobutane

nylon 6

nylon 46

Figure A4.16 Polyesters and polyamides.

polymerizing the lactam of this acid, called *ϵ-caprolactam* (the epsilon refers to the position of the amino group relative to the COOH; the 6 refers to the number of carbons in the amino acid). It is used in fishing lines, reinforced tires, and hoses. *Nylon 46* is a polymer formed from 1,4-butanediamine and adipic (hexanedioic) acid. The 4 and 6 refer to the number of carbons in the respective components. It is used for tire reinforcement.

References and Notes

1. *Proc. Natl. Acad. Sci. U.S.A.*, **88**, 1407 (1991) 1991.
2. The naming of the freons follows a peculiar system. There can be as many as four numbers (but zeros are not written, so often there are fewer): the last is the number of F atoms; the second-last is the number of hydrogen atoms *plus one*; the third-last is the number of carbon atoms *minus one*; the fourth-last is the number of Cl atoms.
3. Tyler, P. E., "Lacking iodine in their diets, millions in China are retarded," *The New York Times,* 4 June 1996, A1, A10.
4. *N. Engl. J. Med.*, **1991**, *324*, 212-8, 260-262.
5. Waldorf, M. M., "The (liquid) breath of life," *Science,* **1989**, *245,* 1043-1045.
6. Riordan, T., "Patents: The man who gave the world vinyl wins a place in the Inventors Hall of Fame," *The New York Times*, 24 July 1995, D2. Semon also invented a new type of synthetic "chewing rubber" which, he said, "looked just like ordinary gum, except that it would blow these great big bubbles." But Goodrich wasn't interested.
7. There are additional benefits to the use of dexibuprofen, including an improved side effect profile. See *Annual Reports in Medicinal Chemistry*, **1990**, *30*, p. 298.
8. Bor, J., "Thalidomide shows that it can heal, too: From deformer of babies to force for good," *The Baltimore Sun*, 2 April 1995, 1A.
9. *Annual Reports in Medicinal Chemistry*, **1990**, *25*, 311.
10. Powell, J. R.; Ambre, J. J.; Ruo, T. I. "The Efficacy and Toxicity of Drug Stereoisomers," in *Drug Stereochemistry: Analytical Methods and Pharmacology,* Wainer, I. W.; Drayer, D. E., Eds. [Vol. 12 of *Clinical Pharmacology*, Weiner, M. Ed.], New York: Marcel Dekker, 1988, pp. 245-270.
11. Reported by Dr. George Preti of the Monell Chemical Sense Center, Philadelphia, at the American Chemical Society National meeting, Washington, August 1990.
12. *Annual Reports in Medicinal Chemistry*, **1990,** *25*, 310. For a clinical comparison with tetracycline, see Blandon, P. T., *Br. J. Dermatol.*, **114**, 493 (1986).
13. Charley, H. "Fruits and Vegetables," in *Food Theory and Applications*, Paul, P. C. and Palmer, H. H., Eds. New York: John Wiley, 1972, pp 310-311.
14. (a) Corey, E. J., *J. Am. Chem. Soc.,* **1978,** *100*, 4620. (b) Woodward, R. B., *J. Am. Chem. Soc.*, **1981**, *103,* 3210.
15. Willett, W.; Sacks, F. M., "Chewing the fat: How much and what kind," *N. Engl. J. Med.,* **1991**, *324*, 121-123.
16. *The Merck Manual*, 15th ed., Berkow, R.; Fletcher, A. J., Eds., Merck: Rahway, NJ, p. 931.

17. Siguel, E. N.; Lerman, R.H., *Metabolism*, August **1994**.

18. *The New York Times*, 24 August 1994.

19. *The New York Times,* 11 September 1994.

20. (a) Mensink, R. P.; Katan, M. B., *N. Engl. J. Med.,* **1990**, *323,* 439-45. (b) Burros, M., "Now what? U.S. study says margarine may be harmful," *The New York Times,* 7 October 1992. (c) "Fat in margarine is tied to heart problems," *The New York Times,* 16 May 1994. But there is a good deal of smoke and heat on this issue, and not much light. In Europe, a method of hydrogenation is employed that does not result in *trans*-unsaturated fats; it may be good to import your margarine from France.

21. Ellison, R. C., "Cheers!" *Epidemiology*, **1990**, *1,* 337-339. See also Stampfer, M.; et al., *N. Engl. J. Med.,* **1991**, *325*, 373-381. The largest and mosr reliable of many studies attribute the benefits simply to alcohol (see Chapter 3). Nonetheless, there are a few studies by French scientists and a group at Cornell that claim only *red* wine (not white wine and not alcohol in general) lowers LDL-cholesterol: Renaud, S.; de Lorgeril, M., *Lancet*, **1992**, *339*, 1523. The active agent could be *resveratrol,* found in the skins only of red grapes, where it may function as an antifungal. Those of us who cherish white wine regret this. The situation is a particular affront to lovers of French sauternes, Hungarian tokays, and German *Trokenbeerauslesen*, all of which (white) varietals owe their peculiar magic to the work of the "noble rot" fungus.

22. For the record: are there any nutritional disorders associated with vegetarianism? If fish is eaten, no. If no fish is eaten, there is a risk of iron deficiency. However, such *ovo-lacto* (only eggs and dairy products) vegetarians live longer than carnivores. *Vegans* consume no animal products whatever and are susceptible to vitamin B-12 deficiency and low levels of calcium, iron, and zinc. *Fruitarians* eat only fruit and can suffer serious health hazards from insufficient protein. See *The Merck Manual*, 15th ed., Berkow, R.; Fletcher, A. J., Eds., Merck: Rahway, NJ, p. 917.

Index

A

B

D

E

F

G

H

I

M

N

Q

R

S

T

U

V

CPSIA information can be obtained at www.ICGtesting.com
Printed in the USA
BVOW09s2338090815

411974BV00011B/44/P